One Small Step:~

The History of Aerospace Engineering at Purdue University

SECOND EDITION

Neil Armstrong Hall of Engineering

One Small Step:~ The History of Aerospace Engineering at Purdue University

SECOND EDITION

BY

A. F. GRANDT, JR.
W. A. GUSTAFSON
L. T. CARGNINO

PUBLISHED BY

SCHOOL OF AERONAUTICS & ASTRONAUTICS
PURDUE UNIVERSITY

For information address:
School of Aeronautics & Astronautics, Purdue University,
701 W. Stadium Avenue, West Lafayette, IN 47907-2045
ISBN 978-1-55753-599-3

Printed and bound in the United States of America
Second edition, October 2010

TABLE OF CONTENTS

LIST OF FIGURES · VII

FOREWORD TO THE FIRST EDITION · XV

FOREWORD TO THE SECOND EDITION · XIX

INTRODUCTION · XXI

1 PREFLIGHT CHECK: · 1
Aeronautical Engineering Before World War II

2 COUNTDOWN: · 23
World War II Programs

3 IGNITION: · 47
The Five-Year Postwar Period 1945–50

4 LIFT-OFF!: · 95
The 1950s

5 GAINING ALTITUDE: · 145
A New Era 1960–73

6 MIDCOURSE CORRECTION: · 189
Transition and Growth 1973–80

7 ALL SYSTEMS GO: · 205
1980–95

8 IN ORBIT: · 237
Golden Anniversary Benchmark 1995

9 IN-FLIGHT REFUELING: · 259
1996–2009

10 MISSION CONTROL: · 285
Alumni Accomplishments

11 RE-ENTRY: · 341
Reflecting on the Past — Looking To the Future

REFERENCES · 344

APPENDICES

A Bachelor of Science Degrees Awarded · 345

B Master of Science Degrees Awarded & Thesis Titles · 387

C PhD Thesis Titles · 427

D Teaching and Research Award Recipients · 451

E Summary of School Titles & School Heads · 453

F Outstanding Aerospace Engineer Award Recipients · 454

G Summary of AAE Faculty · 456

INDEX · 459

LIST OF FIGURES

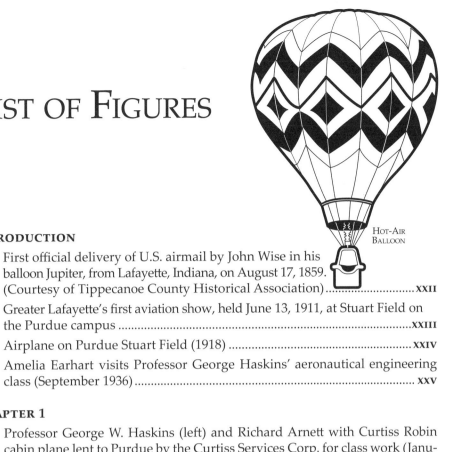

Hot-Air
Balloon

INTRODUCTION

I.1   First official delivery of U.S. airmail by John Wise in his balloon Jupiter, from Lafayette, Indiana, on August 17, 1859. (Courtesy of Tippecanoe County Historical Association) ....**XXII**

I.2   Greater Lafayette's first aviation show, held June 13, 1911, at Stuart Field on the Purdue campus ....**XXIII**

I.3   Airplane on Purdue Stuart Field (1918) ....**XXIV**

I.4   Amelia Earhart visits Professor George Haskins' aeronautical engineering class (September 1936) ....**XXV**

CHAPTER 1

1.1   Professor George W. Haskins (left) and Richard Arnett with Curtiss Robin cabin plane lent to Purdue by the Curtiss Services Corp. for class work (January 1930). Professor Haskins served on the faculty from 1929 to 1937............**3**

1.2   View of aeronautics laboratory showing wind tunnel, Curtiss Robin aircraft, and engine (May 1930)...**5**

1.3   Professors Haskins (right) and Hollister (left) examining twelve-inch wind tunnel (November 1931) **5**

1.4   Aeronautical engineering class. Left to right: P. D. Pruitt (standing), W.H. Rivers, M.G. Haines, M.S. Finch, W.W. Halsteadt, J.L. Fihe, and Professor G.W. Haskins (September 1934) **6**

1.5   Captain Haskins and class of students examining turbo charger for aircraft engine (September 1934). Left to right: D.P. Keller, W.H. Rivers, R. Warren (kneeling), M.G. Haines, and Captain Haskins **6**

1.6   Aeronautics class around aircraft at Purdue airport (October 1934)..............**7**

1.7   Students move glider into position for flight at the Purdue airport (May 1934)...**7**

1.8   Professor G.W. Haskins (second from right) and students examining airplane (October 1936) **8**

1.9   Class of aeronautical engineering students studying construction of Navy aircraft sent to Purdue airport (October 1936) **8**

1.10   Professor Karl D. Wood (1898–1995) **9**

1.11   Professor Joseph Liston (1906–78)..**9**

1.12   Professor Elmer F. Bruhn (1899–1984) **10**

1.13   Senior aeronautics students Frank Crosy (left) and Paul Marshall (right) working on radio-controlled model airplane designed by Professor Wood (February 26, 1938) **13**

1.14 Students testing strength of aircraft wings with bags of sand (July 1940).   **13**

1.15 Purdue wind tunnel showing balance system used to measure aerodynamic loads (February 4, 1939)..**14**

1.16 Professor Liston (second from left) and students examining aircraft engines (circa 1943)..**15**

1.17 Construction of Purdue airport building (June 5, 1934). The Purdue airport was established in 1930 as the first university-owned airport in the nation **17**

1.18 Students with aircraft at Purdue airport (1930s)...**18**

CHAPTER 2

2.1 Collection of photographs from 1945 school pamphlet showing first Aeronautical Engineering building located on main campus, and students working with aerodynamics and flight test facilities..**27**

2.2 Aerial photograph of Purdue campus and airport circa 1943 (facing southwest)...**28**

2.3 Photographs from 1945 pamphlet showing students working in aircraft structures laboratory located in main campus Aeronautical Engineering building...**29**

2.4 Photographs from 1945 pamphlet showing students working in aircraft power plant laboratory located at Purdue airport...**30**

2.5 Curtiss-Wright Cadettes working in various laboratories

    a. Drafting was an important phase of the Curtiss-Wright program. Front row, left: Betty Schaefer; center: Maxine Stevens. Back row, left: Dorothy (nee Wurster) Rout...**36**

    b. Structures laboratory...**36**

    c. Power plant laboratory **37**

    d. Wind tunnel tests...**37**

2.6 Morrow airplane project (1944)

    a. Back row, left to right: unidentified, Professor E. H. Bruhn, Les Schneiter, Dave Mendenhall, Lewis Jones, Ralph Trueblood, and Paul Brink. Second row: unidentified, Walter Smith, Bob Boswinkle, Oscar Lappe, Mr. Morrow, Stanley Meikle (Director of Purdue Research Foundation). Front row: unidentified, Leonard Rose, Bob Pendley, Jack Allen, and Bill Gaugh.....**44**

    b. Students testing Morrow airplane **44**

    c. Completed Morrow airplane examined by (left to right): Professor Bruhn, Mr. Morrow, and unidentified students..**45**

CHAPTER 3

3.1 Organization chart showing relation of Purdue Aeronautics Corporation and School of Aeronautics **52**

3.2 Original air transportation curriculum for new School of Aeronautics........**53**

3.3 Professor Lawrence T. Cargnino (1945)..**56**

3.4 Original aeronautical engineering curriculum for School of Aeronautics....**60**

3.5 Students in air transportation power plant laboratory (circa 1947). Professor Cargnino is standing at rear center in white lab coat, and Professor Briggs is standing front right **71**

3.6   Air transportation laboratory (circa 1947) **72**

3.7   Second-floor corridor connecting Building Units 1 and 2 (circa 1949). Rhea Walker (shown at right) served as head secretary for many years **72**

3.8   Professor Liston, Instructor Crosby, and Mr. Bean in power plant laboratory (circa 1947)..**74**

3.9   Propeller laboratory (circa 1947). Instructor James Basinger is second from left..**74**

3.10 Professor Briggs with students in engine disassembly laboratory (circa 1946) .**75**

3.11 Students in power plant lecture room (circa 1946).......................................**75**

3.12 Various wind tunnels operated by School of Aeronautics during 1940s.

    a. Professor Wood and student with wind tunnel (February 1943) **80**

    b. Low-speed free flight wind tunnel designed by Professor Wood (March 1943)..**80**

    c. Variable-density wind tunnel originally built in Japan **81**

    d. Large subsonic wind tunnel constructed under direction of Professor Palmer (in background). Tunnel was renovated and renamed The Boeing Wind Tunnel in 1992 ....**81**

3.13 Students working in structures laboratory

    a. Conducting wing test..**82**

    b. Professor Bruhn and students (note slide rules). Rear table, left to right: Prof. Bruhn, G. Christopher, R. H. Turner, L. T. Cheung, W. G. Koerner, A. M. Arnold (back to camera), unknown, H. F. Steinmetz. Front table, left to right: R. W. Taylor, R. M. Rennak, R. J. Wingert...................................**82**

    c. Compression test with Tinius Olson test machine **83**

    d. Strain measurements..**83**

3.14 Aircraft metal working laboratory developed by Professor Lascoe of General Engineering Department. Instructor J. Borodavchuk standing at center......**85**

3.15 War surplus C-46. Students Wade E. Mumma and Joseph P. Minton at aircraft windows. Standing left to right: E. F. Bruhn, E. A. Cushman, L. T. Cargnino, W. Briggs, and James Bassinger. Also note P-47 aircraft visible behind C-46 at left.   **86**

3.16 War surplus P-59 jet aircraft (circa 1950)..**87**

3.17 Research project carried out under direction of Professor Hsu Lo (left).......**89**

3.18 Test rig for determining ultimate dynamic energy absorption of basic structural units. PhD program conducted by A. F. Schmitt under direction of Professor Bruhn. L. A. Hromas, top left (circa 1952)..**91**

3.19 Undergraduate enrollment (sophomore through senior) in School of Aeronautical Engineering from 1942 to 1960..**92**

CHAPTER 4

4.1   Professor Milton U. Clauser, school head, 1950–54..**95**

4.2   Professor Harold M. DeGroff (1954), school head, 1955–63...........................**96**

4.3   Professor Maurice J. Zucrow (1952)...**97**

4.4   Professor Angelo Miele (1955)...**98**

4.5   Professor Paul S. Lykoudis (1956) **99**

4.6   Professor Madeline Goulard (1958) **100**

4.7   Professor C. Paul Kentzer (1954) **101**

4.8   Professor Joseph Liston points out features of early jet engine to air transportation student Robert Bass (1954) **110**

4.9   Professors H. M. DeGroff, G. M. Palmer, and M. U. Clauser near subsonic wind tunnel (1953)...**110**

4.10   Professor M. U. Clauser standing in subsonic wind tunnel (1954) **111**

4.11   Air transportation student Joanne Alford, Purdue homecoming queen (1952)...**112**

4.12   Students preparing to board United Airlines DC-6 planes for visit to aerospace industries in the Dallas-Ft. Worth, Texas, and Wichita, Kansas, areas (January 1951)...**114**

4.13   Air transportation students preparing to board flight at LaGuardia Airport in New York City for return to Purdue (January 1950).....................**114**

4.14   Members of Purdue Glider Club studying glider plans (1951) **116**

4.15   Tow car and glider used by Purdue Glider Club (1951)...............................**117**

4.16   Professor Mart I. Fowler, on right shaking hands with Captain Chuck Yeager (first pilot to break sound barrier in 1947). Professor Elmer Bruhn looks on, Captain Massa in doorway (circa 1950) **119**

4.17   Purdue Aeromodelers (early 1950s) **120**

4.18   Purdue Aeromodelers club (Neil Armstrong is fourth from left in second row)...**120**

4.19   Captain Iven C. Kincheloe (BSAE '49) holds a model of the X-2, the jet he flew to a record-breaking height of 24 miles in September 1956 (August 1958 *Redbook* magazine photo)...**129**

4.20   Professor Hsu Lo (1964)..**132**

CHAPTER 5

5.1   Aeronautical and Engineering Sciences Building, 1995 (currently Nuclear Engineering Building)..**147**

5.2   Aerospace Sciences Laboratory (1995) **147**

5.3   Professor John L. Bogdanoff (circa 1971) **151**

5.4   Undergraduate enrollment in Aeronautical and Engineering Sciences......**157**

5.5   Astronauts Grissom (BSME '50), Chaffee (BSAE '57), and White with the Apollo Entry Vehicle (1966)..**159**

5.6   Grissom Hall, 1995 (formerly the Civil Engineering Building) **159**

5.7   Chaffee Hall (circa 1980)..**160**

5.8   Professor Paul E. Stanley (on right) with Dr. McLeod of the Student Health Service (1967)..**172**

5.9   Linda McClatchey Flack, administrative assistant (circa 1975) **183**

5.10   Professor Robert L. Swaim with TR-10 analog computer (1968).................**184**

CHAPTER 6

6.1   Professor Frank J. Marshall (1974) **193**

6.2   Professor George M. Palmer — buildings in the wind tunnel (1970s) **199**

6.3   Rhea Walker, secretary to the head of the School (1961). Ivy-covered building in the background is Old Heavilon Hall Shops **200**

6.4   Professor Ervin O. Stitz (on right) with Dean Emeritus A. A. Potter (1972)..**202**

CHAPTER 7

7.1   Undergraduate enrollment in School of Aeronautics and Astronautics. (Does not include engineering sciences.) **205**

7.2   AAE graduate enrollment from 1961 to 1995...**207**

7.3   Summary of annual sponsored research expenditures **207**

7.4   Calendar year summary of MS and PhD theses completed by the School of Aeronautics and Astronautics **207**

7.5   Professor H. T. Yang advising student (circa 1980) **208**

7.6   Professors A. F. Grandt, W. A. Gustafson, and F. J. Marshall at reception commemorating School's 40th anniversary (July 1, 1985) **209**

7.7   Professor J. P. Sullivan with laser Doppler velocimetry experiment (circa 1980)...**210**

7.8   School of Aeronautics and Astronautics staff (September 1992) **211**

7.9   Professors J. Hancock and T. J. Herrick (July 1, 1985)....................................**213**

7.10 Professors L. T. Cargnino and E. F. Bruhn (circa 1982) **213**

7.11 G. D. Calvert and Professor G. P. Palmer (circa 1987) **214**

7.12 Professors M. M. Stanisic, C. P. Kentzer, and G. P. Palmer (1982) **214**

7.13 G. P. Harston, Professor D. K. Schmidt, and C. R. Malmsten examining circuit board for General Automation computer (circa 1981) **216**

7.14 Professors D. A. Andrisani, K. C. Howell, R. E. Skelton, and D. K. Schmidt (1982) **216**

7.15 Professor K. C. Howell discusses an orbit mechanics problem with graduate students David Spencer (seated) and Hank Pernicka (1985) **216**

7.16 Professor T. A. Weisshaar (center) with R. Benford (left) and S. Stein from Thiokol Corporation with Thiokol SPACE Award plaques (1991)...............**226**

7.17 Students Eric Bennett and Margaret Struckel adjust model rocket in preparation for launch (1991) **227**

7.18 Professor C. T. Sun with materials testing machine in Composite Materials Laboratory (circa 1987) **231**

7.19 Graduate student John J. Rushinsky and Professor Andrisani fly constrained radio controlled airplane in subsonic wind tunnel (circa 1987) **232**

7.20 Students performing AAE 334 wind tunnel experiment (circa 1987)..........**232**

7.21 Graduate student Matt Leddington prepares to make dynamic photoelasticity measurements with ultra-high-speed camera (circa 1988)............................**233**

7.22 Graduate student David P. Witkowski prepares drone for measurements in subsonic wind tunnel (circa 1988)..**233**

7.23 Graduate student Robert Frederick performs high-pressure propulsion measurement (circa 1987) **234**

7.24 Graduate student Jason Scheuring and Professor A. F. Grandt examine fatigue crack growth specimen (1994) **234**

7.25 Graduate student Robert T. Johnston conducts a flow visualization experiment of an aircraft model in the subsonic wind tunnel (circa 1990) **235**

7.26 R. Bateman (BSAE '46), J. Hayhurst (BSAE '69), and R. Taylor (BSME '42) represent The Boeing Company at September 19, 1991, dedication of the Boeing Wind Tunnel..**236**

7.27 Professor S. Schneider works with Ludwieg tube hypersonic wind tunnel (circa 1992) **236**

CHAPTER 8

8.1 1994-95 AAE faculty and research interests **238-239**

CHAPTER 9

9.1 Summary of undergraduate enrollment in School of Aeronautics and Astronautics (excluding freshmen) **260**

9.2 Summary of graduate enrollment in School of Aeronautics and Astronautics from 1961 to 2008..**260**

9.3 Summary of annual externally funded research expenditures by School of Aeronautics and Astronautics from 1970 to 2007 ....**261**

9.4 School of Aeronautics and Astronautics 2006 faculty and staff....................**269**

9.5 Aerial view of the Maurice J. Zucrow Laboratories (Summer 2007) **279**

9.6 Graduate students Brad Wheaton, left, and Peter Gilbert stand near a segment of the Boeing/AFOSR Mach 6 Quiet Tunnel, March 2009 (Purdue News Service photo/Andrew Hancock)..**280**

9.7 Purdue President France A. Córdova, from left, and former Apollo astronauts Neil Armstrong and Eugene Cernan, listen to speakers during October 27, 2007 dedication of the Neil A. Armstrong Hall of Engineering. Mr. Armstrong and Mr. Cernan, the first and last men to walk on the moon, joined 14 other astronauts and former astronauts who are Purdue alumni at the dedication ceremony. (Purdue News Service photo/Dave Umberger) **281**

9.8 Third floor view of atrium of Armstrong Hall showing installation of replica of Apollo I command module. Purdue astronauts Gus Grissom (BSME '50) and Roger Chaffee (BSAE '57) died with Ed White in a training accident when a launch pad burst into flames on October 27, 1967 (Purdue News Service photo/David Umberger)..**282**

9.9 Professor Farris with (clockwise) Faculty Emeriti W.A. Gustafson, G.P. Palmer, and L.T. Cargnino at the Neil Armstrong statue dedication (October 2007)..**283**

CHAPTER 10

10.1 Typical graduation class (1986)..**286**

10.2 Purdue's first class of aeronautical engineers (1943) at their 50th reunion (April 1993). Left to right, back row: M. Howland, R. Herrick, P. Brink, R. Boswinkle, D. Ochiltree, J. Goldman, W. Fleming, I. Kerr. Front row: A. Streicher, C. Hagenmaier, J. Dunn, J. Allen, R. Beebe, R. Pendley **289**

10.3. Purdue alumni selected for space flight. Thirty-seven of the 101 U.S. manned space flights completed by July 1995 had at least one crew member who graduated from Purdue **304-307**

10.4 Alumni astronauts pose with Purdue President France Córdova at the Neil Armstrong dinner, October 26, 2007. Back, left-right: Mark Brown, Jerry Ross, Gregory Harbaugh, Janice Voss, Andrew Feustel, Mark Polansky, David Wolf, John Blaha, Charles Walker, Michael McCulley, Donald Williams; Front: Gary Payton, Neil Armstrong, France Córdova, Gene Cernan, Loren Shriver, Richard Covey (Photo by Vincent Walter)..**307**

10.5 Distinguished Engineering Alumnae Lana Couch (BSAE '63) with (left to right) Dean Yang, Professors Gustafson, Cargnino, Sullivan, and Palmer (April 1994) **320**

10.6 Purdue President S. C. Beering, former faculty member S.S. Shu, Professor A.F. Grandt, and Dean H.T. Yang at the May 1991 commencement prior to awarding the Honorary doctorate to Dr. Shu..**333**

10.7 Honorary doctorate recipient R. Bateman with Professor A.F. Grandt and Dean H.T. Yang at May 1992 commencement **334**

FOREWORD

TO THE FIRST EDITION

WRIGHT FLYER

L. T. Cargnino

A. F. Grandt, Jr.

W. A. Gustafson

This history of the Purdue School of Aeronautics and Astronautics was prepared in the spring of 1995 in connection with the school's golden anniversary celebration. The authors are all current or former members of the school faculty. Professor L. T. Cargnino joined the school in 1945, and taught until his retirement in 1984. Professor W.A. Gustafson joined the school in 1960, and has been its associate head since 1979. He also served as acting head during the 1984-85 academic year and the spring '93 semester. Professor A.F. Grandt joined the school in 1979, and served as head from 1985 to 1993. By coincidence, he was born in July of 1945, the same month the current school was formally established as an independent academic unit in the Schools of Engineering, and two months before Professor Cargnino joined the original faculty.

Much of the material in the first three chapters has been edited from an internal Purdue University report by Professor Elmer F. Bruhn titled *A History of Aeronautical Education and Research at Purdue for Period 1937-50*. Professor Bruhn came to Purdue in January of 1941, and served as first head of the School of Aeronautics from 1945 through 1950. Written during the 1966-67 academic year, this report was Professor Bruhn's last major assignment before retiring from Purdue in June of 1967. Although Professor Bruhn's school history was written nearly 30 years ago, and copies are in the Purdue University libraries, it has not been widely circulated.

Thus, the present authors felt it would be appropriate to include portions of that document here.

Although we have heavily edited Professor Bruhn's report for length and style, and have supplemented his descriptions with additional information in some cases, most of Chapters 1 through 3 retain Professor Bruhn's original organization and interpretation of early school events. In several instances we have quoted directly from Professor Bruhn's writings when it seemed appropriate to give his first-person description of an event or individual. Such cases are clearly identified in the text.

Most of Chapter 4 was prepared in 1988-89 by Professor Cargnino, and was originally published in May of 1989 as a School of Aeronautics and Astronautics report titled *Aeronautical Engineering at Purdue University from 1950 to 1960.* That report was subsequently revised in May of 1991, and both versions have been distributed on an informal basis by the school.

The remaining chapters have been prepared by the authors for this volume. We have relied heavily on internal University documents as well as personal recollections and those of other staff and alumni. The detailed information in Chapter 8 [now Chapter 10] about members of the class of 1943 was assembled by James R. Dunn (BSAE '43) in connection with his class's 50th anniversary reunion.

We apologize for any errors in the text, and would appreciate having them brought to our attention. We welcome comments from readers, and would appreciate receiving other anecdotes about the school's programs, its staff, and its alumni.

We dedicate this book to our current and former students. Truly it can be said that faculty bask in the reflected glow of alumni achievements. We are proud of the accomplishments of former students, and thank them for setting high standards which motivate current students (and faculty) to achieve higher levels of performance.

We wish to acknowledge and thank the many individuals who helped us prepare this volume. Diane Schafer typed much of the manuscript, and Linda Flack and Terri Moore helped assemble information from various school records. Lisa Tally, Engineering Administration, proofread most of the text, and provided some of the statistics about astronaut accomplishments recorded in Chapter 8 [now Chapter 10]. James Dunn gave permission to use much of the information he collected about his 1943 classmates included in Chapter 8 [now chapter 10], and he and fellow classmate Jerry Goldman reviewed several of the first chapters dealing with the School's early programs. Professors J. L. Bogdanoff, S. J. Citron, and A. L. Sweet read parts of the manuscript and made a number of helpful suggestions.

Michele Brost provided some enrollment data from the Registrar's Office. Helen Schroyer, Library Special Collections, and Vincent Walter, Office of Publications, helped locate in the University archives some of the photographs reproduced in the text. Many of the early photographs are from the J. C. Allen & Son collection. The Engineering Productions Office prepared final camera-ready artwork for the book: Matt Harshbarger illustrated the aircraft featured on chapter title pages and Susan Ferringer designed the final text and arranged printing of the book.

Following the original edition in August of 1995, another printing was made in January of 1996. The second printing corrected a few minor typographical errors and omissions, but made no major changes. At the time of the third printing in the fall of 1999, it was also desired to provide an update of school activities. Since the 1995 text represents the status of the school and its alumni at the time of its 50th anniversary, it was decided to relegate subsequent additions (except for correction of minor errors and omissions) to an addendum incorporated at the end of the original volume.

A. F. Grandt, Jr.

W. A. Gustafson

L. T. Cargnino

School of Aeronautics and Astronautics

Purdue University

September 1999

FOREWORD

SPIRIT OF
ST. LOUIS

TO THE SECOND EDITION

    The first edition of this history of the Purdue School of Aeronautics and Astronautics was prepared in spring 1995 in connection with the school's golden anniversary celebration. Two printings of the original version resulted in 400 and 250 copies respectively. (The second printing corrected a few minor errors.) A third printing of an additional 400 copies followed in September 1999 and contained a new addendum intended to update progress that had occurred in the intervening four years.

    This second edition is a more ambitious updating of the many developments that have occurred since the 50th anniversary celebration. The past decade and a half has been a most exciting period in the school's history, marked by substantial growth in academic and research programs, many new faculty, increased recognition, and relocation to a new home in the Neil Armstrong Hall of Engineering. Rather than relegate description of these changes to a separate addendum as in 1999, we decided to reorganize the book.

    Most of the first seven chapters are largely unchanged, except for a few minor additions and corrections. The original benchmark chapter 9 that described the 1994-95 academic year has been renamed as chapter 8; a new chapter 9 has been added to reflect the 1995–2009 period; and the original chapters 8 and 10 have been relocated as chapters 10 and 11. The appendices also have been updated to reflect current data, a new appendix F has been added, and an index has been added.

    We wish to acknowledge and thank the following staff from the School of Aeronautics and Astronautics for helping assemble the information for the second edition: A. Broughton, L. Crain, L. Flack, K. Johnson, P. Kerkhove, J. LaGuire, and T. Moore. We also greatly appreciate the assistance of S. Ferringer, M. Hahn, and A. Roberts for editing and designing the final text for printing. The dedicated efforts of Ms. Roberts to update the original text to incorporate the many changes to this second edition are gratefully acknowledged.

    The authors hope that this updated volume is of interest to the many

Profs. Grandt, Cargnino, and Gustafson, June 2009

students, alumni, friends and supporters of the Purdue School of Aeronautics and Astronautics. As before, we would appreciate learning of errors in the text, and would welcome comments and additional anecdotes from readers.

A.F. Grandt, Jr.
W.A. Gustafson
L.T. Cargnino

Purdue University
School of Aeronautics & Astronautics
Neil Armstrong Hall of Engineering
701 W. Stadium Ave.
West Lafayette, IN 47907-2045
Telephone: (317) 494-5117

July 2010

INTRODUCTION

LOCKHEED ELECTRA 10E

Purdue University has played a leading role in providing the nation with the engineers who designed, built, tested, and flew the many aircraft and spacecraft that so changed human progress during the 20th century. It is estimated that Purdue has awarded 6% of all BS degrees in aerospace engineering, and 7% of all PhDs in the United States during the past 65 years. These alumni have led significant advances in research and development of aerospace technology, headed major aerospace corporations and government agencies, and have established an amazing record for exploration of space. More than one third of all U.S. manned space flights have had at least one crew member who was a Purdue engineering graduate (including the first and last men to step foot on the moon). Originally written on the occasion of the golden anniversary of the Purdue University School of Aeronautics and Astronautics, this book attempts to describe how that organization has developed into one of the world's leading institutions of aerospace education and research.

Although it may surprise some that a small community in northwest Indiana has played such a leading role in developing air and space travel, Greater Lafayette has a long history with air transportation. Lafayette was, for example, the location of the first U.S. airmail delivery on August 17, 1859.[1]* This aviation experiment, which occurred before Abraham Lincoln became president, was accomplished by means of the hot air balloon *Jupiter* (Figure I.1). The balloon pilot was John Wise of Lancaster, Pennsylvania, and his flight was directed by U.S. Postmaster Thomas Wood of Lafayette. This first airmail delivery consisted of 123 letters and 23 circulars, and traveled approximately 25 miles to Crawfordsville, Indiana, where the balloon was forced to land due to lack of buoyancy. The mail was then transferred to another U.S. mail agent, who sent it on by railroad to New York City. It also has been reported that Mr. Wise conducted scientific experiments during this flight to measure the presence of ozone in the upper atmosphere. Thus, 10 years before Purdue University was established in 1869, Lafayette already had a history of experimentation with air travel, and with using that new technology for scientific exploration.

* References listed on page 344.

Figure I.1   First official delivery of U.S. airmail by John Wise in his balloon Jupiter, from Lafayette, Indiana, on August 17, 1859. (Courtesy of Tippecanoe County Historical Association).

Community interest in aviation continued when Purdue University was established across the Wabash River in what was to become West Lafayette. Purdue alumnus George Ade, for example, referred to an aeronautical engineering department at the fictional Indiana Institute of Technology (his pseudonym for Purdue) in his 1908 musical comedy *The Fair Coed*. After opening in Lafayette, this play had a successful run on Broadway, and may have been the first suggestion that aeronautical engineering should play a strong role in Purdue's curriculum.[1] The Purdue Aero Club was organized in 1910 under the direction of Professor Cicero B. Veal of mechanical engineering, and the community's first aircraft demonstration was held on June 13, 1911, during Purdue's Gala Week (Figure I.2). Sponsored by the Purdue Alumni Association and the *Lafayette Journal* newspaper, this Aviation Day attracted an estimated 17,000 people to see Lincoln Beachey and C. C. Witmer land flimsy biplanes on the Purdue athletic field.[1,2] Although Aviation Day was a great success, plans to take George Ade and Purdue President Winthrop E. Stone aloft that day had to be canceled, as strong winds prevented Beachey and Witmer from carrying passengers. (It also has been reported[3] that the Board of Trustees considered flight too dangerous, and

*Figure I.2   Greater Lafayette's first aviation show, held June 13, 1911, at Stuart
Field on the Purdue campus.*

forbade Stone from flying.) At any rate, an unplanned excitement did occur
when Beachey's engine failed during one demonstration, and he was forced
to make a dead stick landing on a nearby Purdue farm field. Other flights to
campus in the next few years continued to draw large crowds (Figure I.3).

The first Purdue graduate to become an aviator was J. Clifford Turpin
(class of 1908), who was taught to fly by Orville Wright. Turpin set an altitude record of 9,400 feet in 1911, establishing an alumni tradition that was
continued 55 years later, when an X-2 aircraft flown by Purdue graduate
Captain Iven C. Kincheloe (BSAE '49) set an altitude record of 126,000 feet
in 1956. (That record was subsequently surpassed by alumni Neil A. Armstrong [BSAE '55] and Eugene A. Cernan [BSEE '56] during their flights to
the moon.) Lieutenant George W. Haskins was the first alumnus to land at
Purdue, as he flew from Dayton, Ohio, in 1919, and landed on the same Stuart
Field that had been the site of the 1911 Aviation Day. Lieutenant Haskins
carried a resolution from the Dayton alumni group proposing that a School
of Aviation Engineering be established at Purdue.[1] As discussed in the following chapter, formal courses in aeronautical engineering finally began
two years later in 1921 in the School of Mechanical Engineering. (Lieutenant
Haskins later returned to campus himself to lead that effort from 1930 to
1937. See Figure I.4.)

In 1930, Purdue became the first university in the United States to offer
college credit for flight training, and it opened the nation's first college-owned
airport[1,2] in 1934. Although somewhat controversial among faculty, the concept of providing academic credit for flight training was actively promoted

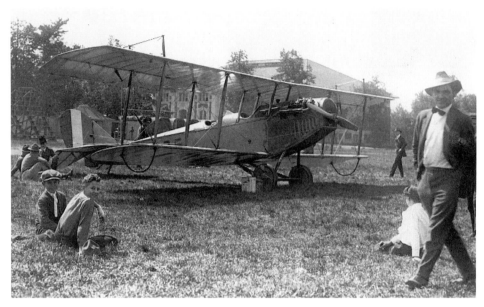

Figure I.3   Airplane on Purdue Stuart Field (1918).

by Purdue President Edward C. Elliott. Elliott became convinced of the future of aviation when, discovering that a snowstorm made area highways impassable, he asked local pilot Lawrence I. Aretz to fly him to Indianapolis to make a speech. Elliott later was responsible for bringing Amelia Earhart to Purdue as a counselor on careers for women, a staff position she held from 1935 until her disappearance in 1937 (Figure I.4). Purdue was also instrumental in providing funds for Earhart's ill-fated Flying Laboratory, the Lockheed Electra that she intended to fly around the world in 1937. Purdue Libraries houses an extensive Earhart collection, which continues to be studied by those seeking to solve the mystery surrounding her final flight.

     Purdue's active involvement in flight training continued during the 1930s, and it was an important military flight training center during World War II. A program for teaching aviation technicians was started in 1954-55, and a two-year professional pilot program was created in 1956. A general aviation flight technology course was created in 1964, and a BS program in professional piloting was approved in 1964. These pilot training and aircraft maintenance programs continue in the aviation technology programs offered by the Purdue College of Technology.

     The focus of this book, however, is on Purdue's aeronautical and astronautical engineering programs. Chapter 1 describes the beginnings of aeronautical engineering at Purdue before World War II, and Chapter 2 traces the war activities that led to the establishment of an independent school in

Figure I.4   Amelia Earhart visits Professor George Haskins' aeronautical
engineering class (September 1936).

1945. The subsequent growth and accomplishments of that program are
chronicled in the remaining chapters.

AERONAUTICAL ENGINEERING BEFORE WORLD WAR II

Although the School of Aeronautics and Astronautics was not formally established as an independent member of the Schools of Engineering until July 1, 1945, aeronautical education and research started two decades earlier with programs in the School of Mechanical Engineering. Beginning with a few courses offered in the early 1920s, the fledgling aeronautical effort grew in depth and scope, leading to a formal four-year curriculum in aeronautical engineering during World War II. The first aeronautical engineering degrees were granted by a combined School of Mechanical and Aeronautical Engineering in August of 1943. This chapter describes Purdue's aeronautical engineering programs before World War II.

1920–30

Purdue began limited education in aeronautical engineering during the 1921–22 academic year with four elective courses offered in mechanical engineering. These courses are listed in Table 1.1 as they appeared in the 1924–25 Purdue catalog and were often referred to as the senior aeronautical option. Professor G. A. Young, head of the School of Mechanical Engineering, offered the new courses in response to strong student interest in aviation. Professor Martin L. Thornburg, a 1915 ME graduate and veteran of the Air Service, was in charge of instruction.[1] (The School of Mechanical Engineering had 25 staff members, 603 undergraduates, and 16 graduate students in 1922.) An aerodynamics laboratory was established in Heavilon Hall and equipped with a fully assembled airplane and operating engines. Professor Thornburg left Purdue in 1924 to become head of mechanical engineering at the University of Arizona, and responsibility for the aeronautical courses was given to Professors Elbert F. Burton and Alan C. Staley. They were followed by Major William A. Bevan (1876–1943) of the Air Service, who joined Purdue in 1926. Promoted to associate professor in 1928, Major Bevan left in 1929 to help establish an aeronautical engineering program at Iowa State College.

The course descriptions given in Table 1.1 are limited in theory or technical analysis. The beginning of this decade was, however, only 12 years after the Wright brothers had delivered the first military airplane to the Army. Performance requirements for that aircraft were a two-person payload, a cruising speed of 40 miles per hour, and a duration of two hours. Furthermore, only two years had elapsed since the end of World War I, and although a great expansion had occurred in military flight, commercial aviation was practically nonexistent. Thus, the courses described in Table 1.1 were realistic relative to the progress of aviation at that time. Although a formal four-year curriculum in aeronautical engineering was not available until the 1940s, many Purdue civil and mechanical engineering graduates took the aeronautical electives and entered the new aeronautical industry during the 1920s and 30s.

Table 1.1   Aeronautical Engineering Courses Offered by the School of Mechanical Engineering During the 1924–25 Academic Year

---

193.   **Aeronautical Engineering:** Sem. 1, 2 + 0, Cr. 2.

Rotary and stationary aviation engines, advantages and principles of operation kind. Theoretical and practical problems. Professors Burton and Staley.

194.   **Aeronautical Engineering:** Sem. 1, 2 + 0, Cr. 2.

Meteorology, principles of mechanical and buoyant flight, and the application of engineering to its various problems, with special attention given to a discussion of the performance and possibilities of approved and proposed types of aircraft and the commercial and military future of aviation. Professors Burton and Staley.

195.   **Aircraft Engine Design:** Sem. 1, 0 + 6, Cr. 2.

The chief characteristics of aviation engines and their requirements for different services are discussed. The elements common to all types are worked out in a group, after which each student completes the calculations and details for a motor to meet certain requirements. Professors Burton and Staley.

196.   **Aircraft Design:** Sem. 2, 0 + 6, Cr. 2.

The various elements of construction and design of aircraft are considered. Suitable materials are selected and the main parts calculated for strength and airfoil. Each student takes up a special type and works out the details. Professors Burton and Staley.

1930–37

As the 1930s began, two new names appeared on the list of mechanical engineering faculty giving courses in aeronautical engineering: Professor George W. Haskins (Figure 1.1) and Mr. Bauers. (Professor Haskins is the same Lieutenant Haskins who had flown to Purdue in 1919 from Dayton, Ohio, becoming the first graduate to land an aircraft on campus.) Subsequent Purdue catalogs show they modified and expanded the aeronautical courses offered as technical electives in mechanical engineering. Relatively few changes were made between 1930 and 1937, however, and the material presented in Table 1.2, taken from the 1930–31 catalog, is quite representative of aeronautical instruction during that period. Comparing the 1930–31 curriculum to that from 1924 –25 still shows heavy emphasis on the aircraft power plant. More treatment of performance, stability and control, and the theory of aerodynamics is, however, evident. Textbooks on aerodynamics were now becoming available, and much test data on airfoil characteristics were being published by the several NACA (National Advisory Committee

Figure 1.1
Professor George W. Haskins (left) and Richard Arnett with Curtiss Robin cabin plane lent to Purdue by the Curtiss Services Corporation for class work (January 1930). Professor Haskins served on the faculty from 1929 to 1937.

for Aeronautics) government research laboratories. Purdue also had its own wind tunnels for aerodynamic measurements,[1] with earlier tunnels having been constructed by Professor Solberg, by Professor "Bridge" Smith of civil engineering, and a third by Professor Haskins. Various photographs of school facilities and student groups from this era are shown in Figures 1.2–1.9.

Table 1.2   Aeronautical Engineering Courses Offered by the School of Mechanical Engineering During the 1930–31 Academic Year

ELECTIVES FOR UNDERGRADUATE STUDENTS

93.   **Aeronautics:** Sem. 1 & 2, 3 + 0, Cr. 3
For students desiring some nontechnical knowledge of aeronautics. Includes description of types of aircraft and essential principles of aerodynamics. Given by Professor Haskins.

94.   **Aeronautics:** Sem. 1 & 2, 3 + 0, Cr. 3.
Continuation of 93 to include further detailed study of aircraft engines, propellers, and navigation. Professor Haskins.

UNDERGRADUATE AND GRADUATE COURSES

193.   **Aeronautical Engineering:** Sem. 2, 2 + 0, Cr. 2.
Rotary and stationary aircraft; advantages and principles of operation of each kind; theoretical and practical problems. Professor Haskins and Mr. Bauers.

193a.   **Advanced Aeronautical Engineering:** Sem. 1 & 2, 3 + 0, Cr. 3.
A continuation of ME 193 with more advanced study of the power plant, including the propeller and its relation to performance. Professor Haskins and Mr. Bauers.

194.   **Aeronautical Engineering:** Sem. 1, 2 + 0, Cr. 2.
Principles of mechanical flight and application of engineering to its various problems; special attention to the discussion of the performance and possibilities of approved and proposed types of airplanes. Professor Haskins and Mr. Bauers.

194a.   **Advanced Aeronautical Engineering:** Sem. 1 & 2, 3 + 0, Cr. 3.
A continuation of ME 194 with more advanced study of the airfoil, stability, control, and performance of the airplane. The layout of an airplane for a definite purpose. Professor Haskins and Mr. Bauers.

195.   **Aviation Engine Design:** Sem. 2, 0 + 6, Cr. 2.
Aviation engine characteristics and their adaptation to flying services. Calculations of details to meet certain requirements. Professor Haskins and Mr. Bauers.

196.   **Airplane Design:** Sem. 1, 0 + 6, Cr. 2.
Elements and construction and design of an airplane to meet government requirements for a particular service. Professor Haskins and Mr. Bauers.

Figure 1.2   *View of aeronautics laboratory showing wind tunnel, Curtiss Robin aircraft, and engine (May 1930).*

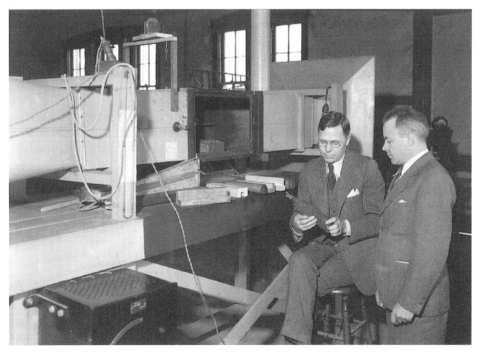

Figure 1.3   *Professors Haskins (right) and Hollister (left) examining 12-inch wind tunnel (November 1931).*

Figure 1.4   *Aeronautical engineering class. Left to right: P. D. Pruitt (standing), W.H. Rivers, M.G. Haines, M.S. Finch, W.W. Halsteadt, J.L. Fihe, and Professor G.W. Haskins (September 1934).*

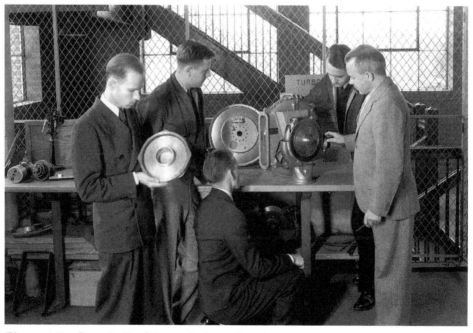

Figure 1.5   *Captain Haskins and class of students examining turbo charger for aircraft engine (September 1934). Left to right: D. P. Keller, W. H. Rivers, R. Warren (kneeling), M. G. Haines, and Captain Haskins.*

Figure 1.6   Aeronautics class around aircraft at Purdue airport (October 1934).

Figure 1.7   Students move glider into position for flight at the Purdue airport (May 1934).

Figure 1.8   Professor G. W. Haskins (second from right) and students examining airplane (October 1936).

Figure 1.9   Class of aeronautical engineering students studying construction of Navy aircraft sent to Purdue airport (October 1936).

Aircraft were still relatively slow in 1930, with a maximum speed of approximately 200 miles per hour. Only three years before, Charles Lindbergh had made his famous 33-hour nonstop flight from New York to Paris in a single engine monoplane, cruising at 105 miles per hour. Mr. Lindbergh's flight gave commercial aviation a big boost, and commercial air transportation was beginning with 10 to 25 passenger airplanes such as the Ford Tri-Motor, the Boeing Tri-Motor, and the Fokker Tri-Motored transport. The famous Douglas DC-3, which was so instrumental in beginning modern air transportation, was still several years in the future. Aeronautical engineers and scientists still believed that airplanes would be limited to the speed of sound, and few, if any, dreamed of the turbojet power plant and propeller-less airplane that were in the near future.

1937-41

Professor Haskins left Purdue and returned to the Air Corps in 1937 and later joined the Civil Aeronautics Board. He was succeeded by three key individuals who were responsible for many important developments in aeronautical education and research at Purdue. In September 1937, Professors Karl D. Wood (Figure 1.10) and Joseph Liston (Figure 1.11) joined the faculty as professor and assistant professor, respectively, while Professor Elmer F. Bruhn (Figure 1.12) joined the school in January of 1941 as an as-

Figure 1.10
Professor Karl D. Wood
(1898–1995).

Figure 1.11 Professor Joseph Liston (1906–78).

Figure 1.12
Professor Elmer F. Bruhn
(1899–1984).

sociate professor.

Professor Wood came to Purdue after many years on the faculty of the Engineering Mechanics Department at Cornell University. The year before coming to Purdue, however, he was a full-time member of the engineering staff of Consolidated-Vultee Aircraft Corporation in San Diego, California. Professor Wood was possibly the first person to publish a comprehensive book on airplane design (K. D. Wood, *Airplane Design, A Textbook on Airplane Layout and Stress Analysis Calculations with Particular Emphasis on Economics of Design*, published by the author, 1st edition, 1934). His original book and its many subsequent revisions were widely used by both students and practicing engineers, and made a valuable, lasting imprint on aeronautical education in the United States and throughout the world. Professor Wood was a strong advocate of education that balanced theory, technical analysis, testing, and design. Although Professor Wood left Purdue in 1944 to become head of the Aeronautical Engineering Department at Colorado University, his dedicated guidance and leadership during his seven-year tenure left a permanent imprint on aeronautical education at Purdue.

Professor Liston came from the University of Oklahoma, where he was a member of the Mechanical Engineering Department. His primary interest at that time was in the field of aircraft power plants. He had previously received his BSME in 1930 and MSME in 1935 from Purdue. Professor Liston remained on the aeronautical staff until 1972, completing 35 years of service to aero education at Purdue. Professor Liston was responsible for the development of outstanding power plant design courses and test facilities during the 1940–50 decade.

Table 1.3 gives the aeronautical engineering curriculum reported in the 1939–40 Purdue catalog. Seniors in mechanical engineering who desired limited knowledge of aeronautics were permitted to take six hours of technical electives in aeronautics courses 83 and 94. Admission to the aeronautical option was limited to students whose grades in mathematics, physics, and applied mechanics averaged at least a B. A fifth year, leading to an MS degree in ME with a major in aeronautical engineering, could be taken with the courses given in Table 1.4.

Table 1.3   Aeronautical Engineering Curriculum Offered by the School of Mechanical Engineering During the 1939–40 Academic Year

AERONAUTICAL OPTION:

Admission to this option is limited to students whose grades in mathematics, physics, and applied mechanics average at least a B.

(3) Airplane Design 194   (3) Airplane Engines 193

(2) Airplane Design 196   (2) Airplane Engine Design 195

CATALOG COURSE DESCRIPTIONS

83.   **Aeronautics:** Sem. 1 & 2, Cr. 1
Elective, open to all sophomores. Required of all students expecting to take senior option in aeronautical engineering. Required of all students taking CAA flight training unless ME 93 or ground school equivalent is taken. Elementary principles and practice of flight, aeronautical nomenclature, aircraft engine, and instrument maintenance. Professor Wood.

93.   **Aerodynamics of the Airplane:** Sem. 1 & 2, 3 + 0, Cr. 3
Required of all juniors planning to take the senior option in aeronautical engineering. Introduction to problems of flight, wing and engine-propeller characteristics, performance, stability, and control of airplane. Professor Liston.

94.   **Aeronautical Navigation and Meteorology:** Sem. 1 or 2, 3 + 0, Cr. 3
Intended primarily for mechanical engineering juniors and seniors. Navigation instruments, course calculation by dead reckoning, practice of celestial navigation, use of radio navigation. Thermodynamics of the atmosphere, atmospheric stability, factors affecting wind circulation, air mass characteristics, meteorological instruments, etc. Professors Binder and Liston.

193.   **Aircraft Engines:** Sem. 2, 3 + 0, Cr. 3
Theory of the principles, design, and limitations of aircraft engines; study of means for increasing performance. Professor Liston.

194.   **Airplane Design:** Sem. 1, 3 + 0, Cr. 3
Principles of airplane layout to meet a given specification; stress analysis of principal parts; design of a few details, with emphasis on economic factors. Professors Wood and Liston.

195.   **Airplane Engine Design:** Sem. 2, 0 + 6, Cr. 2
Design possibilities and limitations; selection and preliminary design of an aircraft engine. Professor Liston.

196.   **Airplane Design:** Sem. 1, 0 + 6, Cr. 2
Laboratory, drafting, and shop work in connection with the building and testing of models and airplanes. The student develops a preliminary design of airplane; supplementary to course 194.

198.   **Advanced Airplane Design:** Sem. 2, 1 + 6, Cr. 3
Continuation of ME 194 involving a more extensive design project. More adequate studies of stability, control, performance, strength, and rigidity. Professor Wood.

Table 1.4   *Fifth-Year Curriculum in Mechanical Engineering Leading to MS Degree with Major in Aeronautical Engineering (1939-40)*

Fifth Year Leading to MS Degree

Advanced courses in aeronautical engineering leading to an MS with a major in aeronautical engineering. Courses available and their catalog descriptions follow:

Catalog Course Descriptions

293.   **Advanced Aerodynamics:** Sem. 1, 3 + 0, Cr. 3.
Brief study of theory of ideal fluids and more extensive study of theory of actual fluids, with technical application. Professor Wood.

294.   **Theory and Design of the Propeller:** Sem. 2, 2 + 3, Cr. 3.
Theory and design of aerial propeller by several methods, including Eiffel's and Modified Drzewiaki system; drawings; analyses of the stresses. Professor Wood.

295.   **Airplane Testing and Development:** Sem. 1, 1 + 6, Cr. 3.
Laboratory work at the airport and the construction of special instruments and actual test flying. Flight test technique and correlation with wind tunnel tests. Professor Wood.

296.   **Airplane Testing and Development:** Sem. 2, 1 + 6, Cr. 3.
Continuation of ME 295. Special lab fee of $10. Professors Liston and Wood.

297.   **Advanced Aircraft Engine Design:** Sem. 1, 1 + 6, Cr. 3.
Design drawing and calculations for aircraft engines. Advanced study of ways for increasing performance. Professor Liston.

298.   **Advanced Aircraft Engine Design:** Sem. 2, 1 + 6, Cr. 3.
Continuation of ME 297.

Comparing this curriculum with those given earlier, one notes that aeronautical education was greatly expanded under Professors Wood and Liston, particularly in the fields of theoretical aerodynamics, airplane design, and aircraft engine design. The curriculum did not, however, have courses in aircraft structures (i.e., materials, stress analysis, structural design, aerodynamic loads, dynamics, flutter, etc.). Furthermore, laboratory courses in aerodynamics and aircraft propulsion were practically nonexistent.

When Professors Wood and Liston joined Purdue in the fall of 1937, laboratory facilities consisted of a small continuous-flow wind tunnel located in old Heavilon Hall (Figure 1.2), plus an assortment of airplane and aircraft engine parts that only could be used as teaching aids for descriptive phases of an aeronautical course. Professors Wood and Liston soon initiated steps, however, to develop laboratory facilities (Figures 1.13–1.16).

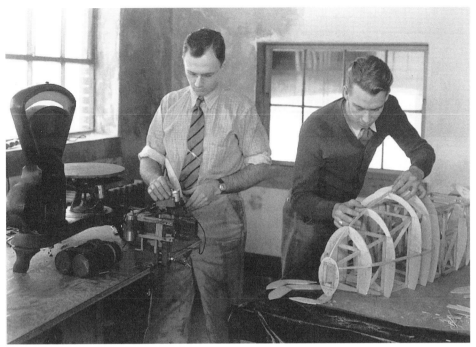

Figure 1.13 Senior aeronautics students Frank Crosby (left) and Paul Marshall (right) working on radio-controlled model airplane designed by Professor Wood (February 26, 1938).

Figure 1.14 Students testing strength of aircraft wings with bags of sand (July 1940).

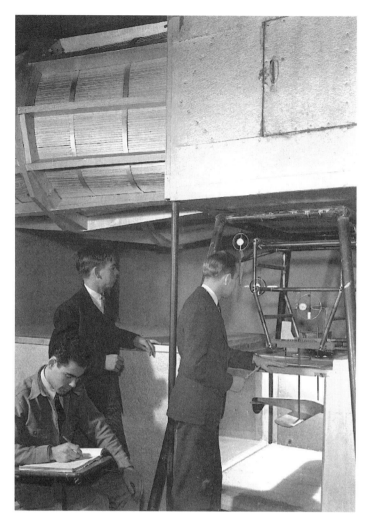

Figure 1.15
Purdue wind
tunnel showing
balance system
used to measure
aerodynamic loads
(February 4, 1939).

The structural-design weakness was ably addressed by the recruitment of Professor Elmer F. Bruhn in January of 1941. Professor Bruhn received a BS in civil engineering from the University of Illinois in 1923 and an MS in 1925 from the Colorado School of Mines. There he taught subjects dealing with mechanics and structures for five years, advancing to the rank of associate professor. He then left academia for a career in the aircraft industry, spending 12 years with North American Aviation and Vought-Sikorsky Aircraft Companies. He was deeply involved in several aircraft design projects for those companies, and at the time he joined Purdue was the project engineer supervising the final design and construction of a long-range flying boat.

Professor Bruhn's industrial experience impressed upon him the need for a comprehensive book on aircraft structural analysis and design, and was a key factor leading to his decision to join the School of Mechanical En-

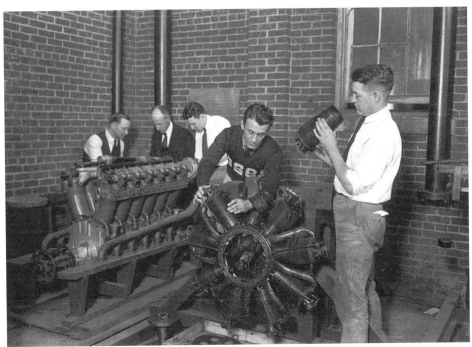

Figure 1.16 Professor Liston (second from left) and students examining aircraft engines (circa 1943).

gineering in January 1941. As he approached retirement in 1967, Professor Bruhn recalled his reaction to a letter from Professor Wood inviting him to visit Purdue to discuss a faculty position.[4]

> *My first reaction to the letter was that I was not interested; however, I had felt for a number of years that a new book on aircraft structural analysis and design better suited for college curricula and for practicing aero engineers was badly needed. I had started to try to write such a book, but the progress had been extremely slow, as time demands of a project engineer on a large project permits very little time for such a task as writing a textbook. Having had five years' teaching experience in a leading technical school, I knew that a teaching position gives far better opportunities for carrying on extracurricular work, such as writing a textbook, as compared to that permitted by a position in industry. I, therefore, decided to accept the invitation to visit Purdue, where I met Dean Potter, Professors Wood and Solberg, and others. I tentatively accepted the offered position of associate professor in the School of Mechanical Engineering, depending on whether I could make satisfactory arrangements for leaving my present position. I obtained an indefinite leave of absence after agreeing to keep in contact with the nearly completed project and returning to the company the following summer to wind up loose ends, etc. I arrived on the Purdue campus*

in January 1941, with the idea that I could try to complete the structures book in two years while teaching, and then if I felt I preferred work in industry to teaching, I would return to industry. It turned out that I never left Purdue to return to industry because of the many interesting and challenging situations that developed in the following years, and the pleasant associations with many faculty members and hundreds of students.

This proved to be a most fruitful decision, for Professor Bruhn played a key role in establishing aeronautical engineering as a separate school at Purdue following World War II. Moreover, his subsequent structural design textbook,[5] first published in 1943, sold more than 100,000 copies, and continues to be a mainstay in the aviation industry.

The improvement and expansion of aeronautical education during the 1937–41 period coincided with the rapid growth and progress of commercial and military aviation during the 1930s. The land-air transportation system, for example, was operating such airplanes as the DC-3, DC-4, and Boeing Stratoliner, while oceanic air travel was well established with Sikorsky, Martin, and Boeing Flying Boats. Military aviation had also greatly expanded with increasing aggression in Europe by Hitler's Germany. Such famous airplanes as the Curtiss P-40 and the Boeing B-17 Flying Fortress were in mass production, and many other new designs of all types of military aircraft were being rushed into production. The increasing world tension caused President Roosevelt to request legislation on May 16, 1940, to allow production of 50,000 military airplanes per year. Although such a number seemed fantastic at that time, by the end of World War II, the United States was producing aircraft well above that rate. It was fortunate that Purdue had begun to expand its aeronautical programs, for universities were soon asked to train men and women for the aviation buildup required by the war. Purdue's participation in that effort is described in Chapter 2.

Purdue Airport Facilities

Although this book deals mainly with aerospace engineering at Purdue, it is appropriate to discuss the involvement of the Purdue airport. In the late 1920s, Purdue President Edward C. Elliott and David Ross, president of the Board of Trustees, became very interested in aviation.[1] In 1930, Mr. Ross made the first of several donations of land to be used for a Purdue airport, and as president of the board of trustees, appointed a committee to formulate plans and policy for its development.

Implementation of these plans fell to G. Stanley Meikle, director of the Purdue Research Foundation. On November 1, 1930, working with representatives from the U.S. Coast and Geodetic Survey and the Department of Commerce, Purdue officials selected and approved an airport site. A wind sock was mounted on a dead tree that day, and a stone ring was put in place to mark the landing site, satisfying government requirements for having an airport designated on U.S. maps. For several years, the airport remained practically a bare field, but then, through relief grants by the Civil Works Administration and further financial support by Mr. Ross, the land was graded and drained, a rotary beacon and boundary lights were installed, and the first hangar-laboratory building erected. On September 4, 1934, the Purdue airport was officially opened for business, and was the first university-operated airport in the United States (Figures 1.17 and 1.18).

Many interesting aviation activities took place between 1934 and 1940. Amelia Earhart joined the Purdue faculty and outfitted a Lockheed Electra transport airplane as a Flying Laboratory. A credit course in flying, ME 96, (Limited Commercial Pilot Training) is shown in the 1935–36 Purdue catalog. The Civil Aeronautics Authority established its civilian-pilot training (CPT) program in 1938, and Purdue, already having an airport and hangar facilities, was the first university to get pilot training underway. More than 500 pilots were trained in the next few years. The CPT program was succeeded by the War Training Service (WTS) program to train pilots to be Naval aviation flight instructors. Additionally, the Inter-American flight training program was

Figure 1.17 Construction of Purdue airport building (June 5, 1934). The Purdue airport was established as the first university-owned airport in the nation.

Figure 1.18 Students with aircraft at Purdue airport (1930s).

conducted to train pilots for Central and South American countries following U.S. entry to the war. Later, the airport and campus became headquarters for MPATI (Mid-West Program for Airborne Television Instruction), circa 1958–1967, as well as the University-owned certified airline called Purdue Airlines from 1953 to 1972.

Thus, when Professors Wood and Liston joined the ME school in September 1937, Purdue had an operating airport and much experience with flight training. These two new faculty, along with President Elliott, Dean of Engineering A. A. Potter, PRF Director Meikle, and Professor G. A. Young (head of the ME school), were intensely interested in expanding the facilities at the airport for aeronautical engineering education and research. This combination of an operating airport and an administration enthusiastic about aviation formed the basis for subsequent expansion of aeronautical engineering educational and research programs.

PROPOSAL FOR AN AERONAUTICAL ENGINEERING LABORATORY

In June of 1938, professors Wood, Liston, and P. C. Emmons submitted an elaborate document titled "A Proposed Aeronautical Engineering Laboratory for Purdue University." The document reviewed the status of the aeronautical industry and of educational programs at Purdue and other universities, and presented detailed plans for a new building and laboratories to be located at the Purdue airport. The building would contain laboratories on a main floor that was roughly twice the size of the present hangar. Second and third floors would contain offices, instrument laboratories, design and computing rooms, lecture rooms, etc. While costs were not given for the new building, equipping the various laboratories would require an estimated $37,878.

Although this proposal was not accepted by the administration, it provides valuable insight into the thinking of Professors Wood and Liston after their first year at Purdue, and gives background for subsequent expansions that did occur during the 1940s and 50s. Thus, it seems appropriate to conclude this first chapter of the Purdue aeronautical engineering story with a review of some of the statistics and ideas presented in that 1938 proposal.

Growth in airplane production and travel is summarized in Table 1.5.

Table 1.5   Growth of Airplane Production and Travel

Year	Number of Airplanes	Airline Passenger Miles (millions)	Aeronautical Engineers in U.S.
1924	500		800
1928	5,000	30	2,000
1932	1,000	120	1,200
1936	3,000	430	3,000
1940 (est.)	6,000	1,000	5,000

The proposal notes that there were approximately 40,000 individuals employed in the aeronautical industry at that time, with 10% holding engineering positions. It was pointed out that *"this high percentage of engineers is due to the multiplicity of technical problems involved in the development of new airplanes and new airplanes are being developed now more rapidly than ever before. As the manufacturing and operating procedure become more standardized, it is not expected that there will be a reduction of engineering personnel but rather that some of the engineers will be transferred to shop and maintenance work, and the vast size of this field will easily permit absorbing a large number of technical men."*

In reviewing the status of aeronautical education in the United States, universities that received endowments from the Daniel Guggenheim Foundation were listed. The Guggenheim endowments were major grants awarded from 1926 to 1930 to stimulate aviation education and research. The Guggenheim schools and their endowments are summarized in Table 1.6.

Table 1.6   Schools Receiving Guggenheim Funds

UNIVERSITY	GUGGENHEIM FUNDS RECEIVED
New York University	$500,000
California Institute of Technology	350,000
Georgia School of Technology	300,000
University of Washington	290,000
Massachusetts Institute of Technology	264,000
University of Akron	250,000
Leland Stanford University	195,000
University of Michigan	78,000
Syracuse University	60,000

When comparing Purdue University's programs with these schools, the authors state:

> Although Purdue has not had the benefit of a Guggenheim grant or of any other endowment of similar size and as a result it is somewhat handicapped by lack of laboratory facilities, its work at present parallels the work given undergraduate students at the Guggenheim schools. Essentially this work consists in courses in fundamental theory of aerodynamics and engines and in the fundamentals of design of planes and engines. In addition, a very limited amount of research work is now in progress. However, there are no facilities for practical training in the fabrication of aircraft structures, the study of various problems connected with sheet metal construction, the study of heat treating and the application of special alloys to aircraft, or the study of the many practical phases of aircraft engine construction and maintenance. Purdue is unique, however, in having excellent ground work for such practical training and extended research work, since it has under its direct control a rather excellent airport.

Although it is pointed out that only limited research was under way, the following comments were made regarding research at Purdue in summer 1938.

> An active program of flight research is at present under way covering the general field of performance, stability and control of airplanes in flight and the effect of changes in design on these items. The principal active projects are designated by Engineering Experiment Station Nos. M-98, M-99, and M-100 with titles as follows:
>
> - M-98 — Flight Testing of a Piper Cub Airplane. Investigators, P. C. Emmons, K. D. Wood, and J. Liston.
> - M-99 — Correlation of Ground and Flight Tests on Airplane Engines.

Investigators, P. C. Emmons, K. D. Wood, and J. Liston.

- *Testing of Radio Controlled Model Airplane. Investigators, E. R. Brown, F. M. Crosby, and K. D. Wood. (See Figure 1.13)*

Each of these projects has resulted in substantial achievement and is reported in the theses submitted in June 1938, which are on file. Other topics using apparatus developed in connection with the above projects and suitable for immediate investigation may be listed as follows:

- *Measurement of propeller torque in flight.*
- *Measurement of drag of an airplane by towing tests.*
- *Characteristics of small wooden propellers determined from flight tests.*
- *Development of apparatus for determination of airplane stability from flight tests.*
- *Flight test measurements of lateral and directional stability.*
- *Flight tests of lateral and directional control.*

Development of these projects will probably be handicapped by lack of space and tools, though a limited amount of progress can be made even with the present equipment and space available at the airport.

With facilities such as we contemplate providing in the building here described, Purdue would be in a position to assume a leading position among the world's aeronautical institutions.

The proposal also includes a lengthy discussion of educational philosophy regarding the balance between theory and practice, and offers the following observations.

In the preparation of an undergraduate curriculum in aeronautical engineering, the question invariably arises as to how much specialization should be given. One point of view is that very little time should be devoted to a specialized subject such as aeronautical engineering, and it is maintained that if such specialization is included it will be at the sacrifice of basic engineering fundamentals, which will result in the student's having too narrow a training. The other group maintains that such a procedure will practically eliminate any chance for undergraduates' obtaining connections in the aviation industry at the completion of their four years' work. They maintain that in order to prepare the young engineer for the highly specialized work of aeronautical engineering, he must be given a rather extended training in strictly aeronautical engineering subjects. In a final analysis, both of these points of view are correct and the net result is that as aviation continues to progress and become more and more highly technical, it becomes increasingly difficult to give a student a thorough background in engineering fundamentals and at the same time give him sufficient specialization to make him of value to the industry. This condition is a natural consequence, but it has been rather reluctantly accepted by many people in technical schools. However, it cannot be ignored regardless of what we may desire in the matter, and the only reasonable alternative is to accept the fact and plan our schedules accordingly. In this we have a very close parallel with the professional fields of medicine and law.

When discussing future developments at Purdue, the authors state:

In our opinion, fundamental theory should continue as a basic point of emphasis. This appears to be in line with the conclusions of practically all of the other important colleges. However, we believe that supplementing of this fundamental theory with even a limited amount of practical training would greatly enhance the value of the courses as now given. Such practical training could not very well be added to the regular school year courses as they are already heavy, but it could be instituted as a required summer course and would then parallel rather closely the required summer courses in civil engineering and other technical lines. ... The other phase of activity at Purdue which merits increased emphasis is graduate research. There can be little dispute of the fact that sound progress in aeronautics is very dependent on research and development. It is believed that this is more important in aviation that in any other industry, and research activity certainly is not outside the fields of endeavor of technical schools. Unquestionably further emphasis should be made in this direction, but at the same time too great an overlapping of the work of other schools should be avoided. ... For graduate research, there is here an unparalleled opportunity for development work involving actual airplanes in flight, the testing and development of actual airplane structures, the testing and development of airplane engines and fuels, and possibly also the development and use of apparatus for improved investigation of meteorological forecasting and the analysis and interpretation of weather data.

Thus, in 1938 we have the vision of Professors Wood and Liston for a sound educational program that seeks balance between theory and application. They recognized that future developments would rest on rigorous preparation in engineering fundamentals, but that exposure to practical problems was needed to provide the context for studying basic principles. (This is, in fact, the basic philosophy behind the Cooperative Education Program started by the school in 1964.) Moreover, the need for an active research program that involves both theoretical and laboratory work is clearly stated. As discussed in the following chapters, World War II quickly brought short-term needs for applications of aeronautical engineering, but later programs developed strong interests in fundamental principles and basic research. This balance between theory and application, and undergraduate and graduate education and research, remains one of the school's strengths and continued objectives.

Chapter 2

Countdown:~

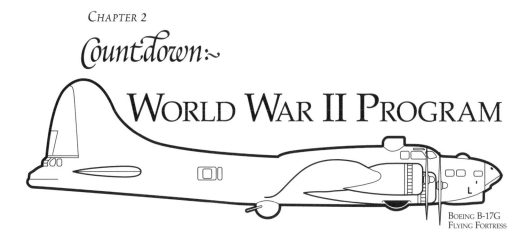

World War II Program

BOEING B-17G
FLYING FORTRESS

Four-Year Undergraduate
and Graduate Curriculum

Although the United States was not yet formally involved in the expanding world war, Dean of Engineering A. A. Potter had decided by mid-1941 that aeronautical engineering would play an important part in the growing war training effort. Since ME had provided courses in aeronautical engineering for 20 years, and had greatly expanded aero coursework after the arrival of Professors Wood and Liston in 1937, Dean Potter recommended that ME immediately develop a four-year BS curriculum in aeronautical engineering. The name of mechanical engineering was changed to the School of Mechanical and Aeronautical Engineering in 1942, and Professor Harry L. Solberg, who had been appointed head of ME in March of 1941 following Professor G. A. Young's retirement, was named head of the combined schools. (Professor Young had served as head of ME from 1912 to 1941, and all of the original aeronautical engineering courses were developed under his administration.)

The initial four-year BS curriculum from the 1941-42 Purdue catalog is given in Table 2.1. Note that this program required $160\frac{2}{3}$ credits. The previous weakness in structures and aerodynamics laboratory courses had been eliminated, with 6 hours of aircraft structures theory, 3 hours of structures laboratory, and 8 hours of laboratory work in aerodynamics, engines, and shop. Design had been increased to 11 hours, and the eighth-semester student was required to choose between airplane or engine design. Students completing the aeronautical engineering curriculum could also obtain a BSME by taking an additional semester of work. The sophomore-year summer program was eliminated the following year, requiring some rearrangement of the original four-year program given in Table 2.1. An MS also was offered, requiring 18 hours in a major area and 12 hours in a minor. The major subject courses listed in the 1941-42 catalog are given in Table 2.2.

The first students to pursue an official aeronautical engineering degree under the new ME-aero program began their studies in 1940. Since the University had gone on an accelerated three-semester-per-year schedule during the war, the first graduates were ready to receive their diplomas in August of 1943. These individuals were truly pioneers, progressing through the first full aeronautical curriculum under very hectic wartime conditions. They served as guinea pigs for the new courses, helping to develop laboratories and facilities and setting the academic standards for subsequent generations of students. A list of the first aeronautical engineering graduates is given in Appendix A, and a detailed description of their professional careers is given in Chapter 10.

Having established a formal aeronautical engineering curriculum, and realizing that World War II would lead to additional training needs, the administration was faced with providing space and laboratories for the expanded program. Since aeronautical engineering was now part of the combined School of Mechanical and Aeronautical Engineering, the logical solution would have been to add a south wing at the east end of the Mechanical Engineering building. Professor Solberg assigned the task of developing such a space layout to Professor Wood, who was assisted by Professors Liston and Bruhn.

Table 2.1   Aeronautical Engineering Curriculum Offered during 1941-42
Academic Year ($160^2/_3$ hours required for graduation)

FRESHMAN YEAR COMMON TO ALL ENGINEERING

Credits	First Semester	Credits	Second Semester
(4)	Chemistry 1 or 21	(4 or 3)	Chemistry 2 or 22
(3)	English 1	(3)	English 10 or Speech 14
(2)	Engineering Drawing II	(2)	Engineering Drawing 12
(0)	Engineering Lecturers	(0)	Engineering Lectures 2
(3)	Mathematics 1	(5)	Mathematics 2
(2)	Mathematics 7	($1^2/_3$)	Military Training 2
($1^1/_3$)	Military Training 1	(2)	Surveying 6 or Shop 34
($^2/_3$)	Personal Living 9	($17^2/_3$)	
(2)	Shop 34 or Surveying 6		
(18)			

Table 2.1   Aeronautical Engineering Curriculum (continued)

SOPHOMORE YEAR

Same as the sophomore year in mechanical engineering except that Aeronautics 1, cr. 1 is added to either semester.

Credits	First Semester	Credits	Second Semester
(3)	English 10, 19, or 27	(4)	Applied Mechanics 21
(3)	Economics or English 31	(3)	Mechanism 11
(4)	Mathematics 3	(4)	Mathematics 4
($\frac{1}{3}$)	Mech. Engr. Problems 90	(5)	Physics 22
(5)	Physics 21	($1\frac{2}{3}$)	Military Training 4
($1\frac{2}{3}$)	Military Training 3	(2)	Shop 35 or 36
(2)	Shop 35 or 36	($19\frac{2}{3}$)	
(1)	Aeronautics 1		
(20)			

SUMMER SESSION AT END OF SOPHOMORE YEAR (BEGINNING SUMMER 1942)

Credits	Summer Session
(4)	Thermodynamics 34
(4)	Applied Mechanics 22
($\frac{2}{3}$)	Testing Materials 31
($1\frac{1}{3}$)	Aeronautical Shop 50
(10)	

JUNIOR YEAR (BEGINNING SUMMER 1942)

Credits	First Semester	Credits	Second Semester
(2)	Descriptive Geometry 16	(3)	English 31 or Economics 1
($3\frac{2}{3}$)	Electrical Engineering 19	($3\frac{2}{3}$)	Electrical Engineering 20
(4)	Fluid Mechanics 105	(3)	Internal Comb. Engines 36
(3)	Industrial Management 110	(1)	Mechanical Engineering Lab 78
(3)	Differential Equations 107	(2)	Aero. Materials & Processes 132
(4)	Aerodynamics 100	(3)	Aircraft Structures 131
($19\frac{2}{3}$)		(3)	Kinetics of Machinery 65
		($18\frac{2}{3}$)	

SENIOR YEAR (BEGINNING SUMMER 1943)

Credits	First Semester	Credits	Second Semester
(5)	Machine Design 66	(3)	Engineering Admin. 103
($3\frac{1}{3}$)	Aircraft Engines 110	(2)	Aero Instruments 105
(5)	Aircraft Structures 133	(3)	Propellers 103
(2)	Airplane Layout & Draft. 104	($4\frac{2}{3}$)	Airplane Design 135
(3)	Non-Technical Electives	or (6)	Engine Design 111
($18\frac{1}{3}$)			plus Technical Elective
		(3)	Econ. of Air Transp. 142
		(3)	Non-Technical Elective
		($18\frac{2}{3}$ or 20)	

Table 2.2   Major Subject Courses for MS Degree in Aeronautical Engineering (1941-42)

Technical Aerodynamics 201 and 202	(6 hours)
Aircraft Engines 211 and 212	(6 hours)
Aircraft Structures 231 and 232	(6 hours)
Heat Transmission 203 and 204	(6 hours)
Hydrodynamics 207 and 208	(6 hours)
Vibrations 215 and 216	(6 hours)
Elasticity 217 and 218	(6 hours)
Research in Aero Engr 299	(6 hours)

The first plan involved adding a complete three-story wing to the ME building to take care of foreseeable growth in aeronautical engineering. Space would be provided for large fluid mechanics, aerodynamics and structures laboratories, a 600-mph wind tunnel, and extensive office research labs and test rooms. This proposal was unrealistic, however, and a second, three-story half-wing plan was proposed. That proposal was also too expensive, and the time of completion was too long relative to the urgent need for new aero facilities. Finally, the small two-story building shown in Figure 2.1 was hurriedly built and completed by the fall of 1942. This building was the campus home for aeronautical engineering until the entire school was moved to the airport in 1948. Figure 2.2 is an aerial photograph of the campus during World War II, and shows the location of the Purdue airport and the campus aeronautical engineering building.

Laboratory exercises played an important educational role during the 1942–47 period, both as a means to substantiate results of theoretical calculations and to provide data for the senior design courses. Figures 2.1 and 2.3 show several photographs of students working in the aerodynamics and structures laboratories in the new campus aero building, in the large design room located in building No. 2 at the airport, and at the flight test laboratory at the airport. Building 2 was completed in 1941 and housed the power plant, flight test laboratories, and design rooms. Figure 2.4 shows students working in the power plant and flight test laboratories.

It is believed Purdue was the first university to offer a course in airplane flight testing. That unique course involved comparing theoretical flight performance and stability with the actual flight characteristics measured in a flying airplane. This course provided the opportunity for many of the 1942–1947 aero graduates to find flight test positions in industry. (A photograph of students preparing for a flight test experiment is included in Figure 2.1.)

The aerodynamic layout and design room at the airport.

The aeronautical laboratory on the main campus.

Engineering cadettes running a model test in the wind tunnel.

Students preparing for a flight-test experiment.

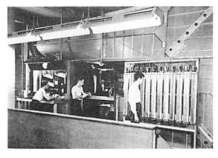

Students making an airplane model test in the 100 mph, 30-by-48-inch-throat wind tunnel.

The quiet, free-flight (inclinable) wind tunnel for classroom demonstration.

Figure 2.1   Collection of photographs from 1945 school pamphlet showing first Aeronautical Engineering Building located on main campus and students working with aerodynamics and flight test facilities.

Following the Pearl Harbor attack in December 1941, the United States became formally involved in World War II. University education and research played a vital part in the subsequent war effort, and Purdue programs in aeronautical engineering are discussed in the remainder of this chapter. To keep the size of this program in perspective, it is useful to note the *total Purdue enrollment* during this period.

- 1940 — 6,966 students

- 1941 — 6,473 students

- 1942 — 6,687 students

- 1943 — 4,102 students

- 1944 — 3,762 students

- 1945 — 5,628 students

Figure 2.2   Aerial photograph of Purdue campus and airport circa 1943 (facing southwest).

Students fabricating structural test units.

Students static-testing a plywood fuselage.

Students measuring strains in an airplane metal wing with electrical strain-gage equipment.

Students testing a wing beam under combined bending and axial loads.

Students making a compression test of a skin-stringer specimen.

Engineering cadettes making a static proof test of horizontal tail surfaces.

Figure 2.3 *Photographs from 1945 pamphlet showing students working in aircraft structures laboratory located in main campus Aeronautical Engineering Building.*

A corner of the disassembly laboratory, showing some of the facilities available for work on aircraft engines.

The power plant and flight-test laboratories, showing stacks for the torque stands at the right.

The C.F.R. fuel octane-rating engine with dynamometer for single-cylinder instruction and research.

Students preparing to test a full-scale, air-cooled aircraft engine in Torque Stand No. 1.

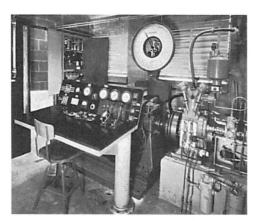

The supercharged E.G.C. test unit showing some of the automatic control equipment, the main control panel (in front of the electronically-controlled A.C. dynamometer), and the special test engine (to the right).

Students installing an aircraft engine and a controllable propeller in the torque stand preparatory to testing.

Figure 2.4   Photographs from 1945 pamphlet showing students working in aircraft power plant laboratory located at Purdue airport.

AIR CORPS CADET PROGRAM

The Air Corps Cadet Aeronautical Engineering Program was an extensive 12-week effort with the curriculum shown in Table 2.3. The program was given twice, once to groups of 50 students who arrived in January of 1941 and again in April of 1941. The rigorous curriculum was based on the fact that all students were engineering graduates with strong academic records. A large majority held mechanical engineering degrees obtained from around 40 universities.

Table 2.3   Air Corps Cadet Curriculum (1941)

	Hours per Week	
Course Name	*Lecture*	*Lab or Computation*
AC 11 Aerodynamics	4	0
AC 12 Airplane Structures	5	6
AC 13 Aircraft Engines	4	3
AC 14 Aircraft Instruments	2	0
AC 15 Materials of Construction	2	0
AC 16 Aircraft Performance	2	0
AC 17 Aircraft Propellers	1	0

The cadets trained for work in Air Corps operations and aircraft maintenance. After completion of the three-month Purdue program, they were sent to Chanute Field at Rantoul, Illinois, for several weeks of intensive practical training, and then commissioned as Army Air Corps officers. Reflecting on this program in the late 1960s [4], Professor Bruhn wrote:

> *My initial major assignment at Purdue was the teaching of the structures course, which involved five hours of lecture work and six hours of computing laboratory. As I look back over the past 25 years at Purdue, I feel that this group of 100 cadets was no doubt the most outstanding group of young men that I had the privilege of working with. They were a dedicated and hard working group and the finest types of young manhood.*
>
> *I remember very vividly one morning near the end of June 1941, when the second group of 50 were only one month away from the end of the program. As I entered one of the larger class rooms in the ME building, the group as a whole appeared to be terribly downcast. When I asked the reason for all the gloom, they replied, "Professor, the way the Hitler war machine is rolling to victories, we are fearful this war will be over before we finish at Chanute Field and get our commission." Little did any of us realize at that time that the Pearl Harbor catastrophe was less than six months away. Years later, I learned that several of them had died in the great Pacific Air campaign.*

CURTISS-WRIGHT CADETTE PROGRAMS

The urgent military demand for tens of thousands of airplanes presented a tremendous engineering design and production task. Although most U.S. aircraft companies expanded their facilities with large new plants built by the government, finding the necessary engineering staff became a critical problem. Purdue conducted two programs between 1943 and 1945 to train young women for technical positions normally held by men at that time: Curtiss-Wright Cadette Programs Number 1 and Number 2.

In late 1942, the Curtiss-Wright Airplane Corporation decided to train 700 to 1,000 young women for work in the engineering departments of their three main plants located in Buffalo, New York; Columbus, Ohio; and Louisville, Kentucky. The approximate cost for this training was $1 million. Cornell, Iowa State, the University of Minnesota, the University of Texas, Rensselaer Polytechnic Institute, Pennsylvania State College, and Purdue were invited to send representatives to a meeting in New York City on the proposed training. As a result of this meeting, all seven schools decided to participate in what became known as the Curtiss-Wright Cadette Training Program.

It was decided to begin training seven weeks later, and students were recruited from graduates of colleges and universities across the United States. Little time was available for recruiting or for student evaluation and screening. As the first 100 cadettes arrived at Purdue on February 12, 1943, Professor E. F. Bruhn later recalled [4]: *"I definitely remember that day as there was over six inches of snow on the ground and still snowing as the 100 cadettes from the far South were seeing their first snow. As this very attractive group of cadettes filled the hallways of the Aero building, it appeared the Aero school men students were very pleased and readily assisted the girls in getting registered and oriented, so to speak. The Aero staff seemed to appear in smiles or be quite happy with the sudden turn of events."*

The first 100 cadettes were graduates from a large number of colleges and universities. The largest group (13) came from the University of Texas, and the next largest group (7) from the University of Minnesota. Cadettes were housed on the upper floors of Wood Women's Residence Hall under the supervision of Helen Ripley. Curtiss-Wright sent Ruth Morrison to campus to serve as personnel supervisor for the cadette group. Her major responsibilities were to keep in close contact with the cadettes, to counsel them, check on class attendance, and pay their salaries. Purdue staff were only responsible for academic issues during the first cadette program, as all personal matters were handled by Miss Morrison. All living expenses and a $10-per-week salary were paid by the Curtiss-Wright Corporation.

The cadette program was offered under the War Training Division at Purdue with Professor C. W. Beese as the director of war training. Since the curriculum involved staff from several schools, Professor K. D. Wood was appointed coordinator for the program, and was assisted by Professor Bruhn. February 12, 1943, marked the first experience of many Purdue staff with teaching all-female classes.

The curriculum for the first cadette program is given in Table 2.4 and was common to all seven participating colleges. The program consisted of two 22-week-long terms and was very heavy in drafting, materials processing, and testing. The schedule included supervised study, and involved the Purdue teaching staff given in Table 2.5. Eighty-three cadettes completed the first program, with most assigned to the Columbus plant, and a few to the Buffalo and Louisville plants. Commencement exercises were held on December 14, 1943, with Dr. Lillian M. Gilbreth giving the commencement address.

Within a few months, increasing war work at Columbus again presented a staffing problem, and that division decided to repeat the cadette training program at Purdue. The second program was shortened to six months with two 12-week terms and a week's vacation between the two terms.

Table 2.4   *First Curtiss-Wright Cadette Curriculum (44 Weeks, Starting February 12, 1943)*

TERM 1		TERM 2	
Course Number/Name	*Class Hours*	*Course Number/Name*	*Class Hours*
CW 1 Engr. Mathematics	6	CW 1 Engr. Mathematics	3
CW 2 Job Term. & Specif.	3	CW 2 Theory of Flight	3
CW 3 Elements of Aircraft Drawing & Standards	9	CW 3 Aircraft Drawing & Descriptive Geometry	12
CW 4 Elem. Engr. Mechanics	6	CW 4 Strength of Materials (11 weeks) and Aircraft Structures (11 weeks)	6
CW 5 Properties and Processing of Aircraft Materials	6	CW 5 Aircraft Materials & Testing	6
CW 6 Supervised Study	10	CW 6 Supervised Study	10
TOTAL hours/week	40	TOTAL hours/week	40

*Table 2.5   Teaching Staff for Curtiss-Wright Cadette Program Number 1
(Coordinator — K. D. Wood)*

Mathematics	J. N. Arnold, C. Holton, N. Little
Aeronautics	E. F. Bruhn, A. D. Johnson, K. D. Wood, W. E. Woodward
Drafting	M. H. Bolds, S. B. Elrod, F. H. Thompson
Applied Mechanics	S. Fairman, E. F. Miller, W. B. Sanders
Materials Processing and Testing	E. F. Bruhn, R. B. Crepps, C. A. Haag, A. D. Johnson, O. D. Lascoe, R. W. Lindley, H. F. Owen, P. J. Panlener, L. F. Schaller, A. J. Vellinger, K. D. Wood
Total Teaching Staff	21, consisting of four professors, three associate professors, five assistant professors, and nine instructors.

Cadette recruiting and selection were directed by Warren Bruner, coordinator of cadette training for the Airplane Division of the Curtiss-Wright Corporation. Several cadettes from the first program assisted in recruiting, and as before, numerous universities and junior colleges were contacted. The age limit was lowered to 18 by the date of plant induction, and a few high school graduates were accepted. Candidates were selected on the basis of their school record and specific information furnished by their teachers.

The July 1944 cadettes were housed in one wing of the Women's Residence Halls. By October of 1944, crowded conditions made it necessary to assign the next group to vacant fraternity houses operated by the University. Each group had a house mother. The cadettes had to buy their own textbooks, slide rules, drawing equipment, and shop clothing. Curtiss-Wright paid all other expenses, including housing and tuition, which entitled cadettes to all the privileges of regular university students (i.e., admission to athletic events, health service, etc.). Each cadette also was paid $10 per week.

The curriculum for the second Curtiss-Wright Cadette Program given in Table 2.6 was revised somewhat to reflect the shorter terms and experience with the first group. As before, the teaching staff included faculty from several schools on campus (see Table 2.7). Figure 2.5 shows cadettes working in the aerodynamics, structures, and metal shop laboratories, and in the drafting room. George Palmer and Paul Homsher, listed in Table 2.7, were senior students in aeronautical engineering, and assisted in teaching the aerodynamics course. Mr. Palmer subsequently joined the faculty of the School of Aeronautical Engineering in 1948, retiring in 1987. Mr. Homsher later became a vice president with McDonnell Aircraft Company, and received the Distinguished Engineering Alumnus Award from the Schools of Engineering in 1980.

Table 2.6   Curriculum for Curtiss-Wright Cadette Program Number 2 (July 1944)

Term 1			Term 2		
	Hours			*Hours*	
Course Number/Name	*Class*	*Lab*	*Course Number/Name*	*Class*	*Lab*
CW 1 Mathematics	6	0	CW 7 Aircraft Drafting	0*	15*
CW 2 Aircraft Drafting	0*	15*	CW 8 Structural Analysis	6	3
CW 3 Materials & Drafting	2	0	CW 9 Aerodynamics	3	3
CW 4 Shop Tools & Practices	1	3	CW 1a Mathematics	4	0
CW 5 Principles of Flight	2	0	TOTAL hours/week	13	21
CW 6 Mechanics (statics)	4	0			
TOTAL hours/week	15	18			

* Lectures, 2–5 hours a week deducted from lab time

*Table 2.7   Staff for the Curtiss-Wright Cadette Program Number 2
(Coordinator — E. F. Bruhn)*

Cadette Teaching Staff

Mathematics	C. D. Olds, H. W. Strand
Drafting	S. B. Elrod, H. H. Bolds, C. E. Mayfield, C. V. Mock, F. H. Thompson, F. Zarrora
Materials and Processes	O. D. Lascoe, P. J. Panlener, C. A. Haag, L. F. Schaller, A. J. Vellinger
Mechanics and Structures	C. S. Cutshall, S. Fairman, N. Little, J. H. Mathews
Terminology-Flight	E. F. Bruhn
Aerodynamics	S. Fairman, P. Homsher, J. Liston, G. Palmer

Curtiss-Wright Officials in Direct Contact with Program

Director of Training, Columbus Plant	Richard Crow
Production Design Engineer, Columbus Plant	Frank Mallett

In the first program, Miss Morrison, a company employee, was directly responsible for all cadette personal problems and contact with the Curtiss-Wright Corporation. Since Curtiss-Wright was in great need of more help, and the company believed that appropriate jobs would be found for all cadettes, academic performance requirements were low. Students quickly learned of this fact, which had a detrimental effect on the performance of a number of the cadettes.

a. Drafting was an important phase of the Curtiss-Wright program. Front row, left: Betty Schaefer; center: Maxine Stevens. Back row, left: Dorothy (nee Wurster) Rout.

b. Structures laboratory.

Figure 2.5   Curtiss-Wright Cadettes working in various laboratories.

c. Power plant laboratory.

d. Wind tunnel tests.

Curtiss-Wright decided to treat cadettes like other Purdue students for the second program, and Miss Morrison did not return to campus. All matters involving contact with the Women's Hall officials, the Health Service, and the Dean of Students, as well as all cadette counseling, were handled by the program coordinator (Professor Bruhn). The teaching staff held frequent meetings during the second program, and maintained continuous contact with company representatives, involving much more administrative work for the second program than for the first.

One hundred and sixteen cadettes started the second Purdue program on July 4, 1944, and 98 finished on December 23, 1944. Of the 19 who did not finish, eight left on account of health, two left of their own accord during the first week, and three more withdrew during the term. Six were removed by faculty and/or company action. Eighty-three more cadettes reported for another session that started on October 2, 1944, and ended on March 24, 1945. From this group, five cadettes left voluntarily, and only one was eliminated because of poor performance.

Writing about this program in 1968, Professor Bruhn states [4]:

> I could write a small book relative to the many interesting and baffling problems or situations that were presented as cadette advisor. In a sense, as part of my job as coordinator of the cadette program, I acted as an assistant to the Dean of Women, with the instructions not to refer cadette matters to the Dean of Women unless it was urgent and beyond my powers to give a proper decision. In several situations, I had to call the dean for help.

> I believe the cadette program as a whole was a success. In talking with Curtiss-Wright officials during the several years following the completion of the program, they gave me the impression that most of the cadettes had worked out well in their various job assignments. The success of the program was due to the dedicated efforts of the teaching staff under difficult conditions. They constantly went out of their way to assist the cadettes, and thus also assist our nation in a vital and critical period of history.

> The senior Aeronautical men students also deserve some credit, as many of them became well acquainted with the cadettes and voluntarily acted as tutors, so to speak. I definitely know that two of the Aero seniors married cadettes soon after the cadette program ended, and I am sure several more cadettes met their future mates on the Purdue campus.

Some appreciation for cadette life and esprit de corps during this wartime training effort may be gained by examining four songs (following) of the many sung about the C-W Cadette program.

C. W. SWEETHEART

Oh, she is my C.W. Sweetheart,
The loveliest girl that I know.
I'll follow her over the world —
Wherever her transport may go.
Her hair is as sleek as aluminum,
Her eyes are as blue as the sky;
Her long and slender fuselage
Draws whistles wherever she flies.

Her chin with its dihedral angle
Is lifted and pointed with vim.
Her in-line nose molded proudly
Would satisfy any man's whim.

Beneath her fuselage streamlined
With curves that are faired and clear,
She moves so fast and gracefully
Her trim little landing gear.

THE GLOBE TROTTER

Oh, she speeds thru the air with the greatest of ease,
That hot little number, the "flap-in-the-
breeze"
Her airfoils are sweptback, and dihedral,
too:
Results of our study at dear ole Purdue.

With nose pointed proudly, and wings
lifted high
She soars on her journey on up in the sky,
She's a ship of the future; of course, you
can tell.
But of this we are certain
She's a bat out of hell.

THANKS FOR THE MEMORIES

Thanks for the memories
Of moment, lift, and drag,
Along with coke and fag,
Of lab reports and drafting sports
And metatarsal sag.
Oh, thank you so much.

Thanks for the memories
Of Women's Residence Halls,
Of midnight smoker brawls,
Of fire alarms and girlish charms
And greasy overalls.
Oh, thank you so much!

Many are the times that we've calculated,
And many are the times that we've
estimated;
But, still we're never dated —
Still we are here without any beer.

Oh, thanks for the memories
Of torsion, torque, and shear,
Of sailors with a leer,
Of walks on Intramural Field
Alone with you, my dear:
Oh, thank you so much!

DON'T SEND MY GIRL TO LOCKHEED

Don't send my girl to Lockheed
The dying mother said.
"Don't send my girl to Consolidated —
I'd rather see her dead.
But send my girl to Douglass,
Or better still to Bell;
But as for dear old Curtiss-Wright!!!!!!
I'd see her first in hell."

NAVY V-12 PROGRAM

The U.S. Navy designed The Navy College Training program, usually referred to as the V-12 program, to solve its own critical staffing problem. The V-12 program began on July 5, 1943, at Purdue, with an enrollment of 1,263 men; successive enrollments raised the total to 2,730 at Purdue. Approximately 400 of these men received BS degrees from Purdue. The V-12 program was conducted in 16-week terms, and the University, by giving a semi-term in the fall of 1943, was able to provide a single calendar for both civilian and Navy students. The aeronautical, civil, electrical, and mechanical schools also adopted the three-term Navy schedule for civilian students, discontinuing the prewar, two-semester-and-summer-school schedule.

Table 2.8 shows the V-12 aeronautical engineering curriculum reported in the Purdue catalog. Note that starting in the sixth term, two options were offered: the structures and the engines options. The structures option required 12 credit hours of aerodynamics, 13 credits in aircraft structural theory, laboratory testing, materials and processes, structural design, and a course in vibrations and flutter. The engine option emphasized courses dealing with theory, testing, and design of aircraft power plants.

Table 2.8   Navy V-12 Curriculum in Aeronautical Engineering

TERM 3	*Credits*	*Course Number/Name*	
	(4)	MS5	Calculus I
	(4)	CN1a	Chemistry Ia
	(3)	BAN1	Economics I
	(3)		Elective
	(3)	PsN1	Psychology I - General
	$(1^2/_3)$	MTs3	Military Training 3
	(1)		Physical Education Men 12
	$(19^2/_3)$		
TERM 4	*Credits*	*Course Number/Name*	
	(4)	MS6	Calculus II
	(2)	CN2a	Chemistry IIa (8 weeks)
	(2)	CN6	Engineering Materials (8 weeks)
	(5)	AN1, AN2	Analytical mechanics I, II
	(3)	BAN2	Economics II
	(2)	MEN 1	Kinematics
	$(1^2/_3)$	MTs4	Military Training 4
	(1)		Physical Education Men 12
	$(20^2/_3)$		

Table 2.8 Navy V-12 Curriculum in Aeronautical Engineering (continued)

TERM 5	Credits	Course Number/Name	
	(3)	AEN1	Aerodynamics I
	(3)	MEN4a	Thermodynamics Ia
	(4)	EEN10	Electrical Engineering I
	(3)	CEN3	Strength of Materials I
	(2)	CEN4a	Materials Laboratory Ia
	(3)	MN7	Calculus III
	(1)		Physical Education Men 12
	(19)		

TERMS 6-8: ENGINES OPTION

TERM 6	Credits	Course Number/Name	
	(4)	EEN11	Electrical Engineering II
	(5)	MEN3	Heat Power I
	(2)	AEN13	Vibration and Dynamic Balance
	(3)	CEN6	Fluid Mechanics
	(3)	CEN7a	Structures Ia — Structural Analysis
	(3)	AEN6	Aircraft Materials and Processes
	(1)		Physical Education Men 12
	(21)		

TERM 7 (Effective July 1945)	Credits	Course Number/Name	
	(3)	AEN4	Airplane Structures I
	(4)	AEN15	Mechanism and Machine Design
	(3)	AEN21	Metal Processing
	(5)	MEN11	Heat Power II — Internal Combustion Engines
	(3)	AEN16	Aircraft Power Plants
	(1)		Physical Education Men 12
	(19)		

TERM 8 (Effective November 1945)	Credits	Course Number/Name	
	(3)	AEN17	Aircraft Engine Design
	(3)	AEN19	Aircraft Engine Laboratory
	(3)	AEN18	Utilization of Fuels and Lubricants
	(2)	GEN5	Contracts and Specifications
	(3)	AEN9	Propellers
	(3)	MEN17	Metallurgy
	(1)		Physical Education Men 12
	(18)		

Table 2.8   Navy V-12 Curriculum in Aeronautical Engineering (continued)

TERMS 6-8: STRUCTURES OPTION

TERM 6	Credits	Course Number/Name	
	(3)	AEN2	Aerodynamics II
	(4)	EEN11	Electrical Engineering II
	(3)	CEN6	Fluid Mechanics
	(2)	AHN20	Wind Tunnel
	(3)	CEN7A	Structures Ia — Structural Analysis
	(3)	AEN6	Aircraft Materials and Processes
	(1)		Physical Education Men 12
	(19)		

TERM 7 (Effective July 1945)	Credits	Course Number/Name	
	(3)	AEN4	Airplane Structures I
	(3)	AEN7	Airplane Design Practice I
	(3)	MEN3a	Heat Power Ia
	(3)	AEN9	Propellers
	(3)	AEN11	Applied Aerodynamics
	(2)	AEN14	Airplane Static Tests
	(1)		Physical Education Men 12
	(18)		

TERM 8 (Effective November 1945)	Credits	Course Number/Name	
	(3)	AEN5	Airplane Structures II
	(3)	AEN8	Airplane Design Practice II
	(2)	AEN10	Aircraft Components
	(3)	AEN3	Aerodynamics III
	(2)	AEN12	Elementary Vibration and Flutter
	(2)	GEN5	Contracts and Specifications
	(3)	AEN16a	Aircraft Power Plants
	(1)		Physical Education Men 12
	(19)		

THE MORROW AIRPLANE PROJECT

Although this project was only indirectly related to the war effort, it provided an interesting educational opportunity for senior students taking Professor Bruhn's structural analysis courses. Moreover, since the project was a good test for the school's structures laboratory facilities during World War II, it is summarized here.

Howard Morrow, one of several brothers who owned a large chain of nut and confectionery stores throughout the United States and Canada, owned and operated an airport with extensive manufacturing facilities at Rialto, California. He became interested in building a small, high-performance, two-place commercial airplane that also might serve as a trainer for the major pilot training programs being carried out by the government. Mr. Morrow proposed using bonded plywood, a nonstrategic material, for the wing, body, and tail surface.

The preliminary aerodynamic design and airplane configuration were developed by engineers from the North American Aviation Company. Although formally known as the Morrow Victory Trainer, the aircraft was essentially a scaled-down version of the North American Aviation Mustang fighter, and was often referred to as the "little mustang." Mr. Morrow started to build the airplane at his Rialto Airport facilities, but the Pearl Harbor attack brought requests from the government to use his facilities for more direct war services, forcing him to discontinue work on the airplane project.

Grove Webster, manager of the Purdue Aeronautics Corporation, met Mr. Morrow in California and discussed the uncompleted airplane. Learning that Purdue had an airport with shop facilities, an aero school with a structures laboratory, and an individual with considerable industrial experience in Professor Bruhn, Mr. Morrow proposed to complete his airplane at Purdue.

Professor Bruhn decided that the job of conducting stress analyses, static testing an airplane to destruction, and writing reports of those tests would be excellent experience for senior students taking his structures course, and agreed to help Mr. Morrow complete the airplane. The class involved two three-hour periods per week. Students who wished to spend more time were paid an hourly wage. Mr. Morrow offered a number of awards and prizes for students. The main two awards, based on leadership relative to planning and completing work on schedule, were a two-week, all-expense-paid vacation at any point in the United States. The two student winners chose Colorado as their vacation location.

Photographs showing students static testing the Morrow airplane and Mr. Morrow visiting the structures laboratory are given in Figure 2.6. Eventually the Morrow aircraft was certified, flown, and sold to a South American country. Upon later reflection on this project [4], Professor Bruhn wrote, "*I am sure the students obtained far more knowledge and experience in this project as compared to the normal course contents. I enjoyed supervising the project and was gratified to know that we had structural test facilities to test a complete airplane to destruction.*"

a. Back row, left to right: unidentified, Professor E. H. Bruhn, Leslie Schneiter, Dave Mendenhall, Lewis Jones, Ralph Trueblood, and Paul Brink. Second row: unidentified, Walter Smith, Bob Boswinkle, Oscar Lappi, Mr. Morrow, Stanley Meikle (Director of Purdue Research Foundation). Front row: unidentified, Leonard Rose, Bob Pendley, Jack Allen, and Bill Gaugh.

b. Students testing Morrow airplane.

Figure 2.6   Morrow airplane project (1944).

PROFESSOR WOOD LEAVES

In January of 1944, Professor Wood obtained a leave of absence from Purdue for the month of February for a consulting project with an airplane company in California. Shortly after his return, he announced his intention to leave Purdue and take a position with the University of Colorado as head of a new department of aeronautical engineering. His departure was a great loss. During his seven years at Purdue, Professor Wood had been responsible for the extensive expansion in aeronautical education and research activities. He held a firm conviction that aero engineering education should be a well-balanced program among theory, technical applied analysis, laboratory testing, and design of the flight vehicle.

Professor Wood died on April 19, 1995, more than 50 years after leaving Purdue. His belief in a rigorous technical program that balanced theory and application remains his Purdue legacy, along with a high regard by his students. He was also, however, remembered for his sense of humor. J. L. Allen (BSAE '43), for example, recalled one of his puns [6]: *"When Professor Wood announced that his newborn daughter would be named Cecibel Maxine, he*

c. Completed Morrow airplane examined by (left to right): Professor Bruhn, Mr. Morrow, and unidentified students.

decided to reduce the name to C_{lmax}. This 'in-the-trade' terminology refers to the aerodynamists' terminology for the maximum lift coefficient of a wing."

Professor Wood left when the second cadette program was about to begin, and the extensive Navy V-12 aeronautical engineering program was just getting under way. Professor Bruhn had acted as an assistant to Professor Wood in a minor way on the cadette program and other matters, and was assigned to assume his administrative duties.

Professor Wood had handled most of the aerodynamic instruction for the school, and since both the Curtiss-Wright Program and the V-12 program required courses in aerodynamics, his departure made it necessary to obtain additional staff members. Fortunately, at that time the aero school was combined with mechanical engineering, and since wartime enrollment in ME had declined, it was possible for Professor Solberg to assign several ME staff, including Professors Binder, Freberg, and Hall, to temporarily help with the Aero program. As pointed out previously, the World War II aeronautical engineering programs relied heavily on the assistance of faculty from other departments. When the war ended, however, these temporary staff returned to their home schools, and it was necessary to hire permanent faculty for the expanded aeronautical engineering program. The next chapter discusses the establishment of postwar programs in aeronautical engineering, and describes the recruitment of faculty, development of facilities, and formation of curricula for the postwar period.

Chapter 3

Ignition:~

The Five-Year Postwar Period 1945–50

Boeing B-47
Stratojet

Planning for Postwar Programs

World War II saw a tremendous buildup in the U.S. aviation industry. In 1943 alone, for example, the nation produced 29,355 bombers, 23,988 fighters, 7,012 transports, 19,339 trainers, and 5,604 aircraft of miscellaneous types. As described in the previous chapter, Purdue played a key role in training the engineering staff needed for this rapid expansion of design and manufacturing activities. Dean of Engineering A. A. Potter had realized the importance of air power at the onset of the war, and had quickly recommended a four-year degree program in aeronautical engineering, leading to the combined School of Mechanical and Aeronautical Engineering established on July 1, 1943.

As the war came to a successful conclusion, however, Dean Potter and Professor Harry L. Solberg, head of the combined Aero-ME school, requested a comprehensive study of the future of postwar aeronautical education at Purdue. This assignment was given to Professor E. F. Bruhn in 1944, who was assisted by Professor Joseph Liston and Grove Webster.

Since the war had served as such a strong stimulus for aviation education across the country, many other long-range studies were undertaken at this time. The American Council on Education for the Civil Aeronautics Administration, for example, had recently polled 1,500 colleges and universities to determine what aviation-related courses had been offered during the 1943-44 academic year, and what plans there were for 1944-45. The survey found that 399 institutions had offered, or were planning to offer, education in aviation, while 844 reported "no aviation courses," and 257 failed to respond. Of the 399 institutions with aviation programs, only 18 offered a BS in aeronautical engineering, and only eight provided graduate work.

Furthermore, only two universities, Purdue and Southern California, had recognized the need to expand beyond the traditional limits of engineering and to consider the broader field of air transportation.[7]

First thoughts at Purdue dealt with estimating student interest for post-war aeronautical education. Since the Air Forces involved more than two million men, it was logical to conclude many would desire to continue in some phase of aviation after the war. Although projecting student enrollment is always difficult, few would have predicted the large impact that returning veterans would have on Purdue's postwar programs. Purdue enrollment was 6,966 students in 1940, dropped to 3,762 at the height of the war in 1944, rose to 5,628 in 1945, doubled the next year to 11,462 in 1946, and peaked at 14,674 in 1948. Enrollments declined to approximately 9,300 students in 1951 and '52, and then resumed a period of steady growth until reaching 30,000 in the late 1970s, leveling off to approximately 40,000 students on the West Lafayette campus in 2008.

The next question was whether there would be demand for work in aviation after the war. One answer appeared to be given by government predictions that more than 500,000 small, privately owned airplanes would be built in the near postwar years. This increase in commercial aircraft would require hundreds of new airports and many individuals educated in airport operations, flight training, and flight administration. Since wartime develop-ment of four-engine bombers and transports had proven that transoceanic travel was safe and economically feasible, commercial airlines also faced a period of tremendous expansion. Thus, it seemed logical to conclude there would be increasing demand for graduates trained as commercial pilots and in flight operations and control; maintenance; and airfreight operations.

At first, postwar demand for aeronautical engineers did not appear as promising, because the greatly expanded aircraft industry would quickly halt much of its production when the war ended. Downsizing would not prove too important for aeronautical engineers, however, because many of the engineers taken on by the aircraft industry came from other areas that curtailed peacetime production for the war. These thousands of transplanted engineers would return to their respective industries when the war ended.

Development of the turbojet engine, which first became public around late 1943, also meant that new airplanes would soon be designed with speeds that were practically twice that of wartime aircraft. Commercial fleets of jet-powered airliners were expected for a worldwide air transportation system. Moreover, aerodynamic research indicated supersonic flight was possible. Thus, there was considerable evidence in early 1944 that aviation had tremendous potential for technical progress and growth.

Assuming that aviation had a promising long-range future, the next question was the role Purdue could play in postwar education and research. The following facts were quite evident:

(1) Purdue had an airport adjacent to the campus, and programs there would not present problems for students relative to class attendance. Moreover, the Purdue airport was located so that future expansion involving longer runways was quite possible.

(2) Purdue had already established the Purdue Aeronautics Corporation with Grove Webster as its director. Before joining Purdue, Mr. Webster had been head of civilian pilot training under the Civil Aeronautics Authority (CAA). At Purdue, he had directed extensive flight training programs sponsored by the U.S. State Department, overseen other large civilian and military training programs, and acted as manager of the Purdue airport and its fixed base operations. His practical experience, broad knowledge of civil aviation, and acquaintance with aviation leaders would be most valuable for long-range, postwar programs in the broad field of air transportation.

(3) Purdue had introduced a four-year bachelor's degree in aeronautical engineering in 1942, and only minor modifications would be needed to meet postwar requirements. By 1944, the tireless efforts of Professor Liston and his associates, Bean and Crosby, had resulted in an excellent aircraft power plant laboratory at the airport, and planning was already under way for expansion into turbojet and rocket-jets.

(4) Design rooms and other laboratories also were available at the airport, while campus aero facilities consisted of a small two-story building adjacent to the ME building. Although additional campus facilities could be provided by adding a full wing and a third floor, as proposed in 1942, the ME school would need such space for its own postwar growth. Furthermore, dividing aeronautical engineering facilities between the airport and campus was not an ideal arrangement, and abandoning the excellent laboratory and educational facilities at the airport did not seem logical.

The initial recommendation from this preliminary study was to establish an additional four-year curriculum in air transportation, and to consolidate all aeronautical engineering and air transportation activities at the Purdue airport. The administration reacted favorably to this recommendation and requested that the proposed program be studied in more detail.

Dean Potter requested input from a broad segment of the aircraft and aircraft engine industry relative to the soundness and adequacy of the two proposed postwar curricula. An elaborate questionnaire regarding the aeronautical engineering plans was sent to 13 leading aircraft companies and 10 aircraft engine companies. Response from this survey clearly showed that industry had great interest in postwar education in aeronautical engineering. Two companies, Douglas and North American Aviation, made composite reports submitted by their presidents. In general, these industrial comments were quite favorable in regard to the postwar engineering plans, and provided several constructive criticisms.

Comments from the air transportation industry also were obtained by contacting the president, vice-president, chief engineer, chief of operations, chief of maintenance, director of long-range planning, etc., of all major airlines. Similar letters were written to airport managers, to a number of well-known fixed base operators, and to top national and state officials associated with civil aviation agencies. Again, response from this broad spectrum of the aviation world was extremely favorable toward the postwar air transportation program, and resulted in many offers of assistance.

In reviewing the many conversations held at that time with other educators and industrial leaders about Purdue's postwar plans, Professor Bruhn recalled two particular conversations in the spring of 1945.[4] The first was a meeting arranged by President Elliott with Dr. Abbot, minister of education for Great Britain. After a lengthy session with Dr. Abbot, Professor Bruhn recalls him stating, *"I have visited practically every leading aero school in the United States to see what their postwar plans were, but this proposed postwar Aeronautical and Aviation Education and Research Center at Purdue is by far the most interesting and comprehensive that we have seen or heard about."* Following a similar meeting that spring with Professor Hunsacker, head of MIT's Aeronautical Engineering Department, Professor Bruhn later recalled, *"I came back to Purdue from this meeting with an aviation pioneer feeling that we were starting to develop and build something really worthwhile for the future of aviation in the broad field of aviation education and research."*

In order to acquaint prospective students and employers with Purdue's long-term plans, Professor Bruhn, Mr. Webster, and Professor Liston developed a 20-page pamphlet titled "Aviation at Purdue." The pamphlet, which was bound in an attractive cover, discussed the coming air age and Purdue's postwar education plans in aeronautical engineering and air transportation. Plans for an aeronautical center located at the Purdue airport were presented along with several pages of photographs illustrating existing airport and campus facilities. Four pages from this pamphlet, which show existing facilities in the aerodynamics, aircraft structures, and power plant areas, were reproduced earlier as Figures 2.1, 2.3, and 2.4 in Chapter 2.

SCHOOL OF AERONAUTICS ESTABLISHED

At its spring 1945 meeting, the board of trustees approved a separate School of Aeronautics effective July 1, 1945, with Professor Bruhn as acting head. The broad title "School of Aeronautics" was selected because degrees were to be offered in both aeronautical engineering and air transportation.

Three new staff had joined the School of Mechanical and Aeronautical Engineering several months previously in anticipation of this official action. These new staff included **Grove Webster**, who had been general manager of the Purdue Aeronautics Corporation, **Dr. Paul Stanley**, who had been director of the ground school for the Purdue Aeronautics Corporation, and **Edward Cushman**, who came from the Allison Company in Indianapolis. Mr. Webster joined the school with the rank of associate, and Dr. Stanley and Mr. Cushman as instructors. An organizational chart showing the relationship of the School of Aeronautics and the Purdue Aeronautics Corporation, as well as the various sections of the Aeronautical Engineering and Air Transportation departments, is reproduced here as Figure 3.1.

The first tasks for the new school were to prepare detailed curricula for both the air transportation and aeronautical engineering programs, and to inform the public of Purdue's postwar aeronautical offerings. The first official pamphlet was published with the air transportation and aeronautical engineering curricula shown in Figures 3.2 and 3.4. The proposed plans for a broad-based aeronautical-aviation education center were mainly due to Professors Bruhn, Liston, and Webster. The proposed postwar curriculum in aeronautical engineering was primarily the thinking of Professors Bruhn and Liston, while the air transportation curriculum was due to Professors Bruhn, Webster, and Stanley. Details of the air transportation and aeronautical engineering programs are discussed in the following sections.

First, however, it may be of interest to note the costs associated with attending Purdue after World War II. University fees in fall 1945 were $54 per semester, plus an additional $75 for nonresidents of Indiana. Books and supplies were estimated to cost $30 to $50 per semester, and refundable deposits for laboratory supplies (i.e. chemistry) and uniforms (all male students were still required to take two years of military training) added $6 to $20 per semester. Special semester fees for aeronautical engineering students included $12 for shop courses, and flight training varied from $328 for Flight 1 to approximately $1,100 for the more advanced flight courses. The G.I. Bill provided veterans with a maximum payment of $500 per academic year for college tuition, course fees, books, and supplies. For comparison purposes, Purdue tuition and fees for fall 2007 were $3,708 per semester, plus an

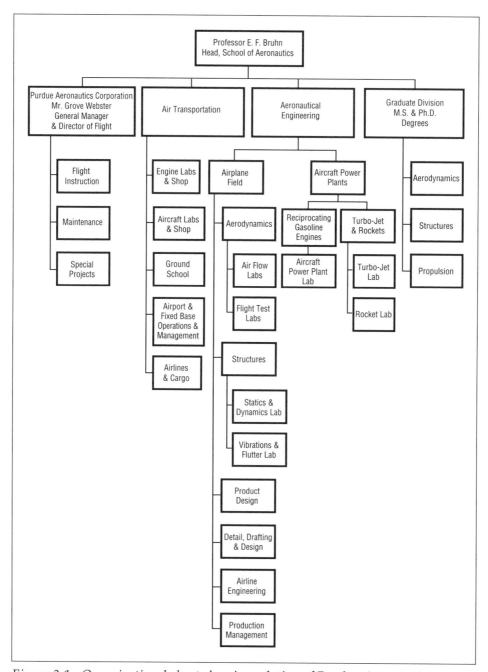

Figure 3.1   Organizational chart showing relation of Purdue Aeronautics Corporation and School of Aeronautics.

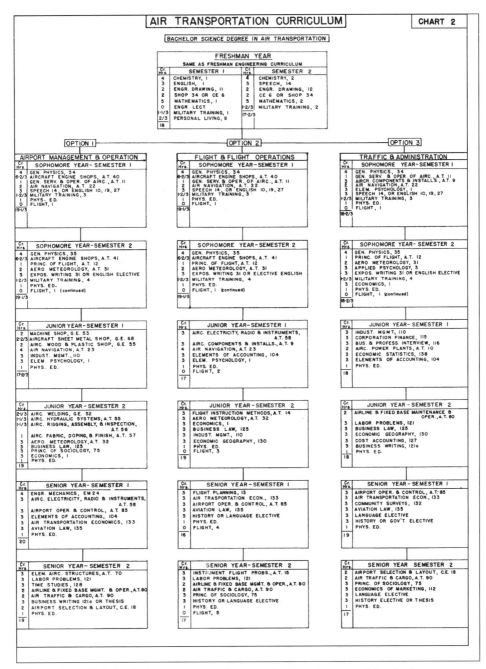

Figure 3.2   Original air transportation curriculum for new School of Aeronautics.

additional $7,404 for out-of-state residents. Books and supplies were esti-mated to cost $525 per semester in 2007-08.

The first official school pamphlet describing the new programs also had the following comments directed to veterans: *"All courses which were given by the various branches of the Armed Forces are being evaluated, and credit for similar courses in the Purdue Aeronautical Curriculum will be given if equiva-lent in content and scope. This applies particularly for the Air Transportation Curriculum for such courses as navigation, meteorology, etc. It will be the policy of the School of Aeronautics to use returning veterans as flight instructors and as-sistants in running the airport. The number of these jobs will depend on the flight enrollment; thus, no promise can be made regarding part time jobs. It is evident that the number of applicants for part time jobs will far exceed the number of jobs available."*

AIR TRANSPORTATION PROGRAM

The initial goal of this new program was to provide the education needed for the rapidly growing air transportation industry. Following the common freshman program required of all engineering students, the air transportation program involved study in three options: airport management and operations, flight and flight operations, and traffic and administration. The original descriptions of these three options as published in the first of-ficial school pamphlet are given below.

Airport Management and Operations Option

In the postwar period hundreds of the present airports will be expanded, and hundreds of new airports will be built throughout the United States and the world. Airports require facilities to take care of the maintenance and storage of airplanes and to provide flight instruction and flight facilities for the public. To manage an airport and operate the flight instruction and maintenance fa-cilities that go along with an airport will require young men and women with special training and education in aircraft maintenance, flying, and business administration. The curriculum outlined in (Figure 3.2) provides this training and education. Particular emphasis is placed on aircraft maintenance, and the shop and laboratory work is extensive enough to qualify a student to pass the requirements for a CAA aircraft and engine mechanic's certificate. Flight 1, which qualifies one for a private pilot's certificate, is required, and the preflight requirements for a commercial pilot's certificate are covered; but additional flying is optional. The curriculum provides a number of basic engineering and business courses.

The Flight and Flight Operations Option

This option is designed for those interested primarily in piloting as a career. Besides giving a basic education in general science and business administration, this option provides all the necessary ground and flight courses leading to all needed CAA flight certificates and ratings for a professional pilot. In addition, the student is qualified for the CAA engine mechanic's certificate. The graduate of this option may engage in many flying operations such as piloting for airlines, industrial firms, or private owners, also production test flying, flight instructor, aircraft sales, etc.

The Traffic Administration Option

This option omits the practical shop work necessary for the maintenance of airplanes as required in the airport management option. However, it provides courses which give a basic understanding of the terminology and uses of the various units of the airplane and its engine.

Worldwide expansion of air transportation means many additional airlines, which in turn mean air terminals and air stations with facilities for carrying on traffic, both passenger and cargo. This means opportunities for young men and young women to start in the traffic business side of air transportation. As noted in (Figure 3.2), this option provides the broad training and education in subjects necessary for this field — economics, finance, law, business administration, human relations, etc. These courses are supplemented by courses in airport operation, traffic, and cargo as required in the other two options.

Once these plans were formulated for the new air transportation program, the next task was to recruit faculty to develop courses and laboratories and teach the new curricula. Purdue was well established as a leading engineering school, and well known to the aircraft manufacturing industry, which had employed hundreds of Purdue graduates in the years before 1945. The University had also earned a national reputation in flight training through the activities of the Purdue Aeronautics Corporation and its manager, Grove Webster. Since Purdue was practically unknown in the airline and the airport operations fields, however, new faculty also had responsibility for publicizing the air transportation program to the aviation world. The initial staff hired for the air transportation program are summarized below.

Edward Cushman joined the School of Mechanical and Aeronautical Engineering in February 1945. Mr. Cushman had been a high school principal and teacher of science courses, but to aid the war effort he became an instructor at the Allison Company Engine Training School in Indianapolis. Mr. Cushman's first Purdue assignment was to obtain laboratory and shop equipment for the air transportation curriculum. Favorable war developments in 1944 caused many industrial training schools to be phased out, and Mr. Cushman was successful in obtaining much surplus equipment for the new curriculum. Many expensive engines, test stands, and elaborate mockups of mechanical, hydraulic, and electrical systems involved in aircraft were obtained for transportation costs only. Mr. Cushman was assisted in the search for surplus equipment by Professor Lascoe of the General Engineering Department and Henry Abbett, purchasing agent for the University. Although Mr. Cushman only remained at Purdue for two years, he was instrumental in obtaining laboratory equipment and organizing course work for the new air transportation program. He later went to the University of Illinois as assistant director of the School of Aviation Technology.

The next step was to obtain additional staff to help Mr. Cushman set up the equipment and work out course details before the first students arrived. Shop and laboratory courses had to satisfy the CAA requirements for a certified program relative to the training of aircraft engine and airframe mechanics.

Lawrence Cargnino, a graduate of Illinois State University, was the next person to join the air transportation staff in September of 1945 (Figure 3.3). Mr. Cargnino possessed exceptionally broad experience in teaching laboratory courses in aircraft engines and airframe systems. He had been an instructor in the Air Force School at Chanute Field in Illinois, chief instructor of aircraft power plants at Seymour Johnson Field in North Carolina, instructor in aircraft power plants at the Willow Run Aircraft Production Plant of the Ford Motor Company, and instructor in the Boeing B-29 Bomber School in Seattle.

*Figure 3.3
Professor Lawrence T.
Cargnino (1945).*

William Briggs joined the air transportation staff in February of 1946. He was a graduate of the University of Michigan in the field of actuarial science. Mr. Briggs also was an instructor in the Ford Willow Run School, and was well known to Mr. Cargnino, who recommended him for the air transportation staff. Both Mr. Cargnino and Mr. Briggs started with the rank of instructor, and while teaching, took graduate work for MS degrees at Purdue. After obtaining their MS degrees in 1948, they advanced to the rank of assistant professor, and later to associate professor.

James Basinger, the next air transportation staff member, was added in fall 1946. He was a graduate of Parks Air College, which had a national reputation as a leading school in the field of training aircraft and airframe mechanics. He joined the Air Forces after graduation, and obtained broad experience in operation and maintenance as well as flying during World War II.

Mart Fowler joined the air transportation staff in fall 1947 with the rank of associate professor to teach the air transportation courses in airline operations. Mr. Fowler held an MS from the University of Pittsburgh, and had done graduate work at the University of California and at Pennsylvania State College. He had wide teaching and industrial experience, coming to Purdue from Capital Airlines, where he served as director of training. In addition to teaching, one of Mr. Fowler's assignments was to help promote the new air transportation program within the commercial and military aviation world.

Professors Cushman, Cargnino, Briggs, and Basinger were responsible for all aircraft engine and airframe laboratory and shop courses in the air transportation program. They worked in close cooperation with the Purdue Aeronautics Corporation, managed by Grove Webster, to provide students with maintenance experience on actual flying airplanes. The power plant option of the aeronautical engineering curriculum also required a six-week summer program involving four laboratory courses in aircraft power plants. These four courses were developed and taught by Professors Cargnino and Briggs during the summer periods. Professor Cargnino subsequently authored a comprehensive textbook on aircraft propulsion engines, which was widely used by aviation technology and Air Force training schools.

After studying staff teaching loads and available laboratory and shop facilities, officials decided to limit enrollment in the air transportation program to 120 new sophomores per academic year. Applicant screening and selection were assigned to Professor Stanley, who had joined the School of Mechanical and Aeronautical Engineering on January 1, 1945, in a temporary appointment with the rank of instructor. On July 1, 1945, he was appointed assistant professor in the new School of Aeronautics.

At the beginning of the July 1945 term, air transportation admitted a few students to test the sophomore laboratories. Since the war with Japan ended on August 15, 1945, there were a large number of applicants for the spring 1946 semester. Dr. Stanley admitted less than one quarter of these first applicants. His admission criteria included consideration of the military rank of the applicant on the argument that if a man could rise from a 2nd lieutenant to major or lieutenant colonel in the military forces, he undoubtedly had considerable ability, and was likely suited for the air transportation program. The 1947 air transportation graduating class, for example, consisted of 21 men, of whom seven were lieutenant colonels. In general, the initial graduates from the air transportation programs obtained good jobs in the aviation world, and great credit is due Dr. Stanley for his excellent work with student selection.

Faculty members from other campus departments were also instrumental in starting the new air transportation program. The economics department, for example, provided two new courses for the air transportation curriculum: Air Transportation Economics 133, given by Professor Mayhill, and Aviation Law 135, given by Professor Ganong. The civil engineering school provided a course in airport selection and layout, and Professor Lascoe and his associates in the general engineering department provided a number of new shop and laboratory courses for both the air transportation and aero engineering curricula.

The airline operations and air cargo courses had been initiated by Professor Mayhill of the economics department before Mr. Fowler joined the air transportation faculty in fall 1947. Professor Mayhill made it a practice to invite speakers from the various airlines to lecture in these courses and to provide material on the airline operations and air cargo fields. In the fall of 1946, the United Airlines Company, showing great interest in the air transportation school, sent eight of its top officials to Purdue to lecture and discuss various phases of airline operations.

Although well known in engineering, Purdue was new to the field of air transportation, and officials decided to sponsor a number of conferences to acquaint the nonengineering world with Purdue's new aviation programs. Since important sources of employment for new graduates would be airports with fixed-base concerns, and national and state aviation agencies, the air transportation school decided to sponsor an annual conference on airport and fixed-base operations. The first two-day conference was offered on February 18–19, 1947, and the second on February 16–17, 1948. They attracted many aviation officials and employees from throughout the United States, many of whom flew to Purdue in their own planes. The success of these conferences was primarily due to the efforts of Professors Stanley, Webster, Cargnino, Briggs, and Basinger.

Since another air transportation goal was to prepare graduates for careers with the world's rapidly expanding airline systems, the staff decided to sponsor an annual air transportation conference. The first of these conferences was held on May 5–6, 1948, and the second on May 5–6, 1949. Professor Fowler was in charge of several staff and student committees charged with planning and carrying out the air transportation conferences, and the school was greatly indebted to his dedicated, continuous efforts in making Purdue's air transportation program known to top officials at all the leading airlines.

Air transportation students played a very important part in all conferences sponsored by the air transportation program. Between conference sessions, one would see small groups of students discussing matters with many of the visitors. These conferences acquainted Purdue to the many top leaders of the aviation world, and helped a considerable number of graduates find good positions in these aviation fields.

Professor Bruhn later recalled [4] the following about these conferences: *"I remember particularly the 1949 Air Transportation Conference because of its large off-campus attendance. At the evening dinner meeting at the end of the first day of the conference, the North Ballroom of the Union building was completely filled. Many top airline officials had arrived during the day to attend this dinner meeting. The principal speaker was Carlton Putnam, chairman of the board of the Chicago and Southern Airlines, later changed to the name Delta Airlines. Professor Fowler had assembled a very interesting entertainment program. I remember the head table extending practically the entire length of the ballroom. I doubt whether such a large representation of the top officials of the airlines of the world had ever before been assembled at one table as at this meeting. The introduction of such a large number of prominent persons from the aviation world was of great interest to our air transportation students."*

AERONAUTICAL ENGINEERING PROGRAM

The goal of the postwar aeronautical engineering curriculum was to provide a well-grounded understanding of how a flight vehicle is designed to meet given performance and operating requirements. The new aeronautical engineering curriculum is shown in Figure 3.4, and is described in the first official school pamphlet at follows.

The "Basic or Airplane Design Curriculum"

This basic or airplane design curriculum, as shown in (Figure 3.4), prepares students for work in the many engineering divisions of the airplane manufacturing industry and in various government aeronautical engineering bureaus and research laboratories, such as the product design, aerodynamics, and structures divisions.

A student, after completing the first five or six terms of the basic curriculum, may select one of three options if he feels that his interests lie in fields of aeronautical engineering other than the analysis, design, and testing phases

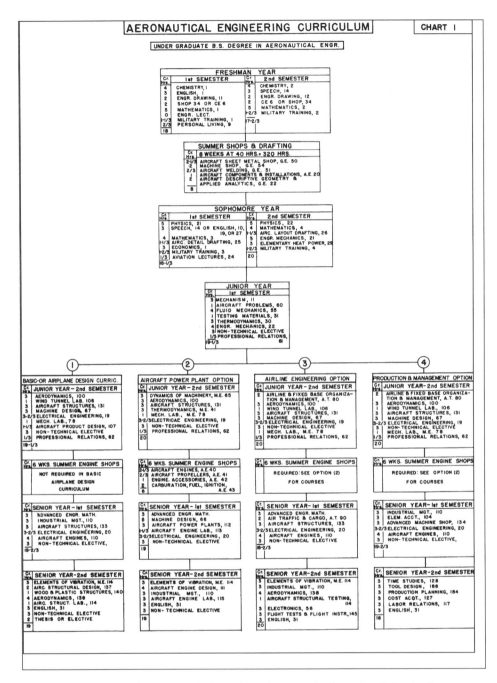

Figure 3.4   Original aeronautical engineering curriculum for School of
Aeronautics.

which are involved in the developing of a new airplane.

1. The Aircraft Power Plant Option

This option (see Figure 3.4) provides exceptional fundamental education and practical training in the field of aircraft power plants. Students who wish to enter the engineering divisions of aircraft engine companies or aircraft power plant accessory manufacturing companies, or government engine research laboratories should follow this option at the end of the fifth term.

2. The Airline Engineering Option

This option provides several courses which are more useful to students who wish to work in the engineering and maintenance divisions of large airlines. The practical shop and laboratory work is confined chiefly to engines and installations common to large airline transports. This option offers a course in flight testing and flight instrumentation and courses on the economics of air traffic, both passenger and cargo.

3. The Production-Management Option

This option (see Figure 3.4) is offered because it is believed that more graduate aeronautical engineers should enter the production-management phase of the aircraft industry. As indicated on (Figure 3.4), the first three years are the same as the basic curriculum, but the fourth year is entirely different and deals for the most part with the principles of production planning, production methods, tool design, labor and personnel problems, etc.

Practical shop and laboratory work plays an important part in the aeronautical curriculum. All students must take an eight weeks intensive shop and laboratory program during the summer after the freshman year. This program as indicated on (Figure 3.4) consists of courses in aircraft materials, aircraft sheet metal operations, aircraft welding, aircraft components and installations, and aircraft descriptive geometry.

Students selecting the aircraft power plant option or the airline engineering option must take an additional six weeks of practical shop work on aircraft engines and engine accessories. This practical background enables the student to obtain a better and more thorough understanding of the courses that follow later which deal with the fundamentals of engineering science and their application in the analysis, design, and production of the airplane.

The curriculum was primarily developed for students who would terminate their formal training with the bachelor's degree and enter the aeronautical industry. For those 10-20% of students with the highest academic record at the end of the junior year, it was customary to recommend senior elective courses in mathematics and advanced courses in fluid flow and elasticity. These few students also were given the option of replacing the senior design courses with additional advanced theoretical courses. Many of these students were employed by one of the NACA government laboratories, while others chose to continue graduate study at Purdue or at other universities.

The new School of Aeronautics officially began on July 1, 1945, two months after the European phase of World War II ended, and six weeks before the Japanese defeat in August. Faculty who had been temporarily assigned to teach courses in the aeronautical war training programs now returned to their own schools. Professors Binder, Freberg, and Hall, for example, returned to ME, and with the exception of several graduate teaching assistants, the aeronautical engineering staff was composed of only two members: Professors Liston and Bruhn. To recruit new faculty for the school's expanded aeronautical engineering programs, Professor Bruhn formulated the following list of desirable qualifications for new faculty:

- Applicants should have an aeronautical engineering degree.

- New faculty should have several years of experience in the aircraft industry or with aeronautical research agencies, such as the NACA.

- Teaching experience in aeronautical engineering subjects would be desirable.

- A master's degree in aeronautical engineering would carry considerable weight in positions of rank below full professor.

- A PhD would be required for the rank of full professor.

- A record of published papers would carry considerable evaluation weight.

Although very few applicants could fill all of these desirable requirements, they gave a good basis for comparison. There were other conditions existing at Purdue, however, that caused considerable difficulty in attracting well-qualified applicants:

- Staff salaries at that time were below those of some other Big Ten universities and other leading schools. Approximate engineering faculty salaries during the 1945–50 period were as follows (with the lower value for 1945 and the higher value for around 1948–50): instructor, $3,200–$3,600; assistant professor, $3,600–$4,400; associate professor, $4,400–$5,000; full professor, $5,000–$5,700.

- The lack of housing for new staff members in the greater Lafayette area.

- The fact that Lafayette was a long distance from major aircraft industry, and that jet-propelled air transportation had not arrived, limited opportunities for staff members to enhance their income by consulting.

Although Dr. Frederick L. Hovde, who arrived as the new president of Purdue in 1946, began to eliminate these unfavorable factors, changes could not be accomplished overnight. An extensive staff housing project was soon under way, however, involving more than 200 separate houses erected along State Street where the married student apartment buildings are now located. Although these houses were of temporary character and rather crude, they served to alleviate the housing emergency until staff could find better facilities as postwar new housing developed. Dr. Lo was one aero faculty member who lived in these houses during his first two years at Purdue.

An extensive recruiting effort was undertaken to locate new faculty for the aeronautical engineering program. Although competition was keen for individuals with the desired academic and industrial backgrounds, several excellent faculty joined the school during the 1945–50 period. Table 3.1 summarizes the aeronautical engineering faculty reported in the Purdue catalog during this period, and indicates additions, promotions, and resignations (graduate teaching assistants are not shown).

Table 3.1   Aeronautical Engineering Staff Reported in Purdue Catalog During the 1945–50 Period

Professor Bruhn in charge for all five years indicated

1945-46	*Professors*	none
	Associate Professors	Liston
	Instructors	Rosenberg, Topinka
1946-47	*Professors*	Liston, Weske, Zucrow
	Associate Professors	none
	Assistant Professors	Herrick, Shanks, Stanley
	Instructors	Crosby, Rottmayer, Rosenberg
1947-48	*Professors*	Liston, Serbin, Zucrow
	Associate Professors	none
	Assistant Professors	Herrick, McBride, Shanks, Stanley
	Instructors	Crosby, Klein, Vanderbilt
1948-49	*Professors*	Liston, Serbin, Zucrow
	Associate Professors	Shanks, Stanley
	Assistant Professors	Augenstein, Herrick, McBride
	Instructors	Klein, Kordes, Misener, Palmer, Robinson, Vanderbilt
1949-50	*Professors*	Liston, Serbin, Zucrow
	Associate Professors	Herrick, Shanks, Stanley
	Assistant Professors	Klein, Lo, Vanderbilt
	Instructors	Palmer, Reese, Robinson, Schmitt

A brief summary of the new staff for the aeronautical engineering portion of the program follows below.

Dr. Paul E. Stanley joined the combined ME-aero school on January 1, 1945, as an instructor. He was promoted to assistant professor on July 1, 1945, and became a member of both the air transportation and aero engineering staffs. Dr. Stanley held a PhD in physics from Ohio State University, and was director of the Purdue Aeronautics Corporation's ground school during its war flight training programs. During the period 1946–50, he acted as an assistant administrator for the school, particularly for the air transportation division. Dr. Stanley later served as interim school head for the two-year period 1964–66. He retired from Purdue in 1976.

Professor Thomas J. Herrick graduated in ME from the University of Illinois in 1936. He came to Purdue in 1936 as a graduate teaching assistant in applied mechanics and was an instructor in applied mechanics from 1938 to 1942. He went on sabbatical leave to the University of Michigan, where he obtained an MS in engineering mechanics in 1940. He also worked toward his PhD at the University of Michigan. He worked at McDonnell Aircraft Corporation as a structural research engineer from 1943 to 1945, and then he rejoined Purdue as an assistant professor of aeronautical engineering in September 1945. He was promoted to associate professor in 1949. Professor Herrick taught all of the basic courses in the aircraft structural area, as well as the first courses in aerodynamics and in airplane design. He was executive assistant to the head of the school from 1968 until his retirement in 1980, serving under three different heads and three different acting heads.

Reinhart Rosenberg received a BS in engineering from the University of Pittsburgh. During the war years, he worked for the Nashville Division of the Consolidated-Vultee Aircraft Corporation, and rose to head of the Vibration and Dynamics Group. He joined the aero staff in September 1945, with the rank of instructor, and took graduate work toward his MS. He taught courses in vibrations, flutter, and aircraft dynamics, and during his two-year stay at Purdue, carried out several sponsored-research projects and published several papers. He obtained his MS at the end of two years, and while working with the Boeing Company during the following summer, became acquainted with the aero staff at the University of Washington. He was offered the position of associate professor, plus the privilege of studying for a PhD, and since Purdue could not match that offer, he joined the University of Washington.

George Topinka was a distinguished aero student who graduated in 1945. He was appointed instructor and helped handle the wind tunnel courses during his one-year stay with the aero school. He resigned to take a position with the long-range planning group of United Airlines.

Dr. Maurice Zucrow joined Purdue in 1946. At that time, he was chief engineer of the Aerojet Company, a leader in the design and production of jet-powered units. Dr. Zucrow had received the University's first PhD in 1928, earning that degree from the School of Mechanical Engineering with a thesis titled "Discharge Characteristics of Submerged Jets." Since power plants had always played an important role in the ME curriculum, the School of Mechanical Engineering proposed to hire Dr. Zucrow to take the lead in developing the jet propulsion field at Purdue. Realizing that knowledge of jet-propelled engines would be essential for aeronautical engineering graduates, Professor Bruhn proposed that Dr. Zucrow be a member of both the ME and aero faculties, and that jet propulsion education and research be a cooperative program between the two schools. This proposal was accepted by the administration, and Dr. Zucrow was a valuable contributor to the School of Aeronautics until he decided in 1953 to spend full time on the ME staff.

Dr. John R. Weske was the first full professor on the aero faculty in the field of aerodynamics. Before coming to Purdue in 1947, he was professor of aerodynamics at Rensselaer Polytechnic Institute, and had wide teaching and research experience, particularly in the hot wire method for measuring airflow characteristics. Dr. Weske's responsibility at Purdue was to coordinate undergraduate and graduate work in aerodynamics, aircraft power plants, and flight testing. Although Dr. Weske did not move his family to Purdue, and decided to leave after only one year, his brief stay on the aero staff was quite fruitful to the initial school programs.

Dr. Merrill Shanks, who had recently joined the mathematics department, became a joint member of both the mathematics and aeronautical engineering faculty during the 1946-47 academic year. Dr. Shanks had gained industrial experience in fluid flow with the NACA Laboratory at Cleveland, Ohio. Since additional faculty members were needed to handle the increasing undergraduate and graduate courses in aerodynamics, and officials realized that future courses in aerodynamics needed a more rigorous mathematical approach, his background was well suited to the new school's needs. Dr. Shanks began with the rank of assistant professor, and continued to be a part-time member of the aero staff, rising to the rank of professor. Although his teaching was confined mainly to courses in aerodynamics, he also acted as mathematical advisor to various curriculum committees.

Frank Crosby was an ME graduate from Purdue and a graduate teaching assistant on the aero staff while working for his master's degree. He was promoted to instructor in 1946 and was primarily concerned with assisting

Professor Liston in building and equipping the aircraft power plant laboratory and handling power plant laboratory courses. He resigned in 1948 to enter industry.

Earl Rottmayer possessed a BSME from Akron University and a 1941 MS in aero from the University of Michigan. He had two years' experience with the Goodyear Aircraft Company and five years with the Dow Chemical Company, where he was in charge of research concerned with use of magnesium alloys in aircraft structures. He joined the aero staff for the 1946-47 year as instructor in the structures area, with the goal of working toward a PhD, but decided to return to industry at the end of one year.

Vern Vanderbilt obtained a BSME from Purdue in 1942, and after spending several years in industry, returned to Purdue as instructor in the aero school. He took graduate work for an MS in aero, which he obtained in 1947, and was then appointed to the rank of assistant professor. He was primarily concerned with teaching aircraft power plant courses and assisting Professor Liston with graduate thesis work. He resigned from the aero staff in 1948 for full-time graduate study for a PhD in electrical engineering. After obtaining that degree, he accepted a research position with the Perfect Circle Company and later formed his own research corporation.

Dr. Hyman Serbin joined the aero staff in 1947 to fill the position of full professor left by Dr. Weske. Dr. Serbin held a BS from Carnegie Institute of Technology, had one year of graduate study at Caltech, a PhD from the University of Pittsburgh (in math, physics, and mechanics), and two years' post-doctorate study at the Princeton Institute of Advanced Study. While working for these advanced degrees, he was an instructor in mathematics. He joined the Fairchild Aircraft Company in 1940, and from 1941 to 1947 was chief of the aerodynamics section, where he was responsible for all aerodynamic performance and stability, flight test analysis and programming, and all vibration, flutter, and heat transfer problems. He was author of a number of published papers before coming to Purdue. Dr. Serbin was responsible for coordinating and guiding undergraduate and graduate instruction and research in aerodynamics at Purdue. He resigned to accept a technical position with the General Dynamics Corporation and later joined the Rand Corporation as a research scientist.

Bruno W. Augenstein received a ScB in physics and math from Brown University in 1943 and an MS in aeronautics from the California Institute of Technology in 1945. He was employed by North American Aviation Company from 1946 to 1948 before joining the Purdue faculty as an assistant professor for the 1948-49 academic year. Mr. Augenstein taught a course on turbomachinery and an advanced class in gas dynamics using largely foreign sources. He left Purdue after one year to accept a position with the Rand Corporation.

Bertram Klein held bachelor's and master's degrees in aeronautical engineering from the Brooklyn Institute of Technology and a bachelor's degree from Brooklyn College. He was chief research assistant to Professor Hoff, head of the aero department at BPI, and was coauthor with Professor Hoff on a series of NACA Technical Notes dealing with stability of stiffened thin wall cylinders. He joined the aero staff in 1947 as instructor and was promoted to assistant professor in 1949. He taught structures courses and assisted in research projects, but left Purdue in 1950 for a research position in industry.

James W. McBride held a BS in engineering from a Canadian university and an MS in aero from MIT. An instructor and research associate for several years at MIT, he had completed all requirements for a PhD except the thesis when the United States entered World War II. During the war, he rose to head of the aerodynamics section of the turbo research division of the Canadian Government Gas Turbine and Jet Development Research Laboratories. His main assignments during his two-year stay at Purdue dealt with teaching internal aerodynamics in turbo engines, directing graduate work, and formulating the design of subsonic and supersonic wind tunnels for the aero school. He left Purdue in 1949 to join the Carrier Corporation as deputy director of the department of technical and fundamental research.

Walter S. Misener was a graduate in engineering from the University of Toronto and worked under Professor McBride during the war. He joined the aero staff at the same time as Professor McBride as instructor in aerodynamics and turbojet engines. While teaching, he took graduate work toward an MS, which he obtained in 1949. He resigned to accompany Professor McBride to the Carrier Corporation.

George Palmer graduated with distinction from Purdue in 1945 with a BS in aeronautical engineering. While a senior, he helped teach the aerodynamic wind tunnel laboratory course in the Curtiss-Wright Cadette war training program. He was granted a scholarship after his junior year by the Consolidated-Vultee Los Angeles Division, where he worked as an engineer in the development section from March to July 1944. Upon graduation from Purdue, he worked for the Consolidated-Vultee Aircraft Corporation for about a year, and then enrolled at the California Institute of Technology, where he obtained an MS in aero with a major in aerodynamics. He spent another year taking the program in jet propulsion and rocketry and received a further degree in aero engineering. At Caltech, he was research assistant in the Supersonic Wind Tunnel Laboratory and later in the Jet-Propulsion Laboratory.

He joined Purdue as an instructor in ME and from 1947 to 1949 was assigned to work under Dr. Zucrow as project engineer on the design and development of the Rocket Research Laboratory. He designed the first liquid-cooled rocket motor for the laboratory. In 1948, he joined the aero staff as instructor, and was promoted to assistant and associate professor in later years. From 1948 to 1950, Mr. Palmer taught courses in applied aerodynamics, designed and supervised the building of the 400-mph wind tunnel, assisted in studies of a supersonic tunnel, and designed and supervised the installation of the first turbojet engine laboratory test unit. He was responsible for developing and giving new courses in applied aerodynamics, flight vehicle design, and laboratory wind tunnel and jet propulsion courses, and was consultant for a number of industrial concerns. Mr. Palmer retired from Purdue in 1987.

Dr. Hsu Lo obtained a BS in aeronautical engineering in China, and an MS and PhD (1946) in aero from the University of Michigan. While enrolled at Michigan, he worked part time as a structural engineer with the Stinson Aircraft Company in Dearborn, Michigan. He joined the structures division of the NACA Langley field laboratories as a research scientist and also taught night school at the University of Virginia. Dr. Lo joined the aero faculty in fall 1949 as an assistant professor, and rapidly rose to the rank of full professor. He developed and taught courses in vibrations, aircraft vibrations and flutter, aircraft dynamics, and aeroelasticity. Dr. Lo served as head of the school from 1967 to 1971, retiring from Purdue in 1979.

Dr. A. F. Schmitt entered Purdue in 1945 and received his BS, MS, and PhD in aero in 1948, 1949, and 1953 respectively. During several summer periods he worked as a stress analyst with the Grumman Aircraft Corporation. He was appointed instructor on the aero staff while working toward a PhD and was promoted to assistant professor in 1953 after receiving the PhD. He taught several senior and graduate courses in structures, aeroelasticity, and aircraft dynamics. He resigned from Purdue in 1955 to join the Ryan Aircraft Corporation as senior dynamics engineer. In 1956, he joined the astronautics division of the General Dynamics Corporation and advanced to manager of the Dynamics and Control Section. In 1963, he joined the Kerrfott Division of the General Precision Corporation as chief engineer and later was made engineering director of a new Aerospace Computer Division, established by the General Precision Corporation, located in New Jersey. Dr. Schmitt received a Distinguished Engineering Alumnus Award from Purdue in May 1967, and was one of the many BS in aero graduates during the 1945–1950 period that advanced to high-level positions in the aerospace industry.

Dr. Bruce Reese received a BSME from the University of New Mexico in 1944, and an MS in ME from Purdue in 1948. He joined the aero school as instructor in 1949 and continued work toward a PhD. He taught a course in jet propulsion and assisted Dr. Zucrow with several research projects. When Dr. Zucrow left the aero school for a full-time position with ME, Dr. Reese accompanied him. When Dr. Zucrow retired in 1966, Dr. Reese was appointed director of the Jet Propulsion Center at Purdue. He later rejoined the School of Aeronautics and Astronautics in 1973, and served as head until his retirement in 1979.

STUDENT ACTIVITIES

Since the air transportation program emphasized supervisory or administrative activities and responsibilities, the air transportation staff encouraged student activities that would promote administrative and writing experience. A few of the activities conducted by the air transportation and aeronautical engineering students are described below.

(a)   **The *Aeroliner* Student Publication**
When the entire School of Aeronautics was moved to the airport in 1947, a few students proposed a weekly publication on school activities to inform air transportation students with what was going on in the aero engineering division of the school and vice versa. The first issues of the *Aeroliner* were four pages long and later increased to six pages. A typical issue included a calendar of coming events, a write-up on a particular staff member, a description of a particular research project, special news on developments in industry, etc. A new editing staff took over each semester and new reporters served from each division of the school. The initial cost of supplies was paid by the school, but later, a voluntary request of 25 cents per student per semester paid for the expenses.

(b)   ***This Week in Aviation*, a Radio Program**
Arrangements were made with WBAA, the Purdue radio station, for a weekly 15-minute program called *This Week in Aviation*, in which different students broadcast news items on Purdue aviation and interesting items about the outside aviation world.

(c)   **The Airfreight Board**
The goal of the Airfreight Board was to acquaint students with actual conditions of the airfreight industry, which was at that time still in its infancy. This goal crystallized in the form of a comprehensive airfreight manual, which was revised and expanded as the semester went by and was used as a reference source in the airfreight course given by Professor Fowler. The board involved

a number of students and committees, and usually met at night, once a week. Professor Fowler was the guiding hand behind this activity and gave much assistance in making the many contacts with industry to obtain information regarding up-to-date operating data, sales methods, etc.

(d) Sigma Alpha Tau Honorary Air Transportation Society

Since students graduating from the air transportation curriculum were not eligible for membership in the aeronautical engineering honorary society, the air transportation students initiated a new honorary society for air transportation students. The new honorary society was named Sigma Alpha Tau, and other chapters were later established at other universities.

(e) Student Chapter of Institute of Aeronautical Sciences

The aeronautical engineering students had a very active chapter of the Institute of Aeronautical Sciences during this period. Most students belonged to the society, and, because of the large attendance, the monthly meetings were usually held in the faculty lounges on the second floor of the Union Building. Most of the guest speakers came from the aircraft industry or from one of the several NACA research laboratories. The student chapter sponsored a two-day meeting in 1947. Student members of IAS chapters at five other Midwestern universities were invited to this meeting, and students were encouraged to present technical papers. Prizes were awarded at a dinner meeting on the evening of the second day.

(f) Inspection Trips to Industry

Faculty members believed that inspection trips to various industrial facilities would have considerable educational value for both students and staff. The air transportation students often made field trips to one of the airlines in Chicago, while the aeronautical engineering students made visits to the Allison Company in Indianapolis, to Wright Field in Dayton, to the NACA Laboratories in Cleveland, etc. The first of several extensive, combined Aero Engineering Air Transportation field trips occurred in January 1950, when 120 students and faculty visited Washington, D.C., Baltimore, and the New York-Long Island area. Travel expenses were exceptionally low due to the cooperation of President Carmichael of Capitol Airlines, who provided two 60-passenger Douglas DC-4 aircraft for the four-day trip. Descriptions and photographs of other field trips conducted during the 1950s are provided in Chapter 4.

Facilities

Until the end of the spring 1946 semester, school facilities consisted of Building Units 1 and 2 at the airport and the campus aero building. The upper floor of Unit 1 was used for the air transportation shops and laboratories, but the first floor was still used by the Purdue Aeronautics Corporation to store and service airplanes. Building Unit No. 2, shown previously in Figure 2.4, housed the aircraft power plant laboratory that Professor Liston and his associates had developed over the past several years (see Figure 3.5). Since the campus aerodynamics and structures laboratories could not be moved to the airport until more space was provided, Purdue administrators allotted approximately $100,000 for new airport facilities. Officials decided to use these limited funds to obtain as much space as possible, and a 150 x 250-foot hangar-type building, with a 20-foot-wide wing along one side, was built. Although occupancy was expected by Christmas 1947, the building was delayed until the 1948-49 year, requiring modification of laboratory courses during the 1947-48 academic year. Before moving the campus equipment to the new airport building, the School of Aeronautics decided to sponsor a dedication dance on its large, open floor. The dance was open to all Purdue students, and was most likely the largest dance held at Purdue relative to the number of couples dancing at the same time in one room.

While the new hangar was constructed, Building Unit No. 1 was also remodeled. The Purdue Aeronautics Corporation vacated the first floor, and the air transportation laboratories and shops located on the second floor were moved to the first floor. The building was extensively remodeled to

Figure 3.5   Students in air transportation power plant laboratory (circa 1947). Professor Cargnino is standing at rear center in white lab coat, and Professor Briggs is standing front right.

provide classrooms, a central school office, design room, library, student lounge, staff offices, and a number of small laboratories. Figures 3.6 and 3.7 show one of the air transportation laboratories and the second floor corridor connecting Units 1 and 2.

After it was decided in 1946 to construct Building Unit 3, administrators requested that the staff work with architect Walter Scholer on future building expansion for a complete aviation education and research center at the airport. When these long-range expansion studies were made in 1946,

Figure 3.6   Air transportation laboratory (circa 1947).

Figure 3.7   Second-floor corridor connecting Building Units 1 and 2 (circa 1949). Rhea Walker (shown at right) served as head secretary for many years.

the school's enrollment was 797 undergraduate students, representing 7% of the entire University enrollment. It was reasonable to assume enrollment would approach 1,500 or more, as the unique aviation education and research center attracted many undergraduate and graduate students from all over the world. These optimistic projections did not, however, come to pass, and Mr. Scholer's proposals for a new aviation center at the airport were not built.

By 1946, Professor Liston had developed an elaborate power plant laboratory to study reciprocating aircraft engines. The laboratory included an engine disassembly and assembly facility and two engineering test cells where students would run tests to determine performance characteristics of reciprocating engines. Professor Liston had also developed an outstanding design course for the reciprocating aircraft engine. Photographs of the power plant and propeller laboratories and staff members Liston, Crosby, Bean, Basinger, Bridges, and students are given in Figures 3.8–3.11.

As previously pointed out, Professor Bruhn had proposed that jet propulsion be a cooperative effort between the ME and AE schools, with assistance from the chemical engineering and metallurgical schools. This proposal was approved by Dean Potter, and during 1947, the main efforts of Dr. Zucrow were devoted to developing the undergraduate and graduate courses in gas turbines and jet propulsion shown in Table 3.2. With the exception of the laboratory course, Dr. Zucrow prepared the initial sets of notes for these courses.

Table 3.2   Jet Propulsion Courses Offered by Schools of Aeronautics and Mechanical Engineering in 1947

Open to Undergraduate and Graduate Students	
ME 139	Steam and gas turbines
AE 115	Gas turbines and jet engines for aircraft
AE 117	Laboratory in gas turbines and jet engines for aircraft
Open to Graduate Students Only	
ME & A E 281	Gas turbines and jet propulsion
ME & AE 282	Gas turbines and jet propulsion
ME & AE 283	Seminar
ME & AE 299	Thesis

Since a comprehensive gas turbine laboratory would take several years to plan, design, and build (if money were available), officials decided that smaller, inexpensive test units would be developed to provide laboratory instruction involving gas turbine principles and operations. The initial

Figure 3.8   Professor Liston, Instructor Crosby, and Mr. Bean in power plant laboratory (circa 1947).

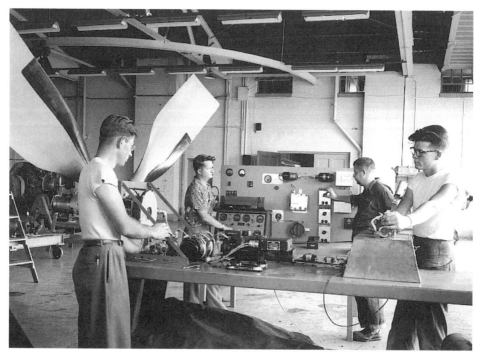

Figure 3.9   Propeller laboratory (circa 1947). Instructor James Basinger is second from left.

Figure 3.10 Professor Briggs with students in engine disassembly laboratory
(circa 1946).

Figure 3.11 Students in power plant lecture room (circa 1946).

step was to obtain several turbojet engines to serve as cutaway models for class lectures and for disassembly and assembly purposes. Professor Liston and his associates designed and installed a turbosupercharger, which was converted to operate as a small gas turbine in a test cell located in the basement of the power plant laboratory (Building Unit 2). By 1950, he and his associates had also completed design plans for a burner test cell, which was also to be located in the basement of Building 2.

Through the efforts of Dr. Zucrow, the School of Aeronautics was able to obtain a small Westinghouse axial flow-model X9.5 B 250-pound thrust jet engine. Professor Palmer and several graduate students, working under the direction of Professor Liston, installed this engine in one of the two engineering test cells, providing aero students with experience in operating and measuring the performance characteristics of a model turbojet engine after the 1949 school year.

It was the consensus of Professors Zucrow, Liston, Weske, and McBride that if Purdue was going to be a leader in graduate instruction and research in the gas turbine field, a comprehensive gas turbine laboratory would be needed. During the period 1947–50, much thought and study were given to such a laboratory, and a number of proposals were submitted for consideration. One proposal placed the gas turbine laboratory near the east end of Building Unit 3, adjacent to a fairly large Quonset hut in which Professor Buttner had installed a large air supply combined to a number of Allison engines with superchargers for use on the Squid research project. Later, Dr. Zucrow expanded that installation for further research in combustion. Professors Liston, McBride, and Misener submitted proposals for space in the basement of Building Unit 2, and proposed construction of a new wing to that building. All of these proposals involved several hundred thousand dollars, and presented a difficult financial problem for the administration. Thus, as the 1940–50 decade ended, the question of a comprehensive gas turbine laboratory was still in the study and discussion stage.

Dr. Zucrow's primary interest was to develop a graduate study and research center in jet propulsion, particularly in the rocket engine field. One of his first sponsored research projects was "A Preliminary Study of the Pressure Distribution in a Coanda Nozzle," and the experimental work on this project was carried out in the basement of aero school Building Unit 2. The year 1947 saw the real beginning of Dr. Zucrow's dream for a research center, however, when the first laboratory building was completed. It was a 40 x 60 foot concrete and concrete-block one-story structure located on University farmland at the northwest corner of the Purdue airport. The cost of the building was around $40,000, with about one half the cost made possible by the Squid Research project sponsored by the Office of Naval Research. This building provided space for two test cells with a central control room, experimental shop facilities and a diesel engine for generating power. A second similar building was completed by 1950 and located adjacent to the west side of Building Unit 1. Thus, by 1950, Dr. Zucrow and his associates were well on their way to developing an outstanding graduate study and research center in the field of rocket engine propulsion.

When the new campus aero building was completed in 1942, Professor Wood and graduate student Alan Johnson designed and supervised construction of a new wind tunnel. It was a twin fan, single return type, having a test cross-section of 38x48 inches, and was powered by a 90-hp Dodge auto engine. It had a maximum speed of about 110 mph, and it featured a three-component balance system using proof rings (Figure 3.12a). The power unit was unsatisfactory, however, because of excessive noise and vibration.

About this same time, Professor Wood designed and built a low-speed free flight wind tunnel. It was a single return type, with a 12x24-inch open throat cross-section that was 36 inches long. The single fan was driven by a 2-hp electric motor that gave a throat speed of approximately 25 mph. The tunnel could be tilted to obtain a wing-gravity component, and was primarily used for classroom demonstrations (Figure 3.12b).

When the new hangar (Building Unit 3) was completed in 1947, all campus aero laboratory facilities and equipment were moved to that airport building, including the two wind tunnels described above. The Dodge auto engine that powered the 110-mph tunnel was replaced by a 25-hp electric motor, and was used for a short time until a larger motor could be found. The school was fortunate to locate a war surplus 50-hp electric motor with Ward-Leonard control systems. Mr. Palmer installed the new power unit and modified the tunnel to obtain better flow characteristics. He added six feet to the tunnel length to obtain a better energy ratio and modified the turning vein design to smooth the airflow. He also improved the measuring balance system by replacing the proof rings with cantilever beams and using strain gauges instead of Ames dials. The remodeled tunnel had a throat velocity of approximately 110 mph and much better flow characteristics; it was used in the required wind tunnel course for aero students. Mr. Palmer was assisted in this redesign project by graduate students Robinson and Novak.

Another small, induced flow, closed throat, atmospheric return type of wind tunnel was completed and operating during the 1947–50 period. The tunnel had a 6x18-inch test section, and a 10-hp electric motor provided a throat velocity of 120 mph. The tunnel was about 10 feet long and could readily be moved. It was used in basic instruction for undergraduate students through measuring pressure distribution on airfoils and for boundary layer measurements.

In the spring of 1947, Professor Weske learned of a Japanese variable-density wind tunnel that the Air Forces had confiscated when U.S. military forces occupied Tokyo. This tunnel was rather advanced because of its variable-density characteristics, and had been dismantled, crated, and shipped to Wright Field. It was still in the crates when Dr. Weske visited Wright Field, and officials there informed him that they would lend the tunnel to Purdue if it agreed to assemble it and place it in operation. In studying the Japanese literature on the tunnel, Dr. Weske decided it would be good for basic research. When the crates were unpacked at Purdue, however, it was discovered that the Japanese had sabotaged some parts of the tunnel. There was a big crack in the steel frame of the 100-hp electric motor and a badly damaged propeller, requiring considerable repair before the tunnel could be put in operation in 1950 (Figure 3.12c).

The aerodynamics group believed that the school should have a better subsonic wind tunnel more suitable for research, and considered two options for an additional subsonic tunnel. One direction was to design and build a tunnel with a 7x9-foot test section with a speed of 200 to 225 mph. This type tunnel would have been quite useful for commercial testing, and the McDonnell Aircraft Corporation indicated that it would give the school considerable testing work. Professor McBride worked out a preliminary design and cost estimate, but after much study and discussion, it was decided that such a tunnel would not best serve the needs of the aero school. It would likely develop into a full-time commercial testing unit, requiring a separate operating staff, and would be too costly for instructional and research purposes.

The other direction was to build a low turbulence wind tunnel with a smaller test section but a faster speed of 350 to 400 mph. Such a tunnel would be good for graduate research projects, as well as for laboratory instruction, and the cost of operation would not be excessive. Professor McBride made a preliminary study for a wind tunnel with a 3x4.5-foot cross section, a throat speed of 400 mph, and a low turbulence factor. The power required was around 400 hp, and since such a large power unit would be quite expensive, the study marked time until a war surplus power plant could be found. After a considerable search, a 400-hp electric motor with dynamic clutch control was located, and could be purchased for about one-tenth of its original $40,000 cost. After much discussion, the staff recommended that the power unit be purchased and that the proposed tunnel be designed and built over a period of several years with small amounts of money that could be appropriated each year.

The administration accepted this recommendation, but shortly there-after Professor McBride left Purdue for a research position with the Carrier Corporation. Mr. Palmer, who had much experience in wind tunnel testing and design, was appointed project engineer, and he completed the detailed design for the tunnel. He kept the same dimensions that Professor McBride had arrived at, but changed the design from a vertical loop with the test section high off the ground to a horizontal loop with the test section at ground level. Mr. Palmer then used student labor to prepare the many detailed fabrication drawings. The plywood tunnel sections were built by the University Wood Shop, but assembly was mainly done with student labor. Design of a proper propeller to go with the power plant was also performed by Professor Palmer. Since money was not available for a new balance system, the balance system used in the older modified 100-mph tunnel was transferred to the new tunnel, and by the beginning of 1950, the tunnel was in operation (Figure 3.12d). Although remodeled in recent years, that basic wind tunnel remains in use, and was still powered by the original war surplus 400-hp motor until it was replaced in 2008.

The 1942 campus aero building included a spacious structures laboratory; it also provided means for applying loads and instrumentation for measuring strains and deflections. The heavy steel superstructure for anchoring large structural test specimens was moved to the airport and mounted at the end of a 30-foot-long, 3-foot-thick concrete slab. The structures laboratory was assigned to the east end of the new airport hangar building. Theoretical structures courses were held in a large classroom adjacent to the structures laboratory, and those classes often were taken to the adjacent laboratory to physically illustrate structural instability or ultimate failure in various types of structural components. The structures and dynamics laboratory in 1948 possessed the equipment summarized in Table 3.3, and several photographs showing students using this equipment are given in Figure 3.13.

Both the air transportation and aeronautical engineering curricula required sheet metal working, machine shop, aircraft welding, and wood and plastics shop courses. A new course in sheet metal forming was needed, since World War II production demands for aircraft had greatly changed metal working methods. The desire for an up-to-date metalworking laboratory course was indicated to the general engineering department, which readily agreed to develop such a laboratory and shop course. A photograph of the aircraft metalworking laboratory is given in Figure 3.14.

a. Professor Wood and student with new wind tunnel located in campus aero building (February 1943).

b. Low-speed free flight wind tunnel designed by Professor Wood (March 1943).

Figure 3.12 Various wind tunnels operated by School of Aeronautics during 1940s.

c. Variable-density wind tunnel originally built in Japan.

d. Large subsonic wind tunnel constructed under direction of Professor Palmer (in background). Tunnel was renovated and renamed The Boeing Wind Tunnel in 1992.

a. Conducting wing test.

b. Professor Bruhn and students (note slide rules). Rear table, left to right: Professor Bruhn, G. Christopher, R. H. Turner, L. T. Cheung, W. G. Koerner, A. M. Arnold (back to camera), unknown, H. F. Steinmetz. Front table, left to right: R. W. Taylor, R. M. Rennak, R. J. Wingert.

Figure 3.13 Students working in structures laboratory.

c. Compression test with Tinius Olson test machine.

d. Strain measurements.

Table 3.3   Structures and Dynamics Laboratory Facilities Available in 1948

Baldwin Southwark Model K Strain Indicator
   Power supply for model K strain indicator
   Stock of A, AR, and C type gages of various sizes
Three Baldwin Southwark Load Cells
   60-cycle power supply for load cells
Brush Direct-Inking Oscillograph (2 channel)
   2 amplifiers for above
   2 Brush amplitude-type pick ups
One Heiland Oscillograph—18 Channel—Photo Recording
   14 galvanometers for Heiland oscillograph
   18 strain gage amplifiers for Heiland oscillograph
   (these are complete except for housing and connections)
One General Radio Vibration Meter Type 761
One General Radio Vibration Analyzer Type 762
Two Consolidated Engine Company Vibration Meters Type 1–110
Two Consolidated Engine Company Velocity Pickups
Two Adjustable Vibrating Reed Type Vibration Pickups
One Frame (multiple fixed reed) Type Vibration Pickup
Two Chronotacs
One Speed Ranger Variable Speed Drive
   6 flexible shafts for above
   3 vibrators—reciprocating mass type for use with above
   1 shake table
One 60,000-lb. Tinius Olsen Testing Machine
One 60,000-lb. Tate Emery Testing Machine
One Rockwell Hardness Tester
Seven Tension Dynamometers, from 300 lbs. to 5,000 lbs. capacity

---

Professor Lascoe of the general engineering department was given the responsibility of providing the shop facilities and developing the course. Professor Bruhn later wrote [4]: *"In this writer's sincere opinion, Professor Lascoe, when it came to the problems of obtaining new and used war surplus machines and material, was the greatest go-getter of all time. In his nationwide search of government warehouses, depots, and wartime plants being placed out of operation, and in his constant contact with many government and military officials concerned with war surplus disposal, he was successful in completely reequipping the various general engineering shops and laboratories with the latest type of machine tools and equipment. These were all practically new and valued at several millions of dollars, but were secured at relatively small cost to Purdue. To provide for our required course in sheet metal forming, he was exceptionally successful in obtaining the latest type of equipment. This writer and the School of Aeronautics are greatly indebted to Professor Lascoe for his continuous fine cooperation with our*

*Figure 3.14 Aircraft metal working laboratory developed by Professor Lascoe
of General Engineering Department. Instructor J. Borodavchuk
standing at center.*

*school. He was of great assistance to our school in working with Mr. Cushman, of
the air transportation staff, in securing war surplus material other than machine
tools, such as electric motors, compressors, motorized lifts, etc."*

The aeronautics staff also believed that surplus aircraft would be of considerable value for studying developments in aircraft structure, power plant installations, and various mechanical, hydraulic, and electrical systems. Three such airplanes were obtained by the school. While visiting Republic Aircraft Corporation at Evansville, Indiana, a plant that produced the famous P-47 Thunderbolt fighter airplane, the school was offered one of the last P-47s to come off the production line. Although this aircraft had cost the government approximately $100,000 with engines, the United States possessed thousands of P-47s at the end of the war, and the surplus airplane was presented to the school without cost. The P-47 was parked on the south side of Building Unit No. 3 for several years. It was then stripped to provide laboratory display specimens and eventually sold to a local Lafayette junkyard.

The school also was able to locate a C-46 Curtiss-Wright twin-engine transport stored at Walnut Ridge, Arkansas. This airplane was the largest two-engine airplane in existence during the war and was widely used to transport troops and supplies. James Basinger of the aero staff was sent to Walnut Ridge to put the airplane in flyable shape. It was flown to the Purdue Airport by aero students Joseph P. Minton and Wade E. Mumma, who had been stationed in Burma during the war, and had flown C-46s over "The Hump" at that time. This airplane, which is shown in Figure 3.15, cost the government around $200,000 but was presented to the school for less than $500. It also was parked south of Unit No. 3 for several years, until an airport operator from Florida visited the school and expressed the desire to buy it. Since the airplane was deteriorating badly, officials suggested that the government sell the airplane to this person. It did so at a price of $5,000, and the Florida airport operator flew the airplane away from the Purdue airport.

Soon after the end of World War II, the Bell Aircraft Corporation delivered the first of several P-59 jet-powered airplanes to the Air Forces. These airplanes were entirely experimental and were mainly intended to demonstrate turbojet power. Soon General Electric and Westinghouse were designing and producing more efficient turbojet engines, and these initial P-59 airplanes became obsolete. The school was successful in obtaining one as a gift in 1946, and it was flown to Lafayette, where the pilot buzzed the field a number of times to give a large number of students their first view of a propellerless airplane. The airplane also was parked on the south side

Figure 3.15 War surplus C-46. Students Wade E. Mumma and Joseph P. Minton at aircraft windows. Standing left to right: E. F. Bruhn, E. A. Cushman, L. T. Cargnino, W. Briggs, and James Bassinger. Also note P-47 aircraft visible behind C-46 at left.

of Building No. 3 and, although it was never flown again, students often fired up the jet engines as a part of their course work on aircraft engines.

Professor Cargnino recalls that the staff drew lots for the honor of starting the turbojet engine for the first time.*"I won that privilege, and although we had very little documentation, was able to enter the cockpit and obtain ignition without too much trouble. The first try was a hot start, however, and shot flames many feet from the rear of the airplane. Although I didn't realize it at the time, these flames came very close to other airplanes parked nearby. Finally, I noticed my colleagues frantically waving for me to shut the engine off before those other aircraft could be damaged. This hot start was quite spectacular, and we repeated it on occasion to impress visitors."* The airplane began to deteriorate badly, however, and the government sold it in 1956 to the Harold Warp Pioneer Village in Minden, Nebraska. (This aircraft was still on display at Pioneer Village in 2008.) A photograph of the P-59 is given in Figure 3.16.

GRADUATE EDUCATION AND RESEARCH

Although World War II ended six weeks after the School of Aeronautics was established in July 1945, it took considerable time to phase out the various war training efforts, and to establish the first graduate programs. The School of Aeronautics section of the 1945-46 Purdue catalog states the following in this regard: *"A new and notable teaching and research staff is now being assembled. Because of the war requirement, it will not be possible to obtain*

Figure 3.16 War surplus P-59 jet aircraft (circa 1950).

the release of future staff members from their present connections until the spring or summer of 1946; consequently, graduate work will not be available before the September semester of 1946."

A graduate MS program in aeronautical engineering was approved in 1946, however, and offered three major fields of study: aerodynamics, aircraft structures, and aircraft power plants. In 1948, graduate work leading to a PhD was approved for study in aerodynamics and jet-propulsion, but a PhD program in aircraft structures was not approved until the early 1950s. Although graduate work in air transportation was not offered before 1950, many AT graduates expressed interest in a master's degree program, and plans for a such a program were developed in the 1949-50 school year.

The first aeronautical engineering graduate degrees were awarded in 1947, with 12 students receiving master's degrees at the February and June commencement exercises. Langdon F. Ayers, with a thesis titled "Design of Turbosupercharger Hot Gas Equipment and Performance Testing of Type B Compressor," and Robert L. Richardson, with a thesis titled "Design of Turbo Supercharger Compressor Test Equipment and Performance Testing of Type B Compressors," earned the distinction of receiving the first graduate degrees from the new School of Aeronautics. The names of other MS recipients in 1947 and subsequent years are recorded in Appendix B, along with a list of other MS thesis titles. By the end of 1949, some half-dozen students were studying toward their PhDs in aeronautical engineering. Navy Commander Richard L. Duncan received the school's first PhD in 1950. His thesis was titled "Analytical Investigation of the Performance of Gas Turbines Employing Air-Cooled Turbine Blades." Subsequent PhD recipients and their thesis titles are recorded in Appendix C.

Although new aero faculty did not arrive until the 1946-47 school year, and were initially busy developing new courses and organizing laboratories, several staff undertook sponsored research projects by the end of the 1949-50 school year. The following is a list of such research projects:

- "Tests of Propellers Under Gyroscopic Loads," PRF 320, sponsored by the U.S. Air Forces. This project was for a contract sum of $30,000, which was quite large for that period. It involved an elaborate test setup to measure propeller strains while rotating a complete P-47 Thunderbolt Fighter about a vertical axis at various speeds while the propellers were rotating. (Directed by Professor Herrick).

- "The Effect of Fuel on the Mass Moment of Inertia and Damping of Oscillating Tanks," PRF 344, sponsored by the Air Forces. (Directed by R. M. Rosenberg and E. F. Kordes)

- "Estimated Frequencies of Rotor Blading," PRF 428, sponsored by U.S. Air Forces. (Directed by Professor McBride)

- "Preliminary Studies to Assist in Drafting Compressor Specifications for U.S. Air Forces," PRF 760, sponsored by U.S. Air Forces.

- "Application of the Response Function to the Calculation of the Flutter Characteristics of a Wing Carrying Concentrated Masses," PRF 570, contract No. NAW-5806. (Directed by Professor Serbin)

- "The Effect of Warping Restraint on Torsional Vibration and Related Aeroelastic Problems," contract No. Nor 433-000. (Directed by Professor Hsu Lo). A photograph of Professor Lo working on this project is given in Figure 3.17.

In addition to these sponsored research projects, faculty and students also carried on theoretical and experimental studies, which often led to research proposals. The titles of some of these first studies and research proposals follow:

- "Research and Development on Guidance and Control Systems for Aircraft and Guided Missiles"

- "Theoretical Analysis of the Dynamics of a Two Control Airplane in Landings in a Yawed Attitude"

Figure 3.17 Research project carried out under direction of Professor Hsu Lo (at left).

- "Investigation of the Possible Use of Corn Stalk Interior Pith Material as a Low Density Core in Aircraft and Missile Sandwich Construction"

- "Theoretical Studies on the Effects of Aspect Ratio on Wing Flutter"

- "A Method of Temperature Determination of High Velocity Gases by Differential Intensity of Two Light Beams of Two Different Wave Lengths"

- "Pressure Method of Measuring Temperature"

- "A Study of Special and Usual Conditions Affecting High Speed Aircraft and Missiles"

- "Experimental Determination of the Effect of Pressure Slots on the Lift and Drag Characteristics of an Airfoil"

Professor Bruhn became interested in aircraft crash safety during 1948-49, studying the dynamic energy absorption characteristics of structural components. His initial research led to a later Office of Naval Research contract dealing with this energy absorption subject. A. F. Schmitt, an instructor on the aero staff working for his PhD, was the principal research engineer on this extensive project. A photograph of the test apparatus employed for this project is given in Figure 3.18. (Dr. Schmitt received his PhD in 1953, and served on the faculty as an assistant professor until resigning in 1955 to pursue a career in industry.)

Professor Bruhn also became interested in auto crash safety during this time, which, in his opinion, required two factors: (1) restraint of car passengers and (2) design of energy absorption into the car to limit the de-acceleration factors that cause internal injuries to the human body. He worked in this area for several years, but eventually lost interest in the subject due to the reluctance of the auto industry to start serious study and research on this difficult safety problem at that time.

1950 PROGRAM STATUS

Figure 3.19 summarizes undergraduate enrollment in both the aeronautical engineering and air transportation programs during the first 15 years of the new school. The peak total enrollment of 736 students (excluding freshmen) in 1947 indicates the large influx that returning GIs had on the new programs; it was the largest undergraduate enrollment in the school's history. Figure 3.19 also notes that although the air transportation program started well, enrollments declined in the 1950s, and the program was discontinued. This point is discussed further in Chapter 4.

*Figure 3.18
Test rig for determining
ultimate dynamic energy
absorption of basic
structural units. PhD
program conducted by
A.F. Schmitt under
direction of Professor
Bruhn. L.A. Hromas,
top left (circa 1952).*

During the 1948-49 academic year, Professor Bruhn asked Dean Potter to be relieved of the headship of the School of Aeronautics and return to full-time teaching and research. He had been involved in extensive administrative efforts since Professor Wood left in early 1944, and was the first head when the new School of Aeronautics was established in July of 1945. Professor Bruhn continued to serve as head, however, until Dr. Milton Clauser was named to succeed him in September of 1950.

The next chapter discusses aeronautical engineering activities during the 1950s. Before proceeding to that era, however, it may be appropriate to give Professor Bruhn's assessment[4] of the school's status at the end of the 1940s.

The school year of 1949-50 saw the completion of the first five years for the new School of Aeronautics. It seems appropriate to summarize what had been accomplished during this five-year period.

Status of Air Transportation Program

The air transportation program was an entirely new educational program for Purdue — "starting from scratch" in 1945. The amazing development and

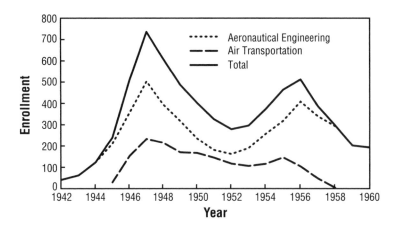

Figure 3.19 Undergraduate enrollment (sophomore through senior) in School of Aeronautical Engineering from 1942 to 1960.

accomplishments of this division of the school ... was due to the dedicated efforts and enthusiasm of the air transportation staff, namely, Webster, Cushman, Stanley, Cargnino, Briggs, Basinger, and Fowler. Under a limited enrollment, this program by 1948 was graduating around 100 students each year. These students were securing good jobs in industry, because as described previously in this report, this group of staff members had carried on a continuous program to acquaint the industrial aviation world (both private and governmental) with what Purdue was developing and providing in this new field of aeronautical education. Thus, as the first five years of the program ended, our school was well known throughout the air transportation world and much cooperation and assistance from it was being given to our school.

Progress of Aero Engineering Program

Curriculum: An undergraduate curriculum had been formulated to give a proper balance between theory, analysis, design, and basic and applied laboratory work. In the senior year the student, if he so desired, could select his technical courses so as to specialize in one of three definite fields of study on aeronautical work; namely, aerodynamics and control, structures and dynamics, and aircraft propulsion.

Senior students who were interested in research or teaching and who had good grades in certain courses were advised to consider going on for graduate study; and, if so interested, were usually advised to select their senior elective courses in the more theoretical or less applied type of course.

A graduate program leading to an MS degree in aero engineering with majors in three fields, aerodynamics, structures, and propulsion, was well developed and had a considerable enrollment. A PhD graduate program with a major in the fields of aerodynamics and jet-propulsion was well underway and a PhD program with a major in the field of structures and dynamics was being submitted for approval as the year 1949-50 ended.

Undergraduate Enrollment: During the years 1948-49, and 50, approximately 120 BS in aero engineering degrees were granted each year, thus making our school one of the largest aero schools in the United States in respect to undergraduate enrollment.

Aero Engineering Staff: As the 1949-50 school year ended, the school possessed the following outstanding staff. In the field of aerodynamics and control: Serbin, Shanks, Palmer, Stanley, and Robinson. In the field of structures and dynamics: Bruhn, Herrick, Lo, Klein and Schmitt. In the field of propulsion: Liston, Zucrow, Palmer, Vanderbilt, and Reese.

Airport Building and Space Facilities

In Building Units 1 and 2:
- *7 average size recitation rooms*
- *1 large lecture room seating more than 100*
- *1 large design room with drafting tables*
- *1 medium size design room*
- *1 62'x102' in size engine and airframe laboratory*
- *5 individual engineering power plant test laboratories*
- *1 medium shop with supplies room*
- *1 well stocked library and study room*
- *1 good student lounge and meeting room*
- *1 school office and good size reception room*
- *14 separate good sized office rooms*

In the Hangar-Type Building Unit 3:
- *This large building contained a main room 100 ft. wide and 250 ft. long. A wing along one side provided three classrooms of 36-student capacity, one classroom of 70-student capacity, and eight large office rooms 20'x20' in size.*

Status of Laboratories

Aircraft Power Plant Laboratory: Professor Liston and his associates had by 1945 developed and had in operation, without doubt, the most complete university educational and research aircraft power plant laboratory in the United States. Soon after 1945, he and his staff associates Zucrow, Palmer, and Vanderbilt began developing additional laboratory installations in the gas-turbine jet field. By 1950 several of these new units — as described previously in this report — were being used for both instruction and research. During this five-year period, much study had been made by the propulsion staff relative to an extensive gas turbine laboratory for future graduate study and research.

Rocket Engine Laboratory: By 1950 Professor Zucrow and his associates from both the Aero and ME schools had completed the first and second units of the rocket engine laboratory located near the west boundary of the Purdue airport, and graduate study and research was well underway. It was evident that the untiring efforts and dynamic leadership of Professor Zucrow would in a few years lead to the most complete university laboratory in the rocket engine field in the United States, thus providing great future opportunities for graduate students in both the Aero and ME schools, since it was to be considered as a joint laboratory of these two schools.

Aerodynamics Laboratory: By 1950, as previously described in this report, five wind tunnels were in operation — the latest one being the 350–400 MPH low turbulence tunnel, which was the result of initial design studies by several staff members, but primarily due to the technical ability of Professor Palmer and his untiring efforts in designing and carrying this large project through to completion and operation. In 1950, the full value of this new tunnel in instruction and research could not be used as funds were not available at that time to include a proper balance system for the tunnel. As pointed out previously in this report, extensive studies relative to a supersonic wind tunnel for the school had been carried out and a proposal submitted for such a tunnel to a government agency for funds; however, by 1950 no such funds became available. Thus, the school possessed no supersonic wind tunnel as the initial five-year period of the school's life ended.

Structures Laboratory: By 1950, the structures laboratory, which was located in the east one-third of Building Unit 3, had load applying equipment and measuring instrumentation to carry on educational and research involving the load spectrum on flight vehicles at that time; it was a laboratory where additional test equipment could readily be added as the flight load spectrum changes as the flight vehicle entered new operating environments.

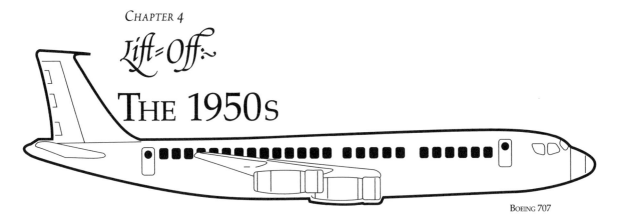

CHAPTER 4

Lift-Off:
THE 1950s

BOEING 707

FACULTY AND STAFF

The School of Aeronautics was well stabilized by 1950, following a five-year growth from the time it was formed in July 1945. By this time the entire operation of the school was located at the Purdue University airport in two large buildings, with separate programs offered in aeronautical engineering and in air transportation. Dr. Elmer F. Bruhn, who had been head of the school since its beginning, decided early in 1949 to return to teaching and research rather than continue in administration. A search was immediately begun for a new head. By midsummer of 1950 Dean Andrey A. Potter announced the appointment of Dr. Milton U. Clauser (Figure 4.1) as the new head of the school, effective September 1, 1950.

Dr. Clauser received his BS and PhD from the California Institute of Technology and came to Purdue from Douglas Aircraft Corporation at El Segundo, California, where he was head of mechanical design. Dr. Clauser was an energetic young man, and while he had no previous experience in academic administration, he soon became very active in the operation and activities of the young school.

The faculty in the program in aeronautical engineering in the fall of 1950 consisted of the following:

Figure 4.1
Professor Milton U. Clauser,
school head, 1950–54.

- **Professors:** E. F. Bruhn, M. U. Clauser, J. Liston, M. U. Serbin, and M. J. Zucrow

- **Associate professors:** T. J. Herrick. M. E. Shanks, and P. E. Stanley

- **Assistant professors:** H. Lo and V. C. Vanderbilt

- **Instructors:** G. M. Palmer, B.A. Reese, R. F. Robinson, and A. F. Schmitt

- **Visiting professor:** L. A. Broglio

The distribution of the faculty in the various areas of specialty was roughly as follows:

- **Structures and dynamics:** Bruhn, Herrick, Lo, and Schmitt

- **Aerodynamics and control:** Clauser, Palmer, Robinson, Serbin, Shanks, and Stanley

- **Propulsion:** Liston, Palmer, Reese, Vanderbilt, and Zucrow

The faculty in the program in air transportation at that time consisted of the following:

- **Professors:** E. F. Bruhn

- **Associate professors:** M. I. Fowler and P. E. Stanley

- **Assistant professors:** W. W. Briggs and L. T. Cargnino

- **Instructors:** J. H. Basinger

- **Associate:** G. Webster

The faculty changed very little in the early years of this decade. Dr. Clauser took his time as new head to adjust to his new assignment and the existing faculty of 18 people. However, one new faculty member was added for the fall 1951 school year. **Dr. Harold M. DeGroff** (Figure 4.2), who received a PhD from the California Institute of Technology in 1950, was added for the fall 1951 school year.

Figure 4.2   Professor Harold M. DeGroff (1954), school head, 1955–63.

Dr. DeGroff was destined to play a leading role in the future of the school, as he later became head and served in that capacity for nearly a decade. Other faculty and staff changes during the 1950s are summarized below.

Professor Vern E. Vanderbilt resigned as assistant professor in 1952 to take a position in industry as director of research for the Perfect Circle Corporation. In the interval between his MS and the time he left the University in 1952, he had obtained his PhD in electrical engineering from Purdue.

Dr. M. J. Zucrow (Figure 4.3) elected to transfer his appointment entirely to the School of Mechanical Engineering during the spring semester of 1953. Dr. Zucrow had held a joint appointment between aeronautical and mechanical engineering since 1946 and had made tremendous progress in developing a first-rate graduate and research program in rocket and jet propulsion. By this time the first and second stages of the rocket engine laboratory had been completed on the west boundary of the Purdue airport and were being used by graduate students in research programs under Dr. Zucrow's direction. His transfer, which became effective with the 1953-54 school year, annoyed

Figure 4.3   Professor Maurice J. Zucrow (1952).

most of the faculty in the aero school because Dr. Zucrow took all of the gas turbine and jet propulsion work, his associates, and graduate students with him, including the new rocket lab on the west end of the airport. This left the aero school with no graduate program in gas turbines or jet propulsion. There was still graduate work in internal combustion engines with Professor J. Liston, but that program was losing popularity with the advent of the jet aircraft, and only a few students were pursuing advanced degrees in that area.

Dr. Clauser announced during the summer of 1954 that he had been granted a one-year leave of absence to work with the Ramo-Woodridge Corporation in Los Angeles, California. He had confided to some of the faculty earlier that he missed the fast pace of industry since coming to Purdue. **Dr. Harold M. DeGroff**, an associate professor, was named acting head during Dr. Clauser's absence by Dr. G. A. Hawkins, dean of engineering. As indicated earlier, Dr. DeGroff had joined the aero faculty in the fall of 1951. He had just completed a two-year assignment at Oak Ridge, Tennessee, on a project conducted by the Fairchild Engine and airplane corporation for the Air Force that dealt with nuclear-powered aircraft. Dr. Clauser elected not to return to Purdue when his leave expired, and Dr. DeGroff was named head of the school effective July 1, 1955.

At this point, several staff members who played key roles in the early development of the school should be mentioned. The head secretary, **Rhea E. Walker** (Figure 3.7), joined the school in 1947 and immediately organized the secretarial staff and all of the school records in a very efficient manner. The school would have had a difficult time without her able assistance during the early years.

George S. Calvert, who supervised the work of the technicians, was another person who contributed much to the early development of the school. Mr. Calvert joined the staff in the fall of 1952 and soon became the indispensable expert in the fabrication and repair of laboratory equipment. The school would have been severely handicapped without his able assistance in maintaining all the laboratory instruments, etc.

Robert L. Conner was another trusted employee who joined the technical staff in 1953 as a machinist. Mr. Conner was an expert machinist who fabricated many of the thesis projects used by graduate students in their research.

By the fall of 1955, the enrollment in aeronautical engineering had climbed to 318 after a low of 162 in the fall of 1952, and additional faculty were sought. **Dr. Angelo Miele** (Figure 4.4), formerly of Naples, Italy, joined the faculty at that time as an associate professor from the Polytechnic Institute of Brooklyn, New York. His area of expertise was flight mechanics and optimization of rocket-powered flight trajectories.

Dr. Hyman S. Serbin, who had been with the faculty since 1947, resigned in August 1955 to take a position with the Convair Division of General Dynamics Corp. in San Diego, California. He had been on a leave of absence the prior year at that facility. His area of specialty was aerodynamics.

Dr. Alfred F. Schmitt left the school in August of 1955 to take a position with

Figure 4.4   Professor Angelo Miele (1955).

Ryan Aeronautical Company in California. Professor Schmitt had obtained a PhD in August 1953 under Professor Elmer Bruhn and had remained with the faculty since that time.

Two new faculty members were added in the fall of 1956. **Dr. R. John Bollard** returned to Purdue University from New Zealand during the summer of 1956 to assume the position of assistant professor of aeronautical engineering. Dr. Bollard had been teaching engineering at Auckland University College since September 1954, when he left Purdue after being awarded a PhD in aeronautical engineering under Professor Hsu Lo.

Paul S. Lykoudis (Figure 4.5), a native of Greece, was appointed an assistant professor in the School of Aeronautical Engineering effective the fall of semester 1956. Dr. Lykoudis did his undergraduate study at the National Technical University of Athens, Greece, receiving his degree there in 1950, with majors in electrical and mechanical engineering. He came to the United States in 1953 to enroll at Purdue, where he was awarded his master's degree in 1954 and his PhD in 1956 in mechanical engineering. Dr. Lykoudis had specialized in the areas of heat transfer, thermodynamics, and fluid flow. He was assigned to teach aerodynamics courses in the aero school, and also offered one of the first graduate

Figure 4.5   Professor Paul S. Lykoudis (1956).

courses in magneto-hydro-dynamics at any university. He also conducted research in a number of allied fields.

During the 1956-57 school year, through the efforts of Dr. Angelo Miele, the aero school was able to get an outstanding authority in the field of aeronautics on a visiting professorship. **Dr. Placido Cicala** came to Purdue on leave from the Polytechnic of Turin, Italy, under the joint sponsorship of the Division of Engineering Sciences and the School of Aeronautical Engineering. He presented a series of lectures on the general theme of "an engineering approach to the calculus of variations."

Dr. Robert J. Goulard was added to the faculty as an assistant professor effective the fall of 1957. He was a native of France and had received his aeronautical engineer's degree from the École Nationale Superieure de L'Aeronautique in Paris in 1949; he had then completed his PhD at Purdue under Dr. H. M. DeGroff. Dr. Goulard's area of specialty was aerodynamics.

Two other new PhD graduates from the school also joined the faculty as assistant professors in aeronautical engineering: **Dr. Leslie A. Hromas** (September 1957) and **Dr. Madeline Goulard** (September 1958, Figure 4.6). Dr. Madeline Goulard, a native of France, had worked under the direction of Lo in the area of aircraft vibrations and aeroelasticity. She received her degree in June 1958, becoming the first woman to earn an aeronautical en-

Figure 4.6   Professor Madeline Goulard (1958).

gineering PhD at Purdue. Dr. Madeline Goulard, who was then the wife of Dr. Robert J. Goulard, had received her aeronautical engineer's degree from the École Nationale Superieure de L'Aeronautique in Paris, France, in 1950. Dr. Hromas had worked under Dr. Harold M. DeGroff, and his area was aerodynamics.

Dr. **Angelo Miele** resigned his professorship during the summer of 1959 and accepted a position on the staff of the University of California at Berkeley in the division of aeronautical sciences. He had been at Purdue since 1955 in the area of flight mechanics. Several graduate students working with him in that area transferred with him.

Dr. **George Lianis**, a native of Greece, was added as an associate professor of aeronautical engineering effective the fall of 1959. Dr. Lianis had received the mechanical and electrical ingeneur degree from the National Technical University of Athens, Greece, in 1953, a diploma from the Imperial College of London in 1954, and a PhD from the same school in 1956. His area of specialty was aircraft structures and the mechanics of solids.

Dr. **C. Paul Kentzer** (Figure 4.7) also joined the faculty in the fall of 1959. He had been a graduate student working under the direction of Dr. Merrill E. Shanks, who had a joint appointment between aeronautical engineering and mathematics. Dr. Kentzer was added as assistant professor of aeronautical engineering with a specialty in aerodynamics. Dr. Kentzer was a native of Poland and in 1948 came to the United States, where he obtained

Figure 4.7   Professor C. Paul Kentzer (1954).

his BS from San Diego State in 1952. He obtained his MS in aeronautical engineering in 1954 at Purdue, followed by a PhD in 1958.

1950 CURRICULUM

The curricula in effect in the fall of 1950 are given in Tables 4.1–4.7. It is interesting to note that the power plant option in aeronautical engineering at that time required $168\frac{2}{3}$ credits. Today, a student can earn both a BS and an MS in engineering with fewer total credits. Military training was required of all able-bodied males (physical education for females) at that time, and those credits (9–12 credits) were included in the totals.

By the fall of 1950, enrollment in most programs at Purdue had declined, primarily as a result of the Korean War. The total enrollment at Purdue dropped from 14,674 in the fall of 1948 to 11,053 in the fall of 1950. This trend continued until 1953, when the enrollment started to increase again. Enrollment in aeronautical engineering decreased from 395 in 1948 to 235 in 1950 (see Figure 3.19). Actually, the enrollment had reached 503 in aeronautical engineering in the fall of 1947. A similar enrollment drop occurred in the air transportation program, from 215 in the fall of 1948 to 167 in the fall of 1950. Total school enrollment was 736 in the fall of 1947, with 233 students participating in the air transportation program and 503 in aeronautical engineering.

Table 4.1   Aeronautical Engineering Curriculum (1950)

Freshman Year

First Semester

- (4)   Chem 1 or 17 (General Chemistry)
- (3)   Engl 1 or 32 (Composition)
- (2)   GE 11 (Engineering Drawing)
- (0)   GE 1 (Engineering Lectures)
- (5)   Math 1 (Algebra, Trigonometry or Math 31 (Elementary Engineering Mathematics)
- (2–3)   Military Training
- ($^2/_3$)   PEM 9 (Personal Living)
- (2)   GE 34 (Welding, Heat Treating, and Casting) or CE 6 (Plane Surveying)

Second Semester

- (4)   Chem 2 or 18 (General Chemistry)
- (3)   Engl 4 (Amer. Books), Engl 10 (Essay), Engl 19 (Drama), Engl 20 (Poetry), or Engl 27 (Fiction) Spch 14 (Principles of Speech)
- (2)   GE 12 (Engineering Drawing)
- (5)   Math 2 (Analytical Geometry) or Math 32 (Elementary Engineering Mathematics)
- ($2^1/_3$–3)   Military Training
- (2)   CE 6 (Plane Surveying) or GE 34 (Welding, Heat Treating and Casting)

($18^2/_3$–$19^2/_3$)   ($18^1/_3$–19)

General Summer Aircraft Shops

The following 8-week summer session must precede the sophomore year:

- ($2^1/_3$)   GE 50 (Aircraft Sheet Metal Shop)
- (2)   GE 54 (Machine Shop)
- ($^2/_3$)   GE 51 (Aircraft Welding)
- (1)   AeroE 20 (Aircraft Components and Installations)
- (2)   GE 16 (Descriptive Geometry)

---

(8)

Table 4.2   Airplane Option of Aeronautical Engineering Curriculum (1950) —
Total Credits: 161 semester hours

SOPHOMORE YEAR

Third Semester

(3)   Spch 14 (Principles of Speech) or
      Engl 4 (American books) or Engl 10
      (Essay) or Engl 19 (Drama) or
      Engl 20 (Poetry)
(5)   Phys 21 (General Physics)
(4)   Math 3 (Calculus I)
(2)   GE 30 (Aircraft Detail and
      Layout Drafting)
(3)   Econ 1 (Elementary Economics)
($2\frac{1}{3}$–3)   Military Training

($19\frac{1}{3}$–20)

Fourth Semester

(5)   Phys 22 (General Physics)
(4)   Math 4 (Calculus II)
(5)   EM 21 (Statics and Kinetics)
(3)   ME 29 (Elementary Heat Power)
($2\frac{1}{3}$–3)   Military Training

($19\frac{1}{3}$–20)

JUNIOR YEAR

Fifth Semester

(3)   ME 11 (Mechanism)
($3\frac{2}{3}$)   ME 55 (Fluid Mechanics)
(3)   ME 40 (Thermodynamics)
(4)   EM 22 (Mechanics of Materials)
(3)   Math 107 (Differential Equations)
      or Math 109 (Numerical Methods
(3)   Govt 30 (Elements of Democracy)
      or Govt 109 (International
      Relations)

($19\frac{2}{3}$)

Sixth Semester

(3)   AeroE 80 (Aerodynamics)
(1)   AeroE 81 (Wind Tunnel
      Laboratory)
(3)   AeroE 71 (Aircraft Structures)
(3)   ME 67 (Machine Design)
($3\frac{2}{3}$)   EE 19 (Direct Currents)
(3)   Hist 1 (Europe since 1914) or
      Hist 5 (U.S. in World Affairs)
(3–$3\frac{2}{3}$)   AeroE 90 (Aircraft Power Plants)

($19\frac{2}{3}$–$20\frac{1}{3}$)

SENIOR YEAR

Seventh Semester

(3)   AeroE 82 (Aerodynamics)
(3)   AeroE 72 (Aircraft Structures)
($3\frac{2}{3}$)   EE 20 (Alternating Currents)
(1)   CE 31 (Testing Materials)
(3)   Psy 74 (Psychology for
      Engineers)
(5-6)   Technical Elective

($18\frac{2}{3}$–$19\frac{2}{3}$)

Eighth Semester

(3)   AeroE 144 (Aircraft Vibrations)
(2)   AeroE 73 (Aircraft Structures
      Laboratory)
(3–$3\frac{1}{3}$)   Technical Elective
(3)   Engl 31 (Expository Writing)
(8–9)   Technical Electives

(19–$20\frac{1}{3}$)

Table 4.3   Aircraft Power Plant Option of Aeronautical Engineering Curriculum (1950) — Total Credits: $168^2/_3$ semester hours

Students who desire to specialize in the field of aircraft power plants will follow the Airplane Option in the third and fourth semesters and may elect the following option in their junior and senior years:

Summer Engine Shops (8 weeks)

Eight weeks of summer engine shops are required. These can be taken after the sophomore or the junior year. If a satisfactory cooperative program can be arranged with the aircraft engine industry, this work in the industry may be substituted for the summer engine shops.

$(4^1/_3)$	AeroE 40 (Aircraft Engines)
$(^2/_3)$	AeroE 41 (Aircraft Propellers)
(1)	AeroE 42 (Aircraft Engine Accessories)
(1)	AeroE 43 (Carburetion, Ignition, Fuel and Oil Systems)
(1)	CE 31 (Testing Materials)

(8)

Junior Year

Fifth Semester		*Sixth Semester*	
Same as Airplane Option		(3)	ME 65 (Dynamics of Machinery)
		(3)	AeroE 80 (Aerodynamics)
		(3)	ME 41 (Thermodynamics)
		(1)	ME 78 (Mechanical Laboratory)
		$(3^2/_3)$	EE 19 (Direct Currents)
		(3)	Hist 1 (Europe Since 1914) or Hist 5 (U.S. in World Affairs)
		(3)	Technical Elective

$(19^2/_3)$

Senior Year

Seventh Semester		*Eighth Semester*	
(3)	ME 67 (Machine Design)	(4)	AeroE 111 (Aircraft Engine Design)
(3)	AeroE 112 (Aircraft Engines)		
(2)	AeroE 113 (Aircraft Engines Laboratory)	(3)	AeroE 115 (Gas Turbines and Jet Engine Power Plants)
(2)	AeroE 157 (Internal Aerodynamics)	(2)	AeroE 117 (Gas Turbines and Jet Engines Laboratory)
$(3^2/_3)$	EE 20 (Alternating Currents)		
(3)	Psy 74 (Psychology for Engineers)	(3)	Engl 31 (Expository Writing)
(3)	Technical Electives	(6)	Technical Electives

$(19^2/_3)$	(18)

Table 4.4   Air Transportation Curriculum (1950) — Leading to the degree of Bachelor of Science in Air Transportation

FRESHMAN YEAR

First Semester		*Second Semester*	
(4)	Chem 1 or 17 (General Chemistry)	(4)	Chem 2 or 18 (General Chemistry
(3)	Engl 1 or 32 (Composition)	(3)	Engl 4 (Amer. Books) or Engl 10
(2)	GE 11 (Engineering Drawing)		(Essay) or Engl 19 (Drama) or
(0)	GE 1 (Engineering Lectures)		Engl 20 (Poetry) or Engl 27 (Fiction)
(5)	Math 1 (Algebra, Trigonometry)		Spch 14 (Principles of Speech)
	or Math 31 (Elementary	(2)	GE 12 (Engineering Drawing)
	Engineering Mathematics)	(5)	Math 2 (Analytical Geometry) or
(2–3)	Military Training		Math 32 (Elementary Engineering
($^2/_3$)	PEM 9 (Personal Living)		Mathematics)
(2)	GE 34 (Welding, Heat Treating	($2\,^1/_3$–3)	Military Training
	and Casting) or CE 6	(2)	CE 6 (Plane Surveying) or GE 34
	(Plane Surveying)		(Welding, Heat Treating and
			Casting)

($18\,^2/_3$–$19\,^2/_3$)   ($18\,^1/_3$–19)

*Table 4.5   Aviation Operations Option of Air Transportation Curriculum (1950) —
Total Credits: $152^2/_3$ semester hours*

Sophomore Year

Third Semester

(4)	Phys 34 (General Physics)
(1)	AT 1 (Private Pilot's Course)
(2)	AT 31 (Aeronautical Meteorology)
(2)	AT 22 (Air Navigation)
(1)	AT 12 (Principles of Flight)
(8)	AT 40 (Principles of Aircraft Power Plant Maintenance)
$(2^1/_3$–3)	Military Training

$(20^1/_3$–21)

Fourth Semester

(4)	Phys 35 (General Physics)
(3)	Econ 1 (Principles of Economics)
(1)	AT 1 (Private Pilot's Course)
$(1^2/_3)$	GE 52 (Aircraft Welding)
(1)	GE 55 (Aircraft Wood and Plastics Shop)
$(2^2/_3)$	GE 68 (Aircraft Sheet Metal Shop)
(5)	AT 53 (Principles of Aircraft Maintenance)
$(2^1/_3$–3)	Military Training

$(20^2/_3$–$21^1/_3)$

Junior Year

Fifth Semester

(3)	Engl 31 (Expository Writing)
(3)	Econ 125 (Business Law)
(3)	Govt 30 (Elements of Democracy)
(5)	AT 59 (Aircraft Power Plant Maintenance)
(4)	AT 60 (Aircraft Maintenance)

(18)

Sixth Semester

(3)	Spch 116 (Business and Professional Interview)
(2)	Econ 135 (Aviation Law)
(3)	Engl 4 (Amer. Books) or Engl 10 (Essays) or Engl 19 (Drama) or Engl 27 (Fiction)
(3)	Psy 1 (Elementary Psychology)
(3)	EM 24 (Engineering Mechanics)
(6)	Electives

(20)

Senior Year

Seventh Semester

(3)	Econ 129 (Government and Business)
(3)	AT 80 (Airline Management and Operations)
(3)	AT 85 (Airport Operation and Control)
(2)	CE 67 (Airport Selection and Layout)
(3)	GE 91 (Elements of Accounting)
(3)	Hist 1 (Europe Since 1914) or Hist 5 (U.S. in World Affairs)
(3)	Elective

(20)

Eighth Semester

(3)	AT 90 (Air Traffic and Cargo)
(3)	Econ 133 (Air Transportation Economics)
(3)	Econ 121 (Labor Economics)
(3)	AT 70 (Elementary Aircraft Structures)
(6)	Electives

(18)

Table 4.6   Flight Operations Option of Air Transportation Curriculum (1950) —
    Total Credits: 155$^2/_3$ semester hours)

SOPHOMORE YEAR
Same as for Aviation Operations Option

JUNIOR YEAR

Fifth Semester

(3)   Engl 31 (Expository Writing)
(3)   Econ 125 (Business Law)
(3)   Govt 30 (Elements of Democracy)
(5)   AT 59 (Aircraft Power Plant
        Maintenance)
(4)   AT 60 (Aircraft Maintenance)
(1)   AT 2 (Advanced Private
        Pilot's Course)

(19)

Sixth Semester

(3)   Spch 116 (Business and
        Professional Interview)
(2)   Econ 135 (Aviation Law)
(3)   Engl 4 (Amer. Books) or Engl 10
        (Essays) or Engl 19 (Drama) or
        Engl 27 (Fiction)
(3)   Psy 1 (Elementary Psychology)
(4)   AT 23 (Air Navigation)
(3)   AT 32 (Aeronautical Meteorology)
(1)   AT 3 (Commercial Pilot's Course)

(19)

SENIOR YEAR

Seventh Semester

(3)   Econ 129 (Government and Business)
(3)   AT 80 (Airline Management
        and Operations)
(3)   AT 85 (Airport Operation and
        Control)
(2)   CE 67 (Airport Selection and Layout)
(3)   GE 91 (Elements of Accounting)
(3)   Hist 1 (Europe since 1914) or
        Hist 5 (US in World Affairs)
(3)   AT 14 (Flight Instruction Methods)
($^1/_2$)   AT 6 (Flight Instructor's Rating)

(20$^1/_2$)

Eighth Semester

(3)   AT 90 (Air Traffic and Cargo)
(3)   Econ 133 (Air Transportation
        Economics)
(3)   Econ 121 (Labor Economics)
(3)   AT 15 (Instrument Flight
        Problems)
(1)   AT 5 (Instrument Rating)
(6)   Elective

(19)

Table 4.7   Aviation Administration Option of Air Transportation Curriculum (1950) — Total Credits: 149⅔ semester hours

Sophomore Year

Third Semester

- (4) Phys 34 (General Physics)
- (1) AT 1 (Private Pilot's Course)
- (2) AT 31 (Aeronautical Meteorology)
- (2) AT 22 (Air Navigation)
- (1) AT 12 (Principles of Flight)
- (3) Engl 4 (Amer. Books) or Engl 10 (Essays) or Engl 19 (Drama) or Engl 27 (Fiction)
- (3) Govt 30 (Elements of Democracy)
- (2⅓–3) Military Training

(18⅓–19)

Fourth Semester

- (4) Phys 35 (General Physics)
- (3) Econ 1 (Principles of Economics)
- (1) AT 1 (Private Pilot's Course)
- (3) AT 9 (Airframe Components and Installations)
- (3) AT 10 (Aircraft Power plants)
- (3) Psy 1 (Elementary Psychology)
- (2⅓–3) Military Training

(19⅓–20)

Junior Year

Fifth Semester

- (3) Engl 31 (Expository Writing)
- (3) Econ 125 (Business Law)
- (3) Hist 1 (Europe Since 1914) or Hist 5 (US in World Affairs)
- (3) Psy 173 (Personnel Psychology)
- (3) Econ 121 (Labor Economics)
- (3) GE 91 (Elements of Accounting)

(18)

Sixth Semester

- (3) Spch 116 (Business and Professional Interview)
- (2) Econ 135 (Aviation Law)
- (3) Econ 138 (Economic Statistics)
- (3) GE 92 (Elements of Accounting II or GE 127 (Cost Accounting)
- (3) Engl 121 (Business Writing)
- (3) Econ 112 (Economics of Marketing) or Econ 130 (Economic Geography), or Econ 141 (World Economics)
- (3) Elective

(20)

Senior Year

Seventh Semester

- (3) Econ 129 (Government and Business)
- (3) AT 80 (Airline Management and Operations)
- (3) AT 85 (Airport Operation and Control)
- (2) CE 67 (Airport Selection and Layout)
- (3) Econ 115 (Corporation Finance)
- (6) Electives

(20)

Eighth Semester

- (3) AT 90 (Air Traffic and Cargo)
- (3) Econ 133 (Air Transportation Economics)
- (3) Spch 106 (Group Discussion) or Spch 114 (Analysis and Argument)
- (9) Elective

(21)

By 1950, both the MS and PhD graduate programs were well established in the School of Aeronautics. As early as February 1947 two students had received masters of science degrees in aeronautical engineering, and the first PhD was granted in 1950 to Commander Richard L. Duncan, U.S.N. His thesis was titled "Analytical Investigation of the Performance of Gas Turbines Employing Air-Cooled Turbine Blades," and his major professor was M. J. Zucrow. (Professor Zucrow had received Purdue's first PhD in 1928.) That same year, eight students received their master's degrees in aeronautical engineering.

Graduate course offerings at that time were as follows. In the area of **Aerodynamics:** Aerodynamics of Perfect Fluids, Aerodynamics of Real Fluids, Supersonic Aerodynamics, Aerodynamics of Turbo Machines, Analysis of Drag, Transonic and Supersonic Aerodynamics, and Rotary Wing Aircraft. In the area of **Structures and Dynamics:** Advanced Aircraft Vibrations, Aircraft Dynamic Analysis, Aircraft Flutter, Advanced Aircraft Structures I, Advanced Aircraft Structures II, Advanced Aircraft Structures Laboratory, and Theory of Plasticity and Aircraft Structures. In the area of **Propulsion:** Advanced Internal Combustion Engines, Advanced Aircraft Power plant Laboratory, Internal Combustion Engine Seminar, Gas Turbines and Jet Propulsion I, Gas Turbines and Jet Propulsion II, Gas Turbine, and Jet Propulsion Seminar and Fundamentals of Rocketry.

FACILITIES

By the early 1950s, the first and second units of the rocket engine laboratory located near the west boundary of the Purdue airport had been completed, and research and graduate study were well under way in that area. This facility was a joint laboratory for both aeronautical and mechanical engineering graduate students.

The aircraft internal combustion engine laboratory was well established and being used by undergraduate and graduate students. This laboratory had been developed during the 1940s by Professor Joseph Liston (Figure 4.8), and consisted of two engine test cells where students could run engineering tests of reciprocating-type aircraft engines. By 1950 one of the test cells had been modified to use a small Westinghouse axial flow turbojet engine — 9.5-inch diameter with a thrust rating of 250 lbs. Also included was a single-cylinder CFR engine in which the compression ratio could be varied during operation for fuel anti-knock testing.

At this time the aerodynamics laboratory had five wind tunnels in operation. The latest of these was the large subsonic tunnel (Figures 4.9 and 4.10) with a capability of 350–400 miles per hour. This tunnel had been

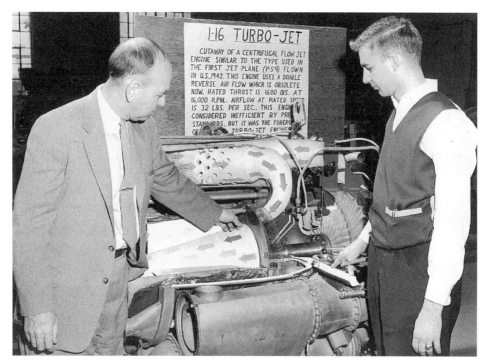

Figure 4.8   Professor Joseph Liston points out features of early jet engine to air transportation student Robert Bass (1954).

Figure 4.9   Professors H. M. DeGroff, G. M. Palmer, and M. U. Clauser near subsonic wind tunnel (1953).

Figure 4.10 Professor M. U. Clauser standing in subsonic wind tunnel (1954).

designed by Professor G. M. Palmer and was built mainly of plywood with student help. The test section was approximately 11.5 square feet, and it used a 400-hp electric motor. This tunnel was upgraded by means of a large grant from The Boeing Company in the early 1990s, and is still in use. The original 440-hp motor was still operating more than 50 years later when it was finally replaced in 2008. Professor Palmer was formally recognized as the designer of the Boeing Wind Tunnel at a special ceremony held on September 25, 2009.

Among the other tunnels in use at that time were a Japanese variable-density wind tunnel that had been confiscated and brought to this country after the war. Its use was rather limited because of its low power (100 hp). It had a 15-inch throat and was capable of speeds up to 300 mph. Another tunnel was a 110-mph tunnel that was originally powered by a Dodge auto engine, but was modified and lengthened by Palmer and driven by a 50-hp electric motor. This tunnel was used primarily in the required wind tunnel courses for aero students.

Two other wind tunnels were small wood facilities that had been brought from campus to the airport in 1948. One was a smoke tunnel. A water table was added in 1953 for demonstrating compressible and viscous flow. Around the same time, a shock tube was added for the study and research in high-speed flow.

The structures laboratory consisted primarily of load-applying equipment and instrumentation involved in measuring the effect of various load spectra on flight vehicles. The following equipment was used: a 60,000-lb. Tinius Olsen testing machine, a 60,000-lb. Tate Emery testing machine, a Rockwell hardness tester, several tension dynamometers, strain indicators, load cells, oscillographs, vibration meters, vibration analyzers, and velocity pick-ups.

Along with the expansion in the graduate programs came an increase in sponsored research. Some of the research came from the aircraft industries, but most came from the Office of Ordnance Research, the Office of Naval Research, the Office of Scientific Research and Development, and the National Advisory Committee for Aeronautics (NACA).

STUDENT ACTIVITIES

By 1952, the undergraduate enrollment had deteriorated even further, reaching its lowest level during the 1950s. In the fall of 1952, the aeronautical engineering enrollment was 162 and the air transportation enrollment was 117 for a total of 289 students. The total University enrollment had decreased

Figure 4.11
Air transportation student
Joanne Alford, Purdue
homecoming queen (1952).

to 9,285 in 1952 from 11,053 in 1950.

There was, however, a bright spot during the fall 1952 semester when one of the sophomores, named Joanne Alford (Figure 4.11), was elected 1952 homecoming queen. Miss Alford was a sophomore in air transportation and was very active in extracurricular activities and in the radio program *This Week in Aviation.* She was a member of Chi Omega sorority. The following year (1953) members of Pi Kappa Alpha fraternity elected her their Dreamgirl and the campus males elected her their homecoming queen. In 1954 she was the national sweetheart of Alpha Chi Rho fraternity. She was a pilot during her student years and captured first place in the spot landing competition at the Midwest Conference of National Intercollegiate Flying Meet in 1954. Miss Alford went on to be named Miss Airpower in both 1956 and 1957 by the USAF Research and Development Corporation. In 1957 she was awarded the Air Force Association's National Citation of Honor for Outstanding Contribution to Air Age Education and National Defense. During her reign as Miss Airpower, in 1956 alone she estimated that she addressed 165,000 children and adults during a tour of the country.

In the early years of the school, including the 1950s, young women were drawn to aeronautics in greater numbers than to the other schools in engineering. The reasons for this trend are not clear, but it was thought by some that the influence of Amelia Earhart might have been present. Ms. Earhart was a familiar sight on campus only a decade or two earlier and had left her imprint on the women on campus. A fellowship award in her name was established in 1938 by Zonta International to be given to women for advanced study in the aerospace sciences.

It was not unusual to find two or three women in a class of about 30 students in those days, and most were able to hold their own academically in what was considered then a male-dominated program. Between 1945 and 1960, 21 women were granted degrees in aeronautical engineering, and five of those later received Amelia Earhart awards for advanced study here or elsewhere. During this same period (1945–57) an additional eight women were granted bachelor's degrees in air transportation for a total of 29. Note that bachelor's degrees in air transportation were not granted after 1958.

Students were very active in extracurricular activities during this decade due primarily to the fact that most of their classes were at the airport, and students felt a closeness almost as though aero were a separate college from the main campus. All of the aero technical courses and laboratories were taught at the airport facilities and only the courses taught by the other departments — such as math and English — were held on campus. This arrangement did not pose any problems for the student since University buses

Figure 4.12 Students preparing to board United Airlines DC- planes for visit to aerospace industries in the Dallas-Ft. Worth, Texas, and Wichita, Kansas, areas (January 1951).

Figure 4.13 Air transportation students preparing to board flight at LaGuardia Airport in New York City for return to Purdue (January 1950).

transported the students to and from the airport on a regularly scheduled basis. Usually the students were scheduled for blocks of courses at the airport, so they would be there for three or four hours (classes) continuously and not have to be shuttled back and forth between the campus and the airport.

During the early 1950s, several inspection trips were planned to acquaint the students with the industrial facilities and the real world of engineering (Figures 4.12–4.13). In January (19–22) 1950, 120 students together with six faculty members spent four days visiting industrial facilities in the Washington, D.C., Baltimore, and New York areas. This group went by air in two DC-4 Capitol Airlines. Eighty-one students were from the air transportation program, while 39 were aero engineering students. The aero students visited the facilities of the Glenn L. Martin Co. in Baltimore and later visited the facilities of Republic Aviation in Farmingdale, New York, and those of Grumman nearby.

The air transportation students meanwhile were visiting the facilities of Capitol Airlines based at Washington International Airport and later the facilities at Idlewild (Kennedy) International Airport, Teterboro Airport, and LaGuardia Airport. Airline installations (American, Pan American, etc.) were visited as well as the airport control facilities. In the evenings all of the students and faculty came together for banquets at the hotel (Annapolis) in Washington D.C. and (Vanderbilt) in New York City. Speakers at the banquets were high-level representatives of the various companies and airports.

A similar inspection trip by air was taken by the students in the School of Aeronautics in 1951. It was a three-day trip to the Chance-Vought Corporation at Dallas; General Dynamics at Ft. Worth; CAA at Oklahoma City; and Boeing, Beechcraft, and Cessna Aircraft Companies at Wichita, Kansas.

The classes were spiced with returning veterans in those early days, helping to create an active group of talented and resourceful students. Some of the activities included the Purdue Glider Club, the Purdue Flying Club, the weekly radio program *This Week in Aviation*, the weekly newspaper *The Aeroliner*, and the Air Freight Board.

The Purdue Glider Club (Figures 4.14–4.15), probably the oldest of these activities, was organized back in 1932 (see Figure 1.7). Despite many setbacks, it managed to survive and stay alive for all those years. In the early years the students used the Purdue airport for their flying activities, but by 1941 the airport had become too busy and the club had to find other places, sometimes cow pastures, or other open fields. Between 1932 and 1942, the club had owned a variety of gliders including an Albatross, a Weaver, a Gull, a Gross, and a Cinema. When Professor Cargnino became faculty advisor in

1945, the club had only two-place Cinemas. At one time the club had four Cinemas and a one-place Franklin which was an experimental soaring plane primarily for ridge soaring.

The Glider Club had developed a woodworking shop in the basement of the Memorial Union and were allowed to use a dope and paint shop at the airport. On nearly any night, several club members could be found working late into the night making wood repairs for gliders that had been damaged in landing or had been flipped over by a wind gust when left unattended on the ground (Figure 4.14). The large wings were made entirely of wood (sitka spruce) and required a lot of handwork repairs.

Most of the gliding and soaring was done on weekends. The club owned a tow car with a stationary winch. The winch was mounted in a trailer and was towed behind a car or truck (Figure 4.15). The winch was powered by an 8-cylinder engine and had 3,000 feet of solid steel wire. Normally the winch would lift the glider into the air at a groundspeed of 45 miles per hour to a height of 700–800 feet. The altitude record for the club at the time was 5,000 feet. The longest flight made by the club then was a two-hour flight. The group attended meets in Dayton and Toledo, Ohio, as well as the annual snowbird meet in Elmira, New York. It was not unusual for the club members to work all night repairing a damaged glider and, with the dope finish on the fabric not yet dry, take off for a meet in the early morning hours. A favorite place to do ridge soaring with the Franklin ship was the high bluffs along the east

Figure 4.14 Members of Purdue Glider Club studying glider plans (1951).

side of Lake Michigan near Benton Harbor, Michigan.

Whenever one of the club members made a flight error or an error in calculation, he was required to carry a trophy shaped like a dumbbell to all his classes and other activities for a period of two weeks. Ribbons were attached to the dumbbell relating the activity that caused the blunder. At Dayton, Ohio, in 1950, the club won the Western Conference Championship trophy for excellence in the spot landing, bomb drop, and paper strafe. The club, which usually numbered 30 to 40 students, was not limited to aero students, but traditionally 35% to 40% of the students were aeronautical engineering or air transportation students.

The Purdue Aero Flying Club was another activity that drew a lot of interest among the aero students in the 1950s. Membership was open to all Purdue students, but usually half of the club members were aero students. Following the Korean War in the mid-1950s, the number of students in the flying club increased greatly due to the number of veterans in school at that time. This was the era when Neil Armstrong was a student in the aero program. Mr. Armstrong returned to the campus in 1952 after flying 78 missions in the Korean War, and he guided the club as its chairman for the next three years. The planes used by the club were usually two-place and four-place Piper cubs, four-place Stinson voyagers, and later Cessna 120s and 140s. Most of the flying was done at Aretz airport because the Purdue airport did not have hangar facilities and provisions for any maintenance outside of its own fleet of planes. The flying club participated in flying meets

Figure 4.15 Tow car and glider used by Purdue Glider Club (1951).

with other Midwest colleges and universities, usually on weekends. Some of the students owned their own planes. Professor W. W. Briggs was the faculty adviser.

A weekly half-hour radio program called *This Week in Aviation* was first introduced during the 1949-50 school year and was continued successfully well past the mid-1950s. It was on the air one evening a week, usually at 7 p.m., over Purdue's radio station, WBAA, and consisted mainly of news concerning aviation events, local and national, that were of interest during that week. It was purely a student activity with the students, mostly from air transportation, gathering and editing the news and then reporting it on the air. Professor Cargnino was faculty adviser, and met weekly with the club members when they selected the news items and also when they were broadcasting. One of the most active members of this club in the early 1950s was Joanne Alford, who as mentioned earlier was the 1952 Purdue homecoming queen. Little difficulty was encountered in attracting students into this activity while she was a member. The group received much help and support at the time from John R. DeCamp, who was program chairman of WBAA and was always there during the broadcasts.

The aero students also developed a weekly newspaper that was very active during the early and middle years of the 1950s decade. It was called the *Aeroliner* and usually consisted of some six to eight pages of newsworthy material, both local and national. The club consisted entirely of students enrolled in aeronautical engineering and air transportation, who wrote the articles and produced the printed copy usually with a mimeograph machine. It was not on the scale of the *Exponent*, but it was a popular newsletter at the time. The faculty adviser most of the time was James H. Basinger, a popular young faculty member. Newsworthy items included such things as new aircraft developments, airline administrative changes, and civil aeronautics administration rulings, as well as campus news such as flying club meets and glider meets.

Another student activity that was very active during this time was called the Airfreight Board. It was formed sometime during the 1949-50 school year by a group of air transportation students under the direction of Professor Mart I. Fowler (Figure 4.16), who taught a course for seniors in air transportation on that subject matter. The board was set up with a chairman, vice chairman, and secretary, and its overall goal was to acquaint the students with the actual present conditions and problems of the air freight industry. The work of the board involved a fairly large number of students in the form of committees, which were given work assignments as they progressed. The students on the board usually met once a week at night

Figure 4.16 Professor Mart I. Fowler, on right shaking hands
with Captain Chuck Yeager (first pilot to break
sound barrier in 1947). Professor Elmer Bruhn looks
on with Captain Massa in doorway (circa 1950).

to discuss their progress. Professor Fowler, who had many contacts in the
industry at the time, gave much of his time and energy to obtaining information and operating data by the students. By the early 1950s, the students
had compiled a very comprehensive airfreight manual that was revised
and expanded in future semesters and was used by Professor Fowler as a
reference source in his course.

Another club that was very popular with the aero and air transportation students was the Purdue Aeromodelers (Figures 4.17–4.18), a student
union-sponsored club that was formed back in 1939 for the purpose of
promoting model airplane building activities at Purdue. The original group
of 10 members included some of the most prominent modelers in the Mid-

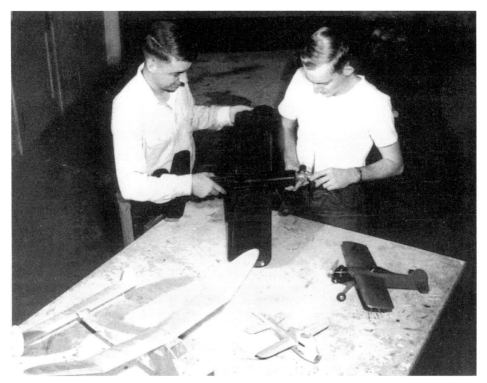

Figure 4.17 Purdue Aeromodelers (early 1950s, Lloyd Hackman, left, BSAE '52).

Figure 4.18 Purdue Aeromodelers club (Neil Armstrong is fourth from left in second row).

west, including James Cahill, who had won the famous Wakefield Trophy in international competition the year before (1938).

Following World War II, the membership grew to more than 75 students. During the school year, the club met once a week in the evenings in the union building where they had a workshop in the subbasement. Early in 1954 a Midwest conference was held to coordinate intra-club contests among about 12 schools in Indiana, Illinois, and Ohio. Intra-club contests were held annually. Types of flying included gliders flown in the union ballrooms, the armory, or outdoors; small speed models flown in the armory during bad weather or outdoors in the intramural fields, and various other types designed by club members. Membership was open to any Purdue student or staff member.

The School of Aeronautics walked off with top honors for its exhibit in the first annual Engineering Open House held May 1, 1954. An estimated 6,000 people viewed the exhibits and demonstrations put on by the various engineering schools. The aeronautics exhibit began with cutaways of turbojet engines in the air transportation lab followed by a movie on fighter-type aircraft by McDonnell Aircraft. Next on the agenda was the power plant lab featuring a full-size engine mounted on a torque stand and outfitted in a test cell instrumented to measure efficiencies. The visitors were then shown the wind tunnel display including the 400-hp, 350-mph wind tunnel. Examples of wing and fuselage construction were shown next, along with structural diagrams. Following the Glider Club's exhibit, which showed how gliders are constructed and operated, came the climax of the tour when the visitors were taken through the fuselage of the B-50 Lucky Lady II, the famous ship that made an around-the-world flight. Probably the most spectacular aspect of the school's open house was the mini Blue Angels flight performance by three Grumman Cougar airplanes over the Purdue airport. Two students, Neil Armstrong and Carl Neidhold, were former Navy pilots and were then in the Navy Reserve program at the Glenview Naval Air station near Chicago. Unbeknownst to most of the school, they, together with another Naval reservist who was not a student, arranged to fly the new Grumman Cougars to the Purdue airport during the open house. Needless to say, it was a very impressive display of high-speed flight maneuvers. Dr. Milton Clauser accepted the plaque awarded to the winning school.

CURRICULUM CHANGES

Shortly after Dr. Clauser arrived as head of the school in 1950, he began to express concern about the curriculum in aeronautical engineering. He believed the curriculum was not rigorous enough and that it was slanted toward a terminal degree program for those going directly into industry. He thought the program should be more theoretical, with fewer laboratories, and that the curriculum at the bachelor of science level should be more a preparation for advanced work and graduate study. At first, some faculty were opposed to his ideas and thought it strange that a man from industry wanted more theoretical work in the curriculum. To some of the other faculty his ideas were perfectly understandable. Dr. Clauser's background and work had been primarily in theoretical areas, and he pushed to get the curriculum changed accordingly. The faculty at that time had a rather diverse background and resisted some of the changes that were planned. However, by the fall of 1953, the faculty approved the addition of a separate option in the curriculum to be called the theoretical aeronautics option. This option was offered for the first time in the 1954-55 school catalog and was open to a select group of entering sophomores for the fall 1954 semester.

The theoretical aeronautics option is given in Table 4.8, and consisted of $147\frac{7}{3}$ semester hours. It is described in the 1954-55 catalog as follows:

> *The theoretical aeronautics option is for the student who has a liking for the more technical and mathematical subjects. It provides an excellent preparation for those who desire to go on for graduate study. Graduates and those with advanced degrees are needed to do the theoretical analysis required in each of the design specialties and for the research work which lays the foundation for progress in aeronautics. The large number of elective courses in the curriculum gives the student a wide latitude in choosing his field of study. Only students with the proper academic standing may take this option.*

Table 4.8   Theoretical Aeronautics Option (1954)

FRESHMAN YEAR

First Semester		*Second Semester*	
(4)	Chem 115 (General Chemistry)	(4)	Chem 116 (General Chemistry)
(3)	Engl 101 (Composition) or Speech 114 (Principles of Speech)	(3)	Speech 114 (Principles of Speech) or Engl 101 (Composition)
(3)	CE 118 (Engineering Graphics)	(4)	Phys 152 (General Physics)
(5)	Math 161 (Mathematics for Engineers and Scientists)	(5)	Math 162 (Mathematics for Engineers and Scientists)
(1)	Engr 100 (Freshman Engineering Lectures)	$(2\frac{1}{3})$	Military Training
(2)	Military Training		
(18)		$(18\frac{1}{3})$	

Table 4.8   Theoretical Aeronautics Option (1954) (continued)

SOPHOMORE YEAR

Third Semester		*Fourth Semester*	
(5)	Phys 240 (General Physics-special section)	(5)	Phys 241 (General Physics—Special Section)
(4)	Math 242 (Calculus II)	(3)	Math 431 (Intermediate Calculus)
(3)	Hist 210 (Europe Since 1914) or Hist 205 (Europe in World Affairs)	(3)	Math 551 (Introduction to Theoretical Statistics)
(3)	Spch 114 (Prin. of Speech) or Engl 202 (Expository Writing)	(3)	ME 206 (General Thermodynamics I)
		(3)	EM 207 (Elementary Mechanics)
$(2\frac{1}{3}$–3)	Military Training	$(2\frac{1}{3}$–3)	Military Training
$(17\frac{1}{3}$–18)		$(19\frac{1}{3}$–20)	

JUNIOR YEAR

Fifth Semester		*Sixth Semester*	
(3)	Math 525 (Applied Theory of Complex Functions)	(3)	Math 533A (Differential Equations)
(3)	Math 523 (Vector Analysis)	(3)	EM 507 (Basic Mechanics III)
(3)	ME 306 (General Thermodynamics II)	(3)	EM 532 (Theory of Material Properties)
(4)	EM 314 (Mechanics of Solids)	(3)	EE 514 (Electronic Circuits)
(4)	EE 412 (Electrical Engineering for Civil Engineers)	(3)	Aero Technical Elective
(3)	Modern Language	(3)	Modern Language
(20)		(18)	

SENIOR YEAR

Seventh Semester		*Eighth Semester*	
(3)	Engl 250 (Amer. Books) or Engl 233 (Essay) or Engl 235 (Drama) or Engl 237 (Poetry) or Engl 238 (Fiction)	(3)	Govt. 301 (Elements of Democracy) or Govt 331 (International Relations)
(3)	Math 515 (Numerical Methods in Analysis)	(15)	Aero Technical Electives
(3)	EE 515 (Electrical Measurement of Non-Electrical Quantities)		
(9)	Aero Technical Electives		
(18)		(18)	

The airplane and power plant options that were offered separately before were now combined as a single curriculum with some required major area courses depending on the choices made. The airplane and power plant option is given in Table 4.9, and is described in the catalog as follows:

> *The airplane and power plant option is for the student who chooses aero-dynamics, structures, or aeroelasticity as his specialty or the student who is interested in the design or installation of the many forms of power plants used by the aeronautical engineer. The first six semesters of this option offer a thorough preparation in the fundamentals of engineering design. This background enables the student to fit into fields of engineering other than aeronautics if he chooses. In the seventh and eighth semesters, a major area group is selected. This group and the elective courses offer a comprehensive background in the design specialty which the student will pursue in his professional career.*
>
> *A large group of electives is available, which permits a wide selection and emphasizes the breadth of interest of the aeronautical engineer.*

Table 4.9   Airplane And Power Plant Option (1954) — Total Credits: $158^2/_3$ semester hours

---

FRESHMAN YEAR

First Semester

(4)	Chem 115 (General Chemistry)
(3)	Engl 101 (Composition) or Spch 114 (Principles of Speech)
(5)	Math 161 (Mathematics for Engineers and Scientists)
(1)	Engr 100 (Fresh. Engineering Lectures)
(2)	Military Training

(15)

Second Semester

(4)	Chem. 116 (General Chemistry)
(3)	Spch 114 (Principles of Speech) or Engl 101 (Composition)
(4)	Phys 152 (General Physics)
(5)	Math 162 (Mathematics for Engineers and Scientists)
$(2^1/_3)$	Military Training

$(18^1/_3)$

SOPHOMORE YEAR

Third Semester

(0)	AE 300 (Aviation Lectures)
(3)	Spch 114 (Principles of Speech) or Engl 202 (Expository Writing)
(5)	Phys 240 (General Physics)
(4)	Math 241 (Calculus I)
(2)	GE 222 (Descriptive Geometry)
(3)	Econ 210 (Elementary Economics)
$(2^1/_3-3)$	Military Training

$(19^1/_3-20)$

Fourth Semester

(5)	Phys 240 (General Physics)
(4)	Math 242 (Calculus II)
(5)	EM 204 (Statics and Kinetics)
(3)	ME 230 (Elementary Heat Power)
$(2^1/_3-3)$	Military Training

$(19^1/_3-20)$

Table 4.9   Airplane And Power Plant Option (1954) (continued)

SUMMER SHOPS

(2)   GE 345 (Machine Tools)
($^2/_3$)   GE 251 (Aircraft Welding)
(4)   AE 381 (Aircraft Power plants)
(2)   AE 365 (Aircraft Components)

($8^2/_3$)

JUNIOR YEAR

Fifth Semester		*Sixth Semester*	
(3)	ME 360 (Mechanism)	(3)	AE 310 (Aerodynamics I)
($3^2/_3$)	ME 310 (Fluid Mechanics)	(1)	AE 320 (Wind Tunnel Lab)
(3)	ME 307 (Elements of Thermodynamics)	(3)	AE 350 (Aircraft Structures I)
		($2^1/_3$)	GE 266 (Aircraft Fabrication)
(4)	EM 312 (Mechanics of Materials)	($3^2/_3$)	EE 315 (Direct Currents)
(3)	Math 421 (Differential Equations)	(3)	Hist 210 (Europe Since 1914)
(3)	Govt 301 (Elements of Democracy) or Govt 331 (International Relations)		or Hist 205 (U.S. in World Affairs) or Elective
		(3)	AE 340 (Aircraft Vibrations)
($19^2/_3$)		(19)	

SENIOR YEAR

Seventh Semester		*Eighth Semester*	
(3)	AE 471 (Aircraft Power Plants I)	(1)	AE 498 (Aero. Engr. Seminar)
($3^2/_3$)	EE 316 (Alternating Currents)	(3)	Engl 250 (Amer. Books) or Engl 233 (Essays) or Engl 235 (Drama) or Engl 237 (Poetry) or Engl 238
(3)	Psy 320 (Psychology for Engineers)		
(3)	Elective, or Hist 210 (Europe Since 1914) or Hist 205 (U.S. in World Affairs)	(3)	Elective
		(9)	Major Area Courses
(7)	Major Area Courses		
($19^2/_3$)		(16)	

TYPICAL MAJOR AREA COURSES FOR AIRPLANE AND POWER PLANT OPTION

Seventh Semester		*Eighth Semester*	
(3)	AE 410 (Aerodynamics II)	(3)	Technical Elective
(3)	AE 450 (Aircraft Structures II) or Technical Elective	(3)	Technical Elective or AE 450 (Aircraft Structures II)
(1)	AE 481 (Aircraft Power Plant Lab)	(3)	Technical Elective
(7)		(9)	

Table 4.9   Airplane And Power Plant Option (1954) (continued)

At least six hours of technical electives should be chosen from the following:

(2)	AE 420 (Aerodynamic Design of Aircraft)	(3)	AE 520 (Stability and Control)
(3)	AE 510 (Theory of Airfoil Sections)	(3)	AE 522 (Aerodynamics of Helicopters)
(3)	AE 530 (Aeroelasticity)		

AIR TRANSPORTATION PROGRAM CANCELLED

As indicated earlier, the air transportation curriculum was first introduced in July 1945 with the formation of the School of Aeronautics. The program was administered by the Schools of Engineering, but the degree was not an engineering degree, but rather a bachelor of science in air transportation. Students were required, however, to complete the freshman program in engineering before entering air transportation.

The fact that the air transportation curriculum was not an engineering program seemed to concern Dr. Clauser from the very beginning when he arrived as head of the school in 1950. He indicated to the faculty in air transportation at several meetings that he believed such a program was out of place in the school and should be terminated or moved to another school.

The air transportation curricula had been developed by Professor Elmer F. Bruhn with the help of Grove Webster and Professor Paul E. Stanley and were first introduced in 1945 with the opening of the new School of Aeronautics. It was a program designed to prepare young men and women for positions in the expanding airline industry, including the large airports and the various governmental agencies involved in the control of air transportation. The graduates had no difficulty being placed in the ever-expanding and dynamic industry at that time.

By the mid-1950s, the enrollment in air transportation had dwindled to nearly half of the peak number reached in 1947: 146 compared to 227. Many of the students in the early years were veterans returning from World War II. During this period a number of students were veterans back from the Korean War. This decrease in enrollment was not unusual, however, since the same changes were occurring in the aeronautical engineering enrollment which was now (the fall of 1955) 318, compared to 497 in 1947.

As indicated earlier, Dr. Clauser was on a leave of absence during the 1954-55 school year and never returned. Dr. DeGroff was named acting head in his absence. Sometime during the spring of 1955, several meetings were held to discuss the future of the air transportation program in the school. These meetings included Dr. DeGroff, the faculty in that program, Dean G. A. Hawkins, Associate Dean P. F. Chenea, and Dr. E. T. Weiler, head of industrial management. The faculty were eventually told that a decision had been reached by the administration to terminate the program in air transportation effective with the fall 1955 semester. The entry in the 1956-57 University catalog was as follows:

Air Transportation

The existing curriculum in air transportation is open only to those students who can complete requirements for the degree by June 1958. The present program in air transportation will be terminated with the class graduating in 1958.

Meanwhile, a new curriculum in transportation studies has been formulated on a broadened and strengthened basis, including the opportunities for students in the field of air transportation.

The new program in transportation studies is in the Department of Industrial Management and Transportation.

The air transportation program lasted only a decade, but left its mark in the form of a number of outstanding graduates. An example of the high achievement of some of these graduates is the fact that two out of the 1948 graduating class have been honored as Distinguished Engineering Alumni: Alvin L. Boyd in 1970, and Martin W. Taylor in 1978. A member of the 1950 graduating class, Richard D. Freeman, was so honored in 1973. A number of the graduates went on to be airport managers in larger cities such as Indianapolis, Baltimore, Dallas, and others. The 1995 manager of the Purdue University Airport (Robert E. Stroud) was an air transportation graduate, class of 1957.

With the deletion of the curriculum in air transportation in 1955, Dean Hawkins envisioned a suitable program would be developed in the Department of Industrial Management and Transportation, which would include air as well as surface transportation. At that time this department was a part of the new School of Industrial Engineering and Management. Professor Mart I. Fowler transferred to that department and continued to teach some of his courses for a few years, but eventually the demand for these courses ended. Professor Fowler had joined the air transportation faculty in the fall of 1947, and had been very influential in the success of that program in the early years. Another faculty member, James H. Basinger, left the University and took a position with the Allison Division of General Motors in Indianapolis. Professor William W. Briggs continued with the School of

Aeronautics on a halftime basis and transferred to freshman engineering halftime, where he did counseling. He had been teaching some courses in aeronautical engineering before these changes. Professor L. T. Cargnino also remained with the school, where he continued to teach some of the courses in aeronautical engineering and later took over the scheduling of classes (schedule deputy) and the counseling of students, which had been done previously by Professor Stanley. Professors Elmer F. Bruhn and Paul E. Stanley continued to teach their regular complement of courses in aeronautical engineering since they had been involved only minimally in the teaching of air transportation courses. Grove Webster, who had been teaching a course in airport management, continued his main duties as manager of the Purdue University Airport, a position he had held since 1942.

NAME CHANGE

Following dissolution of the air transportation program, the faculty decided the name of the school should be changed to more nearly reflect its mission. Accordingly, the name was changed to the **School of Aeronautical Engineering** effective July 1956.

It was about this time, roughly the mid-1950s, when the aeronautical engineering school started producing graduates who would later be involved in the astronaut program that has made Purdue University famous as the "cradle of astronauts." Although the school had no specific program to train astronauts at that time, some visionaries already thought such an idea was possible.

Actually, the first space-pioneer to graduate from the aero school was Iven Carl Kincheloe in 1949 (Figure 4.19). While not a true astronaut in the sense of his being in the astronaut program (which was nonexistent at that time), Mr. Kincheloe did distinguish himself. As an Air Force captain, he established an altitude record of 126,200 ft. (24 miles) in a rocket-powered aircraft in 1956. Captain Kincheloe was killed in July 1958 in a jet plane crash at Edwards Air Force Base in California. Before the crash he was the first man chosen for a trip into outer space. Although Captain Kincheloe's tragic death ended his dream to become the first man in space, his photograph and copies of newspaper clippings describing his career as a space pioneer were carried into orbit by Col. John H. Casper (MSAE '67) aboard Space Shuttle Atlantis on February 28 to March 4, 1990. Those items are on permanent display in Armstrong Hall, along with a certificate verifying that they accompanied Col. Casper into space. When informed of this display, Captain Kincheloe's mother (Frances Kincheloe Crain) expressed the following thoughts in a January 9, 1992 letter to Professor Grandt: *"...I was most interested and most pleased to learn the information regarding the flight of Col. John H. Casper on the Space Shuttle "Atlantis" ... and the material he carried regarding information about my son. I know Capt. Kincheloe would be very honored and proud to be so honored and remembered. ... I truly thank you and your department for thinking of him and so honoring him and helping to put part of his dream into space."* In an earlier letter of November 22, 1991, Mrs. Crain, who was then 85 years old, recalled Captain Kincheloe's lifelong interest in aviation: *"My*

Figure 4.19 Captain Iven C. Kincheloe (BSAE '49) holds a model of the X-2, the jet he flew to a record-breaking height of 24 miles in September 1956 (August 1958 Redbook magazine photo).

son was interested in planes and flying since he was about four years old. Going to an airport and seeing planes and the flying was a big day for him. When he got his pilot license on his 16th birthday and he was able to take a passenger (alone) he asked me to be his first passenger. I also remember after he was in college he took his father and me to the Purdue airport to look at a jet engine they had received for the students to work on. He was so excited and proud of that jet engine. As I look back in time it looked more like a big cigar."

Neil Armstrong, perhaps the school's most famous graduate, first entered Purdue in September 1947 in freshman engineering, but he withdrew in January 1949, after three semesters, for military duty. He was a Naval aviator from 1949 to 1952 and flew 78 combat missions in the Korean War. He returned to Purdue in aeronautical engineering in September of 1952 and received his BSAE in January 1955. Following graduation, he joined NASA's Lewis Research Center in Cleveland, Ohio, and later transferred to the High Speed Flight Station at Edwards AFB, California. He participated in the flight test work as a civilian until he was selected for the second group of astronauts named in September 1962. He was named pilot of the Gemini-8 flight in 1966. Later he was named commander of the Apollo 11 moon landing mission and on July 20, 1969, became the first man to step on the moon.

At about the time that Mr. Armstrong graduated from aero, another young man destined to be an astronaut was midway through his studies. Roger Chaffee entered the School of Aeronautical Engineering as a sophomore in the fall of 1954. A Navy ROTC student, he received his BSAE in June 1957. Following graduation he entered the Navy and was sent to the Naval Air Station at Jacksonville, Florida. In 1963 he attended the Air Force Institute of Technology at Wright-Patterson AFB, where he was selected as an astronaut. His efforts in the space program were cut short by the tragic fire on the Cape Kennedy launch pad in January 1967. Virgil I. (Gus) Grissom, another Purdue graduate (BSME 1950) also perished in that same fire.

The School of Aeronautical Engineering began offering a short course on parachute engineering during the summer of 1956. This was the school's first venture into short-term courses, but the concept was not new, since many such courses were being offered on the Purdue campus at that time, particularly in the field of agriculture. The coordinator of the short course was Professor L. T. Cargnino and the following staff members from the school gave lectures during the two-week program: E. F. Bruhn, H. M. De-Groff, T. J. Herrick, L. A. Hromas, and G. M. Palmer. The following topics were covered: basic theoretical concepts, parachute technology, parachute engineering, textile technology, and special lectures and demonstrations. The program was well attended and drew more than 50 individuals, mostly from industry and the armed services, at a cost of $140 each. The short course offering was considered successful and was offered again during the summer of 1957 (June 3–14). Joint coordinators were used in 1957: L. T. Cargnino and G. M. Palmer. More than 60 individuals attended the second short course on parachute engineering.

During the spring of 1957, faculty members decided that the school should no longer offer several options in aeronautical engineering. Accordingly, the airplane and power plant option was discontinued and the so-called theoretical option given in Table 4.10 was continued as the sole curriculum.

Table 4.10 Aeronautical Engineering Curriculum (1957) — Total Credits: 151 semester hours minimum

FRESHMAN YEAR

First Semester

(4)	Chem 115 (General Chemistry)
(3)	Engl 101 (Composition) or Spch 114 (Principles of Speech)
(5)	Math 161 (Mathematics for Engineers and Scientists)
(1)	Engr 100 (Freshman Engineering Lectures)
(2–3)	Military Training

(15–16)

Second Semester

(4)	Chem 116 (General Chemistry)
(3)	Spch 114 (Principles of Speech) or Engl 101 (Composition)
(4)	Phys 152 (General Physics)
(5)	Math 162 (Mathematics for Engineers and Scientists)
$(2\frac{1}{3}$–3)	Military Training

$(18\frac{1}{3}$–19)

SOPHOMORE YEAR

Third Semester

(4)	Math 261 (Mathematics for Engineers and Scientists III)
(5)	Phys 251 (Gen. Physics)
(3)	ES 205 (Basic Mechanics I)
(2)	AE 265 (Airframe Components and Materials)
(3)	Engl 202 (Expository Writing)
$(2\frac{1}{3}$–3)	Military Training

$(19\frac{1}{3}$–20)

Fourth Semester

(4)	Math 262 (Mathematics for Engineers and Scientists IV)
(3)	ES 206 (Basic Mechanics II)
(3)	ES 315 (Elements of Strength Theory)
(3)	AE 282 (Aircraft Power Plants Components and Materials)
(3)	Econ 210 (Elem. Economics)
$(2\frac{1}{3}$–3)	Military Training

$(18\frac{1}{3}$–19)

Table 4.10 Aeronautical Engineering Curriculum (1957) (continued)

JUNIOR YEAR

Fifth Semester		*Sixth Semester*	
(3)	Math 421 (Differ. Equations)	(3)	AE 340 (Airc. Vibrations)
(3)	AE 313 (Aerodynamics I)	(3)	AE 314 (Aerodynamics II)
(3)	AE 351 (Airc. Structures I)	(3)	AE 352 (Airc. Structures II)
(3)	ME 305 (Gen. Thermodynamics I)	(3)	AE 372 (Airc. Power plants I)
(5)	Phys 262 (Gen. Physics)	(3)	ME 306 (Gen. Thermodynamics II)
(3)	Non-Technical Elective	(3)	Non-Technical Elective
		(1)	AE 398 (Aero. Engr. Seminar)
(20)		**(19)**	

SENIOR YEAR

Seventh Semester		*Eighth Semester*	
(2)	AE 401 (Aeronautical Engr. Lab)	(3)	AE 430 (Intro. to Aeroelasticity)
(3)	AE 417 (Aerodynamics III)	(4)	EE 316 (Electrical Engr.)
(4)	EE 315 (Electrical Engr.)	(2)	IE 470 (Manufacturing Processes)
(6)	Technical Electives	(3)	Non-Technical Electives
(3)	Non-Technical Electives	(6)	Technical Electives
(2)	CE 418 (Aero. Design Drawing)		
(20)		**(18)**	

Figure 4.20 Professor Hsu Lo (1964).

With the advent of the Russian Earth satellite, Sputnik I, launched on October 4, 1957, most engineering educators felt the need to beef up the engineering courses with more math, science, and space-oriented courses. As early as 1952, courses in rocket theory had been introduced by Professor M. Zucrow. By the fall of 1957, a course in flight mechanics was introduced by Professor A. Miele and covered *"a preliminary approach to the problems of the mechanics of terrestrial and extraterrestrial flight; applications in rocket power are emphasized."* By the following year, a course in orbit mechanics was offered by Professor H. Lo (Figure 4.20), which covered *"orbit determination of near-earth satellites and various perturbations; libration and attitude control; orbit transfer and interception; lunar theory and interplanetary orbits; ascending mechanics and reentry."* Graduate-level courses were also upgraded and new courses offered. The 1957-58 list of graduate courses is given in Table 4.11.

Table 4.11 Graduate-Level Courses (1957)

610.	Aerodynamics of Perfect Fluids I. Class 3, cr. 3
611.	Aerodynamics of Perfect Fluids II. Class 3, cr. 3
612.	Aerodynamics of Real Fluids. Class 3, cr. 3
614.	Aerodynamics of Compressible Fluids. Class 3, cr. 3
619.	Hypersonic Aerodynamics. Class 3, cr. 3.
630.	Aircraft Dynamic Analysis. Class 3, cr. 3
631.	Aircraft Flutter. Class 3, cr. 3
651.	Advanced Aircraft Structures II. Class 3, cr. 3
665.	Advanced Aircraft Structures Laboratory. Lab. 6, cr. 2
690.	Advanced Aeronautical Engineering Projects
691.	Aerodynamics Seminar. cr. 1 to 6.
695.	Aircraft Structural Seminar. cr. 1 to 6
697.	Aircraft Propulsion Seminar. cr. 2 to 6
698.	Research MS Thesis
699.	Research PhD Thesis

About this time, Professor Paul Lykoudis developed a new course in astrophysics, which had the following description:

> **Introduction to Astrophysics:** *Observational data concerning stars and galaxies, luminosity, mass, temperature and spectra of stars, motion and distances, the Hertzsprung-Russel diagram; physical principles and mathematical techniques for the study of stellar interiors and stellar evolution, double and pulsating stars; magnetic fields associated with galaxies, interstellar space and stars, with particular emphasis on phenomena associated with the sun and earth, such as solar flares, solar wind, radiation belts, origin of earth's magnetic field, etc. Professors Lykoudis and Lo.*

Professor Joseph Liston, a member of the Purdue engineering faculty since 1937, first in mechanical engineering and then in aeronautical engineering after 1945, decided he would like to teach courses in creative engineering instead of the traditional design-type courses. Professor Liston had been teaching primarily design of aircraft power plants after World War II, and thought that those courses were unrealistic because the students were asked to analyze someone else's ideas. Professor Liston believed that this was not the way an engineer operated in industry. He believed that the classroom emphasis was on the student's ability to absorb and then later recall facts and procedures established by someone else. The student constantly was analyzing problems for which others had already provided the solutions, and he was expected to come up with the same answers. Professor Liston deliberately set out to teach engineering students how to innovate, how to invent, how to come up with a piece of engineering hardware. He named the new course Creative Engineering Synthesis.

Professor Liston enjoyed teaching this type of course but indicated they were the most difficult he had taught because analysis and synthesis are diametrically opposite. In analysis, the approach one must take is almost always patterned. The student generally can find some kind of guidance that will tell him what to do. But when he is told to invent something, all the training he's had in analysis isn't much help, reasoned Professor Liston.

Professor Liston then converted his courses in aircraft power plant design to aircraft power plant evolvement and proceeded to teach them with the creative engineering synthesis approach. Initially, the problem Professor Liston faced was not only the best way to teach such a course, but the unavailability of instructional materials. Professor Liston solved the latter problem by developing his own set of notes, which he incorporated into a textbook a few years later titled *Creative Product Evolvement*.

In those days the students in the aero school referred to Professor Liston, among themselves, as Piston Joe. This was not a derogatory term, but one of respect for his love and knowledge of the piston-type reciprocating engine used in aircraft. In fact, the use of the name Piston Joe was so widespread that when one of his daughters, Dottie, was a student at Purdue in the early 1960s, she was nicknamed Little Piston Joe.

RESEARCH

The research activities in aeronautical engineering had been increasing steadily from about 1955-56, when the total research budget was $24,079, to a high of $57,383 for the academic year 1959-60. Particularly noticeable was the increase to $52,626 in 1958-59 over the prior year 1957-58, when it was $41,515. The individual grants were not large ($28,000–$30,000); nevertheless, they kept the faculty involved in up-to-date activities.

By 1959, six major research programs were underway in aeronautical engineering. These programs were carried on mainly by the following professors: Hsu Lo, John Bollard, Madeline Goulard, Robert Goulard, Leslie Hromas, Paul Kentzer, Paul Lykoudis, Paul Stanley, George Palmer, and Angelo Miele. A complete list of the research, and staff for the academic year 1958-59 is shown in Table 4.12 as they were organized in the six research programs under way at that time.

The research work in flight mechanics was under the direction of Professor Angelo Miele and was supported by the Air Force Office of Scientific Research, the National Aeronautics and Space Administration, and the Army Ballistic Missile Agency at a rate of around $30,000 per year. This work achieved favorable attention both nationally and internationally at the time, with papers being presented at the 8th International Congress of Astronautics in Spain and the 9th International Congress of Astronautics in Holland.

The chemical and radiative aerodynamics research in the School of Aeronautical Engineering was under the direction of Professor Robert Goulard.

Professors Hromas and Kentzer directed the experimental work involving dissociation studies.

All of the research in magneto-fluid-mechanics was under the direction of Professor Paul Lykoudis. His research in this new and exciting field was widely recognized at the time, as indicated by his presence at national and international meetings.

The research in the area of high-speed, high-temperature flight simulation was conducted by Professor George Palmer. A gas plasma jet was developed in 1957, and some early experiments with this device were chronicled in the 1957-58 Purdue alumni movie. During the summer of 1958, an original study of potential surface materials for reentry of the Mercury space capsule was conducted by Professor Palmer for the McDonnell Aircraft Corporation. This work resulted in a contract to continue high-speed, high-temperature simulation for McDonnell.

Professor Palmer, together with two graduate students, investigated both experimentally and theoretically the interaction of electric current and magnetic fields on the heating and propelling of ionized gases. Professors Stanley and Palmer collaborated on an experimental study of radio frequency energy heating of rarefied gases for generating high-speed, high-temperature flow.

The theoretical research in high-temperature structural dynamics was done under the joint supervision of Professors Hsu Lo, John Bollard, and Madeline Goulard. The experimental work was done under the direction of Professor Bollard, who was instrumental in obtaining an experimental facilities grant from the National Science Foundation for the specific purpose of installing an extensive radiant heating facility. The grant was for $50,000 for equipment only. This facility was installed in the spring and summer of 1959.

It is interesting to note that research in aeronautical engineering at Purdue covered in a rather broad fashion many of the basic problems associated with high-speed, high-altitude flight, which at the time coincided closely with the largest proportion of research in aeronautical engineering in the United States.

Research sponsorship was obtained from the following scientific and military establishments: Office of Ordnance Research, Air Force Office of Scientific Research and Development, National Science Foundation, Office of Naval Research, Naval Bureau of Aeronautics, Army Ballistic Missile Agency, National Advisory Committee for Aeronautics, Army Signal Corps, and McDonnell Aircraft Corporation, as well as the Purdue Research Foundation and the Purdue Engineering Experiment Station.

From 1955 to 1959, the faculty in aeronautical engineering participated in national or international meetings or in seminars and colloquia at other institutions on 35 separate occasions, which averaged nearly nine appearances per year. Table 4.13 contains a list of the appearances that the faculty and staff made during that period. Particularly impressive was the participation in international meetings at that time.

The increase in research activities during the latter part of this decade brought an increase in the graduate enrollment. By the spring of 1959, the school had 28 graduate students. Of these 28 graduate students, 11 were working toward PhDs and 17 were working towards master's degrees. Beginning with 1953, when the second PhD in aeronautical engineering was awarded to Al Schmitt, under the direction of Professor Elmer Bruhn, the school granted around one PhD annually; in 1958, three PhDs were granted. The undergraduate enrollment in the spring of 1959 was nearly 300. The undergraduate and graduate enrollment trend for 1955 to 1960 is shown in Table 4.14. The graduate degrees in aeronautical engineering are shown in Table 4.15 for the years 1955 to 1961, both for the MS and the PhDs. Undergraduate degrees granted are shown as well.

The activities in the school, including the curriculum and the individual courses, were being oriented toward a future in space. Since the launch of the Russian Sputnik I in October of 1957, most NASA facilities were being oriented toward space flight, and that was a big influence on all the educational programs in aeronautical engineering.

Table 4.12 Research Staff in the School of Aeronautical Engineering, Purdue University (1959)*

Flight Mechanics

Professor Angelo Miele	PhD, Univ. of Rome
Carlos Cavoti	PRF 1935, ABMA
J.O. Cappellari	PRF 1935, ABMA

Chemical and Radiative Phenomena

Asst. Professor Robert Goulard	PhD, Purdue
Tung Chen	PRF 1717, McDonnell

Experimental Dissociation Studies

Asst. Professor Leslie Hromas	PhD, Purdue
Asst. Professor Paul Kentzer	PhD, Purdue
Leslie Jones	PRF 1717, McDonnell
Harvey Mickelson	PRF 1717, McDonnell

Magneto-Fluid-Mechanics

Assoc. Professor Paul Lykoudis	PhD, Purdue
T.A. Gaichas	EES Fellow
J.R. Barthel	XR Fellow
G.W. Pneuman	PRF 1717, McDonnell

High-Temperature, High-Speed Flight Simulation

Professor Paul Stanley	PhD, Ohio State
Asst. Professor George Palmer	Aero. E. Caltech
F.L. Paris	PRF 1717, McDonnell
W.W. Hill	PRF 1717, McDonnell

High-Temperature Structural Dynamics

Professor Hsu Lo	PhD, Univ. of Michigan
Assoc. Professor John Bollard	PhD, Purdue
Asst. Professor Madeline Goulard	PhD, Purdue
R.L. Swaim	EES Fellow
V.J. Modi	XR Fellow
A.R. Zak	XR Fellow
H. Ohira	XR Fellow
A.L. Seward	XR Fellow
Capt. C.D. Bailey	US Air Force
H.S. Blakiston	Hughes Fellow

* Source of financial support for graduate assistants is in the right-hand column.

Table 4.13 Participation of School of Aeronautical Engineering Staff in Seminars, National, and International Meetings

	Name	Meeting
1955	Miele	American Rocket Society Meeting, Los Angeles, California
	Stanley	Indiana Academy of Sciences, South Bend, Indiana
1956	Lykoudis	Heat Transfer and Fluid Mechanics Institute, Stanford University
1957	Miele	Seminar, Massachusetts Institute of Technology
	Miele	Institute of the Aeronautical Sciences, Los Angeles, California
	Miele	8th International Astronautical Congress, Barcelona, Spain
	Lykoudis	1st National Heat Transfer Symposium, Pennsylvania State University
	Lykoudis	Seminar, Northwestern University
	R. Goulard	American Rocket Society Meeting, New York, NY
	M. Goulard	Midwestern Conference on Solid and Fluid Mechanics, University of Michigan
	Stanley	American Institute of Electrical Engineers, Montreal, Canada
	Lo	Midwestern Conference on Solid and Fluid Mechanics, University of Michigan
	DeGroff	Midwestern Conference on Solid and Fluid Mechanics, University of Michigan
		Seminar, Naval Ordnance Laboratory, Silver Spring, Maryland
	Hromas	American Physical Society Meeting, Lehigh University
1958	R. Goulard	NATO Sponsored Seminars, Laboratories, and Universities in Europe
	Lo	Seminar, Lockheed Missiles Systems Division
	Cavoti	Argentinian Interplanetary Association, Buenos Aires
	Lykoudis	Heat Transfer and Fluid Mechanics Institute, University of California
	Lykoudis	Symposium on Rarefied Gas Dynamics, Nice, France
	Lykoudis	9th International Astronautical Conference, Amsterdam, Holland
	Lykoudis	International Conference on Scientific Information, Washington, D.C.
	Lykoudis	Seminar, University of California
	Lykoudis	Seminar, Johns Hopkins University
	Lykoudis	Institute of the Aeronautical Sciences, St. Louis, MO
	Miele	American Astronautical Society Meeting, New York, NY
	Miele	Aviation Conference, Dallas, Texas
	Miele	Institute for Flight Mechanics, Germany
	Miele	9th International Astronautical Conference, Amsterdam, Holland
	Miele	Seminar, New York University
1959	Lo	Seminar, University of Illinois
	Bollard	Seminar, University of Michigan

Table 4.14 School of Aeronautical Engineering Enrollments

Academic Year	55-56	56-57	57-58	58-59	59-60	60-61
UNDERGRADUATE ENROLLMENT						
BS	318	408	339	293	202	193
GRADUATE ENROLLMENT						
MS	19	17	20	16	13	24
PhD	5	8	9	11	13	18

Table 4.15 Degrees in Aeronautical Engineering

Academic Year	55-56	56-57	57-58	58-59	59-60	60-61
BS	69	91	81	122	68	81
MS	8	9	12	11	5	10
PhD	1	1	3	0	3	3

MERGER WITH ENGINEERING SCIENCES

In March 1960, Professor E. A. Trabant resigned as head of the Division of Engineering Sciences at Purdue to become the dean of engineering at the University of Buffalo in New York. He was scheduled to leave the campus July 1, 1960. Dr. Trabant had been on the Purdue faculty 13 years and had been head of the engineering sciences division since it was formed in 1954. Before that it was called engineering mechanics.

On May 8, 1960, Dean of Engineering G. A. Hawkins announced the merger of the Division of Engineering Sciences with the School of Aeronautical Engineering effective July 1, 1960, with Dr. H. M. DeGroff as head. The name of the new school was to be the **School of Aeronautical and Engineering Sciences**. The faculty in the two schools had mixed feelings about the announcement. It was envisioned at that time by Dean Hawkins that the curricula of these two schools would be integrated and unified in the near future because of the direction of research and graduate studies in each of the schools. Dr. DeGroff echoed these sentiments.

At that time, the School of Aeronautical Engineering was referred to as a professional school by the Purdue faculty, and its aim was to train engineers to work in the aeronautical industries. Engineering sciences, on the other hand, was aimed at grounding students in both engineering skills and basic sciences, expecting them to go into research jobs in engineering industries. Both schools did basic research at the graduate level. For purposes of comparison, a copy of the curriculum in effect for each of the schools in the fall of 1960 is included in Tables 4.16 and 4.17.

The faculty in the School of Aeronautical Engineering at the time of the 1960 merger consisted of the following: **Professors**: R. J. Bollard, E. F. Bruhn, H. M. DeGroff, R. J. Goulard, J. Liston, H. Lo, P. Lykoudis, M. E. Shanks, and P. E. Stanley. **Associate professors**: W. W. Briggs, L. T. Cargnino, T. J. Herrick, L. Hromas, P. Kentzer, and G. Lianis. **Assistant professors**: M. Goulard and G. M. Palmer. During the summer of 1960, following the merger of aeronautical engineering and engineering sciences, Dr. H. M. DeGroff announced the addition of two new faculty members who would join the new school in the fall of 1960, both with the rank of associate professor.

Dr. Francis J. Marshall received his BME from City College of New York in 1948, his MS from Rensselaer Polytechnic in 1950 and a doctor of science in engineering from New York University. Dr. Marshall came to Purdue from the University of Chicago Laboratories for Applied Science, where he was a group leader and senior aeronautical engineer.

Dr. Winthrop A. Gustafson came to Purdue from the Lockheed Aircraft Corporation's Missiles and Space Division in Sunnyvale, California. He received a BSAE in 1950, followed by an MSAE degree in 1954, and a PhD in 1956, all from the University of Illinois. His work at Lockheed was in aerodynamics and high-speed gas dynamics as an associate research scientist.

Table 4.16 Aeronautical Engineering Curriculum (1960)

FRESHMAN YEAR

First Semester		*Second Semester*	
(4)	CHM 115 (General Chemistry)	(3)	CHM 116 (General Chemistry)
(3)	ENGL 101 (Composition) or SPE 114 (Principles of Speech)	(3)	SPE 114 (Principles of Speech) or ENGL 101 (Composition)
(3)	CE 118 (Engineering Graphics)	(4)	PHYS 152 (General Physics)
(5)	MA 161 (Mathematics for Engineers and Scientists)	(5)	MA 162 (Mathematics for Engineers and Scientists)
(1)	ENGR 100 (Fresh. Engineering Lectures)	(2–3)	Military Training
(2–3)	Military Training		

(18–19)   (17–18)

Table 4.16 Aeronautical Engineering Curriculum (1960) (continued)

SOPHOMORE YEAR

Third Semester

(4) MA 261 (Mathematics for Engineers and Scientists III)
(5) PHYS 251 (Heat, Electricity, and Optics)
(3) ESC 205 (Basic Mechanics I)
(3) General Education Elective
(2) Military Training

(17)

Fourth Semester

(4) MA 262 (Mathematics for Engineers and Scientists IV)
(3) ESC 206 (Basic Mechanics II)
(3) ESC 315 (Elements of Strength Theory)
(3) AE 203 (Introduction to Aeronautical Engineering)
(2) Military Training

(15)

JUNIOR YEAR

Fifth Semester

(3) MA 422 (Differential Equations)
(3) AE 351 (Aircraft Structures I)
(3) ME 305 (General Thermodynamics I)
(4) Modern Physics
(3) General Education Elective
(1) AE 305 (Introduction to Digital Computers)

(17)

Sixth Semester

(3) AE 340 (Aircraft Vibrations)
(3) AE 313 (Aerodynamics I)
(3) AE 352 (Aircraft Structures II)
(3) ME 306 (General Thermodynamics II)
(3) AE 398 (Aeronautical Engineering Seminar)
(1) CE 418 (Aeronautical Design Drawing)
(3) General Education Elective

(19)

SENIOR YEAR

Seventh Semester

(2) AE 401 (Aeronautical Engineering Laboratory)
(3) AE 314 (Aerodynamics II)
(4) EE 315 (Electrical Engineering)
(3) Technical Electives
(3) AE 372 (Aircraft Power Plants I)
(3) General Education Elective

(18)

Eighth Semester

(3) AE 417 (Aerodynamics III)
(3) AE 430 (Introduction to Aeroelasticity)
(4) EE 316 (Electrical Engineering)
(3) General Education Elective
(6) Technical Electives

(19)

Table 4.17 Engineering Sciences Curriculum (1960)

FRESHMAN YEAR

First Semester		*Second Semester*	
(4)	CHM 115 (General Chemistry)	(3)	CHEM 116 (General Chemistry)
(3)	ENGL 101 (Composition) or SPE 114 (Principles of Speech)	(3)	SPE 114 (Principles of Speech) or ENGL 101 (Composition)
(3)	CE 118 (Engineering Graphics)	(4)	PHYS 152 (General Physics)
(5)	MA 161 (Mathematics for Engineers and Scientists)	(5)	MA 162 (Mathematics for Engineers and Scientists)
(1)	ENGR 100 (Fresh. Engineering Lectures)	(2–3)	Military Training
(2–3)	Military Training		
(18–19)		**(17–18)**	

SOPHOMORE YEAR

Third Semester		*Fourth Semester*	
(3)	ESC 205 (Basic Mechanics I)	(3)	Basic Mechanics II
(5)	PHYS 251 (Heat, Sound, and Electricity)	(5)	PHYS 262 (Optics and Modern Physics)
(4)	MA 261 (Mathematics for Engineers and Scientists III)	(4)	MA 262 (Mathematics for Engineers and Scientists IV)
(3)	Nontechnical Elective	(3)	STAT 527 (Probability for Engineers)
(2–3)	Military Training	(2–3)	Military Training
(17–18)		**(17–18)**	

JUNIOR YEAR

Fifth Semester		*Sixth Semester*	
(3)	ESC 315 (Elements of Strength Theory)	(3)	ESC 507 (Basic Mechanics III)
(1)	ESC 375 (Experiments in Engineering Sciences I)	(2)	ESC 376 (Experiments in Engineering Sciences II)
(3)	MA 541 (Advanced Calculus I)	(3)	MA 520 (Partial Differential Equations and Applications)
(3)	ME 305 (General Thermodynamics I)	(3)	ME 306 (General Thermodynamics II)
(3)	EE 319 (Electrical Networks I)	(3)	EE 320 (Electrical Networks II)
(3)	AE 313 (Theoretical Aerodynamics I)	(3)	Nontechnical Elective
(3)	Nontechnical Elective		
(19)		**(17)**	

Table 4.17 Engineering Sciences Curriculum (1960) (continued)

SENIOR YEAR

Seventh Semester

(4)   ESC 543 (Continuum Mechanics I)
(3)   ESC 532 (Theory of Material Properties)
(3)   MA 525 (Applied Theory of
        Complex Functions)
(3)   ESC 473 (Projects in Research,
        Design, and Development I)
(3)   Elective
(3)   Nontechnical Elective

(19)

Eighth Semester

(4)   ESC 544 (Continuum Mechanics II)
(3)   ESC 474 (Projects in Research,
        Design, and Development III)
(3)   PHYS 530 (Electricity and Magnetism)
(3)   Technical Elective
(3)   Elective
(3)   Nontechnical Elective

(19)

During the summer of 1960, the faculty and staff of the aero school were moved from the airport to the Engineering Sciences Building on campus after some rearrangements were made for office space. Before the merger, approximately 18 faculty members in Engineering Sciences were housed in the building, but after the merger, the faculty included 33 full, associate, and assistant professors. Essentially, it meant putting two faculty members in one office, but everyone got along. The only facility retained at the airport was the large hangar building that housed the wind tunnels and other laboratory facilities. The other building that formerly housed the aero faculty as well as the classrooms was taken over by the aviation technology department.

The undergraduate enrollment for the fall of 1960 was as follows: aeronautical, 193, and engineering sciences, 144. At the time of the merger, it was estimated that the graduate enrollment for the two programs would be around 100 for the fall. The semester got underway very nicely in spite of the crowded office conditions in the small, one-story Engineering Sciences building. The faculty members in the new School of Aeronautical and Engineering Sciences had a varied background, but possessed resources of extraordinary scope for teaching and graduate research. Many faculty members in both of the two original programs had been heavily involved in sponsored research in a number of diverse areas.

When the two schools were merged, both the dean of engineering, Dr. Hawkins, and the head of the new school, Dr. DeGroff, announced that the ultimate aim of the school would be to combine the two programs into a common undergraduate curriculum. After several faculty meetings during the fall semester, it became rather obvious that a common curriculum could not be achieved very easily because of the disparity in the objectives of the two programs. As described in the next chapter, this difference in objectives, along with other problems, led to the eventual termination of the engineering sciences portion of the program in 1972.

Chapter 5

Gaining Altitude:
A New Era
1960–73

Gemini-7

The fall semester of 1960 marked the beginning of the new School of Aeronautical and Engineering Sciences. Before proceeding with the development of this program, this chapter presents a brief history of the Division of Engineering Sciences so that the effects of this merger can be better understood.

History of Engineering Sciences at Purdue

The Division of Engineering Sciences was created in July of 1954 with a nucleus of faculty from the Division of Engineering Mechanics, which at that time was a part of the School of Civil Engineering. The Department of Applied Mechanics had been formed in 1909 as a service department, without a degree-granting program, for the purpose of teaching basic mechanics to all engineering students. During the following years, the course offerings, both undergraduate and graduate, grew to the point that a degree-granting curriculum seemed appropriate. Thus, in 1944, a Division of Engineering Mechanics was created within the School of Civil Engineering, which offered a separate degree in engineering mechanics.

The mission of the Division of Engineering Sciences as set out in 1954 was to provide a scientific basis to emerging engineering disciplines at both the undergraduate and graduate levels, and to enable graduates of the program to be able to interpret and apply basic science to engineering. These graduates would then be able to become leaders in research and development activities for industries of the future, and faculty research would lead to the infusion of new techniques into some of the existing professional engineering curricula. This program would concentrate on basic sciences such as physics, chemistry, elasticity, thermodynamics, continuum mechanics, and applied mathematics, as well as some new areas including statistical mechanics, solid state physics, and nuclear physics. Hence, the curriculum was designed to provide a broad base of scientific knowledge from which a variety of disciplines could be addressed.

In the fall of 1955, the first group of undergraduate students enrolled, and for the most part they were a select group from the previous year's freshman class. It soon became known among the students that this was a program designed to challenge the best students, and hence it increasingly drew some of the best engineering students that entered Purdue. The undergraduate enrollment from 1955 to 1960 is listed below, while the enrollment after 1960 is included in a graph later in this chapter.

- 1955 — 50 students

- 1956 — 85 students

- 1957 — 85 students

- 1958 — 123 students

- 1959 — 132 students

- 1960 — 144 students

The first head of the division was Dr. Paul F. Chenea, who was succeeded by Dr. E. A. Trabant in 1958, who in turn left the University in 1960. Dr. H. M. DeGroff became the head of the school after the merger. During the period from 1954 to 1960, the enrollment rose rapidly and the following new faculty members were added: P. Feuer in 1954, S. S. Shu in 1955, L. V. Kline in 1956, J. C. Samuels in 1957, M. M. Stansic in 1956, F. Kozin in 1958, and S. J. Citron in 1959. Hence, at the close of the 1959-60 academic year, the Division of Engineering Sciences had a rising undergraduate enrollment of excellent students, a small group of graduate students, and a faculty consisting of a balance of experienced people along with several young faculty members. The faculty members seemed to have a strong commitment to the program, which had recently been accredited, and believed that it was at the leading edge of engineering education.

The School of Aeronautical and Engineering Sciences

As the fall 1960 semester got under way, the School of Aeronautical Engineering and the Division of Engineering Sciences had been merged (Table 5.1), and the two faculties were now located together in the former

Engineering Sciences Building, now renamed as the Aeronautical and Engineering Sciences Building (see Figure 5.1). During the summer of 1960, most of the aeronautical engineering faculty had moved from their airport location (see Figure 5.2) to the new campus location, and were now sharing office space with their engineering sciences colleagues. Although this merger of faculties produced a more crowded environment, everyone seemed to adapt to these new conditions relatively well.

Figure 5.1   Aeronautical and Engineering Sciences Building, 1995 (currently Nuclear Engineering Building).

Figure 5.2   Aerospace Sciences Laboratory (1995).

Table 5.1   Faculty of the School of Aeronautical and Engineering Sciences (1960)

Professor H. M. DeGroff, head; Professor P. E. Stanley, executive assistant

Professors
J. L. Bogdanoff*, R. J. H. Bollard, E. F. Bruhn, C. S. Cutshall*, H. M. DeGroff, A. C. Eringen*, S. Fairman*, R. J. Goulard, J. Liston, N. Little*, H. Lo, P. S. Lykoudis, W. B. Sanders*, M. E. Shanks, S. S. Shu*, P. E. Stanley, E. O. Stitz*

Associate Professors
R. L. Anderson*, W. W. Briggs, L. T. Cargnino, P. B. Feuer*, W. A. Gustafson, J. O. Hancock*, T. J. Herrick, L. A. Hromas**, C. P. Kentzer, L. V. Kline*, G. Lianis, F. J. Marshall, J. C. Samuels*, M. M. Stanisic*

Assistant Professors
S. J. Citron*, M. Goulard, F. Kozin*, G. M. Palmer, J. F. Radavich

*   denotes faculty formerly in engineering sciences
**   on leave and did not return

The administrative decision to merge these two units, while producing a faculty with much greater research potential, caused a number of problems because of the different educational objectives of the two groups. Aeronautical engineering was a professional program dealing with the disciplines important to the design of flight vehicles of various types, while the engineering sciences mission was to develop and bring in new areas of interest to the field of engineering. The clash of these two points of view continued throughout most of this time period.

One significant advantage of the merger was that the library material from each of the two former libraries was combined, thereby resulting in an excellent library covering the general areas of fluid mechanics, solid mechanics, thermodynamics, and applied mathematics. The library continued to grow over the years, and in 1974 contained 21,500 bound volumes of books and journals as well as 53,000 reports. When the Potter Engineering Center opened in the spring of 1977, all of the various school libraries were combined into a single engineering library. This library had 70,000 bound volumes of books and journals, and subscribed to 1,800 journals. Seating was provided for about 400 people.

During the spring of 1961, the school acquired an IBM 1620 digital computer for instruction and research. After a brief period of instruction by a company representative, Professor DeGroff selected several faculty members to teach small divisions of a new computer course to be offered during the fall semester of 1961. A year later this trial course had become a permanent course, AE 305 (two credits), in the aeronautical engineering curriculum. The IBM 1620 was programmed by punching small holes in a paper tape, which meant that when a punching mistake was made, a new tape had to be created. This was a tedious and time-consuming approach, but nevertheless a first step in the use of digital computers in the curriculum. By the fall of 1970, the IBM 1620 had been replaced by an IBM 1130, which could read punched cards, and a teletype terminal was also available to connect to a CDC 6500 in the Computer Science Center. A computer course was required in all of the school's curricula until 1973, at which time computer programming was generally taken during the freshman year.

ADMINISTRATIVE CHANGES

Professor DeGroff had been actively involved with other faculty members in the formation of a private consulting company called the Midwest Applied Science Corporation. This was seen to be a conflict of interest by members of the administration and some faculty, and so in June of 1963 Professor DeGroff began a two-year leave of absence to work full time with the corporation. During this time, Professor Paul E. Stanley became the interim head of the school, with Professor Stephen J. Citron as the executive assistant. After two years, Professor Stanley began a one-year leave of absence to pursue study in bioengineering, and Professor DeGroff returned to the school with the title of executive professor, with Dean Hawkins as fiscal head. Professor DeGroff also was an executive assistant to the dean of engineering. This administrative arrangement proved to be unworkable since many of the same conflicts were still present that had existed in 1963. Hence, at the end of fall 1965, Professor DeGroff moved full time to the dean's office, and an advisory committee was formed with Professor Hsu Lo as chairman to take care of the academic affairs of the school. The other members of the advisory committee were Professors S. J. Citron, R. J. Goulard, P. E. Stanley, and E. O. Stitz. The committee reported to the dean, who was the acting head, but they had considerable responsibility to manage the curriculum and the teaching activities. On February 14, 1966, it was announced in a faculty meeting that Professor Harold Amrine, head of the School of Industrial Engineering, would assume responsibility for certain routine administrative activities including the financial and budgetary functions. This, of course, gave the advisory committee more time to concentrate on academic matters, such as the approaching accreditation inspection, and to devote time to the selection of a new head for the school.

The newly formed advisory committee set about the task of trying to define the goals of the school for the near future. There were three main questions that needed an answer, briefly stated as follows: (1) Should the school have a single head, (2) should the current curricula be retained, and (3) should the school continue to teach the service courses of statics, dynamics, and strength of materials for the other engineering schools? The first question occupied nearly two hours at each of three faculty meetings held on March 4, 7, and 9, 1966, and still there were at least two faculty members who wanted to make additional statements. It was suggested that they provide written statements to the faculty outlining their points of view. The primary aspect of these discussions was whether the school should remain a single academic unit with a single head, or whether it should be split into two units with two heads. The vote on this question, done by mail ballot, resulted in a strong majority for a single head. For the second question, three possibilities were proposed, but the prevailing opinion was to preserve the current three curricula as described in a later section. The faculty also believed that the school should retain the teaching responsibility for the service courses.

During the remainder of the spring semester and the summer of 1966, the advisory committee requested, from the faculty, the names of possible candidates for the position of head of the school. In September, Dean Hawkins, the acting head, assumed a more active role by becoming chairman of the professorial council, thus becoming involved in the promotion process. By October, three people had been selected as candidates for the head's position, and were subsequently invited to come to Purdue for an interview. Of these three candidates, Dr. Hua Lin of the Boeing Company received the most faculty support, and he was extended an offer early in 1967. However, on February 8, 1967, Dean Hawkins announced to the professorial council that Dr. Lin would not accept the position. The head selection process was brought up in the next faculty meeting on February 15, 1967, and the faculty decided not to pursue the other two candidates, but to seek out new candidates from outside the University and to also consider internal candidates. Although Dr. Lin would not be the next head, he did consent to be a visiting professor for the current academic year with the possibility of renewing that appointment in the future. In that capacity, Dr. Lin visited the campus for two days in May 1967, at which time he met with the Aerospace Curriculum Planning Committee and with the Engineering Sciences Curriculum Planning Committee. He again visited the campus in February of 1968 to discuss the future direction of the school.

The head selection process was somewhat slowed down as George A. Hawkins became vice president for academic affairs on July 1, 1967, and Dr. Richard J. Grosh became dean of engineering as well as acting head of the school. An effort had been made to locate additional outside candidates, but few seemed to be available of the stature desired by both the faculty and the dean. Hence, in November 1967, Dean Grosh announced that Professor Hsu Lo would become the next head of the school effective November 27, and Professor John L. Bogdanoff would become the associate head. Professors Lo and Bogdanoff continued their leadership roles until September 1, 1971, at which time Professor Bogdanoff became head (see Figure 5.3). Professor Lykoudis became the associate head on January 1, 1972. Professor Herrick served as an executive assistant to the head beginning in 1968.

Figure 5.3
Professor John L. Bogdanoff (circa 1971).

At a faculty meeting on December 4, 1972, Dean John C. Hancock announced that Professor Bogdanoff wished to resign as head effective January 1, 1973, and that the engineering sciences program would be terminated. Thus, the faculty would again be searching for new leadership to take charge of the school's academic program, which would now concentrate on the aerospace field. During the next several months, George A. Hawkins served as acting head from January 1, 1973, until July 1, 1973, at which time Robert A. Greenkorn, an assistant dean, became acting head. A search committee was immediately appointed with Professor Madeline Goulard as chairperson; after the committee interviewed four candidates, Professor Bruce A. Reese was appointed as the new head on August 20, 1973, retroactive to July 1, 1973. Professor Reese had been director of the Jet Propulsion Center in the School of Mechanical Engineering since 1966, and his arrival brought the school propulsion expertise, which had been lacking since Professor Liston's retirement. The school's name was changed to the School of Aeronautics and Astronautics effective July 1, 1973.

CURRICULUM DEVELOPMENT

One of the objectives of the newly formed School of Aeronautical and Engineering Sciences in the early 1960s was to develop a single new curriculum that would be an acceptable compromise for each of the two existing curricula. In the years preceding 1960, the aeronautical curriculum had become somewhat more science-oriented than it had been earlier, but there were still notable differences between it and the engineering sciences curriculum. The faculty associated with each of the two curricula believed that there were certain unique features about each curriculum that could not easily be discarded. Although the school technically had a single faculty that taught the curricula, there were some fundamental differences of opinion about academic matters between the former engineering sciences faculty and the aeronautical engineering faculty. There were even some differences of opinion within each of the two formerly separate faculty groups. Some of these differences moderated with time, but some persisted throughout the decade. A curriculum committee was appointed in 1960 and worked for about two years attempting to arrive at a single new curriculum that would be acceptable to a majority of the faculty. The resulting proposal was probably the best compromise that could be reached, but it was closer in content to the engineering sciences curriculum than to the aeronautical engineering curriculum. Even so, there was sentiment in favor of the new curriculum, since it would presumably provide a unifying aspect to the new school. At a faculty meeting called in October of 1962 to discuss the new curriculum, Professor Elmer Bruhn, the first head of the School of Aeronautics, rose and gave an eloquent discussion about the tradition of aeronautical engineering at Purdue, and the need to maintain a distinctly aeronautical program in the future. When the vote was taken, the new curriculum was defeated and the two separate curricula were maintained for the future.

During the following two years there were only minor changes in the aeronautical curriculum and in the engineering sciences curriculum. Each curriculum had its own committee with responsibility for making whatever changes seemed to be appropriate. However, by the fall semester of 1964, some significant changes in each curriculum began to take place, which were completed in the spring of 1965, and then implemented for new students entering the school in the fall of 1965. The changes in the aeronautical curriculum came about in part because of the recognition that aerospace was becoming a more scientific industry. Thus, at least some fraction of the employees would be in research and development, and would need a more theoretical background than those who were in the mainstream engineering areas. The following curricula indicate a common sophomore year (Table 5.2) for each of the three curricula, which is followed by an engineering and a science option in the aeronautical and astronautical engineering program (Table 5.3–5.4), and by a separate engineering sciences program (Table 5.5). The school was renamed the School of Aeronautics, Astronautics, and Engineering Sciences in January 1965.

Table 5.2   Common Sophomore Year of the Aeronautical and Astronautical
Engineering and Engineering Sciences Curriculum (1965)

Third Semester		Fourth Semester	
(4)	MA 261 (Mathematics for Engineers and Scientists III)	(4)	MA 262 (Mathematics for Engineers and Scientists IV)
(3)	PHYS 241 (Electricity and Optics)	(3)	PHYS 342 (Modern Physics)
(3)	A&ES 207 (Basic Mechanics I)	(3)	A&ES 208 (Basic Mechanics II)
(3)	STAT 311 (Engineering Statistics and Random Probability)	(3)	A&ES 232 (Introduction to Mechanics of Solids)
(3)	General Education Elective	(3)	General Education Elective
(2)	Military Training	(2)	Military Training
(18)		(18)	

Table 5.3   Engineering Option of the Aeronautical and Astronautical
Engineering Curriculum — Total Credits: 142 semester hours

JUNIOR YEAR

Fifth Semester		Sixth Semester	
(3)	MA 302 (Mathematical Methods in Engineering I)	(5)	A&ES 315 (Aerodynamics I)
(3)	ME 305 (General Thermodynamics I)	(3)	ME 306 (General Thermodynamics II)
(3)	EE 319 (Electrical Networks I)	(3)	A&ES 372 (Propulsion I)
(3)	General Education Elective	(3)	EE 320 (Electrical Networks II)
(3)	A&ES 352 (Structural Analysis I)	(3)	General Education Elective
(2)	A&ES 351 (Digital Computers)		
(17)		(17)	

SENIOR YEAR

Seventh Semester		Eighth Semester	
(3)	A&ES 451 (Design I)*	(3)	A&ES 402 (Aerospace Engineering Laboratory)
(3)	A&ES 464 (Control Systems Analysis)	(3)	A&ES 452 (Design II)*
(3)	A&ES 453 (Structural Analysis II)	(6)	Technical Electives
(3)	A&ES 416 (Aerodynamics II)	(3)	General Education Elective
(3)	A&ES 431 (Aeromechanics I)	(3)	A&ES 417 (Aerodynamics III)
(3)	General Education Elective		
(18)		(18)	

Table 5.4   Science Option of the Aeronautical and Astronautical Engineering Curriculum—Total Credits: 142 semester hours

Junior Year

Fifth Semester		*Sixth Semester*	
(4)	MA 410 (Advanced Calculus)	(3)	MA 523 (Partial Differential Equations)
(3)	MA 422 (Differential Equations)	(3)	A&ES 333 (Fluid Mechanics)
(5)	PHYS 417 (Introduction to Statistical Physics)	(3)	A&ES 352 (Structural Analysis I)
(2)	A&ES 351 (Digital Computer)	(3)	EE 319 (Electrical Networks I)
(3)	General Education Elective	(3)	A&ES 308 (Dynamics of Structures)
		(3)	General Education Elective
(17)		(18)	

Senior Year

Seventh Semester		*Eighth Semester*	
(3)	A&ES 463 (System Analysis)	(3)	General Education Elective
(3)	A&ES 513 (Physics/Chemistry of Gases)	(12)	Technical Electives
(3)	PHYS 530 (Electricity and Magnetism)*		
(3)	Technical Elective		
(3)	General Education Elective		
(15)		(15)	

* May be replaced by technical electives with consent of academic counselor

Table 5.5   Engineering Sciences Curriculum (1965) — Total Credits: 140 semester hours

Junior Year

Fifth Semester		*Sixth Semester*	
(4)	MA 410 (Advanced Calculus)	(3)	MA 523 (Introduction to Partial Differential Equations)
(3)	MA 422 (Differential Equations for Engineering & Physical Sciences)	(3)	PHYS 530 (Electricity & Magnetism)
(3)	PHYS 515 (Heat)	(3)	Area of Specialization
(2)	A&ES 351 (Digital Computers)	(3)	MA 525 (Complex Functions)
(3)	Technical Elective	(3)	A&ES 307 (Advanced Dynamics)
(3)	General Education Elective	(3)	General Education Elective
(18)		(18)	

Senior Year

Seventh Semester		*Eighth Semester*	
(3)	A&ES 461 (Continuum Mechanics I)	(3)	A&ES 462 (Continuum Mechanics II)
(6)	Area of Specialization	(6)	Area of Specialization
(2)	A&ES 533 (Theory of Materials I)	(2)	A&ES 534 (Theory of Materials II)
(2)	CHM 584 (Physical Chemistry I)	(2)	CHM 585 (Physical Chemistry II)
(3)	General Education Elective	(3)	General Education Elective
(16)		(16)	

Five years after the merger of the schools, the curricula had grown in number by the addition of a science option within the aeronautics and astronautics part of the school. It should be noted that this option was oriented towards the aerospace field, unlike the engineering sciences program, and for the first time there was a required course in control theory. This course, A&ES 464, was developed by Professor Madeline Goulard as an elective course that followed A&ES 564, Systems Analysis & Synthesis. During the 1965-66 academic year, the course identification system was undergoing change as some students were still in the old curricula as the new curricula were being phased in. Some service courses for other engineering schools retained their original ESC designation, but the same courses taught for A&ES students had the new designation. A few courses in the engineering option retained their original AE designation.

As the year 1966 began, the school leadership was again in a state of change as discussed in the previous section. The advisory committee was in charge of academic matters, and an accreditation inspection by the Engineers Council for Professional Development (ECPD) was approaching. Several important decisions about the near-term future of the school were made during the early months of 1966, namely, that the school would have a single head, that the recently adopted three curricula would remain in effect, and that the school would continue to offer service courses in statics, dynamics, and strength of materials for the other engineering schools. The resolution of these questions took a considerable amount of time and energy, and sparked some heated discussions.

In October of 1966, it was learned that the ECPD accreditation would continue for another six years. There were, however, several suggestions that required faculty consideration. For the aeronautical curricula, it was suggested that a capstone design course be required, that a course in guidance and control be required, and that an aircraft structures laboratory be developed. A capstone design course had not been in the curricula for several years. In fact, before the merger there were design courses in aerodynamics, structures, and power plants, but none of these were required of all students. A guidance and control course had already been added to the new curricula, but had not been taught since students in the new curricula had not yet become seniors. Comments about the engineering sciences curriculum noted the lack of laboratory courses, particularly in the sophomore and junior years, and the need for continued curriculum development to maintain an innovative program. The decreasing enrollment in the program was noted as a cause for some concern if it persisted. The lack of permanent leadership was cited as an urgent need so that future development could take place.

It was becoming apparent that the aeronautical and astronautical science option, which had attracted only a few students, should eventually be phased out while allowing those students enrolled to complete their program of study. The special courses originally planned for this curriculum could not be taught because of low enrollment, and so other courses were substituted. These changes along with minor changes in the other two curricula were approved in February 1967. In March of 1967, the professorial council sent a list of near-term objectives to Dean Hawkins, outlining the goals of the school until a new head could be found. The following items were mentioned:

- The near-term goals and objectives of the school should be oriented towards the aerospace industry

- The school should strive for a strong science-oriented program, and also a strong engineering program

- Increase efforts to attract students to the science-oriented programs

- Put more motivation and applications in science-oriented courses

- Effect more participation of professors in undergraduate courses and activities

- Create a faculty that is more balanced between science and engineering, reflecting student interest

- Maintain a balance between teaching and research

- Graduate program should continue established research areas, and develop new ones

The undergraduate enrollment in the school during the decade is shown in Figure 5.4, and illustrates rather significant changes, which in turn had some influence on curriculum decisions. The engineering sciences enrollment underwent a continuous decline as the aeronautical enrollment made a large increase followed by a sharp decline. The enrollment summary over the first 65 years is shown on a single graph in Chapter 9, and it is clear that there have been significant fluctuations several times. In the middle of this decade, the rise was undoubtedly due to the strong interest developed by the Mercury, Gemini, and Apollo projects, which resulted in the first Apollo landing on the moon on July 20, 1969. Although interest in aeronautical engineering was relatively high during this period, interest in the science option attracted very few students, and was terminated by a faculty vote on October 28, 1968, with currently enrolled students continued until graduation.

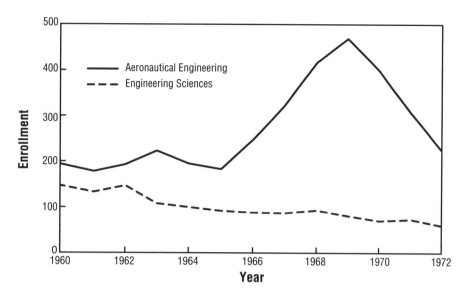

Figure 5.4   Undergraduate enrollment in aeronautical engineering and engineering sciences (1960-1972).

During the spring semester of 1967, both curriculum committees were busy trying to identify the near-term goals for each program as well as to implement recommendations made by the ECPD inspection committee. Part of these considerations involved identifying technical areas in which the school should maintain or develop expertise. Propulsion was an area that provoked considerable discussion, since Professor Liston was the only faculty member actively involved in teaching in that area. Thus, the school was unable to offer many courses other than the required A&ES 372 course in the engineering option, which made it difficult for students to do more advanced work by means of elective courses. Since propulsion was an important component of an aeronautical curriculum, it seemed necessary to find a way to enhance this aspect of the school's program. However, no new faculty members were hired into this area for several years, due to the needs in other areas. A number of other issues also were considered that would improve the various curricula, but which did not change their basic structure. Some of these considerations involved the place of probability and statistics in the curricula, the nature of the general education electives, the languages to be taught in the computer course, the sequences of thermodynamics and electrical engineering courses, and the arrangement of courses and topics in aerodynamics.

The spring semester of 1967 was also a time of considerable activity in preparation for moving to the recently remodeled former Civil Engineering

Building, located just north of the Purdue Memorial Union facing Grant Street. The interior of this building had been completely rebuilt except for some laboratory space on the first floor, which was to remain with civil engineering. The School of Industrial Engineering moved into the second floor, and the School of Aeronautics, Astronautics, and Engineering Sciences occupied the third floor. Since the merger in 1960, the school had occupied the Aeronautical and Engineering Sciences Building, the Aerospace Sciences Laboratory at the airport, and space in the old Heavilon Hall. The latter was a one-story building left standing when the new Heavilon Hall was built in 1956, and contained some laboratory and graduate student office space. It was later destroyed to make room for the new chemistry building, currently named the Brown Laboratory of Chemistry. The relocation involved moving most of the faculty to offices on the third floor in the former Civil Engineering Building, although several faculty members remained in their same offices. The move was carried out over spring break from March 27 to April 1, 1967.

On January 27, 1967, during a test of the Apollo spacecraft, a fire broke out inside the crew compartment, killing the three astronauts on board: Edward M. White, Roger B. Chaffee, and Virgil I. Grissom (see Figure 5.5). The latter two astronauts were Purdue graduates, and so the loss was felt very deeply on campus. A copy of this photograph was taken into space on STS-39 by astronaut Gregory Harbaugh on his April 1991 flight. The University administration decided that a suitable memorial would be to name buildings for Mr. Chaffee and Mr. Grissom. Hence, the former Civil Engineering building, into which the school moved, was named Grissom Hall (see Figure 5.6), and a new building located at the Propulsion Center west of the airport was named Chaffee Hall (see Figure 5.7). Grissom Hall was formally dedicated on May 2, 1968. The school library also was moved to Grissom Hall and occupied the entire south end of the third floor until 1977, when it was incorporated into the engineering library in the Potter building. At that time the library space was converted to office space for faculty and graduate students.

As the fall semester of 1967 got under way, the new dean of engineering, Richard J. Grosh, assumed the position of acting head of the school. He was, of course, well aware of the problems of the school, since he previously served in the dean's office under George Hawkins. The faculty tried to impress on him the need for more faculty in the areas that needed to be strengthened, and particularly the need for stability, which a permanent head could bring to the school. As mentioned in the previous section, the dean decided to name Professor Hsu Lo as head and Professor John Bogdanoff

Figure 5.5
Astronauts Grissom
(BSME '50), Chaffee
(BSAE '57), and White
with the Apollo Entry
Vehicle (1966).

Figure 5.6   Grissom Hall, 1995 (formerly the Civil Engineering Building).

Figure 5.7   Chaffee Hall (circa 1980).

as associate head in November 1967. Thus, the two-year search for a head of the school was ended, which brought with it a promise of leadership that would allow progress towards the academic goals of the faculty.

In 1968, the two curriculum committees of the school continued their deliberations about changes in their respective curricula. The Aeronautics and Astronautics Committee proposed the elimination of the science option, which had never had a large enrollment, and thus this aspect of the school's academic mission would henceforth consist of a single program. This new curriculum was presented to the faculty in May 1968, but was not voted on until October 28, 1968. The Engineering Sciences Curriculum Committee was also busy with innovative ideas that would attract students to combat the continuing downward trend in enrollment. This effort resulted in a proposal for a large-scale systems option in the fall of 1968.

Near the end of spring 1968, Professor Lo thought that it would be helpful to get student input into the school's curricula and course offerings. He apparently had received a number of comments from students during the year, and had given a questionnaire to the seniors graduating in June 1968. The results of the questionnaire were given to the faculty in September 1968, and it was apparent that a significant number of students were unhappy about a variety of things in the curricula. In summary, they perceived a lack of communication between faculty members as well as a lack of communication between faculty and students. They also believed that there was a failure to relate theory to engineering applications, and a rather limited number of elective courses from which to choose. Hence, in the fall semester a public forum was proposed so that students and faculty members could express their opinions. This was held in Fowler Hall in Stewart Center one evening with a few faculty and students on stage and a considerable number of undergraduate and graduate students in the audience, along with several faculty members. Many of the points raised in the previous year's questionnaire came up, and answers were attempted by various faculty members. Whether the discussions effected change is difficult to judge, but probably the faculty were more aware of student concerns after this meeting.

As the fall semester of 1968 began, the school's curriculum committees were structured as follows: aeronautics and astronautics consisted of Professors L. T. Cargnino, M. Goulard, H. Lo, R. A. Schapery, P. E. Stanley, and W. A. Gustafson, chairman; engineering sciences consisted of Professors D. W. Alspaugh, J. L. Bogdanoff, P. B. Feuer, S. S. Shu, A. L. Sweet, and E. O. Stitz, chairman.

The Engineering Sciences Curriculum Committee was engaged in studying several possible curriculum changes. One of these involved significant changes in the mathematics courses by creating a set of four courses to be taught in the sophomore and junior years, thus replacing all mathematics courses after the freshman year. These courses were developed by Professor S. Shu with the objective of providing an integrated sequence of applied mathematical topics with fewer credit hours than would be possible by selecting mathematics courses. These courses were taught on an experimental basis for several years, but there was some opposition from the mathematics department to this concept, and so the courses never acquired permanent course numbers.

Also during the fall of 1968, the Engineering Sciences Curriculum Committee was involved in studying a new curriculum option, which was proposed by Professors J. L. Bogdanoff, R. J. Goulard, and A. L. Sweet, and subsequently adopted as the large-scale systems option. The objective of this option was to develop an appreciation of the social, economic, and engineering variables involved in large complex engineering problems and the interrelation of these variables. Thus, it was necessary to develop the tools needed to assess the merits of various proposed solutions, and to use this knowledge to examine large-scale systems problems such as transportation systems, ecological systems, biological systems, etc. The option consisted of 18 credit hours distributed between the junior and senior years, covering topics in economics, sociology, applied mathematics, computer sciences, and systems analysis. This option was approved by the committee late in the spring semester of 1969, and subsequently by the entire school faculty. The Aeronautics and Astronautics Curriculum Committee also considered the large-scale systems option as a possibility for students in that curriculum, but concluded that it would only be possible if students were willing to add six credit hours to their plan of study. The effort in large-scale systems was later organized as a center, and after a search of several months, Professor A. Alan B. Pritsker was chosen to be a professor of aeronautics, astronautics, and engineering sciences as well as a professor of industrial engineering, and in addition was named the director of the Large-Scale Systems Center effective July 1, 1970. The staff consisted of those named above as well as Professors D. W. Malone and V. Vemuri, who were already on the faculty; two additional faculty were hired later, Professors C. C. Peterson and J. J. Talavage. This option attracted substantial outside research funding, and had about 20 full-time students and several part-time students at the time of peak enrollment in 1972.

In May of 1969, a publication issued by the dean of engineering's office, called the "Engineering News," contained an artist's sketch and description of a proposed new engineering building. The building was to house the Department of Freshman Engineering; the School of Aeronautics, Astronautics, and Engineering Sciences; the School of Materials Science and Metallurgical Engineering; and a new combined engineering library comprised of all the existing school libraries. The proposed building was quite large, consisting of several segments varying from five stories to 12 stories, and was to be built on the site of Michael Golden Hall. The proposed building never got beyond the planning stage, but in 1977 the Potter building was completed, housing the engineering library and a number of interdisciplinary laboratories. Later, in the early 1980s, Knoy Hall was built on the site of Michael Golden Hall, which provided space for several technology departments

The Aeronautics and Astronautics Curriculum Committee began in 1969 by reviewing the curriculum's content in several specific areas. One of these areas was the electrical engineering sequence of courses that students took. There were several meetings devoted to this issue, attempting to arrive at a new sequence that would not be too theoretical and would include some laboratory work. The result of these deliberations was to require a basic circuits course with laboratory, namely EE 201 and EE 207, to be followed by a new course in avionics. This latter course never materialized, and hence an elective to be chosen by the student was substituted. A special committee dealing with laboratory instruction was also formed and made a preliminary report in May 1969. The committee members were Professors A. Ranger, chairman, A. Schiff, and E. Stitz, and their initial recommendation was that the engineering sciences laboratory courses were satisfactory, but that changes were needed in the aeronautics and astronautics courses. The students at that time were required to take a single three-credit-hour course, A&AE 402, which was to include measurement techniques in the areas of aerodynamics, structures, and propulsion. However, with the relatively large enrollment in this program, it became almost impossible to provide equipment and facilities in a single course taught in the senior year. The recommendation was to provide a series of one-credit courses associated with appropriate theory courses in these subject areas, thereby distributing the students among different courses and different semesters. These course changes, along with several other minor changes, were tried on an experimental basis, and subsequently incorporated as permanent curriculum changes. Hence, at the end of the decade, both curricula (Table 5.6–5.7) taught by the school had been changed and are listed below as they appeared in the 1971-72 catalog, which reflected changes made late in May 1970.

Table 5.6   Aeronautical and Astronautical Engineering (1970) — Total Credits: 128 semester hours

SOPHOMORE YEAR

Third Semester

- (4) MA 261 (Multivariate Calculus)
- (3) PHYS 241 (Electricity & Optics)
- (2) ESC 221 (Basic Mechanics I)
- (3) EE 201 (Introduction to Electrical Engineering)
- (1) EE 207 (Electrical Engineering Lab. I)
- (3) General Education Elective

(16)

Fourth Semester

- (4) MA 262 (Linear Algebra & Differential Equations)
- (3) A&ES 208 (Basic Mechanics II)
- (3) A&ES 232 (Introduction to Mechanics of Solids)
- (3) Electrical Engineering Elective
- (3) General Education Elective

(16)

JUNIOR YEAR

Fifth Semester

- (3) MA 302 (Mathematical Methods in Engineering I) or MA 422 (Differential Equations for Engineering and the Physical Sciences)
- (3) ME 305 (Thermodynamics I)
- (3) A&ES 352 (Structural Anal. I)
- (3) A&ES 333 (Fluid Mechanics)
- (2) A&ES 351 (Digital Computers)
- (3) General Education Elective

(17)

Sixth Semester

- (3) A&ES 372 (Propulsion I)
- (3) A&ES 334 (Applied Aerodynamics and Performance)
- (3) ME 306 (Thermodynamics II) or A&ES 361 (Introduction to Random Variables in Engineering)
- (3) A&ES 453 (Structural Analysis II)
- (3) A&ES 340 (Dynamics & Vibration)
- (1) A&ES 350 (Aerospace Structures Laboratory)

(16)

Table 5.6   *Aeronautical and Astronautical Engineering (continued)*

Senior Year

Seventh Semester

- (3) A&ES 416 (Aerodynamics II)
- (3) A&ES 464 (Control System Analysis)
- (3) Elective
- (3) Design Elective
- (3) General Education Elective
- (1) A&ES 450 (Aerospace Aerodynamics and Propulsion Laboratory)

(16)

Eighth Semester

- (12) Technical Electives
- (3) General Education Elective
- (1) A&ES 460 (Physical Measurements and Instrumentation Systems)

(16)

Table 5.7   *Engineering Sciences Curriculum (1970) — Total Credits: 132 semester hours*

Sophomore Year

Third Semester

- (4) MA 261 (Multivariate Calculus)*
- (3) PHYS 241 (Electricity and Optics)
- (3) A&ES 207 (Basic Mechanics I)
- (3) EE 201 (Introduction to Electrical Engineering)
- (1) EE 207 (Electrical Engineering Laboratory I)
- (3) General Education Elective

(17)

Fourth Semester

- (4) MA 262 (Linear Algebra and Differential Equations)*
- (3) A&ES 208 (Basic Mechanics II)
- (3) A&ES 232 (Introduction to Mechanics)
- (2) A&ES 351 (Digital Computers)
- (1) A&ES 290 (Introduction to Engineering Sciences)
- (3) General Education Elective

(16)

Junior Year

Fifth Semester

- (4) Mathematics Elective*
- (3) A&ES 361 (Introduction to Random Variables in Engineering)
- (3) PHYS 515 (Thermodynamics)
- (3) A&ES 333 (Fluid Mechanics) or PHYS 342 (Modern Physics)
- (3) General Education Elective**

(16)

Sixth Semester

- (4) Mathematics Elective*
- (3) A&ES 307 (Advanced Dynam.)
- (4) PHYS 432 (Electromagnetism)**
- (3) A&ES 309 (Statistical Mechanics)
- (3) General Education Elective**

(17)

Table 5.7   Engineering Sciences Curriculum (continued)

SENIOR YEAR

Seventh Semester	*Eighth Semester*
(3) A&ES 461 (Continuum Mechanics I)	(3) A&ES 462 (Continuum Mechanics II)
(3) A&ES 463 (Systems Analysis)	(1) A&ES 476 (Physical Measurements
(1) A&ES 475 (Physical Measurements	Laboratory II)
Laboratory I)	(2) A&ES 534 (Theory of Materials II)**
(2) A&ES 533 (Theory of Materials I)**	(2) CHM 585 (Advanced Physical
(2) CHM 584 (Advanced Physical	Chemistry of Materials)**
Chemistry of Materials)**	(3) A&ES 564 (Systems Analysis and
(3) Technical Elective**	Synthesis)**
(3) General Education Elective	(3) Technical Elective
	(3) General Education Elective
(17)	(17)

*   Can substitute an experimental mathematics sequence for these courses
** Can replace these courses with the large-scale systems option below

The large-scale systems option was created to provide undergraduate students with an opportunity to prepare themselves for the study of large complex problems. Table 5.8 lists the courses that students were required to substitute in the engineering sciences curriculum.

Table 5.8   Large-Scale Systems Option

JUNIOR YEAR

Fifth Semester	*Sixth Semester*
(3) SOC 312 (American Society)	(3) INDM 481 (Elements of
	Industrial Management)
	(3) A&ES 381 (Modeling Tech. for
	Large-Scale Systems I)
(3)	(6)

SENIOR YEAR

Seventh Semester	*Eighth Semester*
(3) A&ES 400 (Modeling Tech. for	(3) A&ES 484 (Problems in Large
Large-Scale Systems II)	Scale Systems II)
(3) A&ES 483 (Problems in Large	(3) Technical Elective
Scale Systems I)	
(6)	(6)

These curricula were no sooner adopted than additional factors came into consideration, producing another round of changes. During the previous decade the school's aeronautical enrollment reached a peak followed by a large drop, while the engineering sciences enrollment made a continuous decline. The undergraduate enrollment in all of the engineering schools is tabulated below.

- 1966 — 4,949 students

- 1967 — 5,206 students

- 1968 — 5,244 students

- 1969 — 5,016 students

- 1970 — 5,035 students

- 1971 — 4,665 students

- 1972 — 4,262 students

- 1973 — 4,215 students

It is seen that engineering enrollment in all fields dropped by about a thousand students in a period of about three years, reaching a minimum in 1973. At the time it was believed that the recent interest in environmental concerns had turned potential students away from engineering, since a number of science and engineering industries were responsible for pollution problems. In an effort to stop declining enrollments, a new Division of Interdisciplinary Engineering Studies was created, which would be very flexible in curriculum requirements by allowing students to formulate their own plans of study. Thus, students could combine, within limits, certain engineering courses along with courses in many other fields of engineering, science, management, and liberal arts to achieve a broad-based education. In addition, most of the traditional engineering disciplines revamped their curricula to make them more flexible and to decrease credit-hour requirements. Following this pattern, the School of Aeronautics, Astronautics, and Engineering Sciences also made further curriculum modifications.

The first change approved by the faculty was in the engineering sciences program, which resulted in an integrated five-year plan leading to the master's degree. The sophomore year of this new program was unchanged from the one above, and the third year was also quite close to the previous junior year. The fourth and fifth years consisted mostly of major and minor area electives so that the plan of study was similar in structure to a usual graduate plan. The completion of 150 hours was required for the master's degree, although at the completion of 123 credits a bachelor's degree could be received. The master's program required that the electives fit into one of three possible options: (1) theoretical and applied mechanics, (2) systems and control, or (3) large-scale systems. This new curriculum was approved in the spring of 1971 and was available to new sophomores in the fall semester. The details of the curriculum first appeared in the engineering catalog for the 1972-73 academic year. This innovative program never developed because of the termination of engineering sciences as described below. The Aeronautics and Astronautics Curriculum Committee was also developing a five-year program during this time period, but finished its work several months later. This program will be described in detail in the next chapter.

The end of this era of the school's history occurred on December 4, 1972, at a faculty meeting presided over by Dean of Engineering John C. Hancock. The announcements made at that meeting were later sent as a memo to the entire faculty of the Schools of Engineering. The first announcement was that the school head, John L. Bogdanoff, wished to resign and return to full-time teaching and research, and consequently George A. Hawkins would assume the position of acting head of the school effective January 1, 1973. In addition, the engineering sciences degree program was terminated and would not accept new students, although those currently enrolled would be able to continue until graduation. Thus, by 1976 all students had either graduated or left the program. The reasons cited for the termination were "decreasing enrollments and a lack of clear and unique purposes for this curriculum." The service courses covering statics, dynamics, and strength of materials, which had been taught by the school's faculty to almost all engineering curricula, were now to be distributed to other schools. Statics and dynamics would become the responsibility of mechanical engineering, and strength of materials would be taught by civil engineering. The large-scale systems program was transferred to industrial engineering. These changes in the school's teaching mission would necessitate the movement of some faculty to different engineering schools. The movement was optional and the choice was up to each faculty member. The relocation of faculty took place after spring 1973 as follows: to mechanical engineering, S. J. Citron, J. Genin, J. H. Ginsberg, and A. J. Schiff; to civil engineering, R. L. Anderson and E. C. Ting; to industrial engineering, C. C. Peterson, A. B. A. Pritsker, A. L. Sweet, and J. J. Talavage; also, V. Vemuri and D. W. Malone left the University.

LABORATORIES

In the fall of 1960, the largest laboratory space was located in the building at the airport, which formerly housed all of the activities of the School of Aeronautical Engineering. There was also additional space in the former Engineering Sciences building. The graduate and undergraduate catalogs for 1961-62 listed the following laboratories:

(1)   *The Aerospace Sciences Laboratory, located at the airport, is the center of experimental research activities of the school. Research programs are presently being conducted in hypersonic aerodynamics, magnetogasdynamics, viscoelasticity, and the behavior of materials and structures at elevated temperatures. The laboratory's equipment includes a plasma jet wind tunnel, a 3.4-meter spectrograph, a unique laboratory for the study of materials to 6000° F and a programmed load capacity of 40,000 pounds. A large subsonic wind tunnel with a test section of approximately 6 x 4 feet and a maximum speed of about 300 mph was available for general testing. The wind tunnel had been designed by Professor George M. Palmer in the late 1940s and had been built by him with student help. This wind tunnel has been used for many different kinds of testing over the years including architectural studies, airplanes, automobiles, trucks, ships, and propellers. It is now nearly 50 years old, but with recent modernization is still playing an important role in research and instruction. Also present in the laboratory was a small instructional blow-down supersonic wind tunnel, a shock tube, and several small subsonic flow facilities.*

(2)   *The computer laboratory contains an IBM 1620 digital computer and an analog computer complex of over 75 stabilized amplifiers.*

(3)   *An Experimental Stress Laboratory located on the campus is equipped for modern stress analysis, including (a) two- and three-dimensional photoelasticity at room and at elevated temperatures, with loading dynamometers, polarized light, and photographic equipment; (b) brittle lacquer survey of stresses in complicated parts; (c) brittle material method of analysis using special metals, plastics, and glass (d) strain measurement using electric resistance strain gauges and indicators.*

(4)   *The Experimental Engineering Sciences Laboratory provides facilities to demonstrate and explore physical laws which are fundamental to engineering. Adequate equipment is available to study heat, light, electricity, and the laws of dynamics.*

(5)   *The Sophomore Aeronautical Laboratory was equipped with several demonstration devices and instrumentation which serve to introduce the student to experimental aeronautical technology.*

By 1963, some of the laboratories mentioned above had increased activity and equipment. In the Aerospace Sciences Laboratory, experimental research was under way on magneto-fluid-mechanics using liquid mercury as a working fluid. Problems being investigated involved natural convection and the structure of shear turbulence in channel flow in the presence of a large electromagnet producing 15,000 gauss at a gap of three inches. This project was developed by Professor Paul S. Lykoudis and his graduate students, with support of the National Science Foundation, and produced a number of PhD theses during the following years.

The 1965-66 catalogs described a new laboratory called the Random Environments Laboratory, which was associated with the Center for Applied Stochastics, and whose facilities included modern equipment for simulation and generation of random environments, and for the analysis of system behavior when subjected to such environments. An analog computer and random shaker with compensating console was available for random vibration experiments. Dynamic phenomena, including accelerations, displacements, and rapidly changing strains, pressures, and temperatures, could be measured electrically. Professor Schiff was in charge of this laboratory after joining the faculty in 1967.

In 1968, a Materials Research Laboratory was added to those located in the Aerospace Sciences Laboratory at the airport. This facility had thermal and mechanical loading equipment for the study of nonlinear materials with memory, and thus allowed for the investigation of viscoelastic materials, and possibly to non-Newtonian fluids. Also on the campus a Systems Laboratory was developed and equipped with three TR-20 analog computers, 30 amplifiers, and electrical and hydraulic servo-systems, to study and design various control and guidance systems.

The Composite Materials Laboratory was added to the research area in 1972 and was dedicated to the testing of particulate and fiber-reinforced composites as well as laminates under controlled environments. Mechanical, electrical, and optical measuring devices were available along with appropriate testing devices for both research and teaching. This laboratory was developed by Professor C. T. Sun and has provided facilities for many PhD theses to the present time.

GRADUATE PROGRAMS AND RESEARCH

The graduate degrees offered in 1960 were the master of science in aeronautical engineering, the master of science in engineering, the master of science, and the doctor of philosophy. The first of these degree titles applied to those students who had a bachelor's degree in aeronautical engineering, the second to those students with a bachelor's degree in engineering sciences or another engineering field, and the third to those who had a bachelor's degree in a field other than engineering. The same degree titles are still in existence for the same purposes, except that the first one is now the master of science in aeronautical and astronautical engineering. A master's degree plan of study had to include a major area and two minor areas, with one of the minors being mathematics. A nonthesis option required 18 credit hours in the major and at least 6 credit hours in each minor along with a total of 33 credit hours for graduation. The thesis option required 12 credit hours of course work in the major exclusive of the thesis, at least 6 credit hours in each minor, and a total of 33 credit hours. The advisory committee for each student was determined by the student and a major professor and consisted of a total of three professors, generally one representing each of the minor areas, and the advisor representing the major area. The student's plan of study had to be approved by the advisory committee and by the chairman of the school's graduate committee.

The doctor of philosophy degree requirements were similar in that the student chose his or her major professor and the other committee members, generally one representing each minor area. The major area required 24 credit hours of course work with another 24 credit hours distributed among the minor areas, but with not less than six credit hours in any one minor. These credit hour totals included those already obtained from a master's degree. A qualifying examination was required before entering the PhD program to ascertain the student's readiness to pursue doctoral studies, and subsequently a preliminary examination was required near the end of the course requirements to assess the ability to successfully complete the proposed research. The final examination was a public presentation of the research results, which presumably would be of sufficient quality to be published in a recognized journal in the field. This degree also required reading proficiency in two languages other than English that were appropriate to the field of study.

The above stated requirements for degrees were modified later in the decade for the various master's degrees and for the doctor's degree. At the beginning of the 1965-66 academic year, the PhD requirements as specified by the graduate school had been modified so that there was no longer a 48-credit hour minimum of course work, but the plan of study was to meet the needs of the student in preparing to pass the examinations and to carry out research. In addition, proficiency in English and one foreign language was required, rather than the two foreign languages previously required. By the following academic year, the minimum of 33 credit hours for the master's degree had also been removed, and by the 1971-72 academic year, the graduate school imposed no foreign language requirements, but individual advisory committees could impose such a requirement if they believed that it was necessary. These guidelines for graduate degrees have remained essentially unchanged up to the present, although the school has generally required 30 credit hours for master's degrees.

Research activities involving professors and graduate students increased significantly during the period 1960–73, reflecting the nature of the aerospace field and the interests of the faculty. The increased emphasis on space flight as well as high-speed aircraft flight brought the need for more research in all the disciplines, which contributed to the development of these flight vehicles. The most representative record of research activity in the school is a listing of graduate theses, both master's and PhD, written during this period (see Appendix B). The record lists 20 MS theses and 120 PhD theses. It is clear from this list that a broad range of interests was held by the faculty at that time, so that it is difficult to classify the theses into a few groups based on subject matter. However, the following broad categories cover most of the thesis topics: continuum mechanics, elasticity, viscoelasticity, wave propagation, magneto-fluid-mechanics, radiative gas dynamics, turbulence, plasma dynamics, astro-fluid mechanics, stochastic phenomena and random processes, control theory, guidance, orbit mechanics, atomization, and gas-particle flows.

It is not possible to give a detailed summary of the research work of each professor in this time period, but a few additional comments concerning some of the senior people may give some indication of the extent and recognition of their work. The emphasis on research sponsored by a government agency or a private company was not as prevalent in this era as in later times, and there are virtually no records remaining that list project titles or expenditures. During this period, the level of support for a research contract was rather small as compared to later years, although there were a few contracts on the order of $100,000. By way of example, a $20,000 contract would probably support a one-half-time research assistant for a year as well as pay a professor full-time during the summer and 10 to 20 percent during the academic year.

The National Science Foundation supported Professor Lykoudis in the area of magneto-fluid-mechanics, Professor Lianis in viscoelasticity, Professor Robert Goulard in radiative gas dynamics, and Professor Bogdanoff in stochastic processes, and all of them took part in extensive lecture trips either in Europe or the Far East. Professors Bogdanoff and Kozin established a Center for Applied Stochastics in the early 1960s to provide courses in probability, statistics, and stochastic processes and to perform research in the use of stochastic processes in predicting the response of structures and systems to random disturbances. The National Science Foundation provided support for many years, and a number of significant results were obtained and applied to understanding earthquake phenomena, which have subsequently been adopted by others doing earthquake research. The center's research program supported a number of graduate students and resulted in several PhD theses. Professor Lo served as a consultant to a number of companies entering the satellite business as a result of his development and teaching of new courses in orbit mechanics. Professor Stanley developed an interest in biomedical engineering, particularly as it applied to electrical phenomena, and made a trip to Scandinavia during the 1965-66 academic year to give lectures and to study. He was later instrumental in establishing the Biomedical Engineering Center within the Institute for Interdisciplinary Engineering Studies at Purdue (see Figure 5.8).

Another indication of faculty interest and activity at the graduate level is the development of new courses, since these courses often represent a current research interest of the faculty members who created them. Table 5.9 lists new courses developed in the period between 1960 and 1973. It should be understood that the original course creator did not necessarily teach the course at all times, since sometimes new faculty would pick up an existing course number and teach it with content modified to their interest; thus, some courses list several faculty members. Courses that appeared in the catalog but were never taught due to little or no enrollment are not included here.

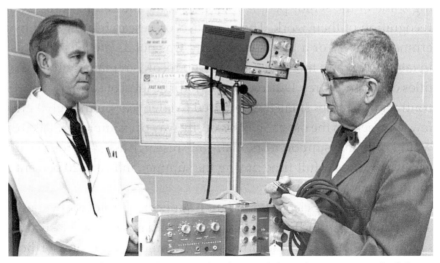

Figure 5.8   Professor Paul E. Stanley (on right) with Dr. McLeod of the Student Health Service (1967).

Table 5.9   New Courses Taught During 1960–73

A&AE Course	Professor	Title
508	Citron, Alspaugh	Systems Optimization Techniques I
516	Kentzer	Computational Fluid Mechanics
517	Lykoudis	Fluid Mechanics of Physiological Systems
519	R. Goulard, Gustafson, Kentzer	Hypersonic Aerodynamics
523	Lykoudis	Introduction to Astrophysics
525	R. Goulard	Physical Optics and Radiative Transfer
535	Feuer	Theory of Material Properties
555	Sun, Koh	Mechanics of Composite Materials
556	Ginsberg	Aeroelasticity
558	Yang	Advanced Matrix Methods in Aerospace Structures
565	M. Goulard, Swaim	Guidance and Control of Aerospace Vehicles
566	M. Goulard	Navigation and Air Traffic Control
567	Bogdanoff, Kozin	Introduction to Applied Stochastic Processes
568	Gersch, Schiff	Stochastics Laboratory
569	Staff	Dynamics of Stochastic Systems
575	R. Goulard, Malone	Methods for Aerospace Design and Operation
580	Pritsker	System Simulation Using GASP II
586	Petersen	Planning & Evaluation of Large Scale Systems
611	Staff	Principles of Fluid Mechanics
606	Citron	System Optimization Techniques
612	Sutton	Aerothermochemistry
613	Gustafson, Hoglund	Viscous Flow Theory
614	Marshall	Theoretical Aerodynamics
615	Gustafson	Rarefied Gas Dynamics
618	DeGroff, Palmer	Applied Aerodynamics
619	Lykoudis	Seminar on Biomedical Engineering Problems
621	Lykoudis	Introduction to Magneto-Fluid Mechanics and Plasma Physics
623	Lykoudis	Advanced Seminar in Astrophysics
625	R. Goulard	Radiative Energy Transfer
626	Stanisic	Mathematical Theory of Turbulence
627	Stanisic	Introduction to Dynamics of the Ocean
633	Feuer	Gas-Surface Interactions
634	Palmer	Aerospace Science Diagnostic Techniques
642	Lianis	Rheology
643	Eringen, Lianis, Koh	Theory of Continuous Media I
644	Eringen, Lianis, Koh	Theory of Continuous Media II
645	Genin	Dynamic Stability of Elastic Systems
646	Eringen, Genin	Elastic Wave Propagation
648	Marshall	Relativistic Mechanics
650	Lianis, Shapery	Inelastic Behavior of Materials & Structures
652	Stanisic	Theory of Plates and Shells
653	Stanisic	Nonlinear Mechanics
664	Bogdanoff, Staff	Application of Stochastic Analysis to Engineering Sciences I
665	Kozin	Application of Stochastic Analysis to Engineering Sciences II
668	Kozin	Methods and Applications of Time Series Analysis

A NEW PROGRAM

During the fall of 1962, correspondence was received from the United States Air Force Academy in Colorado Springs, asking about the possibility of a new master's degree specifically tailored for academy graduates. The proposed program would begin immediately after the cadets graduated in June and would have to be completed by the end of the fall semester in January to accommodate those who would begin flight training in the spring. A faculty committee was appointed to study this proposal and to determine whether such an accelerated degree program was feasible. The program, which was finally approved and forwarded to the graduate council on December 11, 1962, was to consist of a major area called astronautics, and two minor areas designated as electronics and applied mathematics. The six courses taught at Purdue were to be supplemented by a thesis of six credit hours and the transfer of six credit hours of advanced courses from the academy. Thus, for a cadet to qualify, he had to have taken at least six credit hours beyond those needed for the bachelor's degree. The first group of cadets to participate in this program visited Purdue during the spring of 1963 and received an introduction and explanation of the requirements. They also discussed possible thesis topics with the faculty and made a selection of the area in which they wished to work. In June of 1963, about 15 newly commissioned second lieutenants arrived on the campus ready to start work on the master of science degree in astronautics (Table 5.10).

Table 5.10 Master's Degree Curriculum for U.S. Air Force Academy Graduates (1963)

SUMMER SESSION
    EE 595 (Electronics)
    Thesis work

SPECIAL SESSION COURSE (AUGUST TO EARLY SEPTEMBER)
    A&ES 634 (Aerospace Diagnostic Techniques)

FALL SEMESTER
    ASTR 624 (Aerophysics)
    ASTR 630 (Orbital Mechanics and Vehicle Dynamics)
    ASTR 662 (Applied Stochastics)
    ME 651 (Gas Turbines and Jet Propulsion)
    A&ES 698 (MS Thesis)

All students followed the same program during the initial year, but the thesis proved to be a difficult requirement to complete in the short period of time available, and not everyone finished this requirement. When the second group began the program in June 1964, several modifications had been made to the requirements. The thesis had been deleted, and more choices in course selection were possible during the fall semester. Hence, the students took two courses during the regular summer session, immediately followed by an accelerated course from early August to early September. The remaining five courses were taken during the fall semester with freedom to choose, provided the major and two minor areas were reasonably well defined. For a period of several years, the two summer session courses were hypersonics taught by Professor Gustafson, and an electrical engineering course taught by Professor Lindenlaub. The one-month accelerated course just after the summer session was either A&AE 634 taught by Professor Palmer, or ASTR 630 taught by Professor Alspaugh.

This program was sometimes referred to as the "astronaut program," although it was not created to be a guaranteed first step towards becoming an astronaut. However, several graduates from this program did later become astronauts after completing flight training and serving on active duty. They were John E. Blaha and Roy D. Bridges from the class of 1966, John H. Casper from the class of 1967, Loren J. Shriver from the class of 1968, Richard O. Covey from the class of 1969, Guy S. Gardner from the class of 1970, and Gary E. Payton from the class of 1972, who was an Air Force defense payload specialist. There are more details about these and other Purdue graduates who have been astronauts in Chapter 10.

This program continued from 1963 until 1976. In the early years there were about 15 students per year, but by 1970 the number had dropped to about six per year. The special degree originally created for this program, MS in astronautics, was discontinued in 1971, and students were then awarded the MS in aeronautics and astronautics. Beginning in 1973, students were given the entire academic year to complete the degree requirements, and thus graduated in May rather than January. In general, this was considered to be a highly successful program by the faculty, the students, and by the Air Force. The following paragraph is taken from a letter dated April 18, 1966, from General Robert F. McDermott of the USAF Academy to Dean of Engineering George A. Hawkins.

I am sure you will be as pleased as I am to know of the excellent reputation which the opportunity to do graduate work at Purdue has among our students at the Academy. This year, for example, in a class of 470 seniors, we had 56 students who were in strong competition for the 15 opportunities to participate in the cooperative program at Purdue. This challenge has improved the quality of the work of the students involved, and our very best students numbered in this group. As an example of the high regard our people have for this program, I note that two of our graduates who were awarded National Science Foundation Graduate Fellowships elected to use them in this program (Lieutenant Lodge, '64, and Lieutenant Thompson, '65). Additionally, this year one declined the AEC award in favor of attending Purdue in the cooperative graduate program. I have spoken with many Lieutenants who have completed this program and their respect for their Purdue professors and the education they received from them is exactly what every instructor would hope to inspire in his students.

The Cooperative Education Program

The initial development of this program at Purdue University began in the School of Mechanical Engineering in 1955. It was created as an educational program in which work experience and study were to proceed on an alternating schedule after the freshman year. The objective was to find companies that would take relatively young and inexperienced students and provide them with engineering work commensurate with their educational level. Thus, as time went on, students would have the educational and industrial experience to be able to do increasingly responsible engineering work, and by graduation time, they would have five work periods and about 18 months of experience to add to their resumes. This program was begun in the School of Aeronautical and Engineering Sciences in the summer of 1964 with seven students placed in five different companies. Professor Lawrence T. Cargnino was the first coordinator of this program and was responsible for it until his retirement. By 1970, there were 96 students and 22 companies participating, which represented about 20 percent of the undergraduate students. Although there have been many fluctuations in enrollments and participating companies in subsequent years, the cooperative education program continues to play an important educational role for the school's students in 2009.

Staff Departures

During the period of 1960–73 a number of changes in the faculty occurred due to retirement, transfer, resignation, and death. In this section, a chronological listing of departures of full-time permanent faculty members, who had joined the University prior to 1960, will be given followed by a brief resume.

W. Burton Sanders: Professor Sanders died on January 9, 1961, at age

67. He had been a faculty member for more than 40 years since graduating in 1919 with a bachelor of science degree in mechanical engineering. He received a master's degree in 1922. He served as an instructor in applied mechanics and was promoted to assistant professor in 1925, to associate professor in 1931, and to professor of engineering mechanics in 1944. He became a faculty member in the newly created Division of Engineering Sciences in 1955.

Leo V. Kline: Professor Kline resigned from the faculty at the end of the spring semester in 1961 to take a position with IBM in Lexington, Kentucky. He had come to Purdue as an assistant professor of engineering sciences in 1956, and had been promoted to associate professor in 1959. He received a bachelor's degree in mechanical engineering in 1950 from the University of Akron, and then attended the Ohio State University, where he earned a master's degree in 1951 and a PhD in 1954. He had taught courses in solid mechanics and dynamics.

Seibert Fairman: Professor Fairman passed away on June 20, 1961, having served on the Purdue faculty for 40 years. He graduated from Kansas State University in 1919 with a bachelor's degree in mechanical engineering, and obtained a master's degree in the same field from Purdue in 1923. In 1924 he received a professional degree (ME) from Kansas State. He first became a staff member at Purdue in 1921 and rose through the ranks to professor. He was a faculty member in the newly created Division of Engineering Sciences in 1955. He was a veteran of World War I, and received a best teacher award in 1939. He coauthored a textbook with Professor Cutshall on the mechanics of materials in 1953.

R. John H. Bollard: Professor Bollard resigned from the faculty in 1961 to take a position as professor and head of the Department of Aeronautical Engineering at the University of Washington. He had done research in solid mechanics investigating thermal stresses in plates, and, under sponsorship from the National Science Foundation, had prepared a laboratory manual for seniors, which was well received. Professor Bollard had received a PhD in aeronautical engineering from Purdue in 1954 after receiving bachelor's and master's degrees in mechanical engineering from the University of New Zealand in 1948 and 1949. He had returned to Purdue in 1956 as an assistant professor and had rapidly risen to professor in 1960.

Shien-siu Shu: Professor Shu resigned from the faculty in 1963 to accept a position as distinguished professor at the Illinois Institute of Technology. He had been a professor of engineering sciences at Purdue from 1955 to 1963, and had served as an assistant and an associate professor at the Illinois Institute of Technology from 1948 to 1955. He earned a bachelor's degree from the National Tsing Hua University in China in 1935, and a PhD from Brown University in 1947. Professor Shu rejoined the faculty at Purdue in 1968 to develop an applied mathematics sequence of courses for the engineering sciences curriculum. In 1971, he took a leave of absence and became president of National Tsing Hua University in Taiwan. In July of 1973, he also was appointed chairman of the National Science Council of the Republic of China. In 1975, he resigned as president of National Tsing Hua University in order to devote full-time to his work on the National Science Council. He retained this position until 1978, when he retired from Purdue University. Professor Shu was awarded an honorary doctorate by Purdue University in 1991. He died on November 17, 2001.

Chester S. Cutshall: Professor Cutshall retired in June 1964 after serving the University for 44 years. He earned a bachelor's degree in chemical engineering from Purdue in 1919 and was employed at Eli Lilly in Indianapolis for a year before returning to Purdue as a graduate student and instructor in 1920. He received the master's degree in chemical engineering in 1924, and was promoted to assistant professor of applied mechanics in 1926, to associate professor in 1933, and to professor in 1939. He became a faculty member of the Division of Engineering Sciences in 1955. He authored or coauthored three textbooks dealing with mechanics. Professor Cutshall died on February 22, 1994.

J. Clifton Samuels: Professor Samuels resigned from Purdue University in 1965 in order to accept a position as head of the Department of Mechanical Engineering at Howard University in Washington, D.C. He had received a PhD in engineering sciences at Purdue in 1957, preceded by a master's degree in applied mathematics at New York University, and a bachelor's degree in electrical engineering at the Polytechnic Institute of Brooklyn in 1948. His research involved the study of wave propagation and stability of stochastic systems. He was named an assistant professor in 1957, an associate professor in 1959, and professor in 1962.

Neil Little: Professor Little came to Purdue in 1926 after receiving a master of arts degree from the University of Michigan in 1926, which followed a bachelor of arts degree from DePauw University in 1923. He taught courses in solid mechanics and was an original member of the Division of Engineering Sciences. He had the rank of professor at the time of his retirement in 1966. He passed away in 1976.

Elmer F. Bruhn: Professor Bruhn retired in June of 1967 after 26 years on the faculty. He came to Purdue University in 1941 as an associate professor of mechanical engineering after spending the previous 12 years in the aircraft industry. He was with the Berliner-Joyce Aircraft Company (1929–31), the General Aviation Company (1932–33), the North American Aviation Company (1933–35), and the Vought-Sikorsky Aircraft Company (1935–41). During his early years at Purdue, he wrote a textbook titled "Analysis and Design of Airplane Structures," which became the standard reference book on the subject in industry as well as in universities. He began a revision of this book and published volume 1 in 1958 and volume 2 in 1964. During his last year on the faculty, he wrote a history of aeronautics at Purdue up to 1950, which in edited form is the basis for much of Chapters 1–3 of this book. Professor Bruhn received a bachelor's degree in civil engineering from the University of Illinois in 1923, a master's degree in engineering from the Colorado School of Mines in 1925, and a professional degree in structures from the University of Illinois in 1928. He died on April 24, 1984, in West Lafayette after a long and distinguished career in industry and at Purdue, where he served as the first head of the newly created School of Aeronautics in 1945 and was granted an honorary doctorate by Purdue in 1975.

Frank Kozin: Professor Kozin joined the Department of Mathematics in 1956 as an instructor and was promoted to assistant professor in 1957. In 1958, he also received an appointment as assistant professor in engineering sciences, and was promoted to associate professor in 1961 and to professor in 1964. His education was in mathematics; he earned a bachelor's degree in 1952, a master's degree in 1953, and a PhD in 1956, all from the Illinois Institute of Technology. His research was in applied mathematics, statistics, random variables, and stochastic processes. He was an NSF Fellow in statistics in 1961-62 at University College, London, and was also a visiting professor in Japan and Taiwan. He resigned in 1967 to accept a position as professor of electrical engineering at the Brooklyn Polytechnic Institute.

A. Cemal Eringen: Professor Eringen came to Purdue in 1953 as a professor in engineering sciences after employment in several industries and universities. He had obtained a master's degree at the Advanced Engineering School of Istanbul, Turkey, in 1943, and a PhD at Brooklyn Polytechnic Institute in 1948. His teaching and research was in the area of elasticity, wave propagation, molecular theory of gases and liquids, and continuum mechanics. He was a founder of the Society of Engineering Science, and also served as president and editor-in-chief of the *International Journal of Engineering Science.* He resigned in 1967 to accept a position at Princeton University.

Merrill E. Shanks: Professor Shanks had an unusual career at Purdue, arriving in 1946 with a joint appointment in mathematics and aeronautical engineering. He had come from a position as head of theoretical aerodynamics at the NACA Laboratory in Cleveland. His education had been in mathematics. He had received a bachelor's degree from the University of Iowa in 1932, a master's degree from Columbia in 1933, and a PhD from Iowa in 1936. He made many contributions to course development in aerodynamics for both undergraduate and graduate students, and supervised theses in pure mathematics as well as in aeronautical engineering. By the early 1960s his teaching activities diminished in engineering, and he developed a new area of interest in techniques for teaching mathematics to elementary school children. This resulted in textbooks for young students in mathematics. By the end of the 1960s, he had returned full-time to mathematics, and retired in 1978. He passed away on April 27, 1984.

William W. Briggs: Professor Briggs retired in February of 1970 after 24 years on the faculty. He earned a bachelor's degree from the University of Michigan in 1932 and worked for an insurance company until 1942, when he became an instructor in the Ford Airplane School. He came to Purdue as an instructor in air transportation in February 1946. He received a master's degree in education in 1949 and was promoted to assistant professor. Promotion to associate professor occurred in 1960 along with an appointment to the faculty in freshman engineering, where he served until retirement. He died September 6, 1971.

Joseph Liston: Professor Liston retired in 1972 after 35 years of service beginning in 1937. He earned a bachelor's degree in mechanical engineering at Purdue in 1930 and a master's degree in 1935. Professor Liston had been a naval aviator and developed a strong interest in aircraft engines, which he pursued in his academic work. Because of this interest, he was generally known as Piston Joe, a nickname that followed him to retirement. He had developed power plant laboratories at the airport which allowed students to run internal combustion and gas turbine engines on test stands. He later taught courses on aircraft engine design, and collaborated on a book with Paul Stanley titled *Creative Product Evolvement*. Professor Liston passed away on August 7, 1978.

Reidar L. Anderson: Professor Anderson received a bachelor's degree in electrical engineering from Purdue in 1939, and a master's degree in applied mechanics in 1948. From 1939 to 1942, he had been employed in industry and then became an instructor in the War Training school at Purdue. He became an assistant professor of engineering mechanics in 1948, and an associate professor of engineering sciences in 1957. His teaching duties included courses in statics, dynamics, strength of materials, and nuclear engineering laboratory work on a part-time basis with the nuclear engineering department. He transferred to the School of Civil Engineering in 1973.

Stephen J. Citron: Professor Citron joined the faculty in engineering sciences in February 1959 as an assistant professor after completing his PhD at Columbia University earlier that year. He earned bachelor's and master's degrees in engineering from the Rensselaer Polytechnic Institute in 1954 and 1955. His promotions to associate professor and professor came in 1962 and in 1965. He taught courses in applied mathematics and systems optimization, and pursued research in heat conduction, orbit mechanics, and guidance systems. He served the school as executive assistant to the head from 1963 to 1965, as a member of the advisory committee, and later became an assistant vice president for academic affairs. He transferred to the School of Mechanical Engineering in 1973.

John F. Radavich: Professor Radavich had joined the faculty on a one-half-time basis in July of 1960 and still retained his half-time appointment in the physics department. In May of 1973, he transferred to the School of Materials Science and Metallurgical Engineering. He had earned all of his degrees at Purdue in metallurgical engineering in the years 1946, 1948, and 1953. His teaching responsibilities were primarily associated with the school's laboratory courses, and he contributed significantly to their development during this period. He also had a consulting company in which he used an electron microscope for metallurgical investigations of spacecraft materials. He was promoted to associate professor in 1961.

New Faculty

During the period from 1960 to 1973, a number of new faculty members were brought into the school to replace those who left and to facilitate the expansion into new areas

Richard A. Schapery: Professor Schapery earned a bachelor's degree in mechanical engineering from Wayne State University in 1957. He then went to the California Institute of Technology, where he received a master's degree in mechanical engineering in 1958, and a PhD in aeronautics in 1962. He joined the faculty in September 1962 as an assistant professor. He was very active in research in the fields of thermo-viscoelastic phenomena and in composite materials. He was promoted to associate professor in 1965 and to professor in 1968. He resigned and took a position as professor of civil engineering at Texas A&M University effective September 1, 1969.

Emmett A. Sutton: Professor Sutton earned a bachelor's degree in engineering physics and a PhD from Cornell University in 1958 and in 1961. He taught for a year at Hamilton College and came to Purdue as an assistant professor in September 1962. He studied chemical kinetics and chemically reacting flows, and built a shock tube for this purpose at Purdue. Funding for this work was difficult, and so he requested a leave of absence for a year beginning in September of 1965 to continue research at the Avco Corporation in Everett, Massachusetts. He was promoted to associate professor during 1966, but chose not to return.

Kyriakos C. Valanis: Professor Valanis received BSc and MSc degrees from the Imperial College of Science and Technology, London, in 1955 and 1957. He came to Purdue as a graduate student and instructor in February 1961, and completed a PhD in January 1963. He then was appointed an assistant professor and continued his research in thermo-viscoelasticity. He resigned as of September 1, 1964, to accept a position as associate professor of engineering mechanics at Iowa State University.

Richard F. Hoglund: Professor Hoglund joined the faculty in June 1963 as an associate professor and as director of the Aerospace Sciences Laboratory. His education was in mechanical engineering at Northwestern University, where he received a bachelor's degree in 1954, a master's degree in 1955, and a PhD in 1960. He had been a department manager in fluid mechanics at the aeroneutronic division of Ford Motor Company in Newport Beach, California, prior to his arrival at Purdue. His research involved gas-particle flows as applied to rocket propulsion systems, and he was given administrative responsibility for Project SQUID for the Office of Naval Research in 1967. He resigned in January 1969 to become professor of aerospace engineering at Georgia Institute of Technology.

Wilbert M. Gersch: Professor Gersch came to Purdue as an associate professor in September 1963 to work in the area of stochastic processes, control theory, and system identification. He earned a bachelor's degree in electrical engineering at City College, New York, in 1950, a master's degree in mathematics at New York University in 1956, and a Dr. EngSc degree at Columbia University in 1961. He was on leave from Purdue to the department of medicine at Stanford University during the 1966-67 academic year and again beginning in 1968. He resigned his position in 1970 to accept a position as professor of information sciences at the University of Hawaii.

Arnold L. Sweet: Professor Sweet began graduate study at Purdue in September 1960 after receiving a bachelor's degree in mechanical engineering from the City College of New York in 1956, and a master's degree in mechanical engineering from the University of Maryland in 1959. He received a PhD in engineering sciences from Purdue in 1964, and was appointed an assistant professor in September of that year. He was promoted to associate professor in 1968. His research was in the field of stochastic processes and solid mechanics, and was associated with the Large-Scale Systems Center. He transferred to the School of Industrial Engineering in 1973.

Dale W. Alspaugh: Professor Alspaugh entered Purdue in 1957 as a graduate student and received a master's degree in mechanical engineering in 1958, after having earned a bachelor's degree in mechanical engineering in 1955 at the University of Cincinnati. He received a PhD in engineering sciences engineering in 1964 and became an assistant professor in September 1964. His research was in thin shell theory and he taught courses in solid mechanics and orbit mechanics. His promotion to associate professor came in 1968. See later chapters for more details.

Joseph Genin: Professor Genin came to Purdue in 1964 after working for a year at the General Dynamics Corporation in Fort Worth, Texas, as a structural engineer. He obtained a bachelor's degree in civil engineering at the City College of New York in 1952, a master's degree in structural engineering at the University of Arizona in 1957, and a PhD in engineering mechanics at the University of Minnesota in 1963. His work in solid mechanics covered a broad range of subjects including dynamic stability, vibrating systems, rotating systems, and elastic buckling. He wrote textbooks on applied mathematics with J. S. Maybee and a textbook on statics and dynamics with J. H. Ginsberg. His promotion to professor came in 1969, and he transferred to the School of Mechanical Engineering in 1973.

Severino L. Koh: Professor Koh first came to Purdue as a graduate student in 1957, after receiving a bachelor's degree in meteorology at New York University in 1950, followed by a bachelor's degree in mechanical engineering from National University in Manila in 1952 and a master's degree from Pennsylvania State University in 1957. He received a PhD in engineering sciences in 1962 and became a visiting research associate in the period 1962–64. He was an assistant professor in 1964, an associate professor in 1966, and a professor in 1972. His research included studies in thermo-viscoelasticity, composite materials, and micro-elasticity. He transferred to the School of Mechanical Engineering in 1973, and subsequently left the University in 1981.

Linda McClatchey Flack: Miss McClatchey (Figure 5.9) joined the school as a secretary in 1966 just after graduating from high school in Morocco, Indiana. In 1968, she became a secretary for Professor Cargnino in the scheduling office, and has been involved in registration for both undergraduate and graduate students

Figure 5.9
Linda McClatchey Flack,
administrative assistant
(circa 1975).

since that time. She was promoted to administrative assistant in 1981, and as of 2009 has the longest tenure of any support staff member in the school's history.

Anshel J. Schiff: Professor Schiff obtained a bachelor's degree in mechanical engineering from Purdue in 1958, and subsequently a master's degree and PhD also from Purdue in engineering sciences in 1961 and 1967. He became an assistant professor in 1967 and an associate professor in 1972. His research was experimental in nature and oriented towards the study of engineering seismology and building safety and reliability. He transferred to the School of Mechanical Engineering in 1973.

Robert L. Swaim: Professor Swaim joined the faculty as an associate professor in September 1967. He had received bachelor's and master's degrees in aeronautical engineering from Purdue in 1957 and 1959, and a PhD in electrical engineering from the Ohio State University in 1966. He had worked in the aircraft industry and at Wright-Patterson AFB for several years in the area of flight dynamics and control, which he continued at Purdue, and he also developed a new course in air transportation. He was promoted to professor in 1975 and left Purdue in 1978 to accept a position as associate dean of engineering, technology, and architecture at Oklahoma State University in Stillwater (see Figure 5.10).

Chin-Teh Sun: Professor Sun earned a bachelor's degree in civil en-

Figure 5.10 Professor Robert L. Swaim with TR-10 analog computer (1968).

gineering at National Taiwan University in 1962, and received a master's and PhD in theoretical and applied mechanics at Northwestern University in 1965 and 1967. He joined the faculty in 1968 as an assistant professor, and was promoted to associate professor in 1971, and to professor in 1975. As of 2009, Professor Sun continues to be active in research in the area of composite materials, both theoretically and experimentally. He has taught a variety of courses dealing with structures and materials, and has developed new courses in composite materials and fracture mechanics. See additional information in later chapters.

David W. Malone: Professor Malone became an assistant professor in 1968 after completing a PhD in civil engineering at the Massachusetts Institute of Technology. He had previously earned bachelor's and master's degrees in engineering at the University of California at Berkeley in 1961 and 1962. His research was initially in structural analysis, but he became part of the Large-Scale Systems Center and worked on environmental and transportation systems. He resigned in 1973 and joined the Battelle Memorial Institute in Columbus, Ohio.

Arthur A. Ranger: Professor Ranger received his education at the University of Michigan in aeronautical engineering, obtaining a bachelor's, master's, and PhD in 1956, 1961, and 1968, respectively. He joined the faculty in September 1968. His research was experimental in nature and involved using shock tubes to study detonation waves and droplet shattering. He also made notable contributions to the school's laboratory program and helped to formulate a series of one-credit laboratory courses. He spent the 1971-72 academic year as a liaison scientist at the Office of Naval Research in London, and resigned in 1973.

Edward C. Ting: Professor Ting graduated from National Taiwan University in 1962 with a bachelor's degree in civil engineering and was awarded a master's degree in mechanics and materials from the University of Minnesota in 1965. He earned a PhD in applied mechanics from Stanford University in 1968, and joined the faculty later that year. His research area dealt with thermomechanical properties of nonlinear materials with memory, and he taught a variety of courses in solid mechanics. He transferred to the School of Civil Engineering in 1973.

Richard A. Curtis: Professor Curtis came to Purdue in September 1969 after receiving a PhD from Cornell University in 1968 and a master's degree in aeronautical engineering in 1965. He had earlier earned a bachelor's degree in engineering science from Case Institute of Technology in 1964. His research had dealt with wave propagation in ferrofluids and magnetic separation phenomena. He went on leave in 1974 and resigned in 1975.

Jerry H. Ginsberg: Professor Ginsberg joined the faculty in September 1969 as an assistant professor after receiving an EScD and a master's degree from Columbia University in 1969 and 1966, respectively. His bachelor's degree in civil engineering was from the Cooper Union in New York in 1965. He taught courses in solid mechanics and conducted research on forced vibrations in thin shells. He coauthored a textbook on statics and dynamics with Professor Genin, and transferred to the School of Mechanical Engineering in 1973.

Henry T. Y. Yang: Professor Yang became a faculty member in 1969 after receiving his PhD from Cornell University in 1968 and spending a year with a structural consulting company. His master's degree was from West Virginia University in 1965 and his bachelor's degree was from National Taiwan University in 1962, both in the field of civil engineering. His research applied finite element techniques to a variety of structural problems, including the effects of aerodynamic loads. He also wrote a textbook on this subject and worked with many graduate students. His teaching ability was highly regarded by students, who voted him the winner of numerous teaching awards. See additional information in later chapters.

V. R. Vemuri: Professor Vemuri joined the faculty in December 1969 as an assistant professor in the Large-Scale Systems Center to teach and do research in stochastic processes and systems identification. His bachelor's degree in electrical engineering was earned at Andhra University in India in 1958, followed by a master's degree at the University of Detroit in 1963 and a PhD at the University of California at Los Angeles in 1968. He resigned in 1973 to take a position at the State University of New York at Binghamton.

A. Alan B. Pritsker: Professor Pritsker came to Purdue on July 1, 1970, as professor and director of the Large-Scale Systems Center with a split appointment between the School of Aeronautics, Astronautics, and Engineering Sciences and the School of Industrial Engineering. He had obtained a bachelor's degree in electrical engineering in 1955 and a master's degree in industrial engineering in 1956 from Columbia University, followed by a PhD from The Ohio State University in 1961. He had been employed by several companies during the previous five years, and had been involved with research in systems modeling, automation, simulation languages, and operations research. In 1973, he became full-time in industrial engineering with the Large-Scale Systems Center. He later formed the Pritsker Associates Corporation in West Lafayette, which became a very successful consulting company.

William R. Eberle: Professor Eberle was appointed an assistant professor in September 1970, having completed a PhD in the school in May 1970. His bachelor's and master's degrees were obtained in mechanical engineering at Purdue in 1962 and 1964. He taught courses in engineering analysis for the engineering sciences curriculum, developed a new course on air transportation systems, and researched air cushion vehicles. He was on leave from January 1976 to May 1977, at which time he resigned to remain in industry.

Joseph D. Mason: Professor Mason joined the faculty as an associate professor in January 1971 after being employed for four years as a section head in systems simulation at TRW. He received a bachelor's degree in aeronautical engineering from Purdue in 1959, and a master's and PhD in aeronautical engineering from the University of Arizona in 1963 and 1967. His research was in optimization theory with applications to launch and space trajectories. He remained at Purdue for 18 months, and returned to TRW Systems Group in May 1972.

Clifford C. Petersen: Professor Petersen was appointed an associate professor beginning in January 1971, with teaching and research responsibilities in the Large-Scale Systems Center. He earned a bachelor's degree in electrical engineering in 1949 from the Illinois Institute of Technology and was employed in the electronics industry for several years before entering Arizona State University and completing a master's degree and PhD in industrial engineering in 1966 and 1970. He researched systems planning techniques and simulation, and developed new courses for the Large-Scale Systems program. He transferred to the School of Industrial Engineering in 1973.

Blaine R. Butler: Professor Butler had been a graduate student at Purdue in aeronautics, astronautics, and engineering sciences, obtaining a master's and PhD in 1961 and 1965. He had graduated from the United States Military Academy at West Point in 1948 and was a career officer in the United States Air Force, with five years of teaching experience at the Air Force Academy. In the fall of 1972, he came to Purdue with a joint appointment as an associate professor of aeronautics, astronautics, and engineering sciences, and in the freshman engineering department. He was active in counseling freshmen and also taught several courses in aeronautics and astronautics. He went on leave from Purdue in 1983, and resigned in 1985. See additional information in Chapter 7.

John W. Drake: Professor Drake came to Purdue in 1972 as a professor of aeronautics, astronautics, and engineering sciences to develop a program in air transportation. At that time he had just completed the doctor of business administration degree at Harvard University after working at several consulting companies on air transportation problems. He earned a bachelor's degree in electrical engineering at Rensselaer Polytechnic Institute in 1952 and a master of business administration degree from Harvard University in 1954. He developed a sequence of courses in air transportation, which undergraduates could take as major or minor area courses; was active with several graduate students; and carried on an extensive consulting program. See additional information in later chapters.

Joseph J. Talavage: Professor Talavage was hired in August of 1972 as an associate professor of aeronautics, astronautics, and engineering sciences with responsibility for teaching and research in the Large-Scale Systems Center. Previously he had been at Georgia Institute of Technology. He had earned a bachelor's degree in electrical engineering in 1961 from the University of Cincinnati, a master's degree in electrical engineering in 1963 from the University of Pennsylvania, and a PhD from Case Western Reserve University in 1968. His research was related to adaptive pattern recognition and dynamic group behavior. He transferred to the School of Industrial Engineering in 1973.

CHAPTER 6

Midcourse Correction:-

TRANSITION AND GROWTH 1973–80

McDonnell Douglas
(Hughes) AH-64A Apache

At the end of the previous chapter, a major change in the school's structure had just taken place with the termination of the engineering sciences program and the transfer of the service courses to other schools. These changes were accompanied by the movement of faculty. As a result of transfers and resignations, the faculty now consisted of 26 people (Table 6.1). There were 196 undergraduates in aeronautics and astronautics along with 41 students still remaining from the engineering sciences curriculum, who were continuing until graduation. In addition, there were 98 graduate students in the school, of whom 51 were pursuing PhDs.

Table 6.1   Faculty of the School of Aeronautical and Engineering Sciences (1973)

Professor B. A. Reese, head; Professor T. J. Herrick, executive assistant

Professors
J. L. Bogdanoff, J. W. Drake, P. B. Feuer, R. J. Goulard, W. A. Gustafson, G. Lianis, H. Lo, B. A. Reese, S. S. Shu, M. M. Stanisic, P. E. Stanley, E. O. Stitz

Associate Professors
D. W. Alspaugh, B. R. Butler, L. T. Cargnino, M. Goulard, J. O. Hancock, T. J. Herrick, C. P. Kentzer, F. J. Marshall, G. M. Palmer, C. T. Sun, R. L. Swaim, H. T. Y. Yang

Assistant Professors
R. A. Curtis, W. R. Eberle

ADMINISTRATIVE CHANGES

Professor Bogdanoff's resignation as head of the school had been effective as of January 1, 1973, at which time George A. Hawkins was again the acting head until July 1, 1973, when Robert A. Greenkorn took over. On

August 20, 1973, Bruce A. Reese was appointed head on the recommendation of the search committee to Dean Hancock. Professor Reese had come to Purdue as a graduate student in mechanical engineering in 1946 after service in the U.S. Navy from 1944 to 1946. He had received a bachelor's degree in mechanical engineering in 1944 from the University of New Mexico. He earned a master's degree in 1948 and a PhD in mechanical engineering from Purdue in 1953, at which time he joined the faculty as an assistant professor. He was promoted to associate professor in 1955 and to professor in 1958. His areas of interest included heat transfer, gas dynamics, combustion, air-breathing, and rocket propulsion, and he had served on numerous scientific boards in the propulsion field. Just prior to becoming head of aeronautics and astronautics, he had been director of the Thermal Sciences and Propulsion Center in mechanical engineering. Professor Reese took a one-year leave in August of 1979 to be chief scientist at the Arnold Engineering Development Center in Tullahoma, Tennessee, and decided not to return to Purdue at the end of that period.

In February 1975, Associate Professor Robert L. Swaim was appointed associate head of the school, and later that year was promoted from associate professor to professor. Professor T. J. Herrick continued as executive assistant to the head, a position he had held for some time. Professor Swaim left the University in 1978, and Professor Henry Yang subsequently became the associate head. When Professor Reese went on leave in 1979, Professor Yang became the acting head while a search committee chaired by Professor Gustafson began to review candidates for the head's position. The search resulted in Professor Yang being named head as of January 1, 1980.

CURRICULUM DEVELOPMENT

In the last chapter, it was mentioned that the pressure of dropping enrollment in all fields of engineering in the early 1970s had caused each of the Schools of Engineering to formulate new curricula. The changes that were introduced generally took the form of reduced credit-hour requirements and more elective courses. This tended to make the curricula more flexible, with the hope that students would find that attractive. However, the aerospace field was under pressure from several factors in 1973, which tended to make the employment situation rather poor.

The oil embargo initiated by the Mideast oil-producing countries had caused the price of fuel to increase, thus curtailing airline traffic, which in turn caused a drop in orders for new commercial aircraft. In addition, the space program had reduced funding after the last Apollo flight to the moon, and the military aircraft programs were suffering under reduced military spending. Hence, there were large layoffs in the aerospace field and in supporting industries. Even though there was company-supplied data to show that layoffs were greater for non-aerospace engineers employed by aerospace companies, the perception was that aerospace engineering was not a good field to enter. However, by the mid-1970s, the situation for employment had improved, and enrollment began a long period of growth.

The engineering sciences curriculum had just been changed to allow a combined five-year program leading to a master's degree before the decision to terminate the program was made in December 1972. The Aeronautics and Astronautics Curriculum Committee had been considering a similar program, which was adopted late in 1972 and appeared in the 1973–74 catalog (Table 6.2).

Table 6.2 Aeronautical and Astronautical Engineering Curriculum (1972)

COMMON SOPHOMORE YEAR

Third Semester

(4) MA 261 (Multivariate Calculus)
(3) PHYS 241 (Electricity and Optics)
(2) ESC 221 (Basic Mechanics I)
(3) EE 201 (Introduction to Electrical Engineering)
(1) EE 207 (Electrical Engineering Laboratory I)
(3) General Education Elective

(16)

Fourth Semester

(4) MA 262 (Linear Algebra and Differential Equations)
(3) A&ES 208 (Basic Mechanics II)
(3) A&ES 232 (Introduction to Mechanics of Solids)
(3) Electrical Engineering Elective
(3) General Education Elective

(16)

GRADUATION REQUIREMENTS	CREDIT HOURS	
	BS	MS
Freshman and sophomore years	63	63
General education electives (beyond sophomore year)	9	12
Mathematics	3	3
Required introductory courses	18	18
A&ES 333, 334, 352, 372, 464, ME 305		
Additional courses:		
Major area	9	18
Minor area I	6	6
Minor area II (applied mathematics)	0	6
Electives*	15	24
Total Credits	123	150

* Electives could include band or ROTC as well as nontechnical courses provided that there were 3 credits of A&ES laboratory, 3 credits of design, and a total of at least 35 hours of A&ES courses for the BS and 53 hours for the MS.

The format of the curriculum was somewhat different than usual, and a student could be admitted as a graduate student after 100 credit hours had been completed with a 4.8/6.0 grade index. In addition, the credit-hour requirement for both the BS and the MS degrees was reduced from previous requirements, although a graduate student not following the combined program was still required to have 30 credits for graduation. Major and minor areas were defined as: (1) aerodynamics and propulsion, (2) structures, (3) aerospace systems and design, and (4) air transportation.

Table 6.3 shows the courses that were required in each of the respective major and minor areas. The catalog also provided a sample semester layout of courses for each of the four major areas.

Table 6.3   Required Courses (1973)

Aerodynamics and Propulsion
A&ES 416 (Aerodynamics II)
A&ES 450 (Aerodynamics and Propulsion Laboratory)
A&ES 511 (Introduction to Fluid Mechanics)*

Structures
A&ES 340 (Dynamics and Vibrations)
A&ES 453 (Structural Analysis II)
A&ES 553 (Elasticity in Aerospace Engineering)*

Aerospace Systems and Design
A&ES 340 (Dynamics and Vibrations)
A&ES 421 (Stability and Control)
A&ES 451 (Design I)

Air Transportation
A&ES 210 (Introduction to Air Transportation)
A&ES 310 (Analysis of Air Transportation Systems)

* Essential course for a graduate program.

The school's undergraduate enrollment had reached a minimum in 1973, and although a new curriculum was in effect, Professor Reese believed that more changes should be considered. He met with the curriculum committee on September 13, 1973, to present his views, which can be summarized as follows: (1) too many EE courses, including physics, in the sophomore year, (2) too many required courses in mechanics and structures, (3) too much chemistry, (4) more laboratory courses needed, including one laboratory course associated with each basic required course, and (5) introduce aeronautical courses earlier into the curriculum. The curriculum committee members were Professors Cargnino, M. Goulard, R. Goulard, Herrick,

Kentzer, Swaim, Yang, and Gustafson, chairman. Changes began rather quickly. By mid-October, a new sophomore mechanics sequence had been proposed by Professors Lo and Bogdanoff, which had the tentative approval of the flight mechanics and the structures area committees as well as the curriculum committee. The existing sophomore mechanics courses consisted of ESC 221, Statics (two credits); A&ES 208, Dynamics (three credits); and A&ES 232, Introduction to the Mechanics of Solids (three credits); making a total of eight credit hours, while the new proposal included A&ES 203, Aeromechanics I; A&ES 204, Aeromechanics II; and A&ES 204L, Aeromechanics II Laboratory; making a total of seven credit hours.

The rationale for these changes was based on a reorganization of course content so that the first course would contain mostly particle dynamics concepts, with statics considered as a special case of zero acceleration, and a small amount of material on rigid body dynamics. The second course would be primarily devoted to strength of materials, although it would have some statics concepts as needed, and a separate laboratory course would be taken simultaneously. In addition, the curriculum committee recommended that another new course be included in the sophomore year, titled A&ES 251, Introduction to Aerospace Design, which would take the place of the existing electrical engineering elective. The contents of the new design course were proposed by Professor Marshall (see Figure 6.1) to provide an introduction to the design process, in contrast to analysis, by the use of open-ended problems and by developing design modules in disciplines such as aerodynamics, propulsion, structures, etc., which could be made interactive for the design of a flight vehicle. These changes answered some of the comments made by Professor Reese about the curriculum at the beginning of the academic year, and the laboratory committee made proposals on the laboratory courses, which answered the others. Thus, in addition to A&ES 204L, a laboratory course for A&ES 204, the following new courses also were proposed to go along with the corresponding theory courses: A&ES 333L, Fluid Mechanics Laboratory; A&ES 334; Aerodynamics Laboratory; and A&ES 352, Structural Analysis I Laboratory.

On January 28, 1974, a faculty meeting was held to inform the faculty of the various changes being considered by the curriculum committee in the structure of

Figure 6.1
Professor Frank J. Marshall (1974).

the sophomore year. The faculty gave unanimous approval to all the proposed changes, and in April 1974 a faculty document was approved by the faculty and forwarded to the engineering faculty for its approval (Table 6.4). This curriculum had the following format, with the major and minor areas defined as (1) aerodynamics, (2) air transportation, (3) flight mechanics, astronautics, guidance, and control, (4) propulsion, and (5) structures and materials.

Table 6.4   Aeronautical and Astronautical Engineering Curriculum (1974)

SOPHOMORE YEAR

Third Semester		*Fourth Semester*	
(4)	MA 261 (Multivariate Calculus)	(4)	MA262 (Linear Algebra and Differential Equations)
(3)	PHYS 241 (Electricity and Optics)		
(3)	A&AE 203 (Aeromechanics I)	(3)	A&AE 204 (Aeromechanics II)
(3)	EE 201 (Introduction to Linear Circuits)	(1)	A&AE 204L (Aeromechanics II Laboratory)
(3)	General Education Elective	(3)	A&AE 251 (Introduction to Aerospace Design)
		(3)	Elective
		(3)	General Education Elective
(16)		(17)	

GRADUATION REQUIREMENTS	CREDIT HOURS	
	BS	**MS**
Freshman and sophomore years	64	64
General education electives (beyond sophomore year)	9	12
Mathematics	3	3
Required introductory courses	24	24
A&AE 333, 333L, 334, 334L, 352, 352L, 421,372, 464, ME 305		
Additional courses		
Major area	9	18
Minor area I	6	6
Minor area II (applied mathematics)	0	6
Electives*	12	21
Total Credits	127	154

* The 127 credit hours for the BS degree must include a three-hour design course beyond the sophomore year, and the program must include 39 hours of A&AE courses. The 154 credit hours for the MS degree must include at least 57 hours of A&AE courses, and must be a continuous program of five-years' length.

In the years following 1974, the curriculum maintained the same basic structure, but the credit-hour requirements slowly increased. There were also some changes in course content and some additions to the number of required courses; in particular, A&AE 340 (Dynamics and Vibrations) and A&AE 451

(Design I) had all become required courses by 1979 with a total credit-hour requirement of 131. In addition, the combined MS and BS program had been abandoned with the MS degree requirement set at 30 credit hours again.

In order to strengthen the propulsion program in the school, Professor Reese had begun discussions with Professor W. B. Cottingham, head of the School of Mechanical Engineering, about the possibility of a joint program in propulsion between the two schools. In December of 1973, a committee was formed consisting of Professors Reese and Eberle from aeronautics and astronautics, and Professors M. R. L'Ecuyer and J. R. Osborn from mechanical engineering. This resulted in closer cooperation among the two schools, and led to joint listing in 1975 of several propulsion courses, namely, A&AE 372, Jet Propulsion Power Plants, and ME 438, Gas Turbine Engines, along with A&AE 439 (ME 439), Rocket Propulsion; A&AE 538 (ME 538), Air Breathing Propulsion; and A&AE 539 (ME 539), Rocket Propulsion.

The 1975–76 annual report of the school's activities, prepared by Professor Reese for the dean of engineering, mentioned several areas of improvement during the preceding year:*"The undergraduate enrollment had risen from a low of 196 in the fall of 1973 to 222 in the spring of 1976. About 25% of the current students are in the co-op program and are employed by 19 different companies. The implementation of the four new laboratory classes has progressed well with each one being closely associated with a corresponding theory course. For accreditation purposes, the design content of the curriculum is being increased by including design concepts in several of the required courses."* The aerospace industry was recovering well, and a high percentage of students were being employed.

During spring and summer of 1978, faculty were engaged in a study of possible new directions for the school. The faculty identified a number of thrust areas, which appeared to be particularly relevant to the future of the aerospace field as well as to other engineering fields:

- Control of dynamical systems
- Composite materials
- Human-operator systems dynamics and control
- Aeroelasticity
- Failure and damage assessment
- Design
- Transportation systems
- Computational aerodynamics

There was some discussion that these areas could be pursued in a new school with a name such as vehicular dynamics, which would encompass flight vehicles as well as land vehicles. Professor Reese and others visited various private industries in the automotive and aerospace fields to ascertain the level of interest, the viability, and the level of research support that might be available. In addition, discussions were held with government agencies and universities to get their recommendations. The unanimous recommendation was that a school with a new and unfamiliar name would not enhance the current reputation of the School of Aeronautics and Astronautics, which was considerable, and furthermore the prospects for research funding were not promising. The study resulted in an October 1978 report prepared by Professors Reese and Yang titled "New Directions: A Long Range Plan," which outlined the process and objectives of the study and concluded that the pursuit of the thrust areas could best be done with the aerospace field as the unifying factor. Thus, at the end of the decade, the undergraduate curriculum had become well established in terms of requirements, and that structure has persisted with minor additions up to the present time.

LABORATORIES

Undergraduate instructional laboratories made significant progress during this period, in part because of their attachment to corresponding theory courses and the addition of new faculty and equipment. For the first time, students were now exposed to laboratory work in the sophomore year with the new course A&AE 204L, which was taken concurrently with A&AE 204. This course introduced the use of instrumentation and testing machines to do experimental work in solid mechanics.

In the junior year, students had three one-credit laboratory courses, which further developed their skills in experimental work in solid mechanics as well as in fluid mechanics and aerodynamics. The second structures laboratory class, A&AE 352L, complemented A&AE 352 and dealt with stress analysis in aerospace structures. These laboratory classes were organized so that small groups of students performed the tests, evaluated the data, and wrote a report of the results. The other two new laboratory courses, A&AE 333L and A&AE 334L, were held at the Aerospace Sciences Laboratory at the airport because of the need to use various wind tunnels. Both of these courses complemented the corresponding theory courses, and provided continuous exposure to laboratory work during the junior year. In general, the addition of these courses to the curriculum provided a much better experimental background for students, and consequently they were better prepared to take higher-level laboratory courses and to engage in experimental thesis work as graduate students.

GRADUATE PROGRAMS AND RESEARCH

The integrated bachelor's and master's degree program was in existence for about six years during this period, but there were still many graduate students who were not involved, either because they had done undergraduate work elsewhere or because they were already too far along to enter. However, the largest problem concerned the means of support available for graduate students. Before 1973, when the engineering sciences program was still under way, the school had the responsibility to teach service courses in statics, dynamics, and strength of materials for all the Schools of Engineering. This required a large number of graduate students to serve as teaching assistants in these courses, and consequently the school's budget could support a large number of graduate students, who also did research at both the master's and PhD levels. Hence, the need to seek sponsored research money to support graduate students was not as important prior to 1973. Thus, in 1973, as other schools took on the responsibility of teaching the service courses, they also acquired some of the teaching assistant positions. The transfer of teaching assistant positions did not take place immediately, but was phased in over the next one or two years so that current students would still continue to be supported until their degrees were completed. However, by 1976 the number of teaching assistant positions in the school was 16, as compared with 36 such positions in 1972.

In addition to the loss of teaching assistant support in 1973, the 13 faculty who transferred to other engineering schools were responsible for about 75% of the sponsored research spent during the previous year. For example, in the 1972–73 fiscal year, the school spent $434,423 on sponsored programs, and in summer of 1973 the loss of faculty left only $97,300 remaining in the school. However, the 1973–74 sponsored research expenditures totaled $347,044, a noticeable reduction from the previous year, but a significant increase above the $97,300 remaining in the school from the previous year. The faculty clearly made a significant effort to enhance research support, although it is obvious that proposals for much of this research had already been submitted before the summer of 1973. Table 6.5 includes a summary of approximate research expenditures from 1969 to 1980.

During this period, one of the largest single contracts was awarded to Professors Bogdanoff and Lo on "Seismic Resistance of Fossil-Fuel Power Plants" in the amount of $391,796 for a period of a little over two years from the National Science Foundation. However, approximately 20 faculty members acquired some research support from private companies and government agencies such as NASA, NSF, U.S. Air Force, U.S. Army, Department of Transportation, and the Cooper Tire and Rubber Company. Referring to Appendices B and C, which list the theses completed, will give a good summary of the various research areas that were pursued by faculty and graduate students during this time. The number of PhD students completing their degrees during the period from 1973 to 1980 was 52, which reflects a lower graduate student enrollment than had existed during the 1960s, as shown later in Figure 7.2. The undergraduate enrollment had reached a peak in 1969, and had then begun a steep decline until 1973, when it began an increase that lasted for the next 15 years (see Figure 7.1). The reduction in the number of graduate students after 1973 is a reflection of the reduced number of undergraduate students prior to 1973. These enrollment trends were not unique to Purdue, but were common throughout the United States during these years.

Table 6.5   Research Expenditures (1969–80)

Year	Sponsored Research ($)
1969–70	351,000
1970–71	405,000
1971–72	423,000
1972–73	434,000
1973–74	347,000
1974–75	490,000
1975–76	505,000
1976–77	515,000
1977–78	440,000
1978–79	445,000
1979–80	500,000

This period was one of strong research activity, particularly in the area of structures and materials, where Professors Sun and Yang attracted a relatively large number of graduate students, as is evident by examining the theses titles in Appendices B and C. However, the addition of new faculty during the latter years of the 1970s and in the 1980s resulted in a somewhat more balanced research program with more emphasis on aerodynamics, propulsion, dynamics, and control. An interesting area was developed during this period by Professor Palmer involving the aerodynamic testing of new buildings in large cities. Professor Palmer had been responsible for building the large subsonic wind tunnel at the Aerospace Sciences Laboratory in the late 1940s, and now modified the diffuser section to incorporate a large turntable on which could be placed a model of the new building surrounded by existing city buildings (see Figure 6.2). Much of this work was done on a consulting basis for architectural companies, but other work was supported by contracts

*Figure 6.2
Professor George M.
Palmer—buildings
in the wind tunnel
(1970s).*

to study the aerodynamics of cars, buses, trucks, and ocean-going ships.

This was a productive time for course development representing some new research interests of current faculty as well as interests of new faculty. A list of new graduate-level courses and the faculty members who established them follows in Table 6.6.

Table 6.6   New Courses Taught During 1973–1980

A&AE Course	Professor	Title
510	Eberle	Advanced Analysis of Air Transportation Systems
514	Marshall	Matrix Methods in Aerodynamics
515	Gustafson	Aerodynamics of V/STOL Vehicles
518	Curtis	Aerodynamic Noise
522	Swaim	Flight Test Engineering
538	L'Ecuyer (ME)	Air Breathing Propulsion
539	Osborne (ME)	Rocket Propulsion
557	Lianis	Inelastic Behavior of Materials and Structures
570	Drake	Air Transportation Seminar
654	Sun	Fracture Mechanics
690A	Skelton	Model Estimation and Control of Dynamic Systems

FACULTY DEPARTURES

There were a number of departures from the faculty during the period 1973–80, which are listed below. The faculty changes due to the termination of the engineering sciences curriculum were identified at the end of the previous chapter. Several faculty members either retired or resigned during this time period after joining the faculty in the period covered by Chapter 5. They are **A. A. Ranger**, resigned in 1973; **R. A. Curtis**, resigned in 1975; **W. R. Eberle**, resigned in 1977; **S. S. Shu**, retired in 1978; and **R. L. Swaim**, resigned in 1978. Information about them is included in the previous chapter.

Paul S. Lykoudis: Professor Lykoudis came to Purdue in 1953 as a graduate student in mechanical engineering after receiving a diploma in mechanical and electrical engineering at the National Technical University in Greece in 1950. He earned a master's degree in 1954 and a PhD in 1956. He became an assistant professor of aeronautical engineering in January 1956, an associate professor in September 1958, and professor in September 1960. His early research work was directed towards magneto-fluid-mechanics, both theoretically and experimentally, and involved a number of graduate students. He also did research on boundary layers and hypersonic wakes. Some of his research also was directed towards biomedical engineering, particularly the fluid mechanics of the upper urinary tract. His work in liquid metal flow led to problems in nuclear engineering, and he left the school in July 1973 to take the position of head of the Department of Nuclear Engineering at Purdue. He retired in 1992, but remained active in his research interests for several years after that.

Rhea Walker: Miss Walker retired in December 1973 after many years of service to the school as secretary to the head. She first joined the school when Professor Bruhn was head in the late 1940s, and continued in that capacity until retirement. She was a tireless worker with a positive outlook, and was a friend to all (see Figure 6.3).

Paula Feuer: Professor Feuer joined the faculty in 1954 as an instructor in engineering sciences before the merger with aeronautical engineering in 1960. She was subsequently promoted to assistant professor in 1955, to associate professor in 1957, and to professor in 1965. She earned a bachelor of arts degree from Hunter College in

Figure 6.3
Rhea Walker, secretary to the head of the school (1961). Ivy-covered building in the background is Old Heavilon Hall Shops.

1941, followed by a master's degree and PhD in physics from Purdue in 1946 and 1951. Her research was in solid state physics, and she taught courses in statistical mechanics, electromagnetic theory, and material properties. She went on medical leave from the University in 1974, and retired in 1987. It is believed that Dr. Feuer was the first female professor in the engineering sciences program.

Madeline Goulard: Professor Goulard resigned from the University in 1975 after being on leave the previous year. She had received an aeronautical engineer degree at the École Nationale Superieure de l'Aeronautique in France in 1950, and had been employed at ONERA in France for a year and at Canadair in Canada for a year before becoming a graduate student at Purdue in 1955. She received a PhD in aeronautical engineering in 1958 and was appointed assistant professor in September 1958 and promoted to associate professor in 1963. Her interests were in the area of flight dynamics and control, and she developed several new courses in control systems, guidance, navigation, and air transportation. Dr. Goulard was the first female professor of the aeronautical engineering program (before it merged with engineering sciences in 1960). She died on March 14, 2005.

Robert J. Goulard: Professor Goulard resigned from the University in 1975 and accepted a position as research professor at George Washington University. He had received an aeronautical engineering degree at the École Nationale Superieure de l'Aeronautique in France in 1949, and had been employed at Bendix in France, Canadair in Canada, and at General Motors in Detroit before becoming a graduate student in mechanical engineering at Purdue in 1954. He came to the aeronautical engineering school in 1956, received a PhD in 1957, and was appointed assistant professor on October 1, 1957, followed by a promotion to professor in 1960. His research dealt with radiation effects in hypersonic chemically reacting boundary layers, and he initiated several new courses at the graduate level covering these topics. He also was involved in some aspects of the Large Scale Systems program, and served as director of Project Squid. He died on November 3, 2007.

Ervin O. Stitz: Professor Stitz retired in 1975 after 38 years of service to the University as a faculty member. He entered Purdue as an undergraduate student in 1928 and received a bachelor's degree in chemical engineering in 1932. Subsequently he earned a bachelor's degree in mechanical engineering in 1933, followed by a master's degree in applied mechanics in 1944. His first faculty appointment was as an instructor in applied mechanics in 1937, followed by promotions in engineering mechanics to assistant professor in 1944, to associate professor in 1946, and to professor in 1949. Professor Stitz taught courses in solid mechanics including aircraft structures and experimental stress analysis, and was active in photoelasticity research, winning the M. M. Frocht Award in 1972 from the Society of Experimental Stress Analysis. He also won the E. F. Bruhn and A. A. Potter teaching awards in

Figure 6.4   Professor Ervin O. Stitz (on right) with Dean Emeritus A. A. Potter (1972).

1972 (see Figure 6.4). He died in April 1997.

Paul E. Stanley: Professor Stanley retired in 1976 after 30 years of service. He first came to Purdue in 1943 as an instructor in the Purdue Aeronautics Corporation, which was engaged in a civil pilot training program for the Army Air Corps, and then was one of the original faculty members of the School of Aeronautics in 1945. His first appointment was as assistant professor in the air transportation program, and he was subsequently promoted to associate professor in 1948, and to professor in 1956. His education began at Manchester College, where he received a bachelor of arts degree in physics in 1930; he then went on to the Ohio State University, where he received his master's degree and PhD in physics in 1933 and 1937. He taught a number of courses in air transportation, fluid mechanics, and stability and control, and was the interim head of the school from 1963 to 1965, at which time he began a one-year leave to study bioengineering. His work in this area dealt with various aspects of instrumentation and hospital safety, and led to the formation of a Biomedical Engineering Center in 1973 in which he served as the acting director until 1974, when a permanent director was hired. Professor Stanley passed away on May 16, 1983.

George Lianis: Professor Lianis resigned in 1978 after a one-year leave as a visiting professor at a university in Greece. He came to Purdue in 1959 as an associate professor of aeronautical engineering, and was promoted to professor in 1961. His first degree was Dipl.Ing. in mechanical and electrical engineering, obtained at the National Technical University of Athens, Greece, in 1953. In 1954 he received a diploma of Imperial College, mechanical engineering, followed by a PhD in 1956 from the Imperial College in London, England. His research and teaching activities were in plasticity, viscoelasticity, creep buckling, and nonlinear elasticity.

New Faculty

Kenneth W. Kayser: Professor Kayser received all of his degrees from Purdue, earning a BS in engineering sciences in 1969, an MS in 1970, and a PhD in 1973. He was appointed an assistant professor in 1974, and taught courses in dynamics and a physical measurements laboratory course. He left the University in 1977.

Michael P. Felix: Professor Felix joined the faculty in 1974 as an assistant professor, just after completing his PhD and a post-doctoral year at the University of California in San Diego. He had previously completed bachelor's and master's degrees in aeronautics and engineering mechanics in 1965 and 1967 at the University of Minnesota. His research involved the use of lasers to produce stress waves and cracks in materials. He taught courses in dynamics, aircraft structures, and experimental stress analysis while at Purdue, and left in 1977 to take a position at General Dynamics in San Diego.

David K. Schmidt: Professor Schmidt came to Purdue as an assistant professor in 1974 after spending a year at the Stanford Research Institute. He earned a bachelor's degree and a PhD at Purdue in aeronautics and astronautics in 1965 and 1972, and a master's degree in aerospace engineering in 1968 at the University of Southern California. His research interests were in the area of handling qualities and control of flexible aircraft. He taught several courses in the dynamics and control area, including flight dynamics and control, guidance of aerospace vehicles, navigation and air traffic control, and he also developed a laboratory in flight dynamics. He was promoted to associate professor in 1979 and to professor in 1984. He left the University in 1989 to accept a position in aerospace engineering at Arizona State University.

Robert E. Skelton: Professor Skelton joined the faculty in 1975 with the rank of assistant professor. He earned a bachelor's degree in electrical engineering at Clemson University in 1963, a master's degree in electrical engineering at the University of Alabama-Huntsville in 1970, and a PhD in mechanics and structures at UCLA in 1976. He was employed by Lockheed

Missiles and Space Company and by Sperry Rand Corporation in Huntsville, Alabama, between 1963 and 1973, and contributed to the control system design of SKYLAB before entering graduate study at UCLA. His research has been primarily in the area of control theory and its application to the control of large space structures, and he developed several graduate-level courses that reflected those interests. He was promoted to associate professor in 1978, and to professor in 1982.

John P. Sullivan: Professor Sullivan came to Purdue in 1975 as an assistant professor after spending two years at Noctua, Inc., a small company that he cofounded dealing with laser-related electronic and optical systems. His bachelor's degree is in mechanical and aerospace sciences from the University of Rochester in 1967, and his master's degree and Sc.D. in aeronautical engineering from the Massachusetts Institute of Technology in 1969 and 1973. His research and teaching activities have been primarily in the area of experimental aerodynamics, using a laser-doppler velocimeter to make flow field measurements near propellers and in channels. He also has developed new courses in experimental aerodynamics, aerodynamic noise, V/STOL aerodynamics, and design. He was promoted to associate professor in 1979 and to professor in 1984, and served as head of the school from 1993 to 1998.

James F. Doyle: Professor Doyle joined the faculty in 1977 with the rank of assistant professor after completing his PhD work in theoretical and applied mechanics at the University of Illinois. His undergraduate work was done at the College of Technology, Dublin, Ireland, resulting in a Dipl.Eng. degree, and he also earned a master's degree at the University of Saskatchewan, Canada, in 1974. His area of teaching and research has been primarily in the area of experimental solid mechanics involving elastoplastic materials and wave propagation. He was promoted to associate professor in 1981, and to professor in 1988.

Alten F. (Skip) Grandt, Jr.: Professor Grandt came to Purdue in August 1979 as an associate professor after spending eight years in the Materials Laboratory at Wright-Patterson AFB, Ohio. He was promoted to professor in 1983. His education was obtained at the University of Illinois, where he received a bachelor's degree in general engineering in 1968, followed by a master's degree and PhD in theoretical and applied mechanics in 1969 and 1971. His research area involved the study of fatigue and fracture mechanics in a variety of materials and structures, and the problems of aging aircraft. He has taught a variety of courses in the structures area, and has developed new courses in structural design, fatigue, and nondestructive testing. Professor Grandt served as head of the school from July 1985 to January 1993.

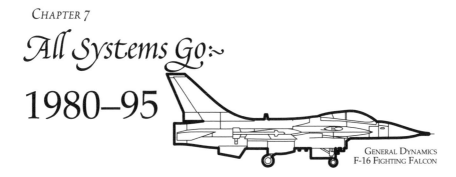

CHAPTER 7

All Systems Go:~
1980–95

GENERAL DYNAMICS
F-16 FIGHTING FALCON

INTRODUCTION

The 1980–1995 period saw significant changes in faculty composition, large fluctuations in undergraduate enrollment, and substantial growth in graduate programs and in sponsored research. Only six of the 21 faculty present in 1979, for example, remained in 1995. The new faculty brought new teaching and research interests to the school, resulting in several program changes. Before discussing these faculty changes and their impact, however, it may be useful to review enrollment between 1980 and 1995.

Figure 7.1 summarizes undergraduate enrollment (sophomore–senior) from the start of the first combined ME-AAE four-year aeronautical engineering curriculum. Note that undergraduate enrollment is character-

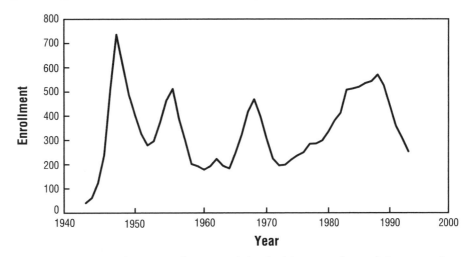

Figure 7.1   Undergraduate enrollment in School of Aeronautics and Astronautics (does not include engineering sciences).

ized by four peaks. The first occurred immediately following World War II, when returning veterans brought total undergraduate enrollment in the new aeronautical engineering and air transportation programs to 736 students in 1947. Enrollment then dropped to 279 by 1952, climbed to 512 in 1956, plunged to 202 in 1959, and reached a low of 178 students in 1961. The late 1960s saw another buildup to 469 in 1969, followed by a drop to 196 students in 1973. Undergraduate enrollments grew steadily for the next 16 years, reaching 570 students in 1989. The downsizing of the aerospace industry associated with the late 1980s saw undergraduate enrollment decrease to 253 by the fall of 1994.

Graduate enrollments are summarized in Figure 7.2. (Prior to 1961, the University did not record graduate students by individual school, and it has not been possible to obtain a complete record before 1960.) Note that the late 1960s saw graduate enrollments of approximately 130, followed by a drop to 56 students in 1979. Part of the decrease in graduate enrollment was associated with disbanding the engineering sciences program in 1972. (It should be noted that the 1960s faculty size was nearly twice that of the 1980s and 1990s.) Graduate enrollments grew substantially during the 1980s, increasing from 56 students (24 MS and 32 PhD) in the fall of 1979 to 148 by the fall of 1994 (74 MS and 74 PhD). Although undergraduate enrollment dropped significantly, graduate enrollment remained strong, near the 140 mark, between 1986 and 1995. (The enrollment and sponsored research data given in Figures 7.1–7.3 are updated in Chapter 9 to reflect changes during the 1995–2009 period.)

As shown in Figures 7.3 and 7.4, the 1980s also saw a significant increase in sponsored research and graduate theses. Annual sponsored research expenditures increased from $580,000 during 1979-80 to $2,341,316 during the 1993–94 academic year. The majority was funded by government agencies (i.e., NSF, DoD, NASA, etc.), although research support by industry increased during the 1990s. This sponsored research brought growth in graduate enrollment, as well as several laboratory and course developments. Figure 7.4 summarizes MS and PhD theses published from 1937 (when aero was still an option in ME) through 2009. Note the strong output of MS theses immediately following World War II, when the new School of Aeronautics was established. Fewer students selected the master's thesis option in the 60s and 70s, and most graduate research was in the form of PhD theses. The late 60s and early 70s saw 10 to 20 PhD theses per year, resulting from the merger with the Division of Engineering Sciences in 1960. When the engineering science program disbanded in 1972 (and faculty size was reduced), PhD output dropped to a low of five in 1979. Both MS and PhD thesis publication then increased steadily during the 1980s.

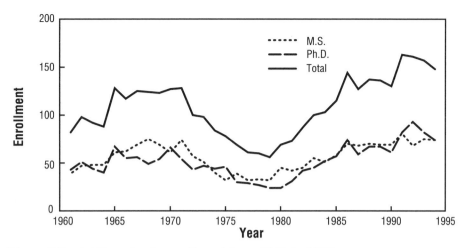

Figure 7.2   AAE graduate enrollment from 1961 to 1995.

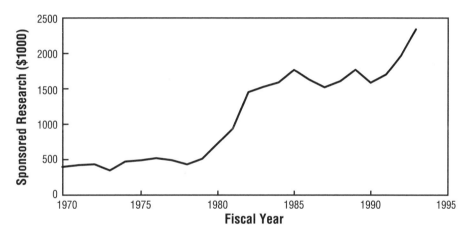

Figure 7.3   Summary of annual sponsored research expenditures.

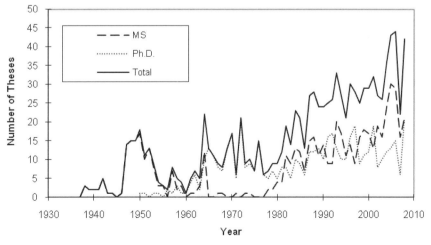

*Figure 7.4   Calendar year summary of MS and PhD theses completed by the
School of Aeronautics and Astronautics.*

ADMINISTRATIVE CHANGES

This 15-year period saw three changes in the head of the School of Aeronautics and Astronautics. Following retirement of Professor Reese in the summer of 1979, **Dr. Henry T. Yang** was named acting head of the school that fall, and head on January 28, 1980. Dr. Yang first joined the faculty as an assistant professor in 1969 and taught courses in the structures and materials area. Finite element structural analysis and its applications were his specialty. He served as school head for five years, until being named dean of the Schools of Engineering on July 1, 1984 (Figure 7.5). He continued as dean for 10 more years, until leaving Purdue in June 1994 to become chancellor of the University of California at Santa Barbara. During his 15-year administrative assignment as school head and dean of engineering, Dr. Yang also was an active member of the school's faculty. He taught each semester during this 15-year period, frequently winning the school's Best Teacher Award, and continued to conduct a large research program with his graduate students. (A list of the school's annual E. F. Bruhn Best Teacher Award winners is given in Appendix D.) Dr. Yang was later named to the National Academy of Engineering and selected as the school's first Neil A. Armstrong Distinguished Professor of Aeronautical and Astronautical Engineering.

Figure 7.5   Professor H. T. Yang advising student (circa 1980).

When Dr. Yang was appointed dean in July of 1984, Professor Winthrop Gustafson was named acting head while a national search was conducted for a permanent replacement. **Professor Alten F. Grandt, Jr.,** was appointed head on July 1, 1985 (Figure 7.6). He had joined the faculty in the fall of 1979, after spending eight years as a research engineer with the Air Force Materials Laboratory located at Wright-Patterson Air Force Base in Dayton, Ohio. Like Dean Yang, Professor Grandt also was a member of the structures and materials group. His area of expertise is in the general field of damage tolerant structural analysis. Professor Grandt served as school head for 7.5 years, stepping aside from those duties on December 31, 1993, to return to full-time teaching and research.

Figure 7.6
Professors A. F. Grandt, W. A. Gustafson, and F. J. Marshall at reception commemorating school's 40th anniversary (July 1, 1985).

Professor Gustafson was once again named acting head in January of 1993 while a national search was conducted for a replacement. **Professor John P. Sullivan** was selected as head in May of 1993, and continued to serve in that capacity in 1995. Professor Sullivan first joined the school as an assistant professor in 1975, and had served as associate head for research and graduate programs for two years prior to becoming head in 1993. Professor Sullivan's technical specialty is experimental aerodynamics (Figure 7.7).

In addition to his two terms as interim head, **Professor W. A. Gustafson** played a key administrative role during this 15-year period (Figure 7.6). After first being named associate head by Dr. Yang on March 28, 1980, he continued in that role with Professors Grandt and Sullivan through 1995. He also performed the duties of undergraduate counselor and co-op coordinator during most of this time, and it is accurate to state that virtually all undergraduate students during this period benefited directly from his dedicated concern and attention to detail. He also had a major influence on graduate programs as he was responsible for teaching assistant and graduate student office assignments during much of this period.

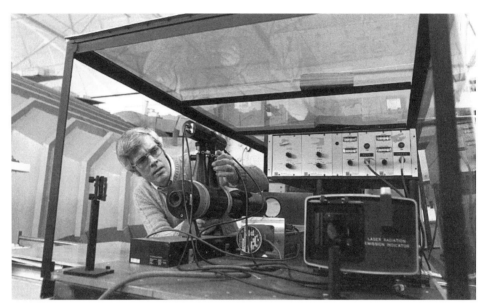

Figure 7.7   Professor J. P. Sullivan with laser Doppler velocimetry experiment (circa 1980).

STAFF DEPARTURES

As can be seen by comparing the list of faculty given in Table 7.1 for the 1980-81 academic year with the 1994-95 faculty list given in Table 7.2, there were significant changes during this period. As the school approached middle age, there were several retirements by faculty who had originally joined the new school in the 1940s. A chronological list of faculty retirements and resignations is given below, along with a few support staff departures. A faculty and staff photograph taken in September 1992 is given in Figure 7.8.

Hsu Lo: Professor Lo retired in December of 1979. He had joined the School of Aeronautics and Astronautics in 1949, and had been with the school since that time, serving as head from 1967 until 1971. Professor Lo was twice voted Best Teacher of the school, and developed several courses dealing with vibrations, aeroelasticity, orbit mechanics, advanced flutter, advanced dynamics, and basic aero mechanics. His orbit mechanics course was the first offered in this country. Professor Lo died on November 3, 1991, in Eugene, Oregon, at the age of 74.

Figure 7.8   School of Aeronautics and Astronautics staff (September 1992).

(1)   Marilyn Engel
(2)   Mario Rotea
(3)   Marie Hathaway
(4)   David Reagan
(5)   Sharon Wise
(6)   Kathleen Howell
(7)   Donald Bower
(8)   Myra Fuqua
(9)   James Doyle
(10) Robert Sanders
(11) Terri Moore
(12) Robert Scott
(13) John Sullivan
(14) C. T. Sun
(15) Linda Flack
(16) Beth Chaney
(17) Stephen Heister
(18) Henry Yang
(19) A. F. (Skip) Grandt, Jr.
(20) Martin Corless
(21) Gus Gustafson
(22) William Bader

(23) Steven Collicott
(24) Terry Weisshaar
(25) Nathan Messersmith
(26) James Longuski
(27) Marc Williams
(28) Robert Skelton
(29) Mark Yost
(30) Gregory Blaisdell
(31) Thomas Farris
(32) Steven Schneider

Not pictured: Dominick
Andrisani II, Horacio
Espinosa, Arthur Frazho,
and Sandy Hall.

Table 7.1   School of Aeronautics and Astronautics Faculty for the 1980-81 Academic Year

Professors
J. L. Bogdanoff, B. R. Butler, J. W. Drake, W. A. Gustafson, F. J. Marshall, J. R. Osborn, M. M. Stanisic, C. T. Sun, H. T. Yang

Associate Professors
D. W. Alspaugh, L. T. Cargnino, A. F. Grandt, Jr., J. O. Hancock, C. P. Kentzer, G. M. Palmer, D. K. Schmidt, R. E. Skelton, J. P. Sullivan, T. A. Weisshaar

Assistant Professors
D. A. Andrisani, J. F. Doyle, A. E. Frazho, M. H. Williams

Table 7.2   School of Aeronautics and Astronautics Faculty for the 1994-95 Academic Year

Professors
J. F. Doyle, T. N. Farris, A. F. Grandt, Jr., W. A. Gustafson, R. E. Skelton, J. P. Sullivan, C. T. Sun, T. A. Weisshaar, M. H. Williams

Associate Professors
D. A. Andrisani, M. J. Corless, A. E. Frazho, S. D. Heister, K. C. Howell, J. M. Longuski, A. S. Lyrintzis

Assistant Professors
G. A. Blaisdell, S. H. Collicott, H. D. Espinosa, N. Messersmith, M. A. Rotea, S. P. Schneider

Thomas J. Herrick: Professor Herrick was an original member of the school's faculty, joining the department in 1945. He retired in May of 1980, after serving 34 years on the faculty (Figure 7.9). Professor Herrick taught all the basic courses in the structures area, as well as the first courses in aerodynamics and airplane design. He was executive assistant to the head of the school from 1968 until his retirement, serving under three different heads and three different acting heads. Professor Herrick died on September 25, 1991, in Lafayette, Indiana, at the age of 78.

Dale W. Alspaugh: Professor Alspaugh left the school in June 1981 to become vice chancellor for academic services at the Purdue North Central Campus in Westville, Indiana. (He later was named chancellor of that campus.) Professor Alspaugh received his PhD from the school in 1965, and had been a member of the school's dynamics and controls group since that time. Dr. Alspaugh died on July 1, 2004.

Figure 7.9
Professors J. Hancock and T. J.
Herrick (July 1, 1985).

Lawrence T. Cargnino: Professor Cargnino was another original member of the school's faculty, having joined the department in September 1945 as a member of the air transportation program. Professor Cargnino retired in May of 1984, after several years of part-time retirement (Figure 7.10).

John Hancock: Professor Hancock retired in May of 1983 after several years of part-time teaching (Figure 7.9). He came to Purdue as a member of the engineering mechanics department in 1936, and had continuous service with the University since that time, with the exception of a few years during World War II. Professor Hancock taught a variety of courses in the structural mechanics area. He died on September 16, 1993, in Lafayette at the age of 80.

Figure 7.10 Professors L. T. Cargnino and E. F. Bruhn (circa 1982).

George D. Calvert: Mr. Calvert was a shop technician who had been with the school for 31 years before retiring in June of 1983. Mr. Calvert had made many contributions to both educational and research projects for the school (Figure 7.11). He died on July 9, 2006.

Figure 7.11 G. D. Calvert and Professor G. P. Palmer (circa 1987).

Claud Tipmore: Mr. Tipmore retired in May of 1984. He joined the school as a shop technician in 1970, and had supervised the machine shop at the aerospace sciences laboratory for several years. Mr. Tipmore died on August 27, 2008 at the age of 91.

Milomir M. Stanisic: Professor Stanisic retired from Purdue in June of 1980, but continued to teach part-time until May of 1985 (Figure 7.12). He had joined the engineering sciences faculty in February of 1956 before that program merged with aeronautical engineering in 1960. Professor Stanisic's teaching and research were in the areas of nonlinear mechanics, theory of continuous media, mathematical theory of turbulence, supersonic aerodynamics, and hydromagnetic turbulence. Professor Stanisic died on March 10, 1991, in West Lafayette at the age of 76. He authored a textbook titled *The Mathematical Theory of Turbulence* in 1985, and was writing a second book at the time of his death to be titled *Nonlinear Phenomena in Mathematical Physics and Engineering Sciences*.

Figure 7.12 Professors M. M. Stanisic, C. P. Kentzer, and G. P. Palmer (1982).

Blain R. Butler: Professor Butler left Purdue in January 1983 to become a department head at Embry Riddle Aeronautical University in Prescott, Arizona. He had held a joint appointment with the School of Aeronautics and Astronautics and the Division of Freshman Engineering since 1972.

John L. Bogdanoff: Professor Bogdanoff retired from the school in December of 1985, following several years on half-time retirement. Professor Bogdanoff joined the engineering sciences faculty in 1950, before that program merged with the School of Aeronautics and Astronautics. He served as associate head of the School of Aeronautics, Astronautics, and Engineering Sciences from 1967 to 1971, and as head from 1971 to 1972. He taught courses in statics, dynamics, strength of materials, vibrations, elasticity, control systems, vectorial mechanics, and probabilistic approaches to cumulative damage. Dr. Bogdanoff was a member of the National Academy of Engineering. Professor Bogdanoff died on July 20, 2003.

George M. Palmer: Professor Palmer retired in June of 1987, following 41 years of service to the school (Figures 7.11–7.12). He graduated with distinction from the School of Aeronautical Engineering in 1945, and had been a part-time instructor with the Curtiss-Wright Cadette War Training Program. He returned to the faculty full-time in 1947, and taught more than 30 different courses during the next 40 years. He was responsible for the required senior design course for approximately 30 years before retiring. He was acting director of the Aerospace Sciences Laboratory from 1969 to 1975, and director from 1975 to 1983.

John R. Osborn: Professor Osborn retired in May of 1989, but returned to teach part-time during the fall 1989 semester. Professor Osborn had transferred from the School of Mechanical Engineering in 1980, and taught courses and conducted research in the propulsion area. Professor Osborn died on December 18, 2009.

David K. Schmidt: Professor Schmidt resigned from the department in May of 1989 to accept a position with another university. Professor Schmidt had received his bachelor's degree from the school in 1961, and his PhD in 1972. He had joined the faculty's dynamics and control group in 1974 (Figures 7.13–7.14).

Francis J. Marshall: Professor Marshall retired in December of 1989 after 29 years on the school's faculty. He was a member of the school's aerodynamics group, and had developed the introduction to design course (AAE 251) in the 1980s (Figure 7.6).

C. Paul Kentzer: Professor Kentzer retired in May 1990 after 37 years on the school's faculty. Professor Kentzer received his PhD from the school in 1958, and joined the aerodynamics group in fall 1959 (Figure 7.12).

Figure 7.13 G. P. Harston, Professor D. K. Schmidt, and C. R. Malmsten
examining circuit board for General Automation computer (circa 1981).

Figure 7.14
Professors D. A. Andrisani,
K. C. Howell, R. E. Skelton,
and D. K. Schmidt (1982).

Figure 7.15
Professor K. C. Howell discusses an
orbit mechanics problem with graduate
students David Spencer (seated) and
Hank Pernicka (1985).

John Drake: Professor Drake retired in May of 1992. He joined the school in 1972 after receiving his doctorate in business administration from Harvard University. He taught and conducted research in the general area of air transportation.

Roberta (Cameron) Copenhaver: Mrs. Copenhaver retired in September of 1990, after serving as an account clerk in the school's business office since 1968. She was one of the first individuals to greet new staff and graduate students, as she arranged their payroll, medical, and retirement plans, etc.

Gene P. Harston: Mr. Harston died suddenly on July 2, 1992, at the age of 52. He was a senior electronics technician and had been employed by the school since 1968 (Figure 7.13).

Mark Yost: Mr. Yost was a laboratory technician who retired in September of 1992 following 15 years with the University, and 9 years with the School of Aeronautics and Astronautics. Mr. Yost's involvement with the school actually began while he was still a high school student in 1944, when he was hired part-time by the School of Mechanical and Aeronautical Engineering to assist with wind tunnel experiments. Mr. Yost died on June 23, 2009.

Henry T. Yang: Dr. Yang resigned from Purdue in June of 1994 to become chancellor of the University of California at Santa Barbara. He first joined the school as an assistant professor in 1968, and served as head from 1979 to 1984. He was dean of the schools of Engineering from 1984 to 1994.

NEW FACULTY

This section summarizes the faculty additions during this 15-year period. As pointed out previously, 16 of the 22 faculty who were with the school in 1995 joined the department after 1980.

Terrence A. Weisshaar: Dr. Weisshaar obtained his BSME from Northwestern University in 1965, an SM from MIT in 1966, and his PhD from Stanford in 1971. He served on the faculty of the University of Maryland, and then at Virginia Polytechnic Institute and State University before joining the school's structures and materials group in the fall of 1980 to teach courses in structural mechanics and design. He joined the school with the rank of associate professor in 1980, and was promoted to professor in 1984.

Dominick A. Andrisani II: Dr. Andrisani obtained his BSAE from Rensselaer Polytechnic University in 1970, his ME from the State University of New York at Buffalo in 1975, and a PhD from SUNY Buffalo in 1979. Prior to joining the school's dynamics and control group in fall 1980, he had worked as an engineer with the Calspan Advanced Technology Center in Buffalo. Dr. Andrisani joined the school with the rank of assistant professor and was promoted to associate professor in 1986 (Figure 7.14).

Arthur E. Frazho: Dr. Frazho also joined the school as an assistant professor in the dynamics and control area in the fall of 1980. He obtained his BSE in computer engineering from Michigan in 1973, an MS in computer information and control engineering from Michigan in 1974, and a PhD in computer information and control engineering from Michigan in 1977. He had served as an assistant professor of electrical engineering at the University of Rochester for three years before joining the School of Aeronautics and Astronautics. He was promoted to associate professor in 1986, and to professor in 1995.

John R. Osborn: Dr. Osborn transferred to the School of Aeronautics and Astronautics from the School of Mechanical Engineering in 1980. He had received his PhD from Purdue in 1957, and had served on the faculty of mechanical engineering until his transfer to the School of Aeronautics and Astronautics. He taught courses in aerospace propulsion, and conducted research in the general field of rocket propulsion.

Marc H. Williams: Dr. Williams obtained his PhD from Princeton in 1974, and remained on the research staff there until joining the school's aerodynamics group as an assistant professor in 1981. He was promoted to associate professor in 1984 and to professor in 1991.

Kathleen C. Howell: Dr. Howell joined the school's dynamics and controls group in January of 1982 as a visiting assistant professor. She completed her PhD in aeronautics and astronautics from Stanford later that year, and became an assistant professor in the fall of 1982. She was promoted to associate professor in 1987 and to professor in 1997. Professor Howell's technical area is in the general area of orbit mechanics (Figure 7.14–7.15).

Martin J. Corless: Dr. Corless joined the school's dynamics and control group as a visiting assistant professor in August of 1984, and as an assistant professor in August of 1985. He received a BE from University College, Dublin, Ireland in 1977, and completed his PhD in mechanical engineering at the University of California, Berkeley in 1984. Professor Corless's research deals with the control of uncertain systems. He was promoted to associate professor in 1989 and to professor in 1996.

Martin Ostoja-Starzewski: Dr. Ostoja-Starzewski joined the structures and materials group as an assistant professor in August of 1985. He received an engineering degree from the Technical University of Cracow, Poland, and a PhD from McGill University, in Montreal, Canada. Dr. Ostoja-Starzewski's research dealt with probabilistic methods in structures and solids. After serving on the faculty for four years, Dr. Ostoja-Starzewski resigned in 1989 to join another university.

Thomas N. Farris: Dr. Farris received his undergraduate degree in 1982 from Rice University and his master's (1984) and PhD (1986) from Northwestern University. He joined the structures and materials group as an assistant professor in August of 1986. He was promoted to associate professor in 1991 and to professor in 1994. His research areas are tribology, manufacturing processes, and fatigue and fracture.

William J. Usab, Jr.: Dr. Usab received his BS from the School of Aeronautics and Astronautics in 1978, and his ScD from MIT in 1984. He joined the school's aerodynamics group as an assistant professor in 1986, after spending two years as an associate research engineer with the United Technologies Research Center. Dr. Usab's specialty was computational fluid dynamics. He left the school in 1993.

Thomas S. Lund: Dr. Lund joined the school as an assistant professor with the aerodynamics group in the fall of 1987 after receiving his PhD from Stanford University. His technical interests were in computational fluid dynamics, hypersonic flow, and V/STOL aerodynamics. He resigned from the faculty two years later to accept a position in industry.

James M. Longuski: Dr. Longuski joined the school's dynamics and controls group in fall 1988 as an assistant professor. He received his undergraduate and graduate education at the University of Michigan, receiving his PhD in 1979. Dr. Longuski was employed by the Jet Propulsion Laboratory in Pasadena, California, from 1979 until 1988. His technical expertise is in the area orbit mechanics, space craft dynamics, control, and orbit decay and reentry. He was promoted to associate professor in 1992 and to professor in 1998.

Lee D. Peterson: Dr. Peterson joined the School of Aeronautics and Astronautics in 1989. He received his undergraduate and graduate education from MIT, and had worked for Sandia National Laboratories for two years before coming to Purdue. He taught in the design area and conducted research dealing with dynamics and control of aerospace structures. He left Purdue after two years to join another university.

Steven P. Schneider: Dr. Schneider joined the school's aerodynamics group as an assistant professor in 1989. He received his undergraduate and graduate education from the California Institute of Technology, receiving his PhD in 1989. His technical specialty is experimental fluid mechanics. He was promoted to associate professor in 1995 and to professor in 2004.

Stephen D. Heister: Dr. Heister joined the school's propulsion group in the fall of 1990. Dr. Heister received his undergraduate (1981) and master's degree (1983) from the University of Michigan, and his PhD from UCLA in 1988. He was employed by Lockheed Aircraft Company from 1981 to 82, and by The Aerospace Corporation from 1983 until joining the school in 1990. While at Aerospace, he served as manager of the Propulsion Technology Section. His research interests include rocket propulsion and liquid propellant injection systems. Dr. Heister was promoted to associate professor in 1994 and to professor in 1999.

Mario A. Rotea: Dr. Rotea received his undergraduate degree from Universidad Nacional de Rosario in Argentina in 1983, his ME from the University of Minnesota in 1988, and his PhD from Minnesota in 1990. He joined the school's dynamics and controls group in the fall of 1990, and his technical interests include multivariable control, optimal control, modeling, and identification. Professor Rotea was promoted to associate professor in

1995 and to professor in 2002.

Gregory A. Blaisdell: Dr. Blaisdell joined the school's aerodynamics group in January of 1991 as an assistant professor. He received his undergraduate (1980) and master's (1982) degrees in applied mathematics from the California Institute of Technology, and his PhD in mechanical engineering in 1991 from Stanford University. His research interests include computational fluid mechanics, transition, and turbulence.

Steven H. Collicott: Dr. Collicott received his BSAE from the University of Michigan in 1983, his MS in 1984, and his PhD from Stanford in 1991. He joined the school's aerodynamics group in January of 1991, where his research interests are in the areas of experimental fluid mechanics, optical diagnostics, and applied optics. He was promoted to associate professor in 1996 and to professor in 2006.

Nathan L. Messersmith: Dr. Messersmith received his undergraduate and graduate education from the University of Illinois, receiving his PhD in mechanical engineering in 1992. He joined the school's propulsion group in January of 1992 as an assistant professor. His technical interests lie in the general areas of air breathing propulsion, gas dynamics, laser-based flow diagnostics, aeroacoustics of turbulent mixing scalar transport in mixing flows, and combustion.

Horacio D. Espinosa: Dr. Espinosa received his degree in civil engineering from Universidad Nacional del Nordeste, Argentina, in 1981, his degree in Ingegneria delle Strutture from Politecnico de Milano, Italy, in 1987, his MS in applied mathematics from Brown University in 1991, and his PhD in applied mechanics from Brown University in 1992. He joined the school's structures and materials group as an assistant professor in January 1992. His research interests are in the areas of micromechanics, constitutive modeling, material instabilities, and wave propagation in advanced materials.

Anastasios S. Lyrintzis: Dr. Lyrintzis received his PhD from Cornell University in 1988. He held faculty positions at Syracuse University, Cornell University, and the University of Minnesota before joining the school as an associate professor in fall 1994. He was promoted to professor in 2002. His technical areas of expertise are computational aeroacoustics, aerodynamics, and traffic flow modeling.

Academic Programs

The undergraduate curriculum remained fairly stable during the 1980–95 period, and is represented by the 1981–83 plan of study reproduced in Table 7.3 from the 1981–83 engineering catalog. Note that 131 credit hours were required for graduation, including 11 hours in mathematics beyond the freshman year and 18 hours in general education electives. In addition, at least 49 hours were needed from the School of Aeronautics and Astronautics, including required credits in aerodynamics (eight hours), dynamics and control (12 hours), propulsion (three hours), and structures and materials (eight hours). These required courses included four hours of laboratory. Other AAE requirements included the sophomore design course (AAE 251) and the senior capstone design course (AAE 451). Students also took nine additional credits in a major and six credits in a minor area selected from the four areas listed above, or from air transportation (see Table 7.4). This curriculum provided students with breadth in the four major aerospace disciplines through the core requirements, along with the opportunity to develop depth in two areas of the student's choice.

This curriculum remained in effect through 1995 with relatively few changes. (The 1995 curriculum is presented in Chapter 9.) One of the more important changes that did occur was discontinuation of the air transportation option following Professor Drake's retirement in 1992. Professor Drake had been the only AAE faculty member working in air transportation for many years, and enrollments in this optional program were small. When he retired, it was decided to use his faculty position to strengthen the propulsion area, which had also been represented by a single AAE faculty member in recent years.

Requirements for the graduate program also remained relatively constant during this period. Both a thesis and nonthesis master's degree were offered. Each required 30 credits, with the thesis accounting for nine hours in that option. Students developed personalized plans of study that included six hours of mathematics, with the remaining hours divided between a major and a minor area. The major and minor were selected from the following: aerodynamics, air transportation, dynamics and control, propulsion, and structures and materials. Again, the air transportation option was discontinued following Professor Drake's retirement in 1992. Also, beginning in 1991, it was possible to minor in manufacturing engineering through an interdisciplinary manufacturing option provided in conjunction with the Schools of Industrial Engineering, Chemical Engineering, Materials Engineering, Electrical Engineering, and Mechanical Engineering.

The PhD required a minimum of 48 hours of graduate coursework (including those taken for an MS). Again, an individualized plan of study was developed with the student's advisory committee subject to general guidelines. That plan of study consisted of a primary area and at least one related area, with at least six hours from outside the School of Aeronautics

Table 7.3   *Plan of Study for Aeronautical and Astronautical Engineering (1981–83 Catalog) — Credit Hours Required for Graduation: 131*

SOPHOMORE YEAR

Third Semester

- (4)   MA 261 (Multivariate Calculus
- (3)   PHYS 241 (Electricity and Optics)
- (3)   AAE 203 (Aeromechanics I)
- (3)   EE 201 (Introduction to Linear Circuits)
- (3)   General Education Elective

(16)

Fourth Semester

- (4)   MA 262 (Linear Algebra and Differential Equations)
- (3)   AAE 204 (Aeromechanics II)
- (1)   AAE 204L (Aeromechanics II Laboratory)
- (3)   AAE 251 (Introduction to Aerospace Design)
- (3)   Elective
- (3)   General Education Elective

(17)

JUNIOR YEAR

Fifth Semester

- (3)   MA 302 (Mathematical Methods in Engineering) or MA 422 (Differential Equations for Engineering and the Physical Sciences)
- (3)   ME 200 (Thermodynamics I)
- (3)   AAE 352 (Structural Analysis I)
- (1)   AAE 352L (Structural Analysis I Laboratory)
- (3)   AAE 333 (Fluid Mechanics)
- (1)   AAE 333L (Fluid Mechanics Laboratory)
- (3)   General education elective

(17)

Sixth Semester

- (3)   AAE 372 (Propulsion I)
- (3)   AAE 334 (Aerodynamics)
- (1)   AAE 334L (Aerodynamics Physical Laboratory)
- (3)   AAE 421 (Stability and Control)
- (3)   AAE 340 (Dynamics and Vibration)
- (3)   Major or minor area electives

(16)

SENIOR YEAR

Seventh Semester

- (3)   AAE 451 (Design I)
- (3)   AAE 464 (Control Systems Analysis)
- (3)   Elective
- (6)   Major or minor area electives
- (3)   General education elective

(18)

Eighth Semester

- (6)   Major or minor area electives
- (6)   Electives
- (3)   General Education Elective

(15)

The 131 credit hours must include at least 49 hours of AAE courses. The required courses and the major and minor area courses may not be taken on a pass/not-pass basis. Engineering Graphics (EG 116) is required in this curriculum.

Table 7.4   Options in Aeronautical and Astronautical Engineering
(1981–83 Catalog)

The school offers five curriculum options for major and minor areas of study in programs leading to the degrees of BSAAE, MSAAE, and PhD.

Aerodynamics. This option emphasizes the study of fluid motion around a body moving through atmospheric air at speeds which range from subsonic to hypersonic. Theoretical and experimental methods are developed to determine forces, moments, and heat transfer, which can be applied to the design of aircraft, missiles, and space vehicles. The basic theory and techniques also find application in other areas such as high-speed ground transportation, hydrofoils, mechanics of blood flow, and noise generation.

Air Transportation. This option addresses the study of the various components of the total air transportation system; aircraft, airplanes, air traffic control, airports; their system design, operation, and interrelationships, and the interfaces between the system and the societal, political, and economic environment. Programs may emphasize such diverse areas as planning, operations, systems, theory, simulation, economics, and environmental impact.

Dynamics and Control. This option involves the study of techniques for aerospace vehicle guidance; systems analysis and control; analysis of flight vehicle trajectories, orbits, and dynamic motion; and system optimization methods. This area deals more with the vehicle as a whole and how the subsystems and related technologies are integrated into the optimal design of a vehicle such that the mission requirements are met.

Propulsion. This option involves the study of propulsion and propulsion devices. It is a study of basic principles of propulsion and the application of gas dynamics, to internal flow and the analysis of the components of the various types of propulsion systems — diffuser, compressor, combustor, turbine, nozzle, feed systems, etc. Various propulsion systems are analyzed: turbojet, fan-jet, ramjet, scramjet, chemical rockets, electric rockets, and combinations. The propulsion option is a cooperative program with the School of Mechanical Engineering.

Structures and Materials. This option primarily involves courses in structural analysis, structural design, and solid mechanics, which deal with the principles of mechanics and analysis techniques necessary to insure structural integrity of a vehicle, primarily an aircraft or spacecraft. Response to a failure of both materials and structures under static and dynamic loads; stationary or transient thermal environments; radiation; and environmental corrosion are investigated theoretically and observed experimentally using photoelasticity, short-duration laser photography, and laser holographic interferometry. The techniques developed in these courses are by no means limited to aerospace applications even though the emphasis is in this area.

---

and Astronautics. All PhD students were required to prepare a research thesis of sufficient importance to merit publication in a recognized and respected technical journal.

Perhaps one of the most significant changes in the school's academic programs during this period was the development of many new courses by the faculty. Table 7.5 shows 43 new courses that were taught for the first time between 1980 and 1995. Table 7.5 lists eight new 400-level courses (undergraduate), 23 new 500-level, and 11 advanced 600-level graduate courses developed between 1980 and 1995. When one considers that senior undergraduates may take 500-level courses along with graduate students, it may be seen that both undergraduate and graduate students benefited significantly from the new course offerings.

The undergraduate design program saw several changes during the 1980–1995 period, including additional faculty involvement in teaching design and increased interactions with industry. For many years the school required two design courses: Introduction to Aerospace Design (AAE 251), developed by Professor F. J. Marshall in the 1970s, and Design I (AAE 451), the senior capstone design course that was taught by Professor Palmer. Following Professor Palmer's retirement in 1987, the capstone course was broken into two sections, one dealing with spacecraft design taught by Professor Gustafson, and a second section dealing with aircraft design taught by Professor Weisshaar. The spacecraft section expanded a working relationship begun in the early 1980s with Lockheed Missiles and Space Company. That interaction involved design projects developed by Lockheed and Professor Palmer (and subsequently Professor Gustafson). Lockheed sent engineers to campus twice each semester: first for a bidder's conference early in the semester to discuss design requirements with students, and again at the end of the semester to help evaluate the final student presentations. The company also provided prize money and a plaque to display names of the winning team members.

The aircraft section of AAE 451 also saw increased industrial involvement during this period, first with the University Space Research Association (USRA), and later through interaction with Thiokol Corporation. The USRA program involved support of a teaching assistant and a working relationship with the NASA Langley Research Center. This program was started in 1986 by Professors Palmer and Grandt through a grant with NASA, and was continued by Professor Weisshaar following Professor Palmer's retirement. Additional industry involvement occurred in 1990 when the Thiokol Corporation established the Thiokol SPACE (Semiannual Purdue Academic Communication Excellence) Award to promote technical communications in design courses (Figure 7.16). The Thiokol SPACE Award recognized written project reports in the sophomore design class (AAE 251) and oral presentations in the aircraft section of AAE 451. Again, Thiokol provided prize money and a plaque to recognize the best designs for both the sophomore and senior classes.

The school introduced three more optional design courses in the fall of 1994 for the senior year. Since the senior capstone course (AAE 451) emphasized conceptual rather than detail design, one goal of these new courses was

Table 7.5   New Courses Taught During 1980–95

AAE Course	First Time Offered	Professor	Title
554	Fall 1980	Grandt	Fatigue of Structures and Materials
590D	Spring 1981	Bogdanoff	Cumulative Damage Model with Applications to Fatigue, Crack Growth, and Wear
421L	Fall 1981	Schmidt	Flight Dynamics and Control Laboratory
517	Fall 1981	Williams	Unsteady Aerodynamics
520	Spring 1982	Sullivan	Experimental Aerodynamics
574	Spring 1982	Andrisani	Digital Flight Control
519	Fall 1982	Gustafson	Satellite Aerodynamics and Planetary Entry
414	Spring 1983	Kentzer	Compressible Aerodynamics
540	Fall 1983	Howell	Spacecraft Attitude Dynamics
513	Spring 1986	Williams	Transonic Aerodynamics
590B	Fall 1986	Ostoja-Starzewski	Probabilistic Methods in Mechanics
590S	Fall 1986	Sun	Advanced Topics in Composites
474	Fall 1987	Andrisani	Experimental Flight Mechanics
412	Fall 1987	Usab	Introduction to Computational Fluid Dynamics
512	Spring 1988	Usab	Computational Aerodynamics
559	Spring 1988	Farris	Mechanics of Friction and Wear
490M	Spring 1988	Sun	Engineering Design Using Modern Materials
519	Fall 1988	Lund	Hypersonic Aerothermodynamics
666	Spring 1989	Corless	Nonlinear Dynamic System Control
590P	Fall 1990	Peterson	Analytical Structural Dynamics and Control
590R	Fall 1990	Rotea	Multi-Variable Control
607	Spring 1991	Longuski	Variational Principles of Mechanics
655	Spring 1991	Sun	Advanced Topics and Composites
490R	Fall 1991	Rotea	Digital Control Systems Analysis
690E	Fall 1991	Usab	Advanced Computational Aerodynamics
590S	Spring 1992	Schneider	Technical Fluid Mechanics
690F	Spring 1992	Frazho	Operational Methods in Control Theory
690R	Spring 1992	Rotea	Multivariable Control
590C	Fall 1992	Collicott	Low Gravity Fluid Dynamics
690P	Fall 1992	Heister	Stability of Free Surfaces
537	Spring 1993	Messersmith	Hypersonic Propulsion
590M	Fall 1993	Messersmith	Systems of Aerospace Combustion
632	Fall 1993	Howell	Advanced Orbital Dynamics
552	Spring 1994	Collicott	Optical Systems Design
590N	Spring 1994	Grandt	Nondestructive Evaluation of Structures and Materials
690E	Spring 1994	Longuski	Optimal Trajectories
415	Fall 1994	Sullivan	Aerodynamic Design
454	Fall 1994	Grandt	Design of Aerospace Structures
590A	Fall 1994	Doyle	Aerospace Structural Dynamics and Stability
690G	Fall 1994	Howell	Astronautical Navigation and Guidance
690S	Fall 1994	Skelton	Dynamical Control of Flexible Structures
590B	Spring 1995	Lyrintzis	Aerodynamic Sources of Sound
490R	Spring 1995	Rotea	Control Systems Design

Figure 7.16 Professor T. A. Weisshaar (center) with R. Benford (left) and S. Stein from Thiokol Corporation with Thiokol SPACE Award plaques (1991).

to provide additional design experience in the specialized technical areas. Aerodynamic Design (AAE 415) was developed and taught for the first time by Professor Sullivan in the fall of 1994. That same semester, Professor Grandt introduced Design of Aerospace Structures (AAE 454). In the spring of 1995, Professor Rotea developed a similar course in the controls area titled Control Systems Design (AAE 490R). All three courses were taught again during the 1995-96 academic year.

Beginning in 1991, Professor Heister developed a rocket launch competition as an integral part of the school's undergraduate rocket propulsion course (AAE 439). The competition used high-power model rockets to demonstrate rocket propulsion and trajectory principles to the class (Figure 7.17). The class was divided into launch teams of 5–7 members as well as a range safety team. Each launch team was responsible for constructing and testing a model rocket with the goal of providing a two-dimensional flight simulation capability, which is utilized on launch day. The range safety team acted as an independent organization responsible for coordination of all launch activities. The project not only exposed students to state-of-the-art experimental techniques for measuring thrust and drag, but also reinforced the importance of teamwork in solving problems. This activity has been quite popular with the students, and has helped them to understand basic propulsion concepts as well as many of the elements of an actual rocket launch. The winning team, based on measurements of landing position and

Figure 7.17 Students Eric Bennett and Margaret Struckel adjust model rocket in preparation for launch (1991).

altitude error, has its picture placed on the "Rocketeers of the Year" plaque located in Grissom Hall.

INDIANA SPACE GRANT CONSORTIUM

The Purdue School of Aeronautics and Astronautics formed the **Indiana Space Grant Consortium** in 1991 in response to the NASA National Space Grant College and Fellowship Program. The original Indiana Consortium comprised Purdue's West Lafayette and Calumet campuses, the University of Notre Dame, and Indiana University, with Ball State University joining in 1995. Professor A. F. Grandt, Jr., served as consortium director from 1991 to 1993, and was succeeded in this position by Professor J. P. Sullivan in 1994. Professors Andrisani (1995–2001) and Caldwell (2002–present) subsequently served as directors of the Indiana Space Grant Consortium.

In addition to $150,000 provided annually by NASA, the consortium received matching funds from its member institutions. The Indiana Space Grant Consortium program provided significant fellowship support for undergraduate and graduate students at the member institutions. Other consortium activities included a combination of precollege, undergraduate, graduate, and faculty efforts directed toward outreach, education, and research in the aerospace field. While more than 25 different programs were sponsored by the consortium between 1991 and 1995, initial emphasis was on outreach activities that employ the mystique of aerospace to encourage academic achievement of younger students. A particular goal was to encourage women, minorities, and those with physical disabilities to pursue education in the aerospace field.

A Taste of Aerospace is another outreach effort developed in 1991 by Professor Dominick Andrisani from the School of Aeronautics and Astronautics to promote interest in mathematics and science among Indiana youth. The program consisted of simple yet insightful experiments that demonstrate key aerospace principles in a manner intended to excite students to learn more about mathematics and science. By the spring of 1995, A Taste of Aerospace had been offered nearly 200 times to more than 8,000 secondary and primary students. Well over 90% of participants were students from underrepresented minority groups, primarily in the Gary and East Chicago area.

A vivid example of the high regard the public has for Purdue University's contributions to space exploration occurred in the aftermath of the January 28, 1986, space shuttle Challenger accident. The campus community was familiar with media attention that accompanied the nation's space triumphs (e.g., the second U.S. manned space flight by Gus Grissom (BSME '50) in 1961, Neil Armstrong's (BSAE '55) first lunar landing in 1969, and Gene Cernan's (BSEE '56) last lunar flight in 1972). It had also shared in the country's sorrow when tragedy occurred, for example Iven Kincheloe's (BSAE '49) death in a 1958 training accident and the Apollo spacecraft fire that killed Mr. Grissom, Roger Chaffee (BSAE '57), and Ed White on January 26, 1967). Few were prepared, however, for the media attention following the Challenger explosion and the loss of its crew. Professor Grandt recalls the events of that day:

> "I was away from the office when the accident occurred (11:38 a.m. EST), and did not return until about 1:30 that afternoon. Although I had been school head for about six months, and had dealt with the press during that time, I was astounded by the number of telephone messages taken by my secretary during the short period I was away from the office. We had dozens of requests from print and broadcast journalists throughout the country seeking the 'Purdue' reaction to the tragedy, and I spent much of the next several days answering media questions and arranging interviews. All of the local and Indianapolis TV stations sent crews to campus to film our faculty and students. Professor J. Robert Osborn, a highly renowned expert on rocket propulsion, granted numerous interviews and appeared on network television to discuss the accident. Several other faculty met with the media to share their thoughts about the accident, and when the nation resumed space flight over two years later with space shuttle Discovery, several television stations returned to film student and faculty reaction to

that successful space flight. (Richard Covey, MSAE '69, was the pilot for that September 29, 1988, launch.) Clearly, the school's faculty and students were at center stage during the weeks following Challenger, and I was most proud of the professional manner in which they patiently responded to repeated media questions about the incident."

Student reaction to this setback took several other forms. Within hours after the explosion, the following neatly handwritten message was anonymously taped to a plaque that dedicates Grissom Hall to the memory of astronaut Gus Grissom, who was killed 21 years earlier with Roger Chaffee and Ed White: *"Michael Smith, Francis 'Dick' Scobee, Sharon Christa McAuliffe, Ronald McNair, Judith Resnik, Ellison Onizuka, Gregory Jarvis. Seven Courageous Americans who gave their all that their country [might] make great strides toward conquering the final frontier. Go west young men and women, we'll miss you. We pray for your souls and want the world to know that your lives and efforts have not been in vain. 28 January 1986."* It was later learned that the author of these touching remarks was Naeem Sayeedi (BSAE '86).

Steven Hertzberg (BSAE '86) was another student who was so moved by the loss of the Challenger crew that he and John Whittenberg, a senior in mathematics, organized the **Purdue Space Shuttle Memorial Fund** to provide scholarships for students interested in the space program. Working with the Purdue Development Office, they helped solicit contributions to establish a scholarship fund that continues to award annual scholarships in memory of the Challenger astronauts. (Several students from the School of Aeronautics and Astronautics have, in fact, received this scholarship.)

FACILITIES

Research facilities at the start of the 1980 decade are described in the following passage reproduced from the 1982–84 graduate school bulletin:

Digital computer facilities are readily available for student use in conjunction with both class and research work. Easy access to the facilities of the University Computing Center is available through several remote terminals maintained for use by students and faculty of this school. As a part of the Engineering Distributed Computer Network, a PDP 11/780 VAX computer is located in Grissom Hall for easy access and use by students and faculty of the school. Additionally, two General Automation minicomputers (one SPC 16/45 and one SPC 16/65 operating in parallel) with 64K memory, floating point processor, dual disk system, and electro-static printer/plotter is available for data acquisition, real time control of experiments, flight simulation, and general purpose computing. (See Figure 7.13).

Laboratories within the school that are available for research activities are listed below.

1.   **The Materials Research Laboratory** *is well equipped to determine conventional mechanical properties of structural materials. Electro-hydraulic test machines and associated equipment are available to measure fracture loads and to study fatigue crack initiation and propagation in metal and polymer test specimens.*

2.   **The Aerospace Sciences Laboratory** *includes a large dual purpose subsonic wind tunnel, which has an open-throat building aerodynamics test section in addition to a conventional test section, a small supersonic wind tunnel, a small shocktube, a low turbulence wind tunnel, and extensive machine shop facilities for model fabrication. The laboratory also includes a computer, a Laser Doppler Velocimeter, and other types of instrumentation.*

3.   **An Experimental Stress Analysis Laboratory** *is equipped for modern methods of stress analysis including (a) two- and three-dimensional photoelasticity, (b) birefringent coatings, (c) brittle lacquer and moiré techniques, and (d) optical and electric resistance strain gages for both static and dynamic strain measurements.*

4.   **A Laser Laboratory** *is equipped with a 200-megawatt pulsed ruby laser, a 250-megawatt pulsed neodymium glass laser, a 30-milliwatt continuous wave helium-neon laser, and two 1-milliwatt helium-neon lasers. This facility enables research and instruction to be conducted in (a) continuous wave holography, (b) pulsed holography and interferometry, (c) laser technology, (d) laser-generated stress waves, (e) ultra-high speed schlieren and polariscope photography, and (f) explosive fracture techniques.*

5.   **The Guidance and Control Laboratory** *contains three TR 20 analog computers with some nonlinear components, X-Y plotters, and oscilloscopes. There are also two servosystems and several gyro instruments. A fixed-base, all digital, flight simulator with force-sensitive control devices and programmable CRT display is driven by the two General Automation minicomputers mentioned previously.*

6.   **The Composite Materials Laboratory** *contains equipment and facilities for general material testing and for curing composite laminates. An autoclave specially designed for curing epoxy-matrix composites is available for laminate fabrication. Two complete MS material and fatigue testing machines (55 kip and 11 kip capacity) and associated equipment are used to perform ultimate strength, stiffness, and fatigue tests on various composite materials. Additional facilities for preparing laminated composites are available as well as impact testing equipment, and an acoustic emission detection device.* (Figure 7.18).

7.   **The Rocket Laboratory** *contains unique equipment, instrumentation, and facilities for basic research on rocket propulsion. Major facilities include a shock tube with an ultrasonic levitator for droplet deforma-*

tion and combustion studies, a servo-positioned strand burner for solid propellant combustion studies, and a forced longitudinal wave motor for combustion instability studies. A large, high pressure air supply is available for air-augmented rocket and ramjet combustion studies. A PDP-11 minicomputer is available for data acquisition/reduction.

*Figure 7.18
Professor C. T. Sun
with materials testing
machine in Composite
Materials Laboratory
(circa 1987).*

The new faculty who joined the department during the 1980s and 90s, along with the growth of sponsored research during this period, led to upgrades of these facilities, and development of new research laboratories. Descriptions of the 1995 laboratories are given in Chapter 8, which deals with the status of the School of Aeronautics and Astronautics in 1995, and will not be repeated here. Photographs of students working in various laboratories are given in Figures 7.19 –7.25.

In 1980, the School of Aeronautics and Astronautics occupied space in Grissom Hall, the Nuclear Engineering Building, and the Aerospace Sciences Laboratory. It also had access to research laboratories in the propulsion area at the Thermosciences and Propulsion Center. Although the school first occupied Grissom Hall in 1968, civil engineering retained some space until completion of an addition to its building in 1988. That area was remodeled and made available to the Schools of Industrial Engineering and Aeronautics and Astronautics at that time. Following this renovation, the School of Aeronautics and Astronautics occupied the entire third floor, portions of the first floor, and part of the basement of Grissom Hall. A new third floor conference room was developed to house the **Edmond F. Ball Aerospace Collection,** a library of historical aerospace volumes donated to the school by Mr. Ball in the late 1980s. A new design resource room

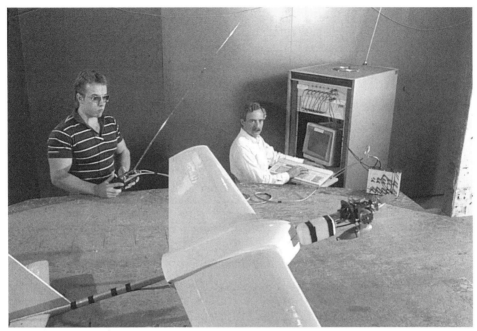

Figure 7.19 Graduate student John J. Rushinsky and Professor Andrisani fly constrained radio controlled airplane in subsonic wind tunnel (circa 1987).

Figure 7.20 Students performing AAE 334 wind tunnel experiment (circa 1987).

Figure 7.21 Graduate student Matt Ledington prepares to make dynamic
           photoelasticity measurements with ultra-high-speed camera (circa
           1988).

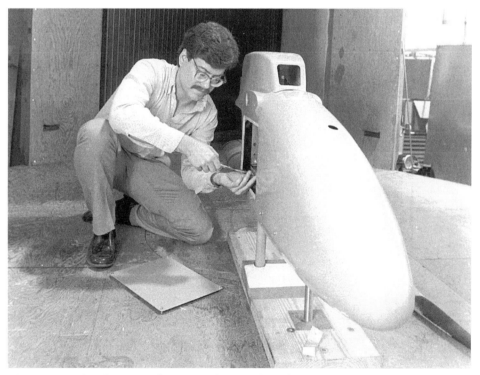

Figure 7.22 Graduate student David P. Witkowski prepares drone for
           measurements in subsonic wind tunnel (circa 1988).

Figure 7.23 Graduate student Robert Frederick performs high-pressure propulsion measurement (circa 1987).

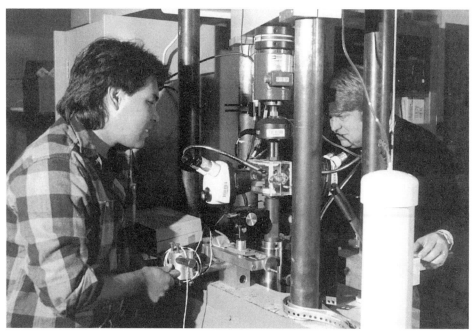

Figure 7.24 Graduate student Jason Scheuring and Professor A. F. Grandt examine fatigue crack growth specimen (1994).

Figure 7.25 Graduate student Robert Johnston conducts a flow visualization experiment of an aircraft model in the subsonic wind tunnel (circa 1990).

was completed on the first floor along with new sites for the dynamics and control laboratory and the guidance and control research laboratory. The fatigue and fracture laboratory moved to the basement of Grissom Hall from its original location in the Aerospace Sciences Laboratory. The schools of Aeronautics and Astronautics and Industrial Engineering also developed a combined computer terminal room on the first floor of Grissom Hall at this time.

The Aerospace Sciences Laboratory also underwent significant modernization during this period. The Aerospace Sciences Laboratory, which was built in 1948 to house the new School of Aeronautical Engineering, had been shared by the schools of Nuclear Engineering and Aeronautics and Astronautics since the 1970s. In early 1980, walls were built to formally separate the areas occupied by those two schools. Also during the 1980s, a new roof was put on the building and new lighting and ventilation systems were installed, along with new windows. The large subsonic wind tunnel that was built in the late 1940s by Professor Palmer was upgraded by means of a large grant from The Boeing Company, and was renamed The Boeing Wind Tunnel (Figure 7.26). Another major addition at the Aerospace Sciences Laboratory included development of a hypersonic test facility through construction of a Ludweig tube designed by Professor Schneider. (Figure 7.27). Again, more detailed descriptions of these facilities are found in Chapter 8.

Figure 7.26 R. Bateman (BSAE '46), J. Hayhurst (BSAE '69), and R. Taylor (BSME '42) represent The Boeing Company at September 19, 1991 dedication of the Boeing Wind Tunnel.

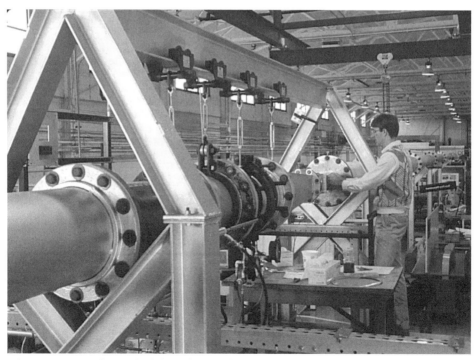

Figure 7.27 Professor S. Schneider works with Ludweig tube hypersonic wind tunnel (circa 1992).

CHAPTER 8

In Orbit:

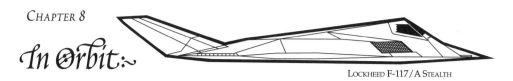

LOCKHEED F-117/A STEALTH

GOLDEN ANNIVERSARY BENCHMARK 1995

The objective of this chapter is to describe the status of the School of Aeronautics and Astronautics during the 1994–95 academic year. It is hoped that this golden anniversary benchmark will demonstrate the progress made during the school's first 50 years and provide a comparison for subsequent developments. Much of the information given here is edited from various annual reports available in spring 1995. Although the given summaries encompass slightly different periods (e.g. calendar year, academic year, etc.), the goal is to provide a snapshot of activities at the school's 50-year point.

STAFF

Faculty members for the 1994–95 academic year are pictured on the next two pages (Figure 8.1). They are grouped by academic specialty, along with a brief summary of their technical interests. In addition to the 22 faculty listed, the school had one faculty opening for the year. Dr. Samir M. Kherat served as visiting assistant professor and taught the introductory sophomore design course (AAE 251). Clerical and technical/professional staff are listed in Table 8.1; many are shown in a group photo taken in September of 1992 (Figure 7.8).

Table 8.1   1984–95 Clerical and Professional Staff

SECRETARIES
M. Engel, M. Fuqua, M. Hathaway, D. Schafer, S. Wise.

PROFESSIONAL/TECHNICAL
W. Bader, D. Bower, M. Chadwell, I. Ellis, L. Flack, T. Moore, D. Halsema, R. Scott, D. Reagan, J. Zachary.

BUSINESS OFFICE
D. Hall, S. Morris.

**AERODY-
NAMICS**

G. A. Blaisdell
Assistant Professor
PhD, Stanford, 1991

Computational fluid
mechanics, transition
and turbulence.

S. H. Collicott
Assistant Professor
PhD, Stanford, 1991

Experimental fluid
mechanics, optical
diagnostics, applied
optics.

W. A. Gustafson
Professor
PhD, Illinois, 1956

Hypersonic aerody-
namics, spacecraft
design.

A. S. Lyrintzis
Associate Professor
PhD, Cornell, 1988

Computational aero-
acoustics, aerody-
namics, traffic flow
modeling.

S. P. Schneider
Assistant Professor
PhD, Caltech, 1989

Experimental fluid
mechanics, laminar-
turbulent transition,
turbulence control.

J. P. Sullivan
Professor and Head
Sc.D., MIT, 1973

Experimental aero-
dynamics, propellers,
laser-Doppler velo-
cimetry.

M. H. Williams
Professor
PhD, Princeton, 1975

Aerodynamics,
computational fluid
mechanics.

**DYNAMICS
AND CONTROL**

D. Andrisani II
Associate Professor
PhD, SUNY–Buffalo,
1979

Estimation, control,
dynamics.

M. J. Corless
Associate Professor
PhD, Berkeley, 1984

Dynamics, systems,
and control.

A. E. Frazho
Associate Professor
PhD, Michigan, 1977

Control systems.

K. C. Howell
Associate Professor
PhD., Stanford, 1983
Orbit mechanics,
spacecraft
dynamics, control
trajectory optimiza-
tion.

J. M. Longuski
Associate Professor
PhD, Michigan, 1979

Spacecraft dynamics,
orbit mechanics, control,
orbit decay and reentry.

M. A. Rotea
Assistant Professor
PhD, Minnesota, 1990

Multivariable control,
optimal control, model-
ing and identification.

R. E. Skelton
Professor
PhD, UCLA, 1976

Dynamics of aero-
space vehicles, control
theory.

PROPULSION

S. D. Heister
Associate Professor
PhD, UCLA, 1988

Rocket propulsion, liquid propellant injection systems.

N. L. Messersmith
Assistant Professor
PhD Illinois, 1992

Compressible mixing layers, air breathing propulsion, combustion, laser-based flow diagnostics.

STRUCTURES AND MATERIALS

J. F. Doyle
Professor
PhD, Illinois, 1977

Structural dynamics, experimental mechanics, photomechanics, wave propagation.

H. D. Espinosa
Assistant Professor
PhD, Brown, 1991

Constitutive modeling, wave propagation.

T. N. Farris
Professor,
PhD, Northwestern, 1986

Tribology, manufacturing processes, fatigue and fracture.

A. F. Grandt, Jr.
Professor
PhD, Illinois, 1971

Damage-tolerant structures and materials, fatigue and fracture, aging aircraft.

C. T. Sun
Professor
PhD, Northwestern, 1967

Composites, fracture and fatigue, computational mechanics.

T. A. Weisshaar
Professor
PhD, Stanford, 1971

Aircraft structural mechanics, aeroelasticity, integrated design.

Figure 8.1   1994–95 AAE faculty and research interests.

STUDENTS

Total enrollment for Purdue's West Lafayette campus during the fall 1994 semester was 34,484, including 27,750 undergraduates, 318 professional students, and 6,416 undergraduates. Other statistical information reported by the registrar is given in Table 8.2. Undergraduate enrollment in the School of Aeronautics and Astronautics in the fall of 1994 was 253 (excluding freshmen). This is the smallest undergraduate enrollment since 1973 (see Figure 7.1) and is indicative of the decline of the aerospace industry in 1995. Graduate enrollment was, however, quite strong, with 75 students enrolled in the master's degree program and 75 pursuing PhDs at the start of the 1994–95 academic year. There were 45 students enrolled in the undergraduate cooperative engineering program. The companies sponsoring co-op students during this academic year are listed in Table 8.3. Other student demographics are given in Table 8.4.

Table 8.2   *Fall 1994 Enrollment Statistics for Purdue University — W. Lafayette Campus (from Office of the Registrar)*

Undergraduate	27,750
Professional	318
Graduate	6,416
Total Enrollment	**34,484**

Marital Status	*Males*		*Females*	
Single	18,317	(53%)	12,995	(38%)
Married	1,673	(5%)	1,499	(4%)
Total	**19,990**	**(58%)**	**14,494**	**(42%)**

Ethnic Group Enrollment			*International Student Enrollment*		
African American	1,202	(4%)	Undergraduate	666	(2%)
American Indian	147	(0%)	Professional	6	(0%)
Asian American	1,409	(4%)	Graduate	1,809	(5%)
Hispanic American	727	(2%)			
Total	**3,485**	**(10%)**	**Total**	**2,481**	**(7%)**

AGE DISTRIBUTION

	Total	*17–19*	*20–21*	*22–23*	*24–25*	*26–30*	*31–35*	*36+*
Undergraduate	**27,750**	10,490	9,844	4,356	1,041	925	445	649
Professional	**318**	0	21	119	84	51	31	12
Graduate	**6,416**	2	88	1,009	1,179	2,198	985	955
Total	**34,484**	**10,492**	**9,953**	**5,484**	**2,304**	**3,174**	**1,461**	**1,616**

Table 8.2   Fall 1994 Enrollment Statistics for Purdue University (continued)

RESIDENCE

	Undergraduate		Graduate		Professional		Total	
Indiana	20,314	(73%)	1,856	(29%)	229	(72%)	22,399	(65%)
Other States	6,315	(23%)	2,474	(39%)	77	(24%)	8,866	(26%)
U.S. Territories	455	(2%)	277	(4%)	6	(2%)	738	(2%)
International	666	(2%)	1,809	(28%)	6	(2%)	2,481	(7%)
Total	**27,750**		**6,416**		**318**		**34,484**	

HOUSING TYPE

On Campus		*Off Campus*	
Undergrad Residence Halls	9,603	Greater Lafayette	14,205
Graduate Houses	1,338	Non-Tippecanoe	2,554
Cooperatives	209	Not Reported	3,142
Fraternities	1,202		
Sororities	902		
Married Student Housing	1,211		
Total	**14,465**	**Total**	**19,901**

Note: Count by housing type reported as of 10/20/94. Total does not equal official enrollment.

Table 8.3   Co-Op Companies Working with the School of Aeronautics and Astronautics (June 1995)

Company	*Location*
Aerospace Corporation	Los Angeles, California
Allison Gas Turbine	Indianapolis, Indiana
Ball Corporation Aerospace Systems	Boulder, Colorado
Hughes Aircraft Space & Comm.	Los Angeles, California
McDonnell Douglas	St. Louis, Missouri
NASA-Ames-Dryden	Edwards, California
NASA-Goddard Space Flight Center	Greenbelt, Maryland
NASA-Johnson Space Center	Houston, Texas
NASA-Langley Research Center	Hampton, Virginia
Rockwell Defense Electronics	Cedar Rapids, Iowa
Structural Dynamics Research Center	Milford, Ohio
Thiokol Corporation	Ogden, Utah
Wright-Patterson Air Force Base	Dayton, Ohio

Table 8.4   *1994–95 Student Demographics for the School of Aeronautics and Astronautics*

Category	Male	Female	International	U.S.	Total
Undergraduate	**206**	**47**	**12**	**241**	**253**
MS	69	6	14	61	75
PhD	71	4	40	35	75
Total Graduate	**140**	**10**	**54**	**96**	**150**

While the downsizing of the aerospace industry resulted in a low undergraduate enrollment during the 1994–95 academic year, student morale remained high, as indicated by the list of "Top Ten Reasons to Be an Aerospace Engineer" reported in Table 8.5. This tongue-in-cheek list, modeled after TV comedian David Letterman's Top Ten lists, was featured on T-shirts sold by the AIAA student section.

Table 8.5   *Top Ten Reasons to Be an Aerospace Engineer (As reported on 1995 AIAA student T-shirt)*

1. McDonald's sounds like McDonnell Douglas.
2. I needed a challenge.
3. A rocket scientist, yes I am!
4. Maximum endurance is only an equation away.
5. I know what skin friction leads to.
6. Aeros get it up and keep it up.
7. I love studying the flow around smooth bodies.
8. Aeros do it with more lift and thrust.
9. Still not sure about my major.
10. Liberal arts was just too damn simple.

FACILITIES

The school's facilities continued to be split between campus and airport locations during its golden anniversary, as it occupied space in Grissom Hall, the Nuclear Engineering Building, and the Aerospace Sciences Laboratory. The school shared Grissom Hall with the School of Industrial Engineering, with its main faculty and staff offices located on the third floor, and laboratories located on the first floor and in the basement. The Nuclear Engineering Building and Aerospace Sciences Building were shared with the School of Nuclear Engineering. Teaching and research laboratories and limited shop facilities were located in the Nuclear Engineering Building. Other research facilities (primarily in the aerodynamics and propulsion areas) were located at the Aerospace Sciences Building (the original Hanger Building, Unit 3, which was first occupied during the 1948–49 academic year) and at the Thermal Science Propulsion Center. The school also shared extensive machine shop facilities at the Aerospace Sciences Laboratory with the School of Nuclear Engineering.

Curriculum

The undergraduate curriculum employed during the golden anniversary year is given in Table 8.6. With a few minor changes, this is the same curriculum that had been in effect for several years. Graduate requirements are summarized in Table 8.7. Courses, instructors, and class enrollments for the 1994–95 academic year are given in Tables 8.8 and 8.9. Table 8.8 indicates that the average undergraduate class sizes for the two 1994–95 semesters were 27.2 and 27.6 students. The fall and spring semester 500-level classes averaged 20.4 and 20.1 students per class, while the 600-level averages were 8.9 and 5.3. For comparison, five years before this time (1989-90 academic year), the fall and spring semester undergraduate class sizes were 52.0 and 58.4, the 500-level sizes were 29.1 and 24.9, and the 600-level course sizes were 9.0 and 6.7. Again, these numbers reflect the decrease in undergraduate enrollment during the 1990s, and are indicative of the fluctuations in class sizes that have occurred during the school's history.

Industrial Affiliates Program

For several years, the School of Aeronautics and Astronautics had conducted an Industrial Affiliates Program in partnership with industry. In return for an annual membership fee, each industrial affiliate participated in an exchange of research information, and receives certain services to assist in development of their manpower needs. Funds received by the school from this program were used to maintain its teaching and research facilities. Industrial affiliates for the 1994–95 academic year are listed below:

- The Boeing Company
- Hughes Aircraft Company
- Lockheed Missiles & Space Company
- McDonnell Douglas Corporation
- Northrop Corporation
- Thiokol Corporation
- TRW

Table 8.6   Undergraduate Curriculum, 1994–95 Academic Year

Sophomore Year

Third Semester

(4) MA 261 (Multivariate Calculus)
(3) PHYS 241 (Electricity & Optics)
(3) EE 201 (Linear Circuit Analysis I)
(3) A&AE 203 (Aeromechanics I)
(3) General Education Elective

(16)

Fourth Semester

(4) MA 262 (Linear Algebra & Differential Equations)
(3) A&AE 204 (Aeromechanics II)
(1) A&AE 204L (Aeromechanics II Lab.)
(3) A&AE 251 (Intro. Aerospace Design)
(3) PHYS 342 (Modern Physics)
(3) General Education Elective

(17)

Junior Year

Fifth Semester

(3) MA 303 or MA 304 (Differential Equations, Including Systems)
(3) ME 200 (Thermodynamics I)
(3) A&AE 352 (Structural Analysis I)
(3) A&AE 333 (Fluid Mechanics)
(1) A&AE 333L (Fluid Mechanics Lab)
(3) General Education Elective

(16)

Sixth Semester

(3) A&AE 372 (Jet Propulsion Power Plants)
(3) A&AE 334 (Aerodynamics)
(1) A&AE 334L (Aerodynamics Lab) or A&AE 352L (Structural Analysis Lab)
(3) A&AE 464 (Control Systems Analysis)
(3) A&AE 340 (Dynamics & Vibration)
(3) General Education Elective

(16)

Senior Year

Seventh Semester

(3) A&AE 451 (Design I)
(3) A&AE 421 (Flight Dyn. & Control)
(1) A&AE 421L (Flight Dyn. & Control Lab.)
(6) Major or Minor Area Electives
(3) General Education Elective

(16)

Eighth Semester

(9) Major or Minor Area Elective
(6) Electives
(3) General Education Electives

(18)

Total   131 Credits
18 credits of General Education Electives
9 credits of Major Area Electives
6 credits of Minor Area Electives
6 credits of Electives

Table 8.7   Graduate Degree Requirements

MS

	Nonthesis Option (credits)	Thesis Option (credits)
Major area	12 (minimum)	9
Minor area	6 (minimum)	6
Mathematics	6	6
Other	6 (electives)	9 (thesis)
Total	**30**	**30**

PhD

- Minimum of 18 credits beyond the MS degree
- Qualifying examinations in major area, minor area, and mathematics (6 credits in minor area may be substituted for minor area qualifying examination)
- Thesis (with preliminary examination) required of all students

NOTES:

All courses must be at 500 level or above.

Major/minor areas to be selected from the following 5 areas:

- Aerodynamics
- Dynamics and control
- Propulsion
- Structures and materials
- Interdisciplinary areas (design, biomechanics, manufacturing, etc.)

Table 8.8   Summarized Class Enrollment Statistics (1994–95 Academic Year)

Semester	Statistic	Three-Credit Courses				One-Credit Laboratory Courses
		200 300 400 Level	500 Level	600 Level	All Level	
Fall 1994	No. of classes offered	18	14	7	39	5
	Total enrollment	489	286	62	837	146
	Average number of students per class	27.2	20.4	8.9	21.5	29.2
Spring 1995	No. of classes offered	18	12	3	33	5
	Total enrollment	497	241	16	754	146
	Average number of students per class	27.6	20	5.3	22.8	29.2

Table 8.9   *Course Enrollments for the School of Aeronautics and Astronautics (1994–95 Academic Year)*

A&AE Course	Most Recent Title	Credit	Fall 1994 Enrollment	Instructor	Spring 1995 Enrollment	Instructor
203 (1)	Aeromechanics I	3	27	Corless	22	Longuski
203 (2)	Aeromechanics I	3	29	Corless		
204	Aeromechanics II	3	24	Farris	41	Grandt
204L	Aeromechanics II Lab.	1	19	Doyle	40	Doyle
251	Intro. Aerospace Design	3	45	Kherat	25	Kherat
333	Fluid Mechanics	3	48	Blaisdell	23	Blaisdell
333L	Fluid Mechanics Lab.	1	56	Schneider	23	Schneider
334	Aerodynamics	3	17	Collicott	46	Williams
334L	Aerodynamics Lab.	1	14	Schneider	31	Schneider
340	Dynamics & Vibration	3	30	Longuski	43	Howell
352 (1)	Structural Analysis I	3	23	Espinosa	21	Espinosa
352 (2)	Structural Analysis I	3	26	Sun		
352L	Structural Analysis I Lab.	1	11	Doyle	18	Doyle
361	Intro. Random Var. Engr.	3			30	Frazho
372	Jet Propulsion Power Plants	3	25	Messersmith	42	Messersmith
412	Intro. Comp. Fluid Dyn.	3	26	Williams		
414	Compressible Aerodyn.	3			26	Collicott
416	Viscous Flow Theory	3			14	Schneider
421	Stability & Control	3	54	Andrisani	30	Andrisani
421L	Flight Dyn. Control Lab.	1	46	Andrisani	34	Frazho
439*	Rocket Propulsion	3	16	Heister	14	Heister
451 (1)	Design I (Aircraft)	3	14	Weisshaar	30	Weisshaar
451 (2)	Design I (Spacecraft)	3	15	Gustafson	15	Gustafson
453	Matr. Meth. Aerosp. Struc.	3			18	Doyle
464	Control Sys. Anal.	3	28	Frazho	46	Rotea
490A	Aerodynamic Design	3	25	Sullivan		
490D	Structural Design	3	17	Grandt		
490R	Control Design	3			11	Rotea
507	Basic Mechanics III	3	21	Longuski		
511	Intro. Fluid Mech.	3	23	Blaisdell		
512	Comp. Aerodynamics	3			17	Williams
517	Unsteady Aerodynamics	3	11	Williams		
519	Satellite Aero. & Entry	3	16	Gustafson		
520	Experimental Aerodynamics	3			20	Sullivan
532	Orbit Mechanics	3	21	Howell		
538*	Air Breathing Propulsion	3	25	L'Ecuyer (ME)		
539	Solid Rocket Propulsion	3			28	Heister
540	Spacecraft Att. Dynamics	3			25	Howell

Table 8.9   Course Enrollments (continued)

A&AE Course	Most Recent Title	Credit	Fall 1994 Enrollment	Instructor	Spring 1995 Enrollment	Instructor
547	Exp. Stress Analysis	3			13	Doyle
553	Elasticity Aerospace Engr.	3	34	Espinosa		
554	Fatigue Struc. & Materials	3	48	Grandt		
555	Mechanics Comp. Materials	3			36	Sun
556	Aeroelasticity	3			23	Weisshaar
558	Finite Element Methods in Aerospace Structures	3	20	Farris		
559	Mechanics Friction & Wear	3			14	Farris
564	System Analysis & Synth.	3	9	Skelton	26	Corless
565	Guidance Aerosp. Vehicles	3			10	Andrisani
567	Intro. Appl. Stoch. Proc.	3	13	Frazho		
590D	Structural Dyn. & Stab.		19	Doyle		
590M	Fund. Aero. Combustion	3	15	Messersmith		
590C	Low Gravity Fluid Dyn.	3	11	Collicott		
590B	Aerodynamic Sound				12	Lyrintzis
590P	Hypersonic Propulsion	3			17	Messersmith
613	Viscous Flow Theory	3	4	Schneider		
626	Math Theory Turbulence	3			6	Blaisdell
655	Adv. Topics Composite	3	17	Sun		
664	Est. Control Uncertain Syst.	3			5	Skelton
690C	Nonlinear Systems	3	14	Corless		
690G	Astro. Nav. & Guidance	3	7	Howell		
690P	Stab. of Free Surfaces	3	8	Heister		
690R	Multi Feedback Control	3	6	Rotea		
690S	Dyn. Control Flex. Struct.	3	6	Skelton		
690B	Variational Principles Mech.	3			5	Longuski
698	MS Thesis Research		31	Staff	32	Staff
699	PhD Thesis Research		70	Staff	70	Staff

* Co-listed with the School of Mechanical Engineering

RESEARCH

Sponsored research projects active during the July 1, 1993, to June 30, 1994, period are summarized in Table 8.10. The school conducted $2,341,316 in sponsored research during that period for the sponsors indicated in Table 8.11. In addition, many other non-sponsored research projects were conducted by the faculty and students during this period. This research led to 62 journal articles or book chapters and 94 conference papers published during calendar year 1993. Seventeen MS and 10 PhD theses were also published during calendar year 1994 (see Appendix B-C for a list of thesis authors and titles).

The **Space Systems Control Laboratory** was an interdisciplinary research effort located within the Institute of Interdisciplinary Engineering Studies. The Space Systems Control Laboratory was established in 1991, with Professor R. E. Skelton as its director. It involved faculty from the School of Aeronautics and Astronautics, the School of Electrical Engineering, the School of Mechanical Engineering, the Department of Computer Science, and the Department of Mathematics. The lab pursued application of control systems science and engineering to space systems. Among its purposes were development of new control concepts, analytical and diagnostic techniques, and computer designs tailored to space system needs. The Space Systems Control Laboratory is a national resource for rapid analysis and design of spacecraft control systems. This is an interdisciplinary activity requiring integration of the fields of identification, dynamic modeling, signal processing, and control design.

Another interdisciplinary research effort was formed in 1993 to address basic research issues associated with aging aircraft. Sponsored by a grant from the Air Force Office of Scientific Research (for four years beginning July 1, 1993), this project seeks to determine how damage forms and grows in older structures. Faculty involved in that effort included A. F. Grandt, Jr. (principal investigator), T. N. Farris, and C. T. Sun of the School of Aeronautics and Astronautics; B. M. Hillberry of the School of Mechanical Engineering; E. P. Kvam of the School of Materials Engineering; and G. P. McCabe of the Department of Statistics.

A general description of the research conducted by the various other academic areas of the school is given below, along with a brief description of laboratory facilities employed for that research.

AERODYNAMICS

Aerodynamics research is directed toward a better understanding of the fundamental laws governing the flow of fluids. Research topics of interest included numerical methods in aerodynamics; computational fluid mechanics; computational aeroacoustics; separated flow around wings and bodies at high angles of attack; diffusion in turbulent flow; aerodynamics of rotors, propellers, and wind energy machines; boundary layers, wakes, and jets in V/STOL applications; aerodynamic noise; aerodynamics of buildings and road vehicles; hypersonic and chemically reacting flows; experimental measurements using the laser systems; and laminar-turbulent transition in high-speed boundary layers.

Experimental facilities included four wind tunnels located at the Aerospace Sciences Laboratory. The first was a large subsonic wind tunnel (the Boeing Wind Tunnel) with a modular test section design. The 1995 configuration included a 4x6-foot closed section with a maximum speed of 250 miles per hour, and a long rectangular boundary-layer test section. The test sections were equipped with a six-component motorized pitch and yaw balance system. Instrumentation included a two-component laser Doppler velocimeter system and a computer data acquisition system. The second wind tunnel was a single-pass, low-turbulence type with a modular plastic tube test section 18 inches in diameter and up to 20 feet long, and a maximum speed of 100 miles per hour. The third tunnel was a small, supersonic, blow-down type, which can operate from Mach 1.5 to Mach 4. This tunnel's air supply also was used for an adjacent gas dynamical flow apparatus designed for nozzle flow studies. The fourth tunnel was a supersonic quiet flow Ludweig tube with a Mach 4 test section. Both of these supersonic tunnels could be operated in pressure-vacuum mode. A high-Reynolds number flow-down water jet apparatus had been recently completed. Various small smoke and calibration tunnels, as well as water tables and tow tanks, also were available.

Dynamics and Control

All modern aerospace vehicles rely upon control systems to improve system performance. Successful control system design requires an understanding of the interactions of dynamic elements, and the trade-offs between vehicle dynamic characteristics, control system properties, and system performance. This is accomplished by a study of dynamics and control.

In 1995, research was divided into the following areas: control and estimation theory, aircraft design for improved handling qualities, modeling and control of aeroelastic aircraft, astrodynamics, and dynamics and control of flexible spacecraft.

Certain research projects in the dynamics and control area required advanced and specialized laboratory facilities. A helicopter experiment was also available for teaching controls. The helicopter consisted of the fuselage, the main motor (for pitch control), and the tail motor (for yaw control). The objective was to model, identify, and control the pitch and yaw axis of the helicopter. The helicopter was connected to a Mac IIsi for the purpose of driving the main and tail motors and measuring the pitch and yaw motion. The Flight Dynamics and Control Laboratory contained several analog computers with some nonlinear components, X-Y plotters, oscilloscopes, and signal-generation devices for studying dynamic systems and evaluating feedback control laws. Control position servos and transducers also were studied. The Flight Simulation Laboratory had fixed-base, digital flight simulation capabilities with force-sensitive manipulators, and CRT display for man-in-the-loop experiments. The simulations were controlled by a 16-bit microprocessor with a 40 MB hard disk for storing experimental data. Here, man/vehicle performance, workload, and vehicle dynamics were studied. An instrumented remotely piloted vehicle, under development, represented a unique research facility upon which to perform many experiments in vehicle dynamics and control. Data communication with a computer-based ground station was provided by a seven-channel telemetry uplink and an eight-channel telemetry downlink.

Propulsion

There are two broad categories in propulsion research: (1) devices that need air to produce thrust, and (2) devices that do not need air to produce thrust. The first category included piston engines, turbojet engines, turbofan/propeller engines, and scramjet and ramjet engines. The second category included various rocket engines. For several years, the emphasis in the School of Aeronautics and Astronautics had been on both of these types of propulsion systems.

The Rocket Laboratory contained unique facilities for basic research on rocket propulsion. Major equipment included a hybrid rocket engine test stand, which was constructed to investigate combustion and performance of these engines. A large, high-pressure air supply was available for air-augmented rocket and ramjet combustion studies. A minicomputer was available for data acquisition/reduction. These facilities were housed in the Thermal Sciences and Propulsion Center.

In addition, propulsion research facilities also were housed in the Aerospace Sciences Laboratory. Facilities were developed to investigate combustion burn-through in turbojet engines using laser sheet flow visualization, thermal paint fluorescence, and pressure paint fluorescence of a jet impinging at various angles to a plate.

Structures and Materials

Structures and materials research includes work in composite materials, computational structural mechanics, damage tolerance analysis, experimental structural analysis, structural mechanics and aeroelasticity, tribology, manufacturing, and wave propagation.

The **McDonnell Douglas Composite Materials Laboratory** contained equipment and facilities for general material testing and for fabrication of composite laminates. An autoclave specially designed for curing epoxy-matrix composites was available for laminate fabrication. A water jet cutting machine was used for specimen preparation. Four complete MS material and fatigue testing machines (55 kip, 22 kip, 11 kip, and 1 kip capacity) and associated equipment were used to perform ultimate strength, stiffness, and fatigue tests on various composite materials. Additional facilities for preparing laminated composites, as well as impact testing equipment and an acoustic emission detection device, were available.

The Dynamic Inelasticity Laboratory contained a 3-inch gas gun and optical instrumentation for wave propagation studies. The facility was designed for the investigation of damage and failure mechanisms in advanced materials. Soft recovery of the impacted targets, which was accomplished by specially designed target fixtures and energy dissipation mechanisms, allowed the identification of stress-induced microdefects and phase changes by means of microscopy studies. Optical instrumentation together with a 1-watt single-mode argon ion laser, and a four-channel oscilloscope with maximum sample rate up to 2 Gs/sec and 1 GHz bandwidth, were used in the measurement of normal and transverse interface and free surface velocities. In addition to the laser interferometric technique, shock stresses in the specimens were measured using piezoresistance gauges.

The Fatigue and Fracture Laboratory was well-equipped to determine mechanical properties of structural materials. Two computer-controlled electro-hydraulic test machines (5 kip and 22 kip capacity) and associated equipment were available to measure fracture loads and to study fatigue crack formation and propagation in test specimens subjected to simulated aircraft or spacecraft load histories. Facilities also were available to artificially corrode specimens in connection with corrosion related research.

The **Photomechanics Laboratory** explored optical methods of stress analysis. Equipment included numerous polariscopes, a 15-mw laser, an isolation table, and a photographic darkroom, allowing research in photoelasticity, Moiré, holography, and interferometry.

The **Structural Dynamics Laboratory** had the latest equipment for recording ultradynamic events. Major equipment included Norland and Nicolet digital recorders, a one-million-frame-per-second dynamic camera, impact gun, and various computer peripherals for data acquisition. The primary research interest was in the impact of structures and the analysis of consequent stress waves.

The **Tribology and Materials Processing Laboratory** was developed in collaboration with the School of Industrial Engineering. Equipment included a 22-kip computer-controlled electro-hydraulic test machine and associated equipment for fretting fatigue testing, infrared sensors for temperature measurements, a friction apparatus for both low- and high-speed sliding indentation, a residual stress analyzer, lapping and polishing equipment, a vibration isolation table, micropositioning stages, rolling contact fatigue testers, Talysurf profilometers, optical microscopes, and a high-pressure pump used for dynamic fracture experiments. Also, access was available to a variety of machine tools, piezoelectric force sensors, and a laser micrometer.

Table 8.10 Sponsored Research Projects Active During the Period July 1, 1993, to June 30, 1994

Sponsor	Project Title	Project Period	Amount ($1000s)	Principal Investigator
NASA	Indiana Space Grant Consortium Project	3/1/91 – 2/28/95	44	Andrisani
NASA	Evaluation of SGS Models for LES of Compressible Flows	5/6/93 – 11/5/94	39	Blaisdell
Lockheed	Graduate Course in Microgravity Fluid Mechanics	6/1/92 – 3/31/94	5	Collicott
ASME	Experimental Study of the Effect of a BowShock on Known Free Stream Disturbances	9/1/92 – 8/31/93	23	Collicott
NASA	Indiana Space Grant Consortium Project	3/1/92 – 4/30/94	5	Collicott
RDL	Implementation of Noise-Reducing Multiple-source Schlieren Systems	1/1/93 – 12/31/93	20	Collicott
NASA	Quiet-Flow Ludweig Tube for Study of Transition in Compressible Boundary Layers	5/1/90 – 11/30/93	17	Collicott
Lockheed	Scale Gravity Probe B Tank Spin Testing	10/7/93 – 6/30/94	18	Collicott
NSF	Presidential Young Investigator Award	8/15/90 – 8/14/94	113	Corless
NASA	Fatigue Crack Growth in Panels Under Aeroacoustic Random Loading	8/3/93 – 8/2/94	29	Doyle
NSF	Correlation Between Microstructural Inelasticity and Macroscopic Response of Advanced Materials: Acquisition of a Multi-Channel Laser Interfer Sample Preparation Equipment Preparation Equipment	7/15/93 – 7/14/94	52	Espinosa
NSF	Micromechanical Study of Inelasticity in Brittle and Amorphous Materials	7/15/93 – 7/15/96	100	Espinosa
NSF	Presidential Young Investigator Award	9/1/90 – 8/31/94	213	Farris
NSF	Engineering Research Center Project	2/1/93 – 8/31/94	106	Farris

Table 8.10 Sponsored Research Projects Active During the Period July 1, 1993, to June 30, 1994 (continued)

Sponsor	Project Title	Project Period	Amount ($1000s)	Principal Investigator
Timken	Controlled Stress CBN Grinding of Bearing Steels	9/1/90 – 8/31/94	130	Farris
AAR	Residual Stresses in Rail Corners and Mechanics of Roller Straightening	3/1/94 – 12/31/94	29	Farris
Cummins Engine	CBN Grinding of M2 Steel: Manufacturing Aspects	11/15/92 – 6/30/94	51	Farris
AAR	Mechanics of Roller Straightening	1/2/93 – 12/31/93	19	Farris
NSF	Mixed H L2 Control Problems	3/1/93 – 2/28/95	59	Frazho
ALCOA	Initiation, Growth, and Coalescence of Small Fatigue Cracks at Notches	3/14/91 – 3/13/94	120	Grandt
GM/ Allison	Analysis of Cracks at Deep Notches	8/1/91 – 1/31/94	59	Grandt
IBMT	Indiana Space Grant Consortium Matching Grant	1/1/92 – 12/31/93	50	Grandt
AFOSR	Materials Degradation and Fatigue in Aerospace Structures	7/1/93 – 6/30/95	1,641	Grandt (Farris & Sun, Co-P.I.s)
NSF	Engineering Faculty Internship with ALCOA Technical Center	9/15/92 – 2/28/94	19	Grandt
AFOSR	Modeling of Liquid Jet Atomization Processes	7/1/92 – 6/30/94	80	Heister
NASA	Indiana Space Grant Consortium Project	3/1/92 – 2/28/95	11	Heister
AFOSR	New Approaches for Modeling Liquid Jet Atomization	6/1/93 – 5/31/96	125	Heister
Cummins	Modeling Injector Cavitation Processes	10/27/93 – 12/31/93	5	Heister
Cummins	Three-Dimensional Injector Cavitation Numerical Modeling	1/1/94 – 7/31/94	9	Heister
NASA	Indiana Space Grant Consortium Project	3/1/94 – 2/28/95	3	Howell
NSF	Analytic Theory of Asymmetric Rigid Bodies Subject to Arbitrary Body-Fixed Torques & Forces	8/1/89 – 7/31/93	70	Longuski
NSF	Analytic Theory and Control of the Motion of Spinning Rigid Bodies	9/1/91 – 8/31/94	180	Longuski

Table 8.10 Sponsored Research Projects Active During the Period July 1, 1993, to June 30, 1994 (continued)

Sponsor	Project Title	Project Period	Amount ($1000s)	Principal Investigator
FAA	Fire Wall Thermal Load Mechanisms in a Combustor Burn-Through Environment with Cross-Flow	8/21/92 – 8/19/94	90	Messersmith
NASA	Indiana Space Grant Consortium Project	3/1/93 – 2/28/95	2	Messersmith
NSF	A State-Space Approach for Multiple Objective Synthesis of Linear Controllers	8/1/91 – 7/31/94	60	Rotea
NSF	NSF Young Investigator Award	10/1/93 – 9/30/94	25	Rotea
NASA	Hot Film Wall Shear Instrumentation for Boundary Layer Transition Research	1/1/91 – 3/31/94	30	Schneider
AFOSR	Laminar-Turbulent Transition in High-Speed Compressible Boundary Layers with Curvature: Controlled Receptivity and Extent-of-Transition Experiments	11/15/93 – 11/14/94	56	Schneider (Collicott Co-P.I.)
NASA	Shakedown and Characterization of the Purdue Quiet Flow Ludwieg Tube	5/1/90 – 12/31/94	91	Schneider
NASA	Supersonic Quiet Tunnel Development for Laminar-Turbulent Transition Research	2/25/94 – 2/24/95	41	Schneider
NASA	Selecting Sensors & Actuators Optimizing Structures for Control Design	8/1/91 – 7/31/94	120	Skelton
NASA	Multiobjective Spacecraft Control Design	4/3/92 – 4/1/94	55	Skelton
ONR	Designing Structures for Controllability	2/1/92 – 11/30/94	179	Skelton
ONR	New Space Structure & Control Design Concepts	3/15/92 – 2/14/94	165	Skelton
ARO	Integrating Structure & Control Design for Mechanical Systems	6/1/93 – 5/31/96	74	Skelton
ONR	Designing Structures for Controllability	12/1/93 – 11/30/94	10	Skelton
NASA	Indiana Space Grant Consortium Project	3/1/93 – 2/28/95	12	Skelton
ONR	Measurement of Surface Pressure, Temperature, & Skin Friction Using Luminescent Imaging	8/1/91 – 9/30/94	208	Sullivan

Table 8.10 Sponsored Research Projects Active During the Period July 1, 1993, to June 30, 1994 (continued)

Sponsor	Project Title	Project Period	Amount ($1000s)	Principal Investigator
NASA	High-Lift Aerodynamics	6/1/93 – 2/28/95	90	Sullivan (Schneider, Co-P.I.)
NASA	Indiana Space Grant Consortium Project	3/1/91 – 2/28/94	452	Sullivan
NSF	Engineering Research Center Project	2/1/93 – 8/31/94	128	Sun
NASA	Modeling Damage Progression in Composite Laminated Structures	8/1/91 – 7/31/94	185	Sun
NASA	Rate Dependent Constitutive Models for High Temperature Polymer Composites	1/9/92 – 1/8/95	153	Sun
CIAC	DOD Ceramics Information Analysis Center	1/1/90 – 8/15/93	46	Sun
ARO	High Velocity Impact of Composite Laminates	1/1/91 – 6/30/94	247	Sun
Cheng Kung	Cooperative Research on Composite Sandwich Structures	7/1/88 – 6/30/94	113	Sun
AFOSR	Mechanics of Actuator-Structure Interfaces Adaptive Structures	8/1/92 – 7/31/95	247	Sun
Galaxy Science Corp.	Evaluations of Current Failure Analysis Methods for Composite Laminates	10/1/92 – 9/23/94	78	Sun
ONR/ ASSERT	Dynamic Response & Failure Thick-Section Composite Cylindrical Shells Subjected to Dynamic Pressure Wave	6/1/93 – 5/31/96	73	Sun
GE	Implementation of Composite Plasticity Model into CSTEM	12/20/93 – 9/30/94	15	Sun
ONR	Effect of Material Nonlinearities on Behavior of Thick-Section Composites under Static and Strong Dynamic Compressive Loading	1/1/90 – 10/31/95	429	Sun
NASA	Interactive Aircraft Flight Control & Aeroelastic Stabilization	11/01/84 – 05/31/94	682	Weisshaar
AFOSR	Aeroservoelastic Tailoring with Adaptive Materials	9/30/91 – 9/29/93	70	Weisshaar (Rotea, Co-P.I.)
USRA	Design Integration in Aeronautical Engineering	6/12/92 – 6/20/94	34	Weisshaar

Table 8.10 Sponsored Research Projects Active During the Period July 1, 1993, to June 30, 1994 (continued)

Sponsor	Project Title	Project Period	Amount ($1000s)	Principal Investigator
NASA	Multidisciplinary Design and Analysis Fellowship Program	10/01/93 – 5/31/94	50	Weisshaar
NASA	Interactive Aircraft Flight Control & Aeroelastic Stabilization	05/01/81 – 5/31/94	732	Weisshaar
NASA	Unsteady Aerodynamics & Aeroelasticity of Advanced Turboprops	5/1/88 – 3/27/95	642	Williams
NSF	Engineering Research Center Project	2/1/93 – 8/31/94	79	Yang
NSF	Active & Passive Controls of Complex Structures with Uncertainties Under Wind Loads	4/1/91 – 3/31/95	136	Yang
NASA	Computational Unsteady Aerodynamics & Aeroelasticity	6/7/83 – 7/31/94	169	Yang (Weisshaar/ Williams, Co-P.I.s)
NSF	Reliability Study on Hybrid Controls of Flexible Uncertain Structures under Wind Loads	9/1/93 – 8/31/95	138	Yang

Table 8.11 Sources of Sponsored Research For 1993-94 Fiscal Year

Source	Percentage of Total
Department of Defense	39.2
NASA	24.8
National Science Foundation	19.5
Industrial	5.8
Other	10.7
Total	**100.0**

The School of Aeronautics and Astronautics conducted $2,341,316 in externally funded research for the period July 1, 1993, to June 30, 1994.

Concluding Remarks

This chapter has served to benchmark the status of the Purdue School of Aeronautics and Astronautics on its 50th anniversary as an independent academic unit. Although aviation courses were first offered by Mechanical Engineering in 1921, and extensive World War II programs led to the first aeronautical engineering degrees awarded by a combined School of Mechanical and Aeronautical Engineering in August of 1943, it wasn't until July 1, 1945 that the separate School of Aeronautics was formally established.

The 50th anniversary of that milestone was celebrated in November of 1995 with a special weekend attended by more than 250 alumni and friends. In addition to an anniversary banquet, tours were given of the school's laboratory facilities, and a tailgate party was held before the football game. The event also was highlighted by two memorial lectures, which featured John B. Hayhurst, BS '69, vice president of product development for Boeing Commercial Airplane Group, and William J. O'Neil, BS '61, Galileo project manager for the Jet Propulsion Lab. Professor and Head John Sullivan noted the occasion with the following remarks: "In its 50-year history, the school has graduated more than 5,500 alumni and has well established itself as a premier provider of aeronautical and astronautical engineering education. We are proud of the accomplishments made by our alumni and gratefully acknowledge the effects they have had on the aerospace industry."

Golden anniversaries are appropriate occasions to reflect on past accomplishments, celebrate current status, and dream of the future. Certainly the Purdue School of Aeronautics and Astronautics had much to be proud of on its 50th anniversary as it faced the 21st century with confidence and optimism. The next chapter summarizes its activities and accomplishments during the subsequent 1996–2009 period.

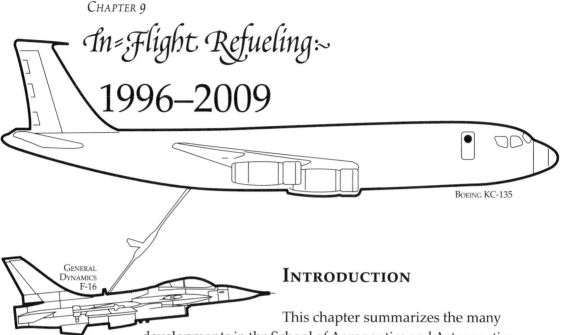

CHAPTER 9

In-Flight Refueling:~ 1996–2009

BOEING KC-135

GENERAL DYNAMICS F-16

INTRODUCTION

This chapter summarizes the many developments in the School of Aeronautics and Astronautics during the 1996–2009 period following its 1995 Golden Anniversary. This productive time saw exceptional growth in academic programs, research, international recognition, facilities, and student/faculty accomplishments. Both the school's undergraduate and graduate programs were, for example, rated fourth nationally in 2009 by *U.S.News & World Report*. Moreover, in fall 2008, *Aviation Week* reported that Aerospace and Defense industry recruiters considered AAE as the top undergraduate program and one of the top three A&D preferred graduate programs.[9]

Purdue University saw record enrollment in fall 2009 with 39,697 students attending its West Lafayette campus. The College of Engineering enrolled 9,549 at this time with 7,035 undergraduate and 2,514 graduate students participating in degree programs. Total School of Aeronautics and Astronautics 2009 enrollment was 904, consisting of 567 undergraduate (excluding freshman) and 337 graduates (the graduate figure is the largest in the school's history). An additional 36 off-campus students were pursuing MS degrees through Engineering Professional Education's distance education facilities. The 2009 undergraduate population included 4.6% minority, 14.5% female, and 8.5% international students. Graduate demographics included 68.5% MS and 31.5% PhD, 85.3 % male and 14.7 % female, and 53% U.S. versus 47% international students. Historical summaries of undergraduate and graduate enrollments in the School of Aeronautics and Astronautics are shown in Figures 9.1 and 9.2. Note that undergraduate attendance is characterized by peaks and valleys that loosely follow employment trends in the aerospace industry.

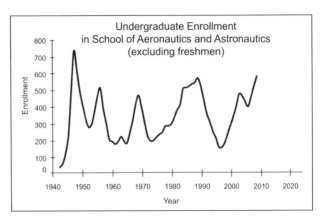

Eighteen new faculty joined AAE from 1996 to 2009, bringing new areas of technical depth and breadth as increased emphasis was placed on interdisciplinary work. The number of female faculty increased from one in

Figure 9.1   Summary of undergraduate enrollment in School of Aeronautics and Astronautics (excluding freshmen).

1996 to five in 2009. Other highlights include the growth in externally sponsored research shown in Figure 9.3. The faculty conducted approximately $7.5 million in sponsored research during the 2007–08 academic year, doubling that from 2001–02.

The spirit of this era is perhaps best characterized, however, by the 2007 move to the Neil Armstrong Hall of Engineering. Some of the activities and accomplishments that culminated with this exciting move are described in the remainder of this chapter.

ADMINISTRATIVE CHANGES

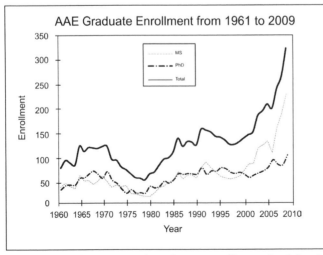

Figure 9.2   Summary of graduate enrollment in School of Aeronautics and Astronautics from 1961 to 2008.

Dr. Martin C. Jischke succeeded Dr. Steven Beering as president of Purdue University in 2000. After receiving his PhD from MIT in 1968 in the area of aerodynamics, Dr. Jishke joined the University of Oklahoma School of Aerospace, Mechanical, and Nuclear

Engineering, eventually becoming school head, dean of the College of Engineering, and finally, interim president of the university in 1985. Dr. Jischke was named chancellor of the University of Missouri–Rolla in 1986, and then president of Iowa State University in 1991. During his seven years as president of Purdue, Dr. Jishke's academic appointment was with the School of Aeronautics and Astronautics. He retired

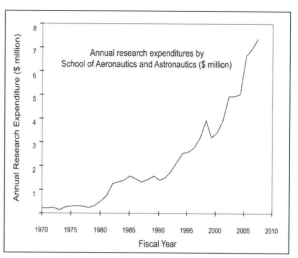

Figure 9.3 Summary of annual externally funded research expenditures by School of Aeronautics and Astronautics from 1970 to 2007.

in 2007 and was succeeded by Dr. France A. Córdova as Purdue's eleventh president on July 16, 2007. Also a member of Purdue's Department of Physics and Astronomy faculty, she served from 1993 to 1996 as chief scientist for NASA, among other positions.

Engineering administration changes during this period include the following sequence of deans of engineering: Henry T. Y. Yang (1984–1994), Richard J. Schwartz (1995–2001), Linda P.B. Katehi (2002–2006), and Leah H. Jamieson (2006–). In 2004, Purdue's Board of Trustees approved a request to rename the "Schools of Engineering" as the "College of Engineering." In announcing this change, Dean Katehi stated:

"Whereas in the past, the Schools of Engineering have emphasized decentralized, independent engineering disciplines, our approach for success in the 21st century, as laid out in our strategic plan, is collaboration, interdisciplinary work, and shared goals among the engineering schools. ... I believe that the change to "College of Engineering" symbolizes Purdue Engineering's commitment to multidisciplinary collaborations in discovery, learning, and engagement."

As described later, this desire for multidisciplinary collaborations was also embraced by the School of Aeronautics and Astronautics.

At the local level, after serving five years as head of the School of Aeronautics and Astronautics, **Professor John Sullivan** resigned in 1998 to return to full-time teaching and research. In 2005, Professor Sullivan was named director of Discovery Park's Center for Advanced Manufacturing, leading an effort to advance high technology production and products for Indiana industry. He also was appointed to the NASA Advisory Council in September 2006 by NASA Administrator Michael Griffin. This six-committee council is the primary source of outside assistance to NASA's role in national space policy. In this regard, Dr. Sullivan was named to the Aeronautics Committee chaired by Neil Armstrong (BSAAE '55).

Professor Thomas N. Farris succeeded Professor Sullivan as head of the School of Aeronautics and Astronautics on July 1, 1998. Professor Farris received his bachelor's degree in Mechanical Engineering in 1982 from Rice University and his master's and doctorate degrees from Northwestern University in theoretical and applied mechanics in 1984 and 1986 respectively. He first joined the school in 1986 as an assistant professor in the structures and materials group with research interests in tribology, manufac-

Thomas N. Farris

turing processes, and fatigue and fracture. He was promoted to associate professor in 1991 and to professor in 1994. Professor Farris served as school head until resigning in June 2009 to become dean of engineering at Rutgers University. His 11 years as AAE head is the longest such term in the school's history. (See Appendix E for a summary of other school heads.)

Professor Kathleen Howell served as interim head following Professor Farris's departure on July 1, 2009 until **Professor Tom I-P. Shih** was named new school head effective August 17, 2009. Professor Shih received his PhD from the University of Michigan in 1981 and held positions with NASA-Lewis Research Center, the University of Florida, Carnegie Mellon University, and Michigan State University before serving as professor and chair of the Department of Aerospace Engineering at Iowa State University from 2003 to 2009. Professor Shih's technical interests include mathematical modeling of fluid mechanics, heat transfer, and combustion problems.

Tom I-P. Shih

Professor W. A. Gustafson retired in June 1998, and his position as associate head was filled on July 1 of that year by **Professor Marc Williams**. Professor Williams had first joined the school's aerodynamics group in 1981. As associate head, he continued Professor Gustafson's earlier responsibilities for undergraduate counseling, co-op coordination, and scheduling.

During his tenure as head, Professor Sullivan established a new Industrial Advisory Council (IAC) in 1996 to help keep the school abreast of changes in the aerospace industry. The council met for the first time on April 29, 1996, and has met regularly each semester since that time. The IAC consists of around two dozen senior members from the aerospace community. The IAC advises the head and faculty on a number of issues (e.g. strategic planning, ABET accreditation, alumni relations, etc.) and helps ensure that students are well prepared for industry upon graduation.

STAFF DEPARTURES

Professor Winthrop A. (Gus) Gustafson retired in June 1998. He first joined the department in fall 1960 after working for several years at Lockheed as an associate research scientist in the areas of aerodynamics and high-speed gas dynamics. Professor Gustafson had been associate department head since 1980 and undergraduate counselor and co-op program coordinator since 1981. In addition to teaching in the aerodynamics area, he also taught the spacecraft section of the senior design course for approximately ten years. Professor Gustafson won the E. F. Bruhn Teaching Award two times, and was presented the Dean M. Beverly Stone Award by the Omicron Delta Kappa National Leadership Honor Society in 1997. He also received the Sagamore of the Wabash Award from Indiana Governor Frank O'Bannon in August 1998. The W. A. Gustafson Teaching Award was established by the School of Aeronautics and Astronautics in 1998 to honor his distinguished career of teaching and service to the school. (Recipients of that award are listed in Appendix D along with winners of the earlier Bruhn Teaching Award.)

Professor Robert E. Skelton resigned in fall 1997 to accept a faculty position at the University of California at San Diego. He first joined the school in 1975 after being employed by industry for several years. Professor Skelton taught in the dynamics and controls area, and his research interests focused on the integration of the modeling and control problems in the analysis and design of dynamic systems. Professor Skelton was director of the Space Systems Control Laboratory, which he helped to establish in 1991. **Professor Nathan L. Messersmith** also left the school in fall 1997 to accept a position with Pratt & Whitney in West Palm Beach, Florida. He joined the department as an assistant professor in 1992, and taught and conducted research in the propulsion area. **Professor Horacio D. Espinosa** resigned in 1999 to take a position with Northwestern University. He had been a member of the structures and materials group since 1992. **Professor Mario Rotea** also resigned in 2007 to take a position with another university. He first joined the dynamics and controls group in 1990.

Finally, **Dean Henry T. Y. Yang** resigned his position as dean of engineering in 1994 to become chancellor of the University of California, Santa Barbara. He first joined the School of Aeronautics, Astronautics, and Engineering Sciences as an assistant professor in 1969, and served as school head from 1979 to 1984 before being appointed dean in 1984. A highly respected teacher, Professor Yang continued to teach one AAE course per semester during his five years as school head and 10 years as dean of engineering. He also maintained an active research program during this period, serving as thesis advisor to many graduate students.

NEW FACULTY

Financed by an annual tuition increase and with gifts and earnings from a new endowment, Purdue initiated a University-wide campaign to increase faculty size and expertise in 2003. The College of Engineering benefited from this effort, growing faculty from 277 in fall 2001 to 348 in 2008, and increasing the number of named or distinguished faculty from 22 to 55. Many of the new engineering faculty were in the following signatures areas intended to reach across traditional boundaries: advanced materials and manufacturing; energy; global sustainable industrial systems; healthcare engineering; information, communications, and sensing systems, intelligent infrastructure systems; nanotechnologies and nanophotonics; perception-based engineering; system of systems; and tissue and cellular engineering.

The School of Aeronautics and Astronautics also enlarged its faculty from 21 to 30 (including joint appointments) from 2000 to 2008. New faculty who joined AAE after its 1995 Golden Anniversary are listed in chronological order below.

Dr. William A. Crossley joined the school as an assistant professor in August 1995. He received his MS and PhD in aerospace engineering from Arizona State University in 1992 and 1995 respectively, and his BS from the University of Michigan in 1990. He was a member of the Advanced Concept Development Group of McDonnell Douglas Helicopter Systems in Mesa, Arizona, while attending graduate school as an MDHS/ASU fellow. His technical interests are in optimal design methods, genetic

William A. Crossley

algorithms and aerospace applications, aircraft and rotorcraft conceptual design, composite, and smart structure design. He was promoted to associate professor in 2001 and to professor in 2009.

John J. Rusek

Dr. John J. Rusek joined the propulsion group in August of 1998 as an assistant professor. He received bachelor's, master's, and doctorate degrees in chemical engineering from Case Western Reserve University. He was the chief engineer of the Energetic Materials Division for the U.S. Naval Air Warfare Center in China Lake, California from 1995 to 1998. Before that, he served as a consultant within that department as well as with other DoD agencies. He also was a research chemical engineer for the U.S. Air Force and an adjunct professor in the School of Engineering for Antelope Valley College in Lancaster, California. He resigned from the school in 2002 to found Swift Enterprises, located in the Purdue Industrial Park, but rejoined the staff in 2003 as an adjunct assistant professor in propulsion, energy conversion, and power generation.

Dr. James L. Garrison joined the school as an assistant professor in 2000, and was promoted to associate professor in 2006. He received his PhD from the University of Colorado at Boulder in 1997, and held positions with the NASA Langley Research Center and the NASA Goddard Space Flight Center prior to coming to Purdue. His technical areas are satellite navigation, GPS, and remote sensing.

James L. Garrison

William Anderson

Dr. William Anderson was named assistant professor with the propulsion group in 2001. He received his PhD from The Pennsylvania State University in 1996, and then held positions with various propulsion companies before coming to West Lafayette. Dr. Anderson was promoted to associate professor in 2007. His technical areas are chemical propulsion and design methodologies.

Dr. Hyonny Kim joined the structures and materials group as an assistant professor in fall 2001. He obtained his PhD in 1998 from the University of California at Santa Barbara, and was a postdoctoral fellow there before coming to Purdue. Dr. Kim's technical areas include composite materials, impact, stability, and adhesive joining. Dr. Kim resigned in 2006 to take a faculty position with the University of California San Diego.

Hyonny Kim

Dr. David L. Filmer joined the school as an adjunct professor in 2002. He received his PhD from the University of Wisconsin in 1961, and served on the faculty of the Purdue Department of Biological Sciences from 1964-2004. His technical interests include satellite design and ground station design for acquisition of satellite data.

David L. Filmer

Dr. Ivana Hrbud joined the school as an assistant professor in 2003. She received her PhD from Auburn University in 1997, and following a year there as a post-doctoral fellow, held positions at NASA Marshall Space Flight Center. Her technical areas are electric propulsion, space power, and advanced in-space propulsion. She resigned from the University in 2009.

Ivana Hrbud

Dr. Charles L. Merkle accepted a joint position with the School of Aeronautics and Astronautics and the School of Mechanical Engineering as the Reilly Professor of Engineering in 2003. He received his PhD from Princeton University in 1969, and held chaired faculty positions with The Pennsylvania State University and The University of Tennessee before coming to Purdue. His technical areas are computational fluid dynamics and mechanics, two phase flows, and propulsion components and systems.

Charles L. Merkle

Dr. Weinong Chen accepted a joint appointment as associate professor with the School of Aeronautics and Astronautics and the School of Materials Engineering in 2004. He was promoted to professor in 2005. Dr. Chen received his PhD from California Institute of Technology in 1995, and held earlier faculty positions at the University of Arizona prior to coming to Purdue. His technical areas are mechanical response of solids and structures under extreme conditions, microstructural effects on mechanical behavior, fatigue behavior of engineering materials, and experimental solid and structural mechanics.

Weinong Chen

Dr. Daniel A. DeLaurentis was appointed assistant professor in 2004. He was promoted to associate professor in 2010. He received his PhD from Georgia Institute of Technology in 1998 and held positions there as research engineer and visiting assistant professor before joining Pur-

Daniel DeLaurentis

due. His areas of interest are design methods, aerospace systems and flight vehicles, and system-of-systems.

Dr. Inseok Hwang was appointed assistant professor in 2004 with the dynamics and controls group after receiving his PhD from Stanford University earlier that year. He was promoted to associate professor in 2010. Professor Hwang's technical areas are hybrid system theory, information inference of complex dynamical systems, safety verification, and their application to the control of multiple-vehicle systems, especially air traffic surveillance and control.

Inseok Hwang

Dr. R. Byron Pipes accepted a joint appointment with the School of Aeronautics and Astronautics, the School of Chemical Engineering, and the School of Materials Engineering as the John L. Bray Distinguished Professor of Engineering in 2004. Dr. Pipes received his PhD from the University of Texas at Arlington in 1972 and held senior academic and administrative positions with the University of Delaware, Rensselaer Polytechnic Institute, and the University of Akron before joining Purdue. His technical areas entail application of nanotechnology to engineering disciplines including aerospace, composite materials, and polymer science and engineering.

R. Byron Pipes

Dr. Alina Alexeenko joined the school as an assistant professor in 2006. She received her PhD from The Pennsylvania State University in 2003, and was a postdoctoral fellow at the University of Southern California from 2004 to 2006. Her technical areas are computational rarefied gas dynamics, kinetic theory of gases, numerical methods for model kinetic equations, direct simulation Monte Carlo techniques, microscale gas flows, coupled thermal-fluid analysis of microdevices, high-altitude aerothermodynamics, and two-phase plume flows.

Alina Alexeenko

Dr. Timothée Pourpoint received his PhD from the Purdue School of Aeronautics and Astronautics in 2005, and after serving two years as a senior research scientist, was named research assistant professor in 2008. His technical areas are aerospace propulsion systems, rocket engine combustors, liquid propellant injection systems, hypergolic propellants, high pressure, and hydrogen storage systems.

Timothée Pourpoint

Dr. Li Qiao was appointed assistant professor in 2007 after receiving her PhD earlier that year from the University of Michigan. Her technical areas are combustion and propulsion (low- and high-speed), experimental flow dynamics, micro-scale power generation, alternative fuels, fire research, and environmental impact of combustion.

Li Qiao

Karen Marais

Dr. Karen Marais joined the school as an assistant professor in January 2009 after serving as a senior lecturer in the Department of Industrial Engineering at the University of Stellenbosch. She received her PhD from the Department of Aeronautics and Astronautics at Massachusetts Institute of Technology in 2005 and also holds a master's degree in space-based radar from MIT. Her technical areas include safety and risk analysis, financial modeling of engineering systems, and environmental impacts of technology.

Dr. Dengfeng Sun joined the school as an assistant professor in fall 2009. He received his PhD from the University of California at Berkeley in 2008 and held a post-doctoral position at NASA Ames before coming to Purdue. His technical interests are in modeling, simulation and optimization of large scale networked systems, control and optimization, studies for National Airspace Systems, air traffic control, and intelligent transportation systems.

Dengfeng Sun

Table 9.1 School of Aeronautics and Astronautics Staff, 2009

Administrative
A. Broughton, T. Moore, L. Flack
Clerical
L. Crain, K. Johnson, P. Kerkhove, J. LaGuire
Professional
S. Meyer, D. Klassen, G. Xia
Technical
M. Chadwell, G. Hahn, J. Kline, M. Kidd, J. Phillips, D. Reagan, R. Snodgrass, J. Younts
Business Office
B. Anthony, J. Jackson, K. Munson, A. Nobile

Figure 9.4 School of Aeronautics and Astronautics 2006 Faculty and Staff

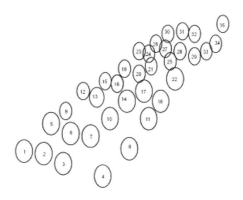

1. T. Yoder	13. A. Broughton	25. S. Heister
2. A. Nobile	14. J. Phillips	26. S. Meyer
3. M. Kidd	15. A. Alexeenko	27. W. Anderson
4. S. Wagner	16. J. Kline	28. I. Hwang
5. D. Andrisani	17. C. T. Sun	29. T. Weisshaar
6. L. Flack	18. T. Moore	30. G. Blaisdell
7. L. Crain	19. J. Longuski	31. T. Pourpoint
8. P. Kerkhove	20. W. Chen	32. S. Schneider
9. J. Garrison	21. A. Lyrintzis	33. M. Corless
10. D. Reagan	22. D. Filmer	34. A. Grandt
11. K. Johnson	23. S. Collicott	35. T. Farris
12. I Hrbud	24. M. Williams	

Dr. Vikas Tomar received his PhD from the Georgia Institute of Technology in 2005. He was an assistant professor at Notre Dame before joining the school's structures and materials group as an assistant professor in August 2009. Dr. Tomar's research focus is on molecular and continuum modeling combined with some degree of characterization of failure in nanostructured as well as microstructured ceramic composites, quantum dots, and ceramic-bio hybrid systems and materials.

Vikas Tomar

OTHER STAFF HIGHLIGHTS

The 2009–10 School of Aeronautics and Astronautics faculty and staff are listed in Tables 9.1 and 9.2. A 2006 group photograph showing many of these individuals is given in Figure 9.4. Table 9.2 lists faculty, with several participating in more than one group, and also includes adjunct faculty and professors from other departments who hold courtesy appoints with AAE.

Table 9.2 2009–10 Faculty by Research Area (including adjunct and courtesy appointments)

Aerodynamics

A. Alexeenko

Assistant Professor; PhD, Penn State, 2003

G. A. Blaisdell

Associate Professor; PhD, Stanford, 1991

S. H. Collicott

Professor; PhD, Stanford, 1991

A. S. Lyrintzis

Professor; PhD, Cornell, 1988.

S. P. Schneider

Professor; PhD, Caltech, 1989

T. Shih

Professor and Head; PhD, Michigan, 1981

J. P. Sullivan

Professor; Sc.D., MIT, 1973

M. H. Williams

Professor and Associate Head; PhD, Princeton, 1975

Aerospace Systems

D. Andrisani

Associate Professor; PhD, SUNY at Buffalo, 1979

B. Caldwell (By Courtesy)

Associate Professor of Industrial Engineering; PhD, University of California-Davis, 1990

W. A. Crossley

Associate Professor; PhD, Arizona State, 1995

D. DeLaurentis

Assistant Professor: PhD, Georgia Institute of Technology, 1998

I. Hwang

Assistant Professor; PhD, Stanford University, 2004

K. Marais

Assistant Professor; PhD, Massachusetts Institute of Technology, 2005

J. P. Sullivan

Professor; Sc.D., MIT, 1973

D. Sun

Assistant Professor; PhD, University of California at Berkeley, 2008

T. A. Weisshaar

Professor; PhD, Stanford, 1971

Astrodynamics and Space Applications

D. Filmer

Adjunct Professor; PhD, Wisconsin, 1961

J. L. Garrison

Associate Professor; PhD, University of Colorado at Boulder , 1997

K. C. Howell

Hsu Lo Professor of Aeronautical and Astronautical Engineering; PhD, Stanford, 1983

J. M. Longuski

Professor, PhD, Michigan, 1979

Dynamics and Control

D. Andrisani

Associate Professor; PhD, SUNY at Buffalo, 1979

M. J. Corless

Professor; PhD, Berkeley, 1984

D. DeLaurentis

Assistant Professor: PhD, Georgia Institute of Technology, 1998

D. Filmer

Adjunct Professor; PhD, Wisconsin, 1961

A. E. Frazho

Professor; PhD, Michigan, 1977

Table 9.2 continued

I. Hwang

Assistant Professor; PhD, Stanford University, 2004

D. Sun

Assistant Professor; PhD, University of California at Berkeley, 2008

Propulsion

W. Anderson

Associate Professor; PhD, The Pennsylvania State University, 1996

J. Gore (By Courtesy)

Vincent P. Reilly Professor of Mechanical Engineering; PhD, The Pennsylvania State University, 1986

S. D. Heister

Professor; PhD, UCLA, 1988

I. Hrbud

Assistant Professor; PhD, Auburn University, 1997

N. Key (By Courtesy)

Assistant Professor of Mechanical Engineering; PhD, Purdue University, 2007

C. L. Merkle

Reilly Professor of Engineering; PhD, Princeton University, 1969

T. Pourpoint

Research Assistant Professor, PhD, Purdue University, 2005

L. Qiao

Assistant Professor; PhD, University of Michigan, 2007

J. J. Rusek

Adjunct Assistant Professor; PhD, Case Western Reserve, 1983

S. Son (by courtesy)

Associate Professor of Mechanical Engineering; PhD, University of Illinois, 1993

Structures & Materials

W. Chen

Professor; PhD, California Institute of Technology, 1995

W. A. Crossley

Associate Professor; PhD, Arizona State, 1995

J. F. Doyle

Professor; PhD, Illinois, 1977

T. N. Farris

Adjunct Professor; PhD, Northwestern, 1986

A. F. Grandt

Raisbeck Engineering Distinguished Professor for Engineering and Technology Integration; PhD, Illinois, 1971

P. Imbrie (By Courtesy)

Associate Professor; PhD, Texas A & M, 2000

R. B. Pipes

John L. Bray Distinguished Professor of Engineering; PhD, University of Texas, 1972

C. Sun

Neil A. Armstrong Distinguished Professor; PhD, Northwestern, 1967

V. Tomar

Assistant Professor; PhD, Georgia Tech, 2005

T. A. Weisshaar

Professor; PhD, Stanford, 1971

(A courtesy appointment allows faculty from another department to advise AAE graduate students.) The 2008–09 faculty included one member of the National Academy of Engineering (Professor Pipes), three distinguished and two named professors, three University Faculty Scholars, and eleven fellows of major societies. By 2008, these 5 female and 25 male professors had published 15 books and more than 1,300 journal papers.

The new position of director of communications and development was established in 1995 to devote increased effort toward fund raising. This full-time position has been held by Nan Clair Ross (1995–1999), Tim Bobillo (1999–2004), Eric Gentry (2004–06), Nathan Wight (2006–2009), and Diane Klassen (2009–present). The development staff meets with school alumni, friends, and aerospace corporations with the goal of keeping these groups involved and connected with the school through personal interaction. Private support helps the school fund professorships, scholarships, and the Neil A. Armstrong Hall of Engineering, along with other initiatives. Around $10 million was raised over a five-year period for construction of Armstrong Hall, for example, and $6.4 million for various school programs during 2008-09.

Professor C. T. Sun was selected as the Neil A. Armstrong Distinguished Professor of Aeronautical Engineering in 1996. He has pioneered research in composite materials, fracture mechanics, structural dynamics, and computational mechanics. Among his many other honors, Professor Sun was recognized in May 2009 by the American Institute of Aeronautics and Astronautics and the American Society for Composites as co-winner of the first J. H. Starnes Award. This new award recognizes *"continued significant contribution to and demonstrated promotion of the field of structural mechanics over an extended period of time emphasizing practical solutions, to acknowledge high professionalism, and to acknowledge the strong mentoring of and influence on colleagues, especially younger colleagues."*

Other named and chaired professorships in AAE during the 1996–2009 period include the Raisbeck Engineering Distinguished Professor of Engineering and Technology Integration (**A.F. Grandt**, 2000), the Hsu Lo Professor of Aeronautical and Astronautical Engineering (**Kathleen Howell**, 2004), the Reilly Professor of Engineering (**Charles Merkle,** 2003), and the John L. Bray Distinguished Professor of Engineering (**R. Byron Pipes**, 2004).

The University Faculty Scholars program was established by Purdue in 1998 to *"recognize outstanding faculty colleagues who are on an accelerated path for academic distinction."* **Professor Stephen D. Heister** was named to the inaugural group of University Faculty Scholars in 1998. Two other AAE professors have received this prestigious designation — **Anastasios Lyrintzis**

(2004) and **Wayne Chen** (2005).

The Purdue University Book of Great Teachers was dedicated on April 23, 1999 to recognize past and present faculty members who have devoted their lives to excellence in teaching and scholarship. Current and former aeronautics and astronautics faculty selected for this honor include Professors **E. F. Bruhn, S. H. Collicott, W. A. Gustafson, K. C. Howell, S. L. Koh, J. M. Longuski, C. T. Sun, and H. T. Yang**.

In order to formally recognize faculty research accomplishments, the **C. T. Sun School of Aeronautics & Astronautics Excellence in Research Award** was established in 2004, with Sun as the first recipient. This award is presented annually to an individual or a team of faculty members in the Purdue University School of Aeronautics & Astronautics to recognize high-quality contributions in science and engineering. Winners of this prestigious school award include Professors Chin-Teh Sun (2004), William Anderson and Stephen D. Heister (2005), Mario Rotea (2006), Steven P. Schneider (2007), Weinong Chen (2008), and Charles L. Merkle (2009).

ACADEMIC PROGRAMS

The AAE faculty formally adopted the following mission statement in 1997:

> To serve the State of Indiana and our Nation by providing degree granting programs — recognized as innovative learning experiences — that prepare students to be exceptional, recognized contributors to aeronautical and astronautical engineering in industry, government laboratories, and universities.
>
> To develop and maintain quality graduate research programs in technical areas relevant to aeronautics and astronautics and to foster a collegial and challenging intellectual environment necessary to conduct enabling and breakthrough research for aerospace systems.

With respect to the second goal, AAE expanded its traditional four technical areas of aerodynamic, dynamics and control, propulsion, and structures and materials in 2006 to include new emphasis areas in **aerospace systems** and **astrodynamics and space applications**. These are intended to broaden the school's instruction and research to keep abreast of the growing needs of industry.

Aerospace systems is related to one of the engineering signature areas mentioned previously, and takes a broad interdisciplinary view of approaches needed to design and operate aerospace systems and products. The goal is to produce innovative, multidisciplinary engineers and researchers who

Table 9.3 Suggested Undergraduate Plan of Study with Aeronautics Concentration, 2008-09 Academic Year

SOPHOMORE YEAR

Third Semester

- (3) AAE 203 (Aeromechanics I)
- (4) MA 261 (Multivariate Calculus)
- (3) Phys 241 (Electricity & Optics)
- (3) MA 265 (Linear Algebra)
- (3) General Education Elective

(16)

Fourth Semester

- (3) AAE 204 (Aeromechanics II)
- (1) AAE 204L (Aeromechanics II Lab)
- (3) AAE 251 (Intro to Aerospace Design)
- (3) ME 200 (Thermodynamics)
- (3) MA 266 (Differential Equations)
- (3) General Education Elective

(16)

JUNIOR YEAR

Fifth Semester

- (3) AAE 333 (Fluid Mechanics)
- (1) AAE 333L (Fluid Mechanics Lab)
- (3) AAE 352 (Structural Analysis I)
- (3) MA 304 (Diff Eqns for Engr & Sci)
- (3) AAE 301 (Signal Analysis for Aerospace Systems)
- (3) General Education Elective

(16)

Sixth Semester

- (3) AAE 334 (Aerodynamics)
- (1) AAE 334L (Aerodynamics Lab) or AAE 352L (Structural Analysis I Lab)
- (3) AAE 340 (Dynamics and Vibrations)
- (3) AAE 372 (Jet Propulsion Power Plants)
- (3) AAE 364 (Control Systems Analysis)
- (3) General Education Elective

(16)

SENIOR YEAR

Seventh Semester

- (3) AAE 421 (Flight Dynamics & Control)
- (1) AAE 364L (Controls Laboratory)
- (3) Technical Elective
- (6) Major or Minor Area Electives
- (3) General Education Elective

(16)

Eighth Semester

- (3) AAE 451 (Aircraft Design)
- (9) Major or Minor Area Electives
- (3) Technical Elective
- (3) General Education Elective

(18)

Table 9.4   Suggested undergraduate plan of study with astronautics concentration, 2008-09 Academic Year

SOPHOMORE YEAR

Third Semester

(3)   AAE 203 (Aeromechanics I)
(4)   MA 261 (Multivariate Calculus)
(3)   Phys 241 (Electricity & Optics)
(3)   MA 265 (Linear Algebra)
(3)   General Education Elective

(16)

Fourth Semester

(3)   AAE 204 (Aeromechanics II)
(1)   AAE 204L (Aeromechanics II Lab)
(3)   AAE 251 (Intro to Aerospace Design)
(3)   ME 200 (Thermodynamics)
(3)   MA 266 (Differential Equations)

(16)

JUNIOR YEAR

Fifth Semester

(3)   AAE 333 (Fluid Mechanics)
(1)   AAE 333L (Fluid Mechanics Lab)
(3)   AAE 352 (Structural Analysis I)
(3)   MA 304 (Diff Eqns for Engr & Sci)
(3)   AAE 301 (Signal Analysis for Aerospace Systems)
(3)   General Education Elective

(16)

Sixth Semester

(3)   AAE 334 (Aerodynamics)
(1)   AAE 334L (Aerodynamics Lab)
       Or AAE 352L (Structural Analysis I Lab)
(3)   AAE 340 (Dynamics and Vibrations)
(3)   AAE 364 (Control Systems Analysis)
(3)   Technical Elective
(3)   General Education Elective

(16)

SENIOR YEAR

Seventh Semester

(3)   AAE 439 (Rocket Propulsion)
(1)   AAE 364L (Controls Laboratory)
(3)   Technical Elective
(6)   Major or Minor Area Electives
(3)   General Education Elective

(16)

Eighth Semester

(3)   AAE 440 (Spacecraft Attitude Dynamics)
(9)   AAE 450 (Spacecraft Design)
(3)   Major or Minor Area Electives
(3)   General Education Elective

(18)

are system thinkers who understand how design decisions impact the many interdisciplinary features of the aerospace system. Topics include requirements definition, functional decomposition, concept synthesis, application of design-oriented analysis methods, and optimization. Initial membership of this group included Professors Andrisani, Crossley, DeLaurentis, Hwang, Sullivan, and Weisshaar, who were joined by Professors Marais and D. Sun in 2009.

The **astrodynamics and space applications** area is intended to provide a comprehensive, well-defined graduate program to support science and space utilization from an engineering perspective. Teaching and research focuses on theoretical and applied celestial mechanics, attitude dynamics, estimation, and filtering. These fundamental principles support applications such as Earth orbital, interplanetary, and libration point missions; trajectory optimization; orbit determination; satellite geodesy; remote sensing; space environment; navigation; spacecraft guidance and station-keeping; communications, telemetry, and data acquisition; constellations and formation flight; and mission operations. This new group included Professors Filmer, Howell, Longuski, and Garrison, who also continued to coordinate their undergraduate teaching with the dynamics and controls group.

Undergraduate curriculum changes during this period were fairly minor and incremental in nature as the school continued its longstanding tradition of requiring aerospace depth, while also providing opportunities for breadth into areas of particular student interest. As indicated in Tables 9.3 and 9.4, the 129 credit hours needed in 2008–09 include required courses in math, science, communication, and basic engineering skills, along with specified AAE courses in laboratories, structures, dynamics and control, aerodynamics, propulsion, and design. In addition to the engineering requirement for 18 hours of general education electives, students also are allowed to select courses from two technical areas as a major and minor. (There are approximately 40 additional AAE courses that may be used as major/minor electives.)

Although there was always the option to emphasize aircraft or spacecraft with the major/minor electives, formal development of separate aircraft and spacecraft concentrations began in 2000 when the senior capstone design class (AAE 451) was separated into AAE 450 (Spacecraft Design) and AAE 451 (Aircraft Design). Design also was recognized as a separate specialty for undergraduate major/minor concentration, with those teaching responsibilities subsequently assumed by the new aerospace systems group. Note that the aeronautics and astronautics concentrations described in Tables 9.3 and 9.4 differ mainly in the choice of the vehicle dynamics course, AAE 421 (Flight

Dynamic and Control) versus AAE 440 (Spacecraft Attitude Dynamics); the propulsion class, AAE 372 (Jet Propulsion Power Plants) versus AAE 439 (Rocket Propulsion); and the senior design class, AAE 451 (Aircraft Design versus AAE 450 (Spacecraft Design).

Graduate study is based on more than 70 AAE graduate courses, with additional offerings in other departments. With the approval of an advisory committee, students tailor their graduate study to major/minor in two of the following areas: aerodynamics; aerospace systems; astrodynamics and space applications; dynamics and control; propulsion; structures and materials; and interdisciplinary areas (such as aeroacoustics, biomechanics, design, and manufacturing).

Students may pursue either thesis or non-thesis MS programs, with the non-thesis option generally entailing 30 credit hours. The thesis MS requires 21 hours plus a thesis on a research topic chosen by the student and advisor. The PhD requires at least 18 credit hours beyond the master's degree or 48 graduate hours beyond the undergraduate degree. The PhD involves written qualifying and oral preliminary examinations, and again includes a research thesis. Lists of all MS and PhD thesis titles published by the school are given in Appendices B and C.

Beginning in 2000, graduate offerings were expanded to working professional engineers through the distance education facilities administered by Engineering Professional Education. Professor Crossley taught the first such AAE course (AAE 550) to 18 campus and 14 remote students in fall 2000, followed by Professor Grandt's AAE 554 in 2001. As the number of distance offerings increased, it soon became possible to obtain the MSAAE degree online by taking many of the same courses offered on campus. (Two distance students, for example, received MSAAE degrees in 2008.) AAE classes taught to distance students during 2000–09 include:

- AAE 507 (Principles of Dynamics), Longuski
- AAE 508 (Optimization in Aerospace Engineering), Longuski
- AAE 512 (Computational Aerodynamics), Merkle
- AAE 514 (Intermediate Aerodynamics), Lyrintzis
- AAE 515 (Rotorcraft Aerodynamics), Lyrintzis
- AAE 550 (Design Optimization), Crossley
- AAE 532 (Orbit Mechanics), Howell
- AAE 552 (Nondestructive Evaluation of Structures and Materials), Grandt
- AAE 554 (Fatigue of Structures and Materials), Grandt
- AAE 555 (Mechanics of Composite Materials), Sun
- AAE 558 (Finite Element Methods in Aerospace Structures), Kim, Tomar

- AAE 564 (Systems Analysis and Synthesis), Corless
- AAE 575 (Intro to Satellite Navigation and Positioning), Garrison
- AAE 590K (System of Systems Modeling and Analysis), DeLaurentis
- AAE 607 (Variational Principles of Mechanics), Longuski
- AAE 690 (Statistical Orbit Determination), Garrison

President Steven C. Beering established the **William E. Boeing Lecture Series** in 1999 to thank The Boeing Company for its generosity to Purdue University over the years and to honor the memory of its founder. Administered by the School of Aeronautics and Astronautics, these annual lectures feature nationally known speakers from the aerospace or air transportation industry. Annual Boeing lecturers have included:

1999:   Michael M. Sears (BSEE 1969, DEA 1999), senior vice president, The Boeing Company, and president of the Military Aircraft and Missile Systems Group

2000:   General Roy D. Bridges (MSAAE'66, DEA'98, OAE'99), director, John F. Kennedy Space Center

2001:   Dr. Jürgen Weber, chairman and chief executive officer, Deutsche Lufthansa AG.

2002:   Major General John L. Hudson (MSAE '74), program executive officer and program director, Joint Strike Fighter Program

2003:   Dr. Mike Howse, director of engineering and technology, Rolls-Royce PLC, Great Britain

2004:   Dr. Paul MacCready, chairman, AeroVironment, Inc.

2005:   Dennis G. Mendoros, founder, owner, and managing director, Euravia Engineering & Supply Co. Ltd.

2007:   Dr. Michael Griffin, administrator, National Aeronautics and Space Administration

2008:   Dr. Sigmar Wittig, chairman emeritus, German Aerospace Center

Purdue Space Day was started in 1996 by AAE student groups to introduce third through eighth grade students to aeronautical engineering and space exploration. **Ann Broughton** became the event coordinator in 2000, and it grew steadily into an independent student organization, which by 2008 had shared the excitement of space with 4,000 elementary students. In addition to highly successful outreach goals, Purdue Space Day also is a professional development program for the many Purdue student volunteers who organize the event and lead the sessions. The interactive day usually is highlighted by presentations from one or more of Purdue's astronaut alumni, and receives substantial media coverage.

The **Outstanding Aerospace Engineer (OAE) award** was established in 1999 to *"recognize alumni who have demonstrated excellence in industry, academia, governmental service, or other endeavors which reflect the value of an aerospace engineering degree."* In addition to honoring these individuals, they also serve as role models for current students. The first OAE banquet was held on October 21, 1999, and recognized 58 alumni who had previously received honorary doctorate degrees or Distinguished Engineering Alumnus awards from the College of Engineering, and astronaut alumni who have flown in space. Subsequent recipients of the annual OAE Award are summarized in Appendix F.

Research Programs

The nature of university research gradually changed at the end of the 20th century, moving from small single investigator programs to larger multi-disciplinary teams collaborating on a particular problem. This shift played a key role in Purdue's strategic plans, and was, in fact, one of the reasons that the College of Engineering chose to invest many of its new faculty positions in the broad signature areas described earlier. One outcome of these larger efforts was the accompanying growth in administrative infrastructure. The Dean of Engineering staff expanded, for example, from one associate dean and two assistant deans during Yang's tenure (1984–94) to 6 associate deans and 1 assistant dean in 2009.

As indicated in the introduction to this chapter, the 1995 –2009 period was a most productive time for research in the school. During the 2006 and 2007 calendar years, for instance, faculty and students published 135 journal articles or book chapters, 288 con-

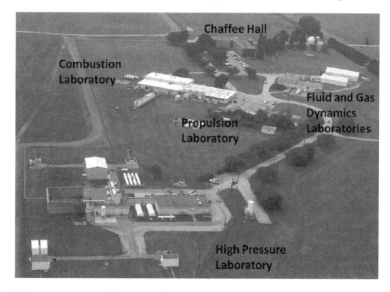

Figure 9.5   *Aerial view of the Maurice J. Zucrow Laboratories (Summer 2007)*

ference papers or technical reports, and 45 MS and 21 PhD theses. Listings of all the research projects and publications produced during this period are given in annual research reports published by the school and posted on its Internet site. Since it is not possible to repeat that information here, readers are encouraged to consult the Web page for details about those research activities. It is perhaps fitting, however, to highlight two of the larger AAE led interdisciplinary research efforts that bracketed the 1996–2009 era: Army-sponsored efforts dealing with damage tolerant armor (1996–99) and gelled rocket propellants (2009–present).

The first example is a three-year Multidisciplinary University Research Initiative (MURI) funded by the U.S. Army in 1996 entitled "Lightweight Layered Materials/Structures for Damage Tolerant Armor." This $3 million project was headed by **Professor C. T. Sun** from the School of Aeronautics and Astronautics and included AAE Professors **J. F. Doyle** and **H. Espinosa**. They were joined by School of Materials Engineering Professors K. J. Bowman and K. P. Trumble and three other researchers from the University of Dayton Research Institute. The team and its students employed advanced experimental and computational methods to understand the response of innovative material systems (e.g., ceramics and composites) and layered structural configurations to high-speed ballistic impact.

The second example is another 2009 Army-sponsored MURI led by **Professor S. D. Heister**, entitled "Spray and Combustion of Gelled Hypergolic Propellants." The $6.25 million, five-year project combined the efforts of nine Purdue engineering professors and food scientists with three faculty from Iowa State University and the University of Massachusetts. In addition to Professor Heister, AAE researchers include **Professors W. E. Anderson, C. Merkle,** and **T. Pourpoint.** Other Purdue team members were mechanical engineering Professors R. P. Lucht, P. E. Sojka, and S. F. Son, agricultural and biological engineering Professor O. H.

Figure 9.6 Graduate students Brad Wheaton, left, and Peter Gilbert stand near a segment of the Boeing/AFOSR Mach 6 Quiet Tunnel, March 2009 (Purdue News Service photo/ Andrew Hancock)

Campanella, and food science Professor C. M. Corvalan. The diverse team combined engineering expertise in propulsion with the knowledge of gels provided by the food scientists to develop innovative gelled propellants. Safer than liquids and easier to control than solid propellants, the hypergolic gels do not require an ignition source, but auto-ignite when mixed with an oxidizer. Testing the highly toxic fuels required special facilities developed at the Zucrow Laboratories. Indeed, one of the key research developments during 1996–2009 is the renovation of Zucrow Laboratories described in the following section.

FACILITIES

The 1996–2009 era saw many advancements in facilities as faculty sought the latest computational and experimental tools to pursue their teaching and research missions. This period also featured relocation of the School of Aeronautics and Astronautics to the new Neil Armstrong Hall of Engineering. While it is not possible to discuss all of these facilities, two particularly notable developments are summarized here — renovation of the Maurice J. Zucrow Laboratories and development of the Mach 6 quiet flow tunnel.

The capability for propulsion system measurements has long been a key aspect of Purdue aerospace efforts. The first aeronautics laboratory located in Heavilon Hall in 1930, for example, featured a fully assembled airplane and operating engine (recall Fig. 1.2). When Professor Joseph Liston joined the school in 1937,

Figure 9.7   Purdue President France A. Córdova, from left, and former Apollo astronauts Neil Armstrong and Eugene Cernan, listen to speakers during October 27, 2007 dedication of the Neil Armstrong Hall of Engineering. Armstrong and Cernan, the first and last men to walk on the moon, joined 14 other astronauts and former astronauts who are Purdue alumni at the dedication ceremony. (Purdue News Service photo/Dave Umberger)

Figure 9.8 Third floor view of atrium of Armstrong Hall showing installation of replica of Apollo I command module. Purdue astronauts Gus Grissom (BSME '50) and Roger Chaffee (BSAE '57) died with Ed White in a training accident when a launch pad burst into flames on October 27, 1967. (Purdue News Service photo/David Umberger)

he developed additional power plant courses and test facilities (Figures 1.16 and 4.8). Indeed, for a period of time in the late 1940s, the newly established School of Aeronautics owned a P-59 airplane, one of the nation's first jet-powered aircraft, and proudly fired up its jet engine for visitors (Figure 3.16).

The first serious jet and rocket propulsion research began, however, when Professor Maurice Zucrow joined the faculty in 1946. Beginning with support from the Office of Naval Research and the Purdue Research Foundation in 1948, Zucrow and his colleagues soon developed the ability to test rocket engines and high pressure gas turbine combustors. The High Pressure Rocket Research laboratory, designed by ME Professor Charles M. Ehresman, was subsequently constructed in the mid-1960s with support from NASA and NSF. These unique facilities are located off-campus on a 24-acre site adjacent to the Purdue airport, and allow testing highly toxic fuels under high pressure conditions.

While employed sporadically over succeeding decades, Zucrow Laboratory saw less frequent use and eventually became obsolete. **Professor Steven Heister** led a major renovation of the Zucrow Laboratory beginning in 2001 with the assistance of a $1 million dollar, two-year grant from the Indiana

21ˢᵗ Century Research and Technology Fund. Additional support came from Rolls-Royce in the form of the nation's first University Technology Center in High Mach Propulsion in 2003. As shown in Figure 9.5, the Zucrow complex consists of five laboratory and an office/classroom building. By 2009, a new spirit of cooperation was flourishing between AAE and ME as this facility was supporting nearly $5 million in sponsored research conducted by approximately 10 AAE and ME faculty and 70 graduate students. **Scott E. Meyer (BSAAE '90, MSAAE '92)** was named managing director of the Zucrow Labororatories in August 2009.

Figure 9.9 Professor Farris with (clockwise) Faculty Emeriti W. A. Gustafson, G. P. Palmer, and L. T. Cargnino at the Neil Armstrong Statue Dedication (October 2007)

Another unique test capability developed during the 1996–2009 period is the Boeing/AFOSR Mach 6 Quiet Tunnel constructed by Professor Steven **Schneider.** Based on a Ludweig tube that provides quiet hypersonic flow for short periods, the concept was first demonstrated by Professor Schneider in summer 1992 (recall Figure 7.27). Then, beginning in 1995, a larger Mach 6 facility was built over a ten-year period with approximately $1 million in support from the Air Force Office of Scientific Research, NASA, Sandia National Laboratories, the Ballistic Missile Defense Organization, and The Boeing Company. At the time of its completion, it was the only facility in the world capable of operating at Mach 6 and was being used to obtain critical experimental data to design new hypersonic aircraft (Figure 9.6).

Perhaps the highlight of the 1995–2009 period was the school's 2007 move from the aging Grissom Hall to the modern, well-equipped Neil Armstrong Hall of Engineering, a wonderful new home that will inspire future generations of faculty and students. The October 27, 2007 dedication of Armstrong Hall was attended by Apollo astronauts Neil Armstrong (BSAAE '55) and Gene Cernan (BSEE '56), the first and most recent men to step on the moon, who were joined by 14 other Purdue astronaut alumni and NASA Administrator Michael Griffin for the ribbon cutting (Figure 9.7). The $53.2 million structure has 210,326 square feet with 20,000 devoted to research labs and more than 50,000 to undergraduate teaching facilities. It is home to the School of Aeronautics and Astronautics, the School of Materials Engineering, the Department of Engineering Education, the Dean of Engineering offices, and several other engineering programs.

In addition to state-of-the-art classrooms and laboratories, Armstrong Hall features a three-story atrium (Figure 9.8), a statue of student Neil Armstrong (Figure 9.9), and a moon rock donated by Martha Chaffee, widow of Roger Chaffee (BSAE, '57). Roger Chaffee and Gus Grissom (BSME '50) tragically were killed along with Ed White in a 1961 fire that engulfed a simulated test of the Apollo I mission. (A replica of that capsule hangs in the Armstrong Hall atrium.) Mrs. Chaffee assigned the lunar sample to Purdue through a NASA program that allows each Gemini, Apollo, or Mercury astronaut, or his survivor, the right to donate a lunar sample to the educational institution of his or her choice.

Many other teaching and research facilities available in the School of Aeronautics and Astronautics are described on the school's Web site. Although it is not possible to describe all of that information here, it should be noted that the school is very well equipped with the modern computational and experimental equipment needed for its teaching and research missions.

Summary

As the School of Aeronautics and Astronautics celebrated its 65th birthday on July 1, 2010, it was in excellent health with little thought about resting on retirement laurels. It enjoyed the largest combined undergraduate and graduate enrollment in its history. It had a dynamic faculty well-versed in many diverse technical areas and armed with the most modern facilities. Moreover, as indicated in the following chapter, its alumni had earned the school an enviable reputation for its ability to graduate leaders who significantly advance the aerospace profession. Certainly, the school was eager and ready to continue its long standing tradition of producing the highest quality graduates and technology needed for the 21st century.

CHAPTER 10

Mission Control:

ALUMNI ACCOMPLISHMENTS

SPACE
SHUTTLE

INTRODUCTION

Alumni achievements are the standard by which any academic program is ultimately judged, and in this regard Purdue has few equals. (Figure 10.1 is a photograph of a typical graduating class [1986], and a list of all graduates from Purdue's aerospace engineering programs is given in Appendices A–C.) The goal of this chapter is to highlight the accomplishments of the school's graduates, and to indicate the varied nature of their careers.

An outstanding academic program needs high-quality students to succeed. Seemingly at odds with this obvious statement, state-supported universities such as Purdue have a special obligation to emphasize admission to qualified state residents, and do not have the option to concentrate on a few select students skimmed from top of the national pool of applicants. This duty has not posed a problem for the School of Aeronautics and Astronautics, however, for it has always attracted strong students to its academic programs. Part of this success may be attributed to the inherent excitement of the aerospace field. Few other disciplines enjoy a student body that has such an avid interest in their profession. Indeed, the aerospace mystique has a powerful motivational influence for learning, and plays a most effective role in outreach activities directed toward encouraging educationally disadvantaged students to higher levels of achievement. Thus, aerospace engineering students tend to be among the most capable and enthusiastic students on campus.

Figure 10.1 Typical graduation class (1986). **Left to right, row 1:** *Troy Rein, Rina Kor, Valerie Schlossberg, Lisa Riddle, Bob Sharp, Rhett Dennerline, Randy Goode, Stephen Wade, Mary Schmitz, Joseph Dick, Lynette Scott, Jeff Bauer, Jennifer Bender, Christine Grandin, Rhonda Thornton.* **Row 2:** *Jackie Johnson, Jeffrey Layton, Dawn Delaney, Kathy Donnellan, Steve Schultz, Kevin Krizman, Tim Moes, Karim Bennis, Michael Messina, Robert Martin.* **Row 3:** *James Passma, John Sanderson, Roy Bruce Hubbard, Khueh Hook Lee, Paul Pontecoro, Neal Jones, Vince Prince, Ross Mohr, Rick Clements, Joseph Speth Jr., Wayne Tygert, Dave Coombs, Lisa Hiday, Professor G. Palmer, Michael Boyle, Steven Hertzberg.* **Row 4:** *Steve Marsh, Scott McNabb, Robert Walker, Chris Pilla, Christopher Franz, Mark Sutherlin, Tim Ewart, Don Monell, Eric VanNorman, Ronald Smith, Todd Rough, Javier Benavente, Professor W. A. Gustafson, Philip Papadakis, Duane Guingrich, Michael Wrobel.* **Row 5:** *Professor F. Marshall, Doug Lee, Eric Wernimont, Douglas Fuhry, Christopher Cleveland, Brent Bates, David Beeman, James Miller, Steven Hiss, Gregory Miller, Michael Phillipis, Todd Ely, James Joyce, Earl Odom III, Professor T. A. Weisshaar, Elliott Keen, Professor J. R. Osborn.*

Beginning in 1955, one student has been selected annually for the *Sigma Gamma Tau Outstanding Senior Award*. Competition for this award involves a two-step process, which considers both academic and service accomplishments and involves input from both faculty and students. First, a list of students whose grades rank among the top 10 percent of seniors graduating in a given year is distributed to faculty for evaluation. The top five students from this faculty balloting are then forwarded to the officers of Sigma Gamma Tau, the student honorary society, who conduct an election among graduating seniors to select the winner. Recipients of the Sigma Gamma Tau Outstanding Senior Award are summarized in Table 10.1.

The rest of this chapter reviews the careers of the first class to receive bachelor's degrees in aeronautical engineering (Class of 1943), as well as those who have received honorary doctorates and Distinguished Engineering Alumnus Awards from the University. A summary of Purdue's astronaut alumni is also provided.

Table 10.1 Sigma Gamma Tau Outstanding Senior Award

Year	Recipient	Year	Recipient
1955	Duane Davis	1983	Craig O. Perry
1956	Robert James	1984	Bradley Hopping
1957	Harry Blackiston	1985	Jonathon Bohlman
1958	William J. Usab	1986	Russ Mohr
1959	Ronald G. Taylor	1987	Christopher P. Azzano
1960	M. Joe Cork	1988	Kendall Yorn
1961	Howard E. Bethel	1989	Jeffrey D. Schultz
1962	Burghard H. Ruterbories	1990	Charissa J. Cosky
1963	James A. Weber	1991	Ronald Cosby
1964	Ronald L. Moore	1992	Markus Heinimann
1965	David Schmidt	1993	Lance Werthman
1966	Gary Fidler	1994	Louis Lintereur
1967	James Green	1995	Victoria Anthes
1968	Steve Zakem	1996	Stephen R. Norris
1969	Neil Walker	1997	R. Sergio Hasbe
1970	Dennis Greene	1998	Kerrie Benish
1971	Stephen T. Montgomery	1999	Abigail Dobbins
1972	George Staab	2000	Nicole Key
1973	Lawrence C. Weeks	2001	Christopher Peters
1974	R. A. Jamrogiewicz	2002	Luca Bertuccelli
1975	Charles R. Stewart	2003	Gina Pieri
1976	Edwin Yarbrough	2004	Brian Ventre
1977	Karen Washburn	2005	George Pollock
1978	Joseph F. Slomski, Jr.	2006	Phillip Boettcher
1979	Randolph Shields	2007	Breanne Wooten & Tim Sisco
1980	Glenn Farris	2008	Amanda Briden
1981	Pamela Alstott	2009	Christine Troy
1982	Joseph G. Myers	2010	Stephanie Sumcad

CLASS OF '43

An interesting case study of alumni accomplishments may be obtained by reviewing the careers of Purdue's first class of aeronautical engineering graduates. As indicated previously, these students completed their BS degrees in August 1943 through the School of Mechanical and Aeronautical Engineering. The information given here is edited from a booklet prepared by James R. Dunn to commemorate the 50th anniversary of his class's graduation. The preface to that booklet is reproduced below.

> *This booklet commemorates the first class of aeronautical engineers at Purdue on the occasion of the 50th anniversary of their graduation. It is intended to provide historical information on the careers of each since graduation. Several of the class members did not graduate until after their military service, but nevertheless participated in the creation of the Aero school. Most of the class contributed in one way or another in building the wind tunnel, setting up the engine test labs at the airport, helping edit Professor Bruhn's structures book and in general, test and form the curriculum for a BS in aero engineering.*
>
> *Over the 50 years, members of the class of '43 have led or contributed to the rapid development of the aerospace industry and some astonishing achievements undreamed of in 1943. Then no one had flown an aircraft through the sonic barrier, and space flight was only accomplished by Buck Rogers in the comics. In fact, air travel was in its infancy. The new DC-3 landing at Purdue's airport was an exciting event. Airplanes were not pressurized for high altitude so that flying through weather was always uncomfortable. Speed and payload were limited by propulsion with reciprocating internal combustion engines and by propeller tip speed. Professor Wood made the point that "if you had enough power, you could fly a brick." The Navy was testing rocket or jet assisted takeoff. The Germans were delivering bombs with a pulse jet engine called a buzz bomb. The V-2 rocket they were developing, more often than not, was blowing up on the launch pad. Radar and Loran were in their infancy and only available for guidance detection on the latest military equipment.*
>
> *Since then, the members of the class have participated in the broad range of technical evolution, which has revolutionized travel and virtually every aspect of living in the 1990s. Our graduates have helped engineer the major developments in aircraft jet engines, rocket engines, and their accessories. They have done research and design in aerodynamics and structures, which have produced the finest military and civilian aircraft in the world and space achievements of greater breadth and depth than all the others of the world. Other members of the class have used their engineering in other fields with great success, and several have made teaching their careers.*

Twenty-six of the 32 original class members were still living and in many cases still active professionally in 1993. Fourteen members attended their golden anniversary reunion held at Purdue on April 24, 1993. A class photo taken at that time is given in Figure 10.2. The following comments about the 1943 class members are edited from J. R. Dunn's reunion booklet.

John L. Allen went into the Army Air Corps after graduation and spent part of the war in the 13th Air Force in the South Pacific. When he was discharged, he joined the NACA Lewis Labs in Cleveland, Ohio, where he worked until retirement in 1986.

Robert L. Beebe was called up by the Army three months before graduation, and instead of getting his BSAE, was given a BSME with an aero major. He was commissioned a lieutenant and assigned as a B-29 flight engineer. He returned to Purdue in 1946 for a graduate degree in management, and then he went to work for Kimberly-Clark Corporation in production standards and industrial engineering. Over the next 32 years, he was promoted several times, eventually becoming corporate vice president. Since retiring in 1978, he has been an adjunct professor at the National University in San Diego, teaching in the management school.

Robert W. Boswinkle went to Langley Research Center after graduation and remained there for 35 years until retiring in 1979. He worked in theoretical and experimental aerodynamic research, in aeroelasticity, flutter,

Figure 10.2 Purdue's first class of aeronautical engineers (1943) at their 50th reunion (April 1993). Left to right, back row: M. Howland, R. Herrick, P. Brink, R. Boswinkle, D. Ochiltree, J. Goldman, W. Fleming, I. Kerr. Front row: A. Streicher, C. Hagenmaier, J. Dunn, J. Allen, R. Beebe, R. Pendley.

and structural dynamics. He participated in the development and implementation of a large and highly complex wind tunnel with a 16-foot transonic and dynamic throat. He became assistant director for aeronautics in 1964 and managed the entire aeronautics program at Langley. He received his MS in aeronautical engineering from the University of Virginia and later got into theoretical physics. After retirement in 1979, he began teaching in a public high school.

Paul L. Brink joined the research department of the Curtiss-Wright Corporation at Buffalo, New York, after graduation, and participated in a flight test program for the Navy conducted at Van Nuys, California. In 1944, he enlisted in the U.S. Navy and received a direct commission. He then served as an experimental officer at the U.S. Naval Ordnance Test Station at China Lake, California, in work associated with rocket and missile development. He was discharged from the Navy in June 1946, and joined the Naval Avionics Facility at Indianapolis, Indiana, where he became director of the Applied Research Department. His own technical work related to gunfire control systems, missile guidance systems, aircraft radar development, and many other types of avionic equipment. He retired from the Naval Avionics Facility in 1978, and then joined the faculty of the School of Engineering and Technology at Indiana University-Purdue University at Indianapolis (IUPUI). He served as assistant dean for research and sponsored programs until retiring from IUPUI in 1984. He also served as a consultant to the Corporation for Science and Technology at Indianapolis after 1984.

Sherwood H. Brown went to the Ames Laboratory of the NACA after graduation, along with Leonard Rose, where both were assigned to the 7x10 foot wind-tunnel division. In 1945, he was commissioned as ensign but continued to do the same work. After being discharged from the Navy in 1945, he returned to Purdue to get a master's degree in aeronautical engineering in 1947. He then accepted a job with Northrop Aircraft Corporation and soon led aerodynamics work on the F-89 and the X-4 aircraft. Most of his 36 years at Northrop were spent as an aerodynamicist on preliminary and advanced designs of missiles, rockets, and aircraft. While working on laminar flow control for the X-21 system, he and his group invented the concept of high bypass ratio turbofan engines, which was adopted first by General Electric for the C-5 and C-5A and subsequent transport aircraft.

James R. Dunn joined Wright Aeronautical Corporation after graduation; he spent a year in the engineering and service departments assisting in the development of the R-3350 engine for the B-29. Then he spent two years in the South Pacific in the Army Air Corps, 20th Air Force, before joining General Electric. There he worked in the development of the first axial-flow gas turbine for aircraft. This engine eventually powered the B-52s. After nine years with gas turbine applications, he moved to California in 1955 to join Ramo Wooldridge Corporation. There he was responsible for the technical direction to the developer of the rocket engines for the Titan ballistic missile and later other rocket engines for space, including the Apollo Lunar Excursion Module Descent Engine. In 1968, Mr. Dunn moved from aerospace work for the application of space technology to other industries, principally petroleum and large fluid handling equipment. In 1970, he became vice president of TRW responsible for pump manufacturing both in the United States and Europe. Following retirement from TRW in 1977, he was cofounder of Energy Management Corporation, where he served as CEO until his retirement in 1989.

Robert H. Essig went to NACA's Lewis Laboratory in Cleveland, Ohio, where he spent eight years. While there, he did graduate studies at Ohio State University. In 1951, he entered the private sector, where his focus was sales and marketing. His products included biosatellites and various aircraft accessories such as hydraulic actuators, landing gear, pumps, compressors, valves, and secondary flight controls. He also was occupied with research and development in the field of material sciences.

William A. Fleming also went to the NACA Aircraft Engine Lab (Lewis) in Cleveland and quickly became involved with the altitude wind tunnel and testing of the Westinghouse turbojet. For the next 15 years he directed research and testing of the turbojets and ramjets in the various altitude facilities at Lewis. He ran the first turbojet afterburner at altitude conditions, which gave General Electric the technology to design the first operational afterburner for the F-86 in Korean duty. In 1960 he moved to NASA headquarters in Washington, where he gravitated to program management. He directed the team that established the program plan for Apollo used as a guide for placing a man on the moon. During his 14 years at NASA headquarters, he held a number of executive positions, including assistant administrator for programs, director of program review, and senior technical officer. He retired from NASA in 1973 and spent the next 12 years as a consultant. After retiring from active business in 1985, he devoted considerable time at the Air and Space Museum on the history of the gas turbine engines, and worked with the curator for air propulsion as a coauthor of the history of the small aircraft gas turbine industry in the United States.

Robert J. Gatineau completed his BS in aero following the war (1946), and continued on for an MS in 1948.

William J. Gaugh went to New York University after graduation. He was an instructor there for a year, and then he went to MIT as a doctoral candidate, again serving as a staff instructor. While at MIT he accepted an offer to head the operation of a new wind tunnel at Texas A&M. However, he later discovered the wind tunnel was not powered and would not be for five years. Hence, he joined the Northrop Corporation in 1947 and remained there until retirement in 1988. While at Northrop, he contributed to a variety of projects, including the design and launching of the Mariner 69 in conjunction with the Jet Propulsion Laboratory. He also was active in the Snark missile, the Flying Wing, and several other significant advances in the aerospace industry. Among his wide range of projects was the direction of the technical analysis, which confirmed the feasibility of the conversion of the B-29/cargo airplane to the Guppy configuration for transport of the shuttle. In his years as part of the Northrop Preliminary Design Group, he was on the team that designed the B-2 bomber.

Jerome M. Goldman was called to active duty by the Army Air Corps before he could get his aeronautical engineering degree. He returned to campus after his discharge and actually graduated in January 1947. While in the service he flew B-29s in the South Pacific, and was stationed on Saipan and Guam in the Mariana Islands with the 20th Air Force. After graduation, he went to New York as a pilot for United Air Lines until 1953. He went back to Purdue to help start Purdue Airlines, and served as chief pilot and director of operations, and then as vice president for operations and maintenance. The airline operated DC-3, DC-6, and DC-9 aircraft. The operation was shut down in December 1972 under his direction, and an attempt to restart the airline with the help of Eli Lilly was unsuccessful. He also earned his MS in E.D. degree while in the Purdue area, and then set up a Lear Jet recurrent audio visual training program for the International Learning Systems group in Greenwich Connecticut. He later joined an air cargo company and served as a line captain until joining the Federal Aviation Administration in New York. He was an air carrier operations aviation safety inspector until retirement from the FAA in 1990. He continued to serve as technical consultant and expert witness for law firms and insurance carriers.

John C. Harrison went to Wright Aeronautical Corporation in the service engineering department. His assignments were to keep the B-29 engines running and to feed back any failure data that could help improve the life of the R-3350. After a year, he was called into the Navy, where he served in the South Pacific on a rocket-carrying landing craft and participated in the battle for Okinawa. Following cessation of hostilities, he volunteered for duty at the Bikini Atoll test site for nuclear bomb tests, and was the first person to enter the lagoon after the tests. As a result, he unknowingly exposed himself to radiation sickness and had to spend six months in Bethesda Naval

Hospital. He was plagued with that ailment for the rest of his life. He was discharged from the Navy and joined a very successful family construction company, where he operated as vice president until retirement in 1985. He died in March 1991.

Robert N. Herrick was drafted into service with the Army Air Corps after graduation and became an aircraft crew chief. He was stationed in the South Pacific and flew between the islands of New Guinea and the Philippines. He was discharged in December 1945 and joined the Chance Vought engineering department. When the company decided to move from Connecticut to Texas, he stayed in Connecticut and bought and managed a hardware store for five years. He went back into engineering at an Avco Lycoming engine plant and was layout design engineer before transferring to the new aerospace division, where he was involved in the design of an early re-entry vehicle nose cone. In 1956, the aerospace division moved to Lawrence, Massachusetts, and he became assistant section chief of the model design section for wind tunnel testing. He joined The Boeing Company in Seattle, Washington, in 1966 and was senior design engineer for the Supersonic Airplane Structure, the MX missile, the Cruise Missile, and many other projects. He retired in 1983. (His brother, Thomas J. Herrick, served on the faculty of the Purdue School of Aeronautics and Astronautics from 1945 to 1980).

Merville C. Howland stayed at Purdue as an instructor of Curtiss Wright Cadettes for almost a year after graduation. He joined the Navy in 1944 and was in charge of airplane maintenance, but preferred engine system development. Upon leaving the Navy, he spent three years in the design and development of automatic packaging machinery. In 1950, he joined Thompson Products (now TRW) to develop aircraft accessories such as a centrifugal fuel pump rotating at 25,000 rpm. Then in 1955, he went to Bendix Aviation to do research and development of engine controls systems for several aircraft, including the F-104 and the B-70. In 1959, he moved to Woodward Governor Company to develop very precise speed controls for aircraft jet engines, diesel locomotives, cruise missiles, etc., and with a significant background in control theory, moved into full authority digital electronic controls.

Lewis P. Jones received his BS in 1943. He died on July 23, 1985. Although little is known about his career after he left Purdue, he was chief of program reviews and reports with NASA in 1968. The 1982 alumni directory lists him as head, Audio-Visual Department, Saudi Arabian National Center for Science and Technology, Riyadh, Saudi Arabia. He is quoted in a 1969 *Purdue Alumnus* magazine article [6] with the following recollections about his student days: *"The 1943 class in aero engineering was unique from the standpoint that we were the smallest school on campus and the first to be awarded the BSAE degree. I recall vividly how we sweated during the closing months of our program as we were paced in our aircraft design courses by the rebuilding of the wind tunnel. It was not available for testing our design models until nearly the end of the school term. We had only one month before graduation to schedule and run wind tunnel tests, compile results, perform necessary engineering data analysis and calculations, and then prepare individual final reports of findings and conclusions—all in the space of about three weeks."*

H. Irving Kerr stayed at Purdue after graduation as an instructor in aeronautical engineering until December 1943. He then joined Douglas Aircraft Company until called to the Army Air Corps/Signal Corps as an instructor. After discharge from the service, he rejoined Douglas (now McDonnell Douglas) and was still there in 1993. He has participated in the design and construction of the DC-5, DC-7, DC-8, DC-9, DC-10, the KC-10, MD-11, MD-12, and the C-17. He moved to the Astronautics Company from 1952 to 1969 and worked on several projects there. He retired from McDonnell Douglas in 1986, but was immediately put on a consulting contract.

Stan H. Lowy joined the Army Air Corps and was an engineering officer as well as a flight engineer on a B-29. On discharge from active duty, he enrolled at the University of Minnesota to get his master's degree in aeronautical engineering. Then in 1947 he joined the Allison Division of General Motors in Indianapolis as a test engineer. In the same year, he became an instructor in mechanical engineering at Oregon State College, where he remained for three years. In the next eight years, he had several jobs in industry, including Hughes Aircraft Company, and also did consulting engineering. He then spent five years at the University of Oklahoma as an assistant professor in aeronautical engineering. He moved to Texas A&M University in 1964 as a professor of aeronautical engineering and later as assistant dean of engineering. He retired to halftime in 1986 but remained as a professor and dean emeritus until fully retiring in 1991. He received numerous honors and awards, including Outstanding Aeronautical Engineer Educator from AIAA and ASEE in 1977.

John I. Nestel was called to active duty before he finished his requirements for the aeronautical engineering degree. He founded Consolidated Airborne Systems in Long Island, New York, a company that produced various airborne instrumentation. He received the Purdue Distinguished Engineering Alumnus Award in 1974. After a distinguished flying career during World War II, he returned to Purdue after discharge and completed his work in 1947.

David W. Ochiltree went to NACA Langley Field after graduation and was assigned to the structures laboratory. In 1946, he was transferred to NACA headquarters in Washington, D.C., to become secretary to the Aircraft Structures Committee. In 1947, he changed course in his career and joined the construction company, Ceco Corporation. He eventually became vice president of its eastern operations, and retired in 1987.

E. C. Olsen (unknown)

Robert E. Pendley went to Langley Field in Virginia after graduation, where he spent nine years in research on transonic and supersonic aerodynamics, high-speed propellers, jet engine inlet and exhaust ducts, and transonic wind tunnel techniques. In 1954, he joined the Douglas Aircraft Company as supervisor of aerothermodynamics, aerodynamic design, and performance of engine installations for the F-4D, F-5D, A-3, and A-4 programs. From 1959 to 1967, he was assistant chief, power plant section, where engine installation and fuel design were made for fighter aircraft and for the DC-8. In 1967, he became manager of advanced design and research programs for the power plant engineering subdivision, and in this capacity managed the NASA contract for the design and flight demonstration of a turbofan noise reduction system for the DC-8. He was promoted to director, acoustic engineering subdivision, where he conducted research, design, demonstration and certification efforts related to interior and exterior noise for advanced and production programs developing and counseling flight crews on noise reduction procedures. From 1984 to retirement in 1987, he was director of aircraft configuration performance and director of regulatory affairs.

Leonard M. Rose joined the NACA Ames Laboratory at the Naval Air Station in Moffett Field, California. He was commissioned an ensign and continued to operate in the 7x10-foot wind tunnel division. After discharge from the Navy, he joined the North American Aviation Corporation in 1951 as an aerodynamics engineer. He held a number of project assignments on the F-100A Supersabre, F-107A, X-15, F-108, and the XB-70. He was appointed vice president of research and engineering in 1968 at the Columbus, Ohio, division of North American Rockwell. After leaving what became known as Rockwell International, he went to Northrop, where he also headed up research for some eight years. He retired in 1989 and died in 1990. He received the Distinguished Engineering Alumnus Award from the Purdue Schools of Engineering in 1970.

Ralph Salisbury died in an airplane crash in 1955.

Leslie E. Schneiter (deceased, 1962).

Paul M. Schroeder (deceased January 4, 1981).

Kenneth R. Scudder was taken out of school by the Navy V-12 program before graduation when he lacked only a semester toward his aeronautical engineering degree. Although he was permitted some leeway, he finished with an ME degree instead. He went to Annapolis on his way to serving 12 years in the Navy. During the war, he was chief engineer on patrol craft. He later attended the Naval Post-Graduate school in Monterey, California, and was assigned to the Office of Naval Research. He saw duty at the David Taylor Model Basin and became assistant design officer for the construction of the USS Forestall aircraft carrier. After leaving the Navy in 1957, he joined the Harry Diamond Laboratories, which was an offshoot of the National Bureau of Standards, to assist in the development of the science of fluidics. His achievements were, among others, the making of a no-moving-parts gyro, and precise fuel control on jet engines as a function of turbine blade temperature

Walter D. Smith worked for Consolidated Vultee Aircraft Corporation on the XB-36 and the XC-99 before joining the Navy, where he was assigned to the Office of Naval Intelligence. After discharge from the Navy, he joined the Glenn L. Martin Company in Baltimore. There he worked on commercial aircraft including the Martin 202 and 303 transports. Later he participated in the design of the XB-51, the XP-6M-1, and the AM-1, and also was involved with the Viking Research Rocket. He headed the technical staff that was responsible for missile development and field test activities. He became chief engineer for the Canaveral division when the cape was set up for testing. He was appointed chief engineer for the Gemini Launch Vehicle development and was program manager for the last eight flights. He also was Martin Marietta program manager for the Titan IIIC and the IIIM. He moved to General Electric Company to become general manager of the re-entry vehicle department for the space program and then general manager of the operations re-entry systems division, from which he retired in 1983. He was given the Purdue Distinguished Engineering Alumnus Award in 1971, as well as awards from NASA and the Air Force Systems Command.

Wilbur A. Spraker left Purdue for the war in March 1943, a semester short of a degree. He served in the Army Air Corps as communications officer running control towers and crypto centers. He came back to Purdue in 1946 to finish his work for a BSAE and stayed on to get a BSME and an MS in aeronautical engineering. He joined Pratt & Whitney in 1948 in the compressor design group, and worked on the T-34, J-48, and second and third J-57 compressors, which were the engines for the B-52 and 707 aircraft. He then went to Convair Ft. Worth early in the B-58 program and from there to Battelle Memorial Institute to work on pumps, fans, blowers, and aircraft instrumentation, as well as heat and mass transfer. In 1971, he moved to Creare in Hanover, New Hampshire, to design and develop compressors and turbines. He then went to Schwitzer in Indianapolis to manage aerodynamic design of turbosuperchargers. He retired as chief engineer in 1988, but continued to consult for Schwitzer and others on radial compressor, pump, and fan design.

Albert H. Streicher was called into the service, before he could graduate, in order to become part of an intensive training group commissioned as Army officers. He went to the University of Alabama and eventually ended up in the Philippines. When discharged, he returned to Purdue and received his degree in 1947. He joined the Glenn L. Martin Company and for a year worked in the pilotless aircraft section on the Viking earth satellite. He also was involved with the MX-774, the Matador, and the Mace rockets. After several assignments in the Chicago area, he joined Bauer & Black Company doing various machine design work. In 1957, he went to the University of California Lawrence Radiation Labs in Livermore, California, where he performed engineering tasks in various support equipment projects. When he retired in 1985, he was in charge of stockpile surveillance of all Navy weapons, including Polaris, Poseidon, Terrier, etc.

Ralph B. Trueblood went to Aeronca Aircraft in Ohio for six months before being called into the Navy. When discharged in 1946, he went back to Aeronca, but moved to Fairchild Personal Plane Division in Kansas. When this division was closed in 1948, he joined the sponsored research staff at MIT and remained there until retirement in 1976. His MIT efforts involved design of aircraft and missile navigation and guidance equipment as well as control devices, and he was head of the Army projects division of the C. S. Draper Laboratory when he retired. He served as associate director of research pro-tem for Clarkson Institute of Technology until it located a permanent director. In 1980, he moved to New Hampshire and set up a consulting business in the field of automatic control systems. During his career, he served on blue-ribbon panels reviewing various weapons as well as the SST. He participated in NATO activities in Europe and New York. He was awarded a gold medal by the New York Academy of Sciences for a paper on traffic control for VTOL aircraft.

Frank E. West went to NACA Langley Field, Virginia, when he graduated and was placed in the Army Air Corps enlisted reserves. During the war, he worked on aerodynamics associated with aircraft development, weapons characteristics, and in-flight failures. After the war, he did applied research and exploratory development involving flow phenomena, aerodynamic loads, stability, and control. These studies were initially at transonic speeds, but this group concentrated on subsonic and hypersonic speed ranges. In 1961, he moved to the Navy's Taylor Research Center as head of the transonic division. Later he became assistant to the laboratory director when emphasis shifted from advanced development to applied research and exploratory development. He retired in 1978.

Charles M. Zimney went to the NACA laboratory in Virginia for a short time and then into the Army Air Corps Air Technical Service Command. Here he began a career in atmospheric investigations and instrumentation. After service, he returned to Purdue and received his master's degree in aeronautical engineering, and then joined the faculty at the University of Minnesota. He then went to the Jet Propulsion Laboratory in Pasadena, California, and worked on the LOCI and other potential spacecraft. He joined the Cooper Corporation for a short while in 1958 before founding the Zimney Corporation. As head of that company, he guided the design and manufacture of many atmospheric test packages and upper atmosphere scientific vehicles. His company produced the only product that was fired into the mushroom cloud of a nuclear bomb and retrieved a sealed gas sample. His products also established altitude records for atmospheric instruments. He later joined the Hughes Helicopter Company and later still, the Telatlantic Company, from which he retired in 1990.

Astronaut Alumni

Purdue alumni have made substantial contributions to the design, construction, and flight of the many spacecraft, which have provided the opportunity to explore the frontiers of space during the past 50 years. As mentioned previously, alumni interest in pushing flight technology to its limits traces back to the beginning of the century when J. Clifford Turpin (class of 1908) set an altitude record in 1911 by flying an aircraft to 9,400 feet. Although this achievement pales with more recent astronaut accomplishments, it does indicate the start of a long tradition in which Purdue graduates have extended the boundaries of flight.

Purdue's involvement in the modern space age begins, perhaps, with Captain Iven C. Kincheloe (BSAE '49). (A photograph of Captain Kincheloe was given previously in Figure 4.19.) An ace during the Korean War, Captain Kincheloe was then assigned by the Air Force to be a test pilot. He flew an X-2 aircraft to an altitude of 126,000 feet (23.9 miles) in 1956, reaching the threshold of space. He was then selected to fly the X-15 into outer space sometime in 1959, but was killed in a training accident on July 26, 1958. The X-15 approach to putting a man into orbit changed with the advent of the

Russian Sputnik and the subsequent space race in the late 1950s and 1960s. The new effort involved the use of manned capsules propelled into orbit by multi-stage rockets, and led to the U.S. Project Mercury and to the later Apollo missions.

Virgil I. "Gus" Grissom (BSME '50) was one of the original seven astronauts selected for the U.S. Mercury program. Aboard his Mercury 4 capsule on July 21, 1961, he became the second American to fly into space. He orbited the earth in Gemini 3 on March 23, 1965, but was then killed in a flash fire during a simulated launch of the Apollo spacecraft on January 27, 1967. Also killed with Mr. Grissom in that fire were school graduate Roger B. Chaffee (BSAE. '57) and Ed White. (Grissom Hall, location of the main offices of the School of Aeronautics and Astronautics from 1967 to 2007, and Chaffee Hall, home to the Thermal Sciences and Propulsion Center, were renamed in honor of these two alumni.) Subsequent alumni to become astronauts were Neil A. Armstrong (BSAE '55) and Eugene A. Cernan (BSEE '56), both of whom had space flights during the Gemini program in 1966, and later made lunar landings. Neil Armstrong was the first person to set foot on the moon on July 20, 1969, aboard Apollo 11, and Eugene Cernan has the distinction of being the last man to leave the moon during his Apollo 17 flight in December 1972. Mr. Cernan also circled the moon aboard Apollo 10 in May 1969.

Following the lunar flights of 1969 and the early 1970s, the nation's manned space effort took a different course with the Space Shuttle missions. Nineteen alumni have been selected for Space Shuttle flights, and all but one of those individuals had flown into space by 2009. In total, Purdue has had 23 astronaut alumni, including Roger Chaffee, who was killed in training. A space-flight log that summarizes alumni space travel is given in Table 10.2. It is interesting to note that by mid-2009, Purdue alumni had flown on 35 percent (55/108) of all U.S. manned space flights, and that 24 percent (38/158) of U.S. space flights had at least one crew member who graduated from the School of Aeronautics and Astronautics. Eight U.S. space missions have flown two Boilermakers as crew members. Photographs of the Purdue astronaut alumni are given in Figure 10.3. Purdue alumni have spent more than 1.9 years in space, believed to be the most time in space accumulated by graduates of any U.S. university.

It is fitting to close this section about Purdue's astronaut alumni with the following 1995 statement from Purdue President Steven C. Beering: *"Having been associated with the U.S. space program since 1960, I am particularly proud that so many Purdue graduates are serving as astronauts. A number of their children are now enrolled here and are preparing themselves to be leaders of the next generation of space explorers. At least three of these students are actively involved*

Table 10.2 Purdue Alumni Space-Flight Log
(source: http://www.nasa.gov/missions/index.html)

Mission	Crew (Purdue alumni in boldface)	Date	Elapsed mission time hrs:min:sec	Cumulative Purdue alumni hrs
PROJECT MERCURY				
Mercury-Redstone 4	**Grissom**	July 21, 1961	00:15:37	00:15:37
GEMINI PROGRAM				
Gemini-Titan III	**Grissom,** Young	March 23, 1965	04:52:31	5:08:08
Gemini-Titan VIII	**Armstrong,** Scott	March 16, 1966	10:41:26	15:49:34
Gemini-Titan IX-A	Stafford, **Cernan**	June 3-6, 1966	72:20:50	88:10:24
APOLLO PROGRAM				
Apollo-Saturn 10	Stafford, Young, **Cernan**	May 18-26, 1969	192:03:23	280:13:47
Apollo-Saturn 11	**Armstrong,** Collins, Aldrin	July 16-24, 1969	195:18:35	475:32:22
Apollo-Saturn 17	**Cernan,** Evans, Schmitt	Dec. 7-19, 1972	301:51:59	777:24:21
SPACE TRANSPORTATION SYSTEM (SPACE SHUTTLE)				
STS 41-D Discovery	Hartsfield, Coats, Resnik, Hawley, Mullane, **C. Walker**	Aug. 30-Sept. 5, 1984	144:56:04	922:20:25
STS 51-C Discovery	Mattingly, **Shriver,** Onizuka, Buchli, **Payton**	Jan. 24-27, 1985	73:33:23	1,069:27:11
STS 51-D Discovery	Bobko, **Williams,** Seddon, Hoffman, Griggs, **C. Walker,** Garn	April 12-19, 1985	167:55:23	1,405:17:57
STS 51-F Challenger	Fullerton, **Bridges,** Musgrave, England, Henize, Acton, Bartoe	July 29-Aug. 6, 1985	190:45:26	1,596:03:23
STS 51-I Discovery	Engle, **Covey,** van Hoften, Lounge, W. Fisher	Aug. 27-Sept. 3, 1985	170:17:42	1,766:21:05
STS 61-B Atlantis	Shaw, O'Connor, Cleave, Spring, **Ross,** Neri-Vela, **C. Walker**	Nov. 26-Dec. 3, 1985	165:04:49	2,096:30:43
STS 26 Discovery	Hauck, **Covey,** Hilmers, Lounge, Nelson	Sept. 29-Oct. 3, 1988	97:00:11	2,193:30:54
STS 27 Atlantis	Gibson, **Gardner,** **Ross,** Shepherd, Mullan	Dec. 2-6, 1988	105:05:37	2,403:42:08
STS 29 Discovery	Coats, **Blaha,** Buchli, Springer, Barian	March 13-18, 1989	119:38:52	2,523:21:00
STS 28 Columbia	Shaw, Richards, Adamson, **Brown,** Leestma	Aug. 8-13, 1989	121:00:08	2644:21:08

Table 10.2 Purdue Alumni Space-Flight Log, continued

Mission	Crew (Purdue alumni in boldface)	Date	Elapsed mission time hrs:min:sec	Cumulative Purdue alumni hrs
STS 34 Atlantis	**Williams, McCulley,** Lucid, Baker, Chang-Diaz	Oct. 18-23, 1989	119:39:20	2,883:39:48
STS 33 Discovery	Gregory, **Blaha,** Musgrave, Carter, C. Thornton	Nov. 22-27, 1989	120:06:49	2,003:46:37
STS 36 Atlantis	Creighton, **Casper,** Mullane, Helmers, Thuot	Feb. 28-March 4, 1990	106:18:22	3,110:04:59
STS 31 Discovery	**Shriver,** Bolden, Hawley, McCordless II, K. Sullivan	April 24-29, 1990	121:16:06	3,231:21:05
STS 38 Atlantis	**Covey,** Culbertson, Springer, Meade, Gemar	Nov. 15-20, 1990	117:54:31	3,349:15:36
STS 35 Columbia	Brand, **Gardner,** Hoffman, Lounge, Parker, Durrance, Parise	Dec. 2-10, 1990	215:05:08	3,564:20:44
STS 37 Atlantis	Nagel, Cameron, Apt, **Ross,** Godwin	April 5-11, 1991	143:32:44	3,707:53:28
STS 39 Discovery	Coats, Hammond, **Harbaugh,** McMonagle, Bluford, Hieb, Veach	April 28-May 6, 1991	199:22:23	3,907:15:51
STS 43 Atlantis	**Blaha,** Baker, Low, Lucid, Adamson	Aug. 2-11, 1991	213:21:25	4,120:37:16
STS 48 Discovery	Creighton, Reightler, **Brown,** Buchli, Gemar	Sept. 12-18, 1991	128:27:38	4,249:04:54
STS 46 Atlantis	**Shriver,** Allen, Hoffman, Chang-Diaz, Nicollier, Ivins, Malerba	July 31-Aug. 8, 1992	191:15:03	4,440:19:57
STS 54 Endeavour	**Casper,** McMonagle, Runco, **Harbaugh,** Helms	Jan. 13-19, 1993	143:38:19	4727:36:35
STS 55 Columbia	Nagel, Henricks, **Ross,** Precourt, Harris, Walter, Schlegel	April 26-May 6, 1993	239:39:59	4,967:16:34
STS 57 Endeavour	Grabe, Duffy, Low, Sherlock, Wisoff, **Voss**	June 21-July 1, 1993	239:44:54	5,207:01:28
STS 58 Columbia	**Blaha,** Searfoss, Seddon, McArthur, **Wolf,** Lucid, Fettman	Oct. 18-Nov. 1, 1993	336:12:32	5,879:26:32

Table 10.2 Purdue Alumni Space-Flight Log, continued

Mission	Crew (Purdue alumni in boldface)	Date	Elapsed mission time hrs:min:sec	Cumulative Purdue alumni hrs
STS 61 Endeavour	**Covey,** Bowersox, Musgrave, K. Thornton, Nicollier, Hoffman, Akers	Dec. 2-12, 1993	259:58:33	6,139:25:05
STS 62 Columbia	**Casper,** Allen, Thuot, Ivins, Gemar	March 4-18, 1994	335:16:41	6,474:41:46
STS 63 Columbia	Weatherby, Collins, Harris, Foale, **Voss,** Titov	Feb. 2-11, 1995	198:28:15	6673:10:01
STS 71 Atlantis	Gibson, Precourt, Baker, **Harbaugh,** Dunbar, Solovyev, Budarin, Thaggard, Dezhurov, Strekalov	June 26-July 7, 1995	235:22:17	6,908:32:18
STS 70 Discovery	Henricks, Kregel, Thomas, Currie, **Weber**	July 13-22, 1995	214:20:05	7,122:52:23
STS 74 Atlantis	Cameron, Halsell, **Ross,** McArthur, Hadfield	Nov. 12-20, 1995	196:31:42	7,319:24:05
STS 77 Endeavour	**Casper,** Brown, Bursch, Runco, Garneau, Thomas	March 19-29, 1996	240:39:18	7,560:03:23
STS 79 Atlantis	Readdy, Wilcutt, Akers, Apt, Walz, **Blaha** (depart)	Sept. 16, 1996	3077:27:55 *(includes Blaha's time aboard Space Station Mir)*	10,637:31:18
and STS 81 Atlantis	Baker, Jeff, Grunsfeld, Ivins, Wisoff, Linenger, **Blaha** (return)	Jan. 22, 1997		
STS 82 Discovery	Bowersox, Horowitz, Lee, Hawley, **Harbaugh,** Smith, Tanner	Feb. 11-21, 1997	239:37:09	10,877:08:27
STS 83 Columbia	Halsell, Still, **Voss,** Thomas, Gernhardt, Crouch, Linteris	April 4-8, 1997	95:13:38	10,972:22:05
STS 94 Columbia	Halsell, Still, **Voss,** Thomas, Gernhardt, Crouch, Linteris	July 1-17, 1997	376:44:34	11,349:06:39

Table 10.2 Purdue Alumni Space-Flight Log, continued

Mission	Crew (Purdue alumni in boldface)	Date	Elapsed mission time hrs:min:sec	Cumulative Purdue alumni hrs
STS 86 Atlantis and	Wetherbee, Bloomfield, Titov, Parazynski, Chretien, Lawrence, **Wolf** (depart)	Sept. 25, 1997	3079:00:50 (includes Wolf's time aboard Space Station Mir)	14,428:07:29
STS 89 Endeavour	Wilcutt, Edwards, Dunbar, Anderson, Reilly, Sharipov, Thomas, **Wolf** (return)	Jan. 31, 1998		
STS 88 Endeavour	Cobana, Sturckow, Currie, **Ross,** Newman, Krikalev	Dec. 4-15, 1998	283:18:47	14,711:26:16
STS 99 Endeavour	Kregel, Gorie, Kavandi, **Voss,** Mohri, Thiele	Feb. 11-22, 2000	269:39:41	14,981:05:57
STS 101 Atlantis	Halsell, Horowitz, **Weber,** Williams, **Voss,** Helms, Usachev	May 19-29, 2000	236:36:00	15,217:41:57
STS 98 Atlantis	Cockrell, **Polansky,** Curbeam, Jones, Ivins	Feb. 7-20, 2001	309:21:00	15,527:02:57
STS 110 Atlantis	Ochoa, **Ross,** Walheim, Smith, Morin	April 8-19, 2002	259:42:44	15,786:45:41
STS-112 Atlantis	Ashby, Melroy, **Wolf,** Sellers, Magnus, Yurchikhin	October 7-18, 2002	259:58:44	16,046:44:25
STS-116 Discovery	Oefelein, Higginbotham, **Polansky,** Curbeam, Patrick, Williams, Fuglesang	December 9-22, 2006	308:45:16	16,355:29:41
STS-125 Atlantis	Massimo, Good, Johnson, Altman, McArthur, Grusfeld, **Feustel**	May 11-24, 2009	309:37:09	16,665:06:50
STS-127 Endeavor	**Polansky,** Hurley, Cassidy, Payette, **Wolf,** Marshburn, Kopra, Wakata	July 15-31, 2009	376:42	17,041:48:50

Neil A. Armstrong

BS '55, Aeronautical Engineering (Purdue)

MS '70, Aerospace Engineering (University of Southern California)

Honorary doctorate '70 (Purdue)

Command pilot, Gemini 8, March 16, 1966

Commander, Apollo 11, July 16–24, 1969 (first manned lunar landing)

Eugene A. Cernan

BS '56, Electrical Engineering (Purdue)

MS '64, Aeronautical Engineering (U.S. Naval Postgraduate School)

Honorary doctorate '70 (Purdue)

Pilot, Gemini 9, June 3–6, 1966

Lunar module pilot, Apollo 10, May 18–26, 1969

Commander, Apollo 17, Dec. 7–19, 1972 (last manned lunar landing)

Virgil I. Grissom

BS '50, Mechanical Engineering (Purdue)

Pilot, Mercury 4, July 21, 1961

Command pilot, Gemini 3, March 23, 1965

Roger B. Chaffee

BS '57, Aeronautical Engineering (Purdue)

Grissom and Roger Chaffee were killed on Jan. 27, 1967, during a simulated launch of their Apollo spacecraft.

Figure 10.3 Purdue Alumni Selected for Space Flight. Thirty-seven of the 101 U.S. manned space flights completed by July 1995 had at least one crew member who graduated from Purdue.

John E. Blaha

BS '65, Engineering Science (U.S.A.F. Academy)

MS '66, Astronautics (Purdue)

Pilot, STS 29 (Discovery), March 13-18, 1989

Pilot, STS 33 (Discovery), Nov. 22-27, 1989

Commander, STS 43 (Atlantis), Aug. 2-11, 1991

Commander, STS 58 (Columbia), Oct. 18-Nov. 1, 1993

Mission Specialist, STS 79 (Atlantis) and STS 81 (Atlantis), Sept. 16, 1996-Jan. 22, 1997 (includes time aboard Space Station Mir)

Roy D. Bridges, Jr.

BS '65, Engineering Science (U.S.A.F. Academy)

MS '66, Astronautics (Purdue)

Pilot, STS 51-F (Challenger), July 29-Aug. 6, 1985

Mark N. Brown

BS '73, Aeronautical & Astronautical Engineering (Purdue)

MS '80, Astronautical Engineering (U.S.A.F. Institute of Technology)

Mission specialist, STS 28 (Columbia), Aug. 8-13, 1989

Mission specialist, STS 48 (Discovery), Sept. 12-18, 1991

John H. Casper

BS '66, Engineering Science (U.S.A.F. Academy)

MS '67, Astronautics (Purdue)

Pilot, STS 36 (Atlantis), Feb. 28-March 4, 1990

Commander, STS 54 (Endeavour), Jan. 13-19,1993

Commander, STS 62 (Columbia), March 4-18, 1994

Commander, STS 77 (Endeavour), May 19-29, 1996

Richard O. Covey

BS '68, Engineering Science (U.S.A.F. Academy)

MS '69, Aeronautics & Astronautics (Purdue)

Pilot, STS 51-1 (Discovery), Aug. 27-Sept. 3, 1985

Pilot, STS 26 (Discovery), Sept. 29-Oct. 3, 1988

Commander, STS 38 (Atlantis), Nov. 15-20, 1990

Commander, STS 61 (Endeavour), Dec. 2-12, 1993

Andrew J. Feustel

BS '89, Solid Earth Sciences (Purdue)

MS '91, Geophysics (Purdue)

PhD '95, Queen's University

Mission specialist, STS 125 (Atlantis), May 11-24, 2009

Guy S. Gardner

BS '69, Engineering Sciences, Astronautics, and Mathematics (U.S.A.F. Academy)

MS '70, Aeronautics & Astronautics (Purdue)

Pilot, STS 27 (Atlantis), Dec. 2-6, 1988

Pilot, STS 35 (Columbia), Dec. 2-10, 1990

Gregory J. Harbaugh

BS '75, Aeronautical & Astronautical Engineering (Purdue)

MS '86, Physical Science (University of Houston)

Mission specialist, STS 39 (Discovery), April 28-May 6, 1991

Flight engineer, STS 54 (Endeavour), Jan. 13-19, 1993

Mission specialist, STS 71 (Atlantis), June 26-July 7, 1995

Mission Specialist, STS 82 (Discovery), Feb. 11-21, 1997

Figure 10.3 Purdue Alumni Selected for Space Flight (continued).

Michael J.McCulley

BS '70, MS '70, Metallurgical Engineering (Purdue)

Pilot, STS 34 (Atlantis), Oct. 18–23,1989

Gary E. Payton

BS '71, Astronautical Engineering (U.S.A.F. Academy)

MS '72, Aeronautics & Astronautics (Purdue)

Defense payload specialist, STS51-C (Discovery), Jan. 24–27, 1985

Mark L. Polansky

BS '78, MS '78, Aeronautical and Astronautical Engineering (Purdue)

Pilot, STS 98 (Atlantis), Feb 7–20, 2001

Commander, STS 116 (Discovery), Dec 9–22, 2006

Commander, STS 127 (Endeavour), July 15–31, 2009

Jerry L. Ross

BS '70, MS '72, Mechanical Engineering (Purdue)

Mission specialist, STS 61-B (Atlantis), Nov. 26–Dec.3,1985

Mission specialist, STS 27 (Atlantis), Dec. 2–6, 1988

Mission specialist, STS 37 (Atlantis), April 5–11, 1991

Mission specialist, STS 55 (Columbia), April 26–May 6, 1993

Mission Specialist, STS 74 (Atlantis), Nov 12–20, 1995

Mission Specialist, STS 88 (Endeavour), Dec 4–15, 1998

Mission Specialist, STS 110 (Atlantis), April 8–19, 2002

Commander, STS 127 (Endeavour), July 15–31, 2009

Janice Voss

BS '75, Engineering Science (Purdue)

MS '77, Electrical Engineering (MIT)

PhD. '87, Aeronautics/ Astronautics (MIT)

Mission specialist, STS 57 (Endeavour), June 21–July 1, 1993

Mission specialist, STS 63 (Discovery), February 2–11,1995

Payload Commander, STS 83 (Columbia), April 4–8, 1997

Payload Commander, STS 94 (Columbia), July 1–17, 1997

Mission Specialist, STS 99 (Endeavor), Feb 11–22, 2000

Loren J. Shriver

BS '67, Aeronautical (U.S.A.F. Academy)

MS '68, Astronautics (Purdue)

Pilot, STS 51-C (Discovery), Jan. 24–27, 1985

Commander, STS 31 (Discovery), April 24–29, 1990

Commander, STS 46 (Atlantis), July 31–Aug. 8, 1992

Scott Tingle

BS, Southeastern Massachusetts University (University of Massachusetts Dartmouth)

MS '88, Mechanical Engineering (Purdue)

Selected for astronaut training starting August 2009

Charles D. Walker

BS '71, Aeronautical & Astronautical Engineering (Purdue)

Payload specialist, STS 41-D (Discovery), Aug. 30–Sept. 5, 1984

Payload specialist, STS 51-D (Discovery), April 12–19, 1985

Payload specialist, STS 61-B (Atlantis), Nov. 26–Dec. 3, 1985

Figure 10.3 Purdue Alumni Selected for Space Flight (continued).

Mary E. Weber

BS '84, Chemical Engineering (Purdue)

PhD '88, Physical Chemistry (University of California, Berkeley)

Mission specialist, STS 70 (Discovery), July 13–22, 1995

Mission Specialist, STS 101 (Atlantis), May 19–29, 2000

Donald E. Williams

BS '64, Mechanical Engineering (Purdue)

Pilot, STS 51-D (Discovery), April 12–19, 1985

Commander, STS 34 (Atlantis), Oct. 18–23, 1989

David A. Wolf

BS '78, Electrical Engineering (Purdue)

MD '82 (Indiana University)

Mission specialist, STS 58 (Columbia), Oct. 18–Nov. 1, 1993

Mission Specialist, STS 86 (Atlantis) and STS 89 (Endeavour), Sept 25, 1997–Jan 31, 1998 (includes time aboard Space Station Mir)

Mission Specialist, STS 127 (Endeavour), July 15-31, 2009

Figure 10.3 Purdue alumni selected for space flight (continued).

Figure 10.4 Alumni astronauts pose with Purdue President France Córdova at the Neil Armstrong Hall dedication dinner, October 26, 2007. **Back,** *left-right: Mark Brown, Jerry Ross, Gregory Harbaugh, Janice Voss, Andrew Feustel , Mark Polansky, David Wolf, John Blaha, Charles Walker, Michael McCulley, Donald Williams;* **Front:** *Gary Payton, Neil Armstrong, France Córdova, Gene Cernan, Loren Shriver, Richard Covey (Photo by Vincent Walter)*

in planning the mission to Mars. Furthermore, our food scientists, under NASA sponsorship, are evaluating and testing methods for growing and recycling food in space. Purdue astronauts were the first and last men on the moon (Armstrong and Cernan), and I predict will be the first on Mars."

OUTSTANDING AEROSPACE ENGINEER

As described previously in Chapter 9, the Outstanding Aerospace Engineer (OAE) Award was established in 1999 to recognize AAE alumni whose professional achievements reflect the value of an aerospace engineering degree. Recipients of this award are summarized in Appendix F. As of October 2009, 101 of the school's approximately 7,000 alumni have been selected for this award.

DISTINGUISHED ENGINEERING ALUMNI

The Purdue Schools of Engineering initiated the Distinguished Engineering Alumnus Award (DEA) in spring 1964 to recognize special alumni accomplishments. Discussing this award in the 1995 DEA program brochure, Purdue President Steven C. Beering states:*"There is something very special about Purdue engineers. Their reputation precedes them, and their accomplishments set them apart. The men and women who have earned the right to call themselves Distinguished Engineering Alumni include leaders in their profession, in business, industry, education, science, and government. They have founded corporations, built cities, and walked in space. They prove that a Purdue degree is only the beginning of a lifetime of education and high accomplishment. Each year the Schools of Engineering look for graduates whose achievements are truly outstanding. There is never a shortage of qualified candidates. In recognizing them, Purdue reminds itself and its students that our standards must always be high and that the future is ours to build."*

A total of 423 of Purdue's 81,000 living engineering alumni had received this award by 2010. The following graduates from the School of Aeronautics and Astronautics have received this highly prestigious honor. Award recipients are listed in chronological order, along with their title and a biographical sketch prepared at the time of the award.

1965 **Walker M. Mahurin** *(BSAE '49), Colonel, USAF Retired, Assistant Division Director, Advanced Programs Development, Space and Information Division, North American Aviation Corporation.* He left Purdue in 1942 to join the Army Air Corps and became the leading ace in Europe until he was shot down and captured. He returned for his degree at

Purdue, then became assistant executive officer to the secretary of the Air Force. Again on combat duty, he flew 67 missions in Korea while pioneering and perfecting the F-86 dive-bombing technique, was shot down and held prisoner 16 months in solitary confinement. *[Editors' note: The first American to greet Colonel Mahurin upon his release from Korean captivity was Marine Colonel John Glenn, who subsequently became the first American to orbit the earth, and a U.S. Senator.]* Colonel Mahurin received many high military honors from U.S., British, French, and Belgian governments. He now shares administrative responsibility for the leading U.S. space contractor. (Col. Mahurin died on May 11, 2010.)

1966 **Walter J. Hesse** *(BSME '44, MSME '48, PhD '51), Vice President and Program Director, V-STOL Program, Vought Aeronautics Division, Ling-Tempco-Vought Aerospace Corporation.* He served with Test Pilot school, U.S. Navy, from 1949 to 1955, training pilots and doing research on flight testing of high-performance aircraft, meanwhile completing work for the doctorate. He joined Chance Vought in 1956 as chief of advanced development planning, became manager of advanced engineering in 1959, program director of nucleonic systems in 1961 and of advanced missile systems in 1964, moving into his present position in 1965.

1967 **Neil A. Armstrong** *(BSAE '55), NASA Astronaut.* He joined the NASA manned space program in 1962 and on March 16, 1966, he commanded the Gemini 8 mission (Figure 10.3). He received the NASA Exceptional Service Medal after this flight. A licensed pilot at 16, he was a naval aviator from 1949 to 1952. He flew 78 missions in the Korean War and was awarded the Air Medal with clusters. After graduation from Purdue, he joined NACA Lewis Flight Propulsion Laboratory, then transferred to NASA High Speed Flight Station, and did test work on the X-1 rocket airplane. He was an X-15 project pilot, flying that aircraft to over 200,000 feet and approximately 4,000 mph and receiving the Octave Chanute Award of the Institute of Aerospace Sciences for outstanding contributions to development and testing of an adaptive control system in the X-15.

1967 **Alfred F. Schmitt** *(BSAE '48, MSAE '49, PhD '53), Director of Aerospace Digital Computers, General Precision Corporation.* He left the Purdue faculty in 1955 to become a senior dynamics engineer at Ryan Aeronautical Company; then was for six years with General Dynamics as design specialist, group engineer, and section manager, responsible for flight control systems design for the Atlas ICBM weapon system, and for space boost vehicles, contributing to numerous space vehicle and satellite programs. He joined General Precision in 1963 as a chief engineer, responsible for design and support of aerospace digital computers supplied to NASA for inertial guidance and to the USAF for navigation of the C-141 global logistics transport aircraft. He assumed his present position in 1965.

1968   **Richard L. Duncan** *(BSEE '37, MSAE '48, PhD '50), Manager, Advanced Planning, Pratt & Whitney Aircraft Division United Aircraft Corp.* He entered the U.S. Navy in 1937 and served as a combat pilot in the Pacific and a project pilot at the Naval Air Test Center. He returned to Purdue after the war for postgraduate studies and then was assigned to the Office of Naval Research, where he worked with advanced propulsion systems for high-speed aircraft and missiles. Later, as assistant director, nuclear propulsion division of the Bureau of Aeronautics, he directed work on the Navy's aircraft nuclear power plant and other advanced nuclear systems. As a captain, he retired from the Navy in 1961 and joined United Aircraft as assistant to the chief scientist, later assuming his present responsibilities. *[Editors' note: In 1950, Dr. Duncan received the first PhD granted by the School of Aeronautical Engineering. By coincidence, his PhD advisor, Professor Maurice Zucrow, was the first recipient of a Purdue PhD, received in 1928 from the School of Mechanical Engineering. Dr. Duncan died on December 29, 1990, at 76 years of age.]*

1968   **Lester W. Smith** *(BSAE '48, MSE '50, PhD '52), General Manager, Ross Engineering Division, Midland-Ross Corporation.* He joined the Combustion Engineering Corporation in 1952 and began working with a group analyzing the nuclear reactor and steam generating system for the submarine Sea Wolf. In 1954, he recommended design criteria for a reactor vessel in Shippingport, Pennsylvania, and later, as assistant chief engineer, he had complete responsibility for stress and thermal analysis of the Enrico Fermi Reactor Vessel at Monroe, Michigan. He was manager of engineering, surface combustion division, from 1962 to 1967, and became assistant general manager in September 1967. He was then promoted to his present position.

1970   **Alvin L. Boyd** *(BSATR '48), Divisional Vice President-Fiscal Management, McDonnell Douglas Corporation, St. Louis, Missouri.* A veteran of World War II, Mr. Boyd graduated with highest distinction and was employed by the State of Missouri as an airport engineer before joining McDonnell as a technical writer. A year later, he was assigned duties in the contracts division and has advanced to his present position. He has successfully directed negotiations for some of the largest firm fixed-price contracts in history. He presently has the responsibility for all contract management and negotiations as well as accounting and treasury functions of the St. Louis division of the corporation.

1970 **Leonard M. Rose** *(BSAE '43), Vice President-Research and Engineering, Columbus (Ohio) Division, North American Rockwell Corporation.* Joining North American in 1951 as an aerodynamics engineer, Mr. Rose has held a number of project assignments on the F-100A Supersabre, F-107A, X-15, F-108, and the XB-70. He heads a new centralized research and engineering organization, which includes related activities of flight operations and aircraft sciences.

1971 **Walter D. Smith** *(BSAE '43), General Manager, Space Re-Entry Systems Department, Re-Entry and Environmental Systems Division, General Electric Company.* Mr. Smith is responsible for directing programs in military space re-entry, recoverable scientific satellites, space defense, and development of applications of space/defense technology for non-space uses. His work involves contracts totaling $140 million. Mr. Smith and his organization have successfully recovered from orbit more spacecraft and vehicles than the remainder of the free world combined.

1973 **William E. Cooper** *(PhD '51), Vice President and Technical Director, Teledyne Materials Research, Waltham, Massachusetts.* Dr. Cooper received his PhD in engineering mechanics in 1951. He has particular competence in the effect of material properties on pressure-retaining equipment, design criteria for water and sodium cooled nuclear power components, and analytical and experimental stress analysis.

1973 **Richard D. Freeman** *(BSATR '50), Vice President, Business Development, Autonetics Division, North American Rockwell Corporation, Anaheim, California.* Mr. Freeman worked his way up through engineering and management levels with several companies before joining Autonetics in July 1972. In this new position, he is responsible for developing strategic market plans based on current and future business environment.

1974 **Robert E. Bateman** *(BSAE '46), Vice President-Washington Operations, The Boeing Co., Washington, D.C.* His early work on engine nacelles and nozzles, his initiation and guidance in hydrofoil design, and, more recently, his financial management of the $1 billion 747 project have won much favorable attention for Mr. Bateman. His leadership in raising funds to build homes for and improve the environment of South Vietnamese Navy families resulted in increased morale and helped expedite the withdrawal of U.S. forces. For this, the Navy awarded Mr. Bateman the Distinguished Public Service Citation.

1974 **John I. Nestel** *(BSME-BSAE '47), President, Consolidated Airborne Systems, Long Island, New York.* A much-decorated World War II flier, Mr. Nestel recently graduated from the United States Navy Test Pilot school so that he could see and use, under actual Navy testing conditions, those instruments produced by his company, which began 17 years ago in his home basement. Today, Consolidated employs more than 200 skilled engineers and technicians in the design and production of aerospace, marine and ground control systems, commercial and industrial television, and temperature measurement and control equipment. Mr. Nestel is an authority on the technical needs of airborne instrumentation.

1976 **Richard M. Patrick** *(BSAE '50, MSAE '52), Vice President, AVCO-Everett Research Laboratory, Inc. Everett, Massachusetts.* Dr. Patrick, who earned a PhD from Cornell University in 1956, joined AVCO-Everett as a principal research scientist and has made many outstanding research contributions in high-temperature gas dynamics and plasmas, including development of a magnetic, annular, arc-driven shock tube as a thruster for space flight. He was appointed to his present position of vice president for advanced products in 1972, and is responsible for the development of lasers for commercial use. An outstanding recent project has been the application of high-power CO_2 laser systems to industrial metalworking and welding. Dr. Patrick has published extensively in professional journals and has been awarded five patents for his innovative work in plasma physics.

1977 **Jack D. Daugherty** *(MSESc '62, PhD '66), Vice President of AVCO-Everett Research Laboratory, Inc., Everett, Massachusetts.* Dr. Daugherty joined AVCO as a principal research scientist and has made outstanding theoretical and experimental research and development contributions in many areas, such as plasma physics, high-voltage electron devices, gas discharge physics, lasers, and laser kinetics. His work with laser projects such as Big Bang and Humdinger has greatly increased the state-of-the-art in carbon dioxide high-energy laser systems. In his current position at AVCO-Everett, he directs all laser research activities sponsored by governmental agencies. Dr. Daugherty has shown creative leadership in the varied applications of high-energy lasers and physical gas dynamics.

1978 **Wendell S. Norman** *(MSAE '58, PhD '61), Director, von Karman Gas Dynamic Facility, ARO Incorporated, AEDC Division, Arnold Air Force Station, Tennessee.* Dr. Norman began his career as a professor at the Air Force Academy. He then joined ARO as an aerodynamicist and has directed aerodynamic studies on such vehicles as the space shuttle, the X-24 research plane, the Trident missile and the cruise missiles. He has developed a number of research and test facilities, such as HIRHO, which will be a high-density, high-speed shock tunnel; and a high Reynolds number transonic test facility, which will be capable of studying high-speed, realistic flows. He was active in applying ground test data and analytical techniques to the analysis of re-entry vehicle motion. Dr. Norman also is a part-time professor of mechanical engineering at the University of Tennessee Space Institute.

1978 **Martin W. Taylor** *(BSATR '48), Vice President, Maintenance, Continental Airlines, Los Angeles, California.* Mr. Taylor has devoted his entire career to air transportation. While at Pan Am, he was in charge of the development and operation of this country's first airline jet engine overhaul facility. He joined Continental Airlines in 1968 and is currently responsible for the maintenance activities of that airline as well as Air Micronesia. Continental has one of the highest rates of aircraft utilization, which is largely credited to the extraordinary maintenance efficiency supervised by Mr. Taylor. Also under his direction are the labor relations activities and the evaluation of new aircraft.

1979 **James D. Raisbeck** *(BSAE '61), President, The Raisbeck Group, Seattle, Washington.* Mr. Raisbeck has devoted his career to the application of aerodynamic research to transport and private aircraft. Through his leadership and understanding of the modern aerodynamics of wings and high-lift technology, the firms he led or founded have succeeded in making it economical to replace existing wings with his innovative designs to markedly improve the safety and efficiency of a wide variety of general aviation aircraft.

1980 **Paul T. Homsher** *(BSAE '45), Vice President, F-15 Saudi Arabian Operations, McDonnell Aircraft Company, St. Louis, Missouri.* Mr. Homsher is responsible for a program consisting of 60 F-15 airplanes, as well as all support, maintenance, training, and construction items. This development project, integrated and coordinated by

Mr. Homsher, has been one of the most successful programs in the history of American aviation. It was conducted on schedule, on cost, and met or exceeded more than 95 percent of all specification requirements. The F-15 aircraft is considered the world's finest air superiority aircraft with a top speed of Mach 2.5. Mr. Homsher has experience involving project engineering, flight testing, and program management, including involvements with the Gemini Manned Spacecraft Vehicle, the F-111 Crew Module, and the Navy's VFX program, as well as the F-15 project.

1981 **Richard H. Petersen** *(BSAE '56), Deputy Director, National Aeronautics and Space Administration, Langley Research Center, Hampton, Virginia.* Mr. Petersen is a leader in analytical and experimental aerodynamics with the Langley Research Center. In addition to his current managerial duties in research and development in aerospace technology, he has made fundamental contributions in the development of the Mercury Space Capsule, several missile systems, and B-70 aircraft. He planned and directed studies of the mission performance of airbreathing launch vehicles, supersonic aircraft, hypersonic aircraft, and future civilian and military aircraft systems, including STOL aircraft for short-haul transportation. He is widely known for his achievements in developing aircraft technology by integrating the disciplines of aerodynamics, internal flow, stability and control, and aeroelasticity.

1982 **Robert D. Hostetler** *(BSES '64, MSIA '65), President and Chief Executive Officer, CTS Corporation, Elkhart, Indiana.* Mr. Hostetler is responsible for directing the long-range profit structure of the corporation, establishing its objectives, plans, and policies, and representing its interests. He has been affiliated with CTS since 1965. In 1968, he was named assistant general manager for Knights, Inc., a CTS company in Sandwich, Illinois. In 1973, he was named president of Knights. In 1977, Mr. Hostetler returned to corporate headquarters as vice president, and in 1978 he became group vice president of the high technology group of CTS companies. In the latter capacity, he oversaw the development of products for the data processing, telecommunications, and medical electronics markets. Since 1979, he has served in successively responsible administrative positions.

*Robert D. Hostetler
(BSES '64, MSIA '65).*

1982 **Bruce A. Reese** *(MSME '48, PhD '53), Chief Scientist, Arnold Engineering Development Center.* Dr. Reese has been chief scientist at the Arnold Engineering Development Center (AEDC), one of the world's largest and most advanced aerodynamic test facilities, since 1979. He succeeded to his post as principal assistant and advisor to the commander after a notable career as student and faculty member at Purdue. Entering Purdue in 1946 as research fellow and instructor, he was appointed professor of mechanical engineering in 1957, director of the Jet Propulsion Center in 1966, and head of the School of Aeronautics and Astronautics in 1973. On leave from Purdue in the early 1960s, he was technical director of the Nike Zeus and Nike X Projects, with responsibility for technical supervision of all the U.S. Army's R&D activities in ballistic missile defense. Recognized for his service on the scientific advisory boards of all three military services and twice awarded the Army's Outstanding Civilian Service Medal, he has assisted in development of most of the present weapons systems in the U.S. *[Editors' note: Technically, Dr. Reese's DEA was awarded by the School of Mechanical Engineering, but since he was head of the School of Aeronautics and Astronautics from 1973 to 1979, he is included here among AAE's Distinguished Alumni.]*

Bruce A. Reese (MSME '48, PhD '53).

1983 **Arthur B. Greenberg** *(BSME '50, BSAE '50, MSAE '52, PhD '55), Vice President and General Manager of Government Support Operations, Programs Group, Aerospace Corporation, Los Angeles, California.* Dr. Greenberg is responsible for a diversified set of technologies and programs, which cover rapidly advancing technological issues in areas such as solar and nuclear energy, civilian applications of satellites, energy conservation technology, electric and hybrid automobiles, offshore structural analysis techniques, energy resource development opportunities, and the strategic petroleum reserve, all of which have major national implications. He consults with government officials on policy and technical matters. Dr. Greenberg serves as chairman of the Aerospace Committee, whose goal is to

Arthur B. Greenberg (BSME '50, BSAE '50, MSAE '52, PhD '55).

develop mobility opportunities for employees and expand their talents. He also is a member of both the Professional Development and Aerospace Sponsored Research Committees.

1984   **R. E. "Jeff" Kasler** *(BSAE '46), Chairman of the Board, Kasler Corporation*. Mr. Kasler assumed the position of president in 1961, and since that time his company has been awarded projects totaling more than $900 million, including bridges, highways, freeways, and water control and distribution, as well as military and defense construction. The corporation is one of the leaders in concrete paving technology, using on-site central mixing plants and electronically controlled slipform paving equipment. Kasler Corporation was the first contracting firm to successfully slipform four layers of freeway in a single pass, and in 1973

R. E. "Jeff" Kasler (BSAE '46).

it set a world record for paving 3.37 miles of a two-lane highway in a single day. Several equipment design, modifications, and paving methods first used by Kasler are now considered industry standards.

1986   **Garner W. Miller** *(BSAE '50), Senior Vice President for Maintenance and Engineering, USAir, Pittsburgh, Pennsylvania*. Mr. Miller became senior vice president for maintenance and engineering and a member of USAir's Executive Committee in 1979. In that capacity, he serves as USAir's chief engineer. USAir employs 12,500 people, serving approximately 70 cities coast-to-coast with 137 modern jet aircraft. As a senior officer of USAir, Mr. Miller is responsible for a staff of 2,500 people working in engineering, maintenance, planning, and administration. He has primary responsibility for engineering, tech-

Garner W. Miller (BSAE '50).

nical, and financial input for selection of new aircraft, engines, systems, and related support equipment, including facilities. Several recent new aircraft acquisition programs were coordinated through his office, and he has been instrumental in implementing advanced communication and navigation systems, which have resulted in major improvements in the USAir fleet.

1988 **Ronald L. Kerber** *(BSES '65), Vice President of Advanced Systems and Technology for McDonnell Douglas Corporation, St. Louis, Missouri.* Until February of this year, Dr. Kerber was the deputy undersecretary of defense for Research and Advanced Technology, where he was responsible for planning and reviewing all research and development programs in the Department of Defense Science and Technology Program. Before joining the Department of Defense in 1985, Dr. Kerber was for 16 years a faculty member at Michigan State University, and was associate dean for graduate studies and research from 1980. He has researched and taught on a variety of subjects, including laser physics, plasma chemistry, and thermodynamics, and has served as a consultant to numerous government agencies. In 1987, he made nine testimonial appearances before Congress, providing expertise on various subjects.

Ronald L. Kerber (BSES '65).

1989 **John B. Hayhurst** *(BSAE '69), Vice President, Marketing, Boeing Commercial Airplanes, Seattle, Washington.* Mr. Hayhurst began his career with Boeing as a customer support engineer, drafting service bulletins for the Boeing 747 during its initial certification process. He received a master's degree in business administration from the University of Washington in 1971. Since then, he has held several positions at Boeing, including marketing manager for the Boeing 747 and 757, and regional director of domestic sales. He was named to his present position in 1987. As vice president for marketing, Mr. Hayhurst is responsible for all of Boeing's commercial airplane marketing efforts, including product strategy, sales support, market research, public relations, and advertising. He leads a diverse group of engineers, economists, mathematicians, and others who form one of the largest market analysis groups in the field.

John B. Hayhurst (BSAE '69).

1990 **Hsichun Hua** *(MSAE '65, PhD '68), Executive Vice President and General Manager, Chung-Shan Institute of Science and Technology.* Dr. Hua came to Purdue for his master's degree following a distinguished military career in both the R.O.C. and U.S. Air Forces. After receiving a PhD from Purdue, Dr. Hua worked with Cessna and Lockheed Aircraft Companies before returning to Taiwan in 1970. He has since played an instrumental role in the development of Taiwan's aerospace industry, in adjunct faculty positions at National Cheng Kung and Tunghai Universities, and in research leadership and management positions in the country's aerospace industry. In his current position, in which he also serves as general manager of the Aero Industry Development Center, Dr. Hua supervises more than 5,500 engineers and technicians in aeronautical research as well as aircraft and engine development and

Hsichun Hua (MSAE '50, PhD '68).

production. The center's notable achievements include the F5-E and AT-3 fighters, as well as the Ching-Kuo, a supersonic lightweight air defense fighter scheduled to begin production this year.

1991   **William W. Brant** *(BSAE '55, MSE '56), Vice President and General Manager, Strategic Operations, Thiokol Corporation.* After serving with the Air Force, Mr. Brant joined Thiokol Corporation in 1958 as head of the BOMARC missile propulsion department. He has held a number of positions with the company, including manager, proposal department, and vice president, Navy strategic programs. Until April of last year, he was vice president, operations, responsible for all production operations in Thiokol's Strategic Operations Division. In his current position, Mr. Brant is responsible for all aspects of the division, which has been a leading supplier of solid rocket propulsion systems in ballistic missiles for the past 30 years. Propulsion programs under Mr. Brant's direction include those for the Peacekeeper, Trident, and Small ICBM missile programs, as well as those for ordnance and composite programs.

William W. Brant (BSAE '55, MSE '56).

1992   **William C. Kessler** *(BSAE '64, MSAE '65), Director, Manufacturing Technology Directorate, Wright Laboratory, Wright-Patterson Air Force Base.* In 1965, Dr. Kessler began a 10-year career at McDonnell Douglas in engineering and management, earning a doctorate in chemical engineering from Washington University in 1974. Dr. Kessler joined the Air Force in 1975 as a senior materials research engineer, and in 1981 was appointed special assistant to the director of the Air Force Materials Laboratory. Since 1988, Dr. Kessler has headed the Manufacturing Technology Directorate at Wright-Patterson Air Force Base, a program that pioneers manufacturing developments for all Air Force products. He is the Air Force's foremost authority on advanced manufacturing and production technology. Dr. Kessler also has served as an adjunct professor at the University of Dayton since 1975, teaching graduate-level classes in numerical methods and transport phenomena in the School of Chemical Engineering.

William C. Kessler (BSAE '64, MSAE '65).

1993 **David O. Swain** *(BSAE '64), Senior Vice President, McDonnell Douglas Aerospace, Long Beach, California.* Mr. Swain was appointed senior vice president of transport aircraft for McDonnell Douglas in April 1991. He is responsible for business, technical, and production operations of all McDonnell Douglas business dealing with airlift and tanker aircraft. In January 1991, he was named vice president-general manager of the C-17 program, a $36 billion program, which is the biggest part of the Defense Department's current five-year procurement plan. From 1989 to 1991, he served as vice president of strategic business and technology development, Douglas Aircraft, and from 1987 to 1989, he served as vice president of strategic business development of McDonnell Douglas Astronautics. In 1986, he was general manager of McDonnell Douglas Electronics, effectively serving as president for that 2,000-person organization.

David O. Swain (BSAE '64).

1994 **Lana M. Couch** *(BSAE '63), Director, National Aero-Space Plane Office, NASA Langley Research Center, Hampton, Virginia.* During a career with NASA that has spanned 30 years, Ms. Couch has made significant contributions toward advancing the state-of-the-art in aeronautics and space technology, and in particular, bringing those two technologies and their technical communities together to enhance the management and development of hypersonic and transatmospheric technologies. Named to her present position in 1990, Ms. Couch is director of the National Aero-Space Plane (NASP) Office at Langley Research Center in Hampton, Virginia. She leads the NASA technical activities in the NASP program. In addition, she provides the program leadership at Langley in the hypersonic research and technology base program. NASP is a multi-billion-dollar, joint NASA/Department of Defense (DOD) program to develop the technology for an airbreathing-engine pow-ered airplane with a goal of flying to orbit after the turn of the century. The hypersonics program is a fundamental research and technology program that addresses key technologies necessary for operational hypersonic systems in the 21st century and beyond.

1995 **James P. Noblitt** *(BSAE '57), Vice President and General Manager, Missiles and Space Division, Boeing Defense and Space Group.* Mr. Noblitt oversees the operations of a diverse aerospace concern whose approximately 4,000 employees in Kent, Washington; Huntsville, Alabama; and Houston, Texas; perform vital work for the nation's space and missile programs. In addition, Mr. Noblitt serves as president, Boeing Com-mercial Space Company. Mr. Noblitt is responsible for Boeing's work as prime contractor on NASA's International Space Station program, the Air Force/NASA Inertial Upper Stage booster rocket program,

Figure 10.5   Lana Couch (BSAE '63) with (left to right) Dean Yang, Professors Gustafson, Cargnino, Sullivan, and Palmer (April 1994).

research and development programs for the
Department of Defense Ballistic Missile Defense Organization, and developmental efforts
on a variety of civil and military space systems.
He also is responsible for production of the
U.S. Army's AVENGER air defense system and
development of various domestic and export
derivatives. Other tactical and strategic missile
programs also come under his purview.

1996 **Robert L. Strickler** *(BSAE '60, MSAE '62,*   *James P. Noblitt*
PhD '68), President and General Manager, TRW   *(BSAE '57).*
Environmental Safety Systems, Inc. (TESS). Dr.
Strickler oversees a wholly owned subsidiary of TRW that serves
as the management and operating contrac-
tor for the Department of Energy's Civilian
Radioactive Waste Management program. He
also leads TRW's market area activities across
energy, environmental, and nonproliferation
business lines. In this capacity, he is responsible
for strategic planning for the market area and
coordinates internal and external business development and project performance activities
for the company. Prior to joining TESS in 1995,
Dr. Strickler was vice president and general   *Robert L. Strickler*
manager of the TRW Ballistic Missiles Division   *(BSAE '60,*
in San Bernadino, California, and was respon-   *MSAE '62, PhD '68)*
sible for TRW's largest systems engineering
project, the Intercontinental Ballistic Missile program.

1997 **William J. O'Neil** *(BSAE '61), Manager, Project*
Galileo, Jet Propulsion Laboratory. Mr. O'Neil has
been employed by the Jet Propulsion Laboratory since 1963, and has made many significant
contributions to the exploration of the moon,
Mars, and Jupiter. He currently is the project
manager for Project Galileo, the first spacecraft to perform asteroid flybys, to probe the
atmosphere of Jupiter, and to orbit Jupiter.
He oversees more than 335 full time and 150   *William J. O'Neil*
contractual employees and an annual budget   *(BSAE '61).*

of more than $75 million. Mr. O'Neil is responsible for identifying technical threats to Project Galileo and for isolating its most difficult challenges. He is the leading expert on the project and is the primary spokesperson for national and international media requests.

1998   **Roy D. Bridges** *(MSAE '67), Director, Kennedy Space Center, National Aeronautics and Space Administration.* Mr. Bridges currently leads the largest and most prestigious spacecraft launch facility in the world, managing an installation of nearly 140,000 acres and a team of approximately 2,000 NASA civil servants and 14,000 contractors. In addition, he oversees the integration and test- ing of the International Space Station. Before joining NASA in 1997, he served 30 years with the United States Air Force, retiring with the rank of major general. His USAF assignments included management of various missile and flight test centers, serving as a flight instructor, as a test pilot, and flying more than 225 combat missions in Vietnam. Mr. Bridges also was a NASA astronaut from 1980 to 1986, and was the pilot on the Spacelab-2 mission (STS 51-F) in 1985.

Roy D. Bridges (MSAE '67).

2000   **Michael T. Kennedy,** *BSAE '70, Vice President, EELV/Delta IV Program, The Boeing Company.* As vice president, EELV/ Delta IV Program for The Boeing Company, Mr. Kennedy is responsible for the total per- formance of this evolution of launch vehicles. After receiving his bachelor's degree from Purdue in 1970, Mr. Kennedy spent his career with McDonnell Douglas, which merged with Boeing in 1997. Initially advancing through the structural-mechanics analysis area from a junior engineer to senior engineer of a large group, he shifted in the mid-1980s from pure technology into more of a project/program management role. He has worked on a variety of strategic defense initiatives programs and the International Space Station program and spent the last four years on development of the new Delta Rockets program.

Michael T. Kennedy (BSAE '70).

2001 **Kenneth G. Miller,** *BSAE '66, Vice President, Air Force Programs An-teon Corp.* Mr. Miller received his bachelor's degree in aeronautical engineering from Purdue in 1966 and a master's degree in systems management at the University of Southern California in 1970. Mr. Miller served in the United States Air Force for 30 years, retiring in 1995 as a brigadier general. He is currently vice president, Air Force Programs, Anteon Corporation. His career included accelerated advancement through many management and leadership positions, highlighted by his appointment as the director of the C-17 aircraft program. In this position, he executed budgets in excess of $2 billion and in-stituted a team approach that resulted in major production improvements and the first flight of the C-17. Mr. Miller's military decorations include the Defense Superior Service Medal, the Legion of Merit, and the Meritorious Ser-vice Medal. He also has received the Freedom Foundation's Award of Merit.

Kenneth G. Miller (BSAE '66).

2002 **Mark K. Craig,** *BSAE '71, Acting Director, Stennis Space Center.* Mr. Craig is acting director of Stennis Space Center, NASA's lead center for rocket propulsion testing and remote sensing applications. He graduated from Purdue in 1971 with a bachelor's degree in aeronauti-cal engineering. Mr. Craig's NASA career began when he was a 19-year-old co-op student at the height of the Apollo program. He has served the space program through his technical ex-pertise, outstanding management skills, and strategic planning. Mr. Craig led negotiations that brought Europe, Japan, and Canada into the Space Station. He was the creator of NASA's strategic enterprise concept and led President Bush's Space Exploration Initiative to integrate robotic and human exploration. Mr. Craig re-ceived the Federal Engineer of the Year Award in 1991 and the NASA Medal for Outstanding Leadership in 1992. He is a two-time recipient of the President's Meri-torious Executive Award and was named an Outstanding Aerospace Engineer by Purdue in 2000.

Mark K. Craig (BSAE '71).

2003   **Major General John L. Hudson,** *MSAA '74, JSF Program Director, USAF.* Major General Hudson received his bachelor of science in astronautical engineering at the United States Air Force Academy in 1973 and his master of science in aeronautics and astronautics at Purdue in 1974. He is program executive officer and program director, Joint Strike Fighter Program, Office of the Assistant Secretary of the Navy for Research Development and Acquisition, Arlington, Virginia. This joint program will develop and produce the next generation strike warfare weapon system for the U.S. Navy, Marine Corps., and Air Force; Royal Navy and Royal Air Force; and other countries under partnership and foreign military plans. Among Major General Hudson's numerous awards and commendations are the Defense Distinguished Service Medal with oak leaf cluster, Defense Superior Service Medal, Defense Meritorious Service Medal with oak leaf cluster, Meritorious Service Medal with two oak leaf clusters, and the Air Force Commendation Medal.

John L. Hudson (MSAA '74).

2004   **Hank Queen,** *BSAE '74, Vice President of Engineering-Product Integrity, Boeing Commercial Airplane Group.* Mr. Queen received his bachelor's degree from the School of Aeronautics & Astronautics in 1974 and currently serves as vice president of engineering and manufacturing at Boeing's Commercial Airplane Group. He began at Boeing as an engineer after graduation and worked in management and leadership roles in several of the company's divisions. In 2001, Mr. Queen was honored with an Outstanding Aerospace Engineering Award from Purdue University.

Hank Queen (BSAE '74).

2004   **Christopher Whipple,** *BSES '70, Principal, Environ International.* Mr. Whipple received a bachelor's degree in engineering science in 1970. In the more than 30 years since, he has performed risk assessments and environment analyses, gauging the risks associated with energy production, fuel emissions, and radioactive waste. He has chaired the International Atomic Energy Agency and the National Academy of Sciences Board on Radioactive Waste Management. Since 2000,

Mr. Whipple has worked as a principal with Environ International. He was elected to the National Academy of Engineering in 2001.

2005 **Paul Bevilaqua,** *MSAAE '68, PhD '73, Chief Engineer, Advanced Development Projects. Lockheed Martin Skunk Works.* Dr. Bevilaqua is an accomplished aircraft design engineer with numerous patents and awards. He studied at Notre Dame as an undergraduate before enrolling in Purdue's School of Aeronautics and Astronautics, where he earned master's and doctoral degrees in aeronautics and astronautics in 1968 and 1973. He then served as a captain in the U.S. Air Force and as deputy director of the Energy Conversion Laboratory at Wright-Patterson Air Force Base in Ohio. From 1975 to 1985, he was manager of advanced programs at Rockwell International, where he led the design of a Navy VSTOL aircraft, a carrier support aircraft, and a tactical transport aircraft. He's been with the Lockheed Advanced Development Company since 1985, where he is currently manager of advanced development projects.

Christopher Whipple (BSES '70).

Paul Bevilaqua (MSAAE '68, PhD '73).

2006 **Allen S. Novick,** *BSAE '65, MSAE '67, PhD '72, Vice President Marketing Intelligence & Support, Rolls-Royce Corporation.* Dr. Novick joined the Detroit Diesel Allison Division of General Motors in 1972 after receiving his bachelor's, master's, and doctorate degrees in aeronautical engineering from Purdue University. He progressed through assignments, including research and technology, preliminary design, advanced engines, engine development, program management, business development, commercial business, customer operations, and supply chain management. He later served as vice president, cost excellence, where he established the Rolls-Royce Cost Excellence philosophy and organizational structure. In his current position as vice presi-

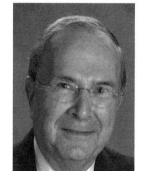

Allen S. Novick (BSAE '65, MSAE '67, PhD '72).

dent, marketing intelligence and support, Dr. Novick assesses and analyzes external market issues and dynamics that impact Rolls-Royce Corporation's Capability Center and Customer Facing Businesses. Dr. Novick served as a member of the General Aviation Manufacturers Association Board of Governors and was chairman of the Association's Technical Policy Committee. He also was a founding member and chairman of the Regional Airlines Association, Associate Member Council, and formerly served as a member of the Board of Directors for Purdue Engineering Alumni Association from 1988 to 1992. He also was a cofounder of the Indiana Advanced Aerospace Manufacturing Alliance (IAAMA) in 2005.

2007   **William H. Gerstenmaier,** *BSAAE '77, Associate Administrator for Space Operations, NASA.* Mr. Gerstenmaier received a bachelor's degree in aeronautical engineering from Purdue University in 1977 and a master's degree in mechanical engineering from the University of Toledo, Ohio, in 1981. As associate administrator for the Space Operations Mission Directorate at NASA Headquarters, Mr. Gerstenmaier directs NASA's human exploration of space and also has programmatic oversight for the International Space Station, Space Shuttle, Space Communications, and Space Launch Vehicles. Mr. Gerstenmaier began his NASA career in 1977 at the Glenn Research Center in Cleveland, performing aeronautical research; he also was involved with wind tunnel tests on the Space Shuttle. In 1980, he joined the Space Shuttle program as propulsion flight controller, and in 1992, Mr. Gerstenmaier got his first managerial assignment for the Orbital Maneuvering Vehicle project at JSC.

William H. Gerstenmaier (BSAAE '77).

2008   **Debra Haley,** *BSAAE '78, Special Assistant to the Commander (Ret.), Aeronautical Systems Center, Wright-Patterson Air Force Base.* Ms. Haley recently retired as special assistant to the command of the Aeronautics Systems Center at Wright-Patterson Air force Base in Dayton, Ohio. Her responsibility and accomplishments were recognized with a presidential award for meritorious service from President George W. Bush in 2005. She received her BSAE from Purdue in 1978 and also holds master's degrees in management from the

Massachusetts Institute of Technology and the Air Force Institute of Technology. Ms. Haley led several organizations at Wright-Patterson AFB. The two most challenging assignments were as chief information officer and executive director for Materiel Systems. As CIO, she and her team operated and maintained the $4 billion communications infrastructure of Air Force Materiel Command, enabling 100,000 people to communicate effectively and securely around the world. As executive director, she led a team of people responsible for buying and implementing software to run the business processes of the Air Force.

Debra Haley (BSAAE '78).

2008 **Michael J. Cave,** *BSE '82, Senior Vice President Business Development and Strategy, The Boeing Company.* After leaving Purdue, Mr. Cave spent two decades in the aviation industry, first with McDonnell Douglas, where his appointments included heading the team that merged commercial airplane units of The Boeing Company and McDonnell Douglas. In 2000, he was named vice president and chief financial officer for Boeing Commercial Airplanes and faced perhaps the greatest challenge of his career — the turnaround of Boeing Commercial Airplanes in the aftermath of September 11, 2001. The company had not launched a new airplane in a decade, was challenged by Airbus, and was said to be no longer competitive in the commercial airplane business. In 2001, the industry's customer base faced the worst downturn in its history. Through Mr. Cave's financial and managerial leadership, the company regained its leadership in innovation, profitability, and sales, and secured US leadership in the industry for a long time to come.

Michael J. Cave (BSE '82).

2009 **Michael H. Campbell,** *BSAAE '83, Executive Vice President and Chief Operating Officer, Fair Isaac Corporation.* After graduation from Purdue in 1983, Mr. Campbell co-founded General Optimization with two professors from the University of Chicago that aimed to "bring optimization down to the microcomputer and make it widely available."

His entrepreneurial endeavors later expanded to include Campbell Software, which he founded in 1989. That company would go on to develop the leading workforce management solution for the labor-intensive retail marketplace and become the top-selling provider of workforce management applications in the United States retail industry. Campbell Software was sold to SAP Americas in 1999 for an undisclosed amount. In 2005, Mr. Campbell joined Fair Isaac Corporation and is now executive vice president and chief operating officer. In this role, he is responsible for sales, global product line P&L functions, product management, professional services, and partner management for Fair Isaac's diverse portfolio of analytic solutions and software. Fair Isaac applies high-level math to study fraud, determine credit worthiness and predict consumer behavior.

Michael H. Campbell (BSAAE '83).

2010 **Darryl W. Davis,** *BSAAE '78, President, Phantom Works, Boeing Defense, Space and Security.* Upon graduating from Purdue, Mr. Davis accepted an offer to work in McDonnell Douglas's propulsion department. In 1997, McDonnell Douglas merged with The Boeing Company. After holding several positions of increasing responsibility in both McDonnell Douglas and Boeing, Mr. Davis took his current position as president of Phantom Works for Boeing Defense, Space, and Security. In this position, he leads an organization of approximately 2,400 employees in five major business elements: Advanced Boeing Military Aircraft (ABMA), Advanced Network & Space Systems (ANSS), Advanced Global Services & Support (AGS&S), Advanced Modeling & Simulation (AMS), and Strategic Development & Experimentation (SD&E).

Darryl W. Davis (BSAAE '78).

HONORARY DOCTORATES

It is appropriate to conclude this chapter of alumni achievements by reviewing the careers of those who have received honorary doctorates from Purdue. The honorary doctorate is the highest recognition that the University can bestow on a former student or staff member. The following alumni and faculty have been honored by the School of Aeronautics and Astronautics in this manner. They are listed in chronological order along with a biographical description prepared at the time the degree was awarded.

1953 **Donovan R. Berlin,** *BSME '21, ASVP General Manager, Boeing VERTOL, Philadelphia, Pennsylvania.* Mr. Berlin was a gifted engineer who helped develop a number of important airplanes over the first half of the 20th century. He began his career with Douglas Aircraft, then moved on to Northrop Aircraft Companies, where he was chief engineer. He then joined Curtiss-Wright Corporation in 1934, becoming chief engineer and director of engineering in the years before World War II. He designed the P-36 and the P-48 aircraft. The P-40, also known as the Hawk 75, was considered one of Mr. Berlin's greatest designs, and was one of the most widely used airplanes in World War II. After the war, Mr. Berlin continued his distinguished career by working for a variety of other aeronautics corporations on projects such as the Navy's FH-1 Phantom and the Army's CH-47 Chinook helicopter. He retired in 1979 and died in 1982. [8]

1967 **Milton U. Clauser**, *Director of International Development, Communications Satellite Corporation.* Dr. Clauser was an engineer with the Douglas Aircraft Company from 1937 until 1950, when he became head of the Purdue School of Aeronautics. He served as head of the school from 1950 to 1954, and then resigned to take various positions in industry. He was director of aeronautics and propulsion with the Ramo-Wooldridge Corporation from 1954 to 1955, and was vice president at Space Technology Laboratories from 1955 to 1960. He then founded his own consulting company and was active in this capacity until becoming director of research and engineering at the Institute for Defense Analysis in 1965. He assumed his current position with the Communications Satellite Corporation in 1966.

1969 **Richard Lee Rouzie,** *BSME '28, Vice President-Flight Test and Service, Commercial Airplane Division, The Boeing Company.* During his 39-year career with The Boeing Company, Mr. Rouzie has made significant contributions to aeronautical engineering leading to the jet age, especially in civilian aviation. He was appointed chief project engineer commercial, in June 1958, and directed design activities for all versions of the Boeing 707 and 720 model aircraft. He became chief engineer production in July 1960 and directed engineering design activities for all commercial and military versions of the Boeing 707, 720, and 727. He then became director of engineering on July 12, 1962, and was elected vice president of engineering for the commercial airplane division on February 1, 1966. His duties there encompassed overall direction of the engineering department activities and flight testing and customer service dealing with production of the 707, 720, 727, 737, and 747 aircraft. His title was changed to vice president for flight test and service in July 1968.

1970 **Neil A. Armstrong,** *BSAE '55.* Mr. Armstrong's honorary degree is awarded for courageous and dedicated personal contributions to the technology of supersonic flight and lunar exploration (Figure 10.3); and for the finest exemplification of those priceless human attributes of character, competence, and rigorous self-discipline demanded of all men who would help mankind reach for the stars.

1973 **Richard W. Taylor,** *BSME '42, vice president and general manager of aeronautical and information systems division, The Boeing Company.* Mr. Taylor began his aerospace career with Boeing in 1946 as a design engineer, designing instrumentation for use in flight testing the first model of the B-50 bomber and the Stratocruiser commercial airliner. Today, he directs most of the firm's military aircraft programs and has responsibility for determining potential new markets in this field, with emphasis on military derivatives of Boeing's commercial jetliners. Mr. Taylor's division presently consists of some 3,000 employees, about two-thirds of whom are engineers. In addition, he holds responsibility for more than 1,000 commercial airplane group employees who support the programs he directs. During his Boeing career, Mr. Taylor has received numerous awards, including selection as a Purdue Distinguished Engineering Alumnus in 1968. He also received the Society of Experimental Test Pilots James H. Doolittle Award for excellence in technical management that year. This award is given annually to an individual who not only has excelled in test flying but also has achieved success in top management with industry.

1975 **Elmer F. Bruhn,** *Professor Emeritus, Purdue University.* Professor Bruhn has distinguished himself as an engineer and an engineering educator. He was responsible for the basic structural design of the AT-6 Texan, a classic military trainer. He joined the School of Mechanical Engineering in 1941 as an associate professor. In 1945, when the School of Aeronautical Engineering was formed, he was named professor and acting head of the new school, and from 1947 to 1949, he served as head. As an engineering educator, he organized a new school that has grown to become one of the largest and most widely respected in the country. He authored a textbook on aircraft structures, which has sold over 100,000 copies, and he has been an extremely important influence in the design of all modern aircraft structures.

1977   **Richard E. Adams** *(BSME '42, Aero option)*. Mr. Adams is vice president of the General Dynamics Corporation and general manager of the Fort Worth division of that corporation. The division he manages is recognized as an excellent and important national production facility, and he currently manages one of the nation's most important military aircraft programs: the F-16 air-combat fighter. Mr. Adams joined General Dynamics as an assistant project engineer in 1951, and he rose rapidly through the positions of project engineer, chief of preliminary design, chief of advanced design, director of advanced programs, and vice president. His career as one of the nation's top engineering managers encompasses a number of major aircraft programs, including the B-58 supersonic bomber, holder of 19 world speed and altitude records, and the F-111, a versatile fighter-bomber and the first operational variable-sweep aircraft. As vice president of engineering at the Convair division, he was responsible for the design of the fuselage of the DC-10 transport and the F-16. Under the leadership of Adams, the Fort Worth division won an intensive competition for the United States Air Force lightweight air-combat fighter. The United States Air Force contract, combined with the decision of four NATO consortium countries to build the F-16, will result in the largest international aircraft program ever undertaken, potentially amounting to $15 billion.

1979   **Herbert F. Rogers**, *BSAE '49*. Mr. Rogers currently is division vice president, deputy general manager, and F-16 program director at the Fort Worth division of General Dynamics Corporation. Named manager of advanced aircraft systems for the division in 1967, he assumed responsibility for the management of all Air Force-oriented program studies, including the early advanced Fighter (FX) studies. During 1969 and 1970, he was program director for the division's B-1 program effort. In late 1970, Mr. Rogers was transferred to the Convair division, where he assumed the duties of chief engineer, and became director of the Convair Space Shuttle Booster Program. In 1972, he served as director of advanced programs at Convair and later became vice president of marketing at the Fort Worth division. As the F-16 program director, he is responsible for executive direction of the entire F-16 Air Combat Fighter Program, a highly complete weapon system of unprecedented technical and managerial scope and magnitude. A somewhat unusual dimension of Mr. Rogers' management responsibility is the multinational aspect of the production program. Under his direction, the F-16 program consists of a co-production effort never before attempted by the participating countries. Co-production includes the manufacture of the F-16 by industry in the United States, Belgium, Denmark, the Netherlands, and Norway for delivery to air forces in all five countries. The F-16 multinational production is an innovative program that allows each participating country's industry and citizens to benefit by actually taking part in producing an airplane that is a vital part of its defense. Mr. Rogers has expanded the basic knowledge from his academic training with manufacturing experience, design engineering, and advanced technological analysis. His current responsibilities are a direct result of this varied education and wide range of experience, combined with his unusual perception, initiative, leadership, and judgment.

1980 **Edmund V. Marshall,** *MSAES '47.* Mr. Marshall has had an illustrious career in space systems. During his professional career, he has been responsible for innovations in aircraft systems design, and he holds five patents related to that field. He contributed to the original engineering studies for the U.S. Lunar Landing Program, and was industrial manager of the program for the design and development of the life support system for the Apollo astronauts' backpacks and Lunar Excursion Module. He has been involved in the management of a variety of programs and products including research on artificial heart pumps, kidneys, and respiration devices, and has managed a department producing biomedical devices. Most recently he has been responsible for the general management of the largest firm in the field of military aviation gas turbines. Mr. Marshall's professional appointments have been just as impressive. They have included, among others chief, aircraft design, and manager of space flight vehicles at the Chance Vought Corporation; division vice president, space systems and biomedical systems department, and executive vice president for the Hamilton standard division of United Technologies Corporation;

vice president and general manager of Pratt and Whitney Aircraft, Florida Research and Development Center, and most recently, president of Pratt and Whitney Aircraft Government Products Division in West Palm Beach, Florida.

1986 **Richard H. Petersen**, *BSAE '56.* As director of the National Aeronautics and Space Administration's Langley Research Center in Hampton, Virginia, Mr. Petersen is responsible for the center's aeronautical and space research programs, as well as the center's facilities, personnel, and administration. For the five years prior to assuming his current duties in 1980, he was the center's deputy director. Earlier, Mr. Petersen was chief of the aerodynamics division at NASA's Ames Research Center, Moffett Field, California, where he directed a 200-man team conducting research in aerodynamics and fluid mechanics. He worked at Ames most of his career, joining NASA there in 1957. Mr. Petersen received a bachelor of science in aeronautical engineering, with highest distinction, from Purdue in 1956, and a master's degree in aeronautics from the California Institute of Technology in 1957. He was a Sloan Executive fellow at the Stanford Graduate School of Business in 1972-73. Mr. Petersen is a fellow of the American Institute of Aeronautics and Astronautics and

Richard H. Petersen (BSAE '56).

Figure 10.6   Purdue President S. C. Beering, former faculty member S. S. Shu, Professor A. F. Grandt, and Dean H. T. Yang at the May 1991 commencement prior to awarding the Honorary doctorate to Shu.

a member of its board of directors. He received the Presidential Rank of Meritorious Executive in 1982 and NASA's Outstanding Leadership Medal in 1984. He was chosen in 1980 for Purdue's Distinguished Engineering Alumnus Award.

1991 **Shien-siu Shu.** Born in Wenchow, China, Professor Shu (Figure 10.6) received a Tsing Hua fellowship for advanced study in the United States. He earned his doctorate in applied mathematics at Brown University. He taught at the Illinois Institute of Technology, and during the periods of 1955–63 and 1968–79, he was professor of aeronautics astronautics and engineering science at Purdue University. In 1970, Professor Shu went on leave from Purdue to serve as president of the National Tsing Hua University in Taiwan, where he established a modern engineering college and significantly expanded enrollment. From 1973 to 1981, he served as chairman of the National Science Council of Taiwan. He established a scientific infrastructure, which paved the way for a number of research initiatives. Largely because of his vision, the world-renowned Hsin Chu Science Park was founded. The park is now known as the cradle of the high-tech industry in Taiwan. He served as chairman of the board of the Industrial Technology Research Institute of Taiwan from 1979 to 1988. Among the honors he has received is the plaque of the Order of Brilliant Star presented by the president of the People's Republic of China.

Figure 10.7   Honorary doctorate recipient R. Bateman with Professor A. F. Grandt and Dean H. T. Yang at May 1992 commencement.

1992 **Robert E. Bateman** *(BSAE '46)*. Mr. Bateman (Figure 10.7) graduated from Purdue with a bachelor's degree in aeronautical engineering. He joined the Boeing Company as an aerodynamist; and during the next 42 years held a variety of management assignments including senior engineering representative at Strategic Air Command head-quarters, manager of Advanced Space Systems, general manager of the Turbine Division, 747 program executive, and general manager of the Systems Division. He initiated Boeing's programs in hydrofoil systems and was vice president/general manager of Boeing Marine Systems. When Mr. Bateman retired in 1988, he was responsible for managing and coordinating governmental affairs of the Boeing Company at the international, national, state, and local level. He has been deeply involved in a variety of civic and charitable activities, serving as a trustee of the Naval Aviation Museum Association, the U.S. Navy Memorial Foundation, the Atlantic Council, the Naval War College Foundation, and the Helping Hand Foundation. He is chairman of the board of trustees of the Museum of Flight Foundation and chairman of the advisory board of the Seattle World Affairs council. Among his many honors and awards, he has been both a Distinguished Engineering Alumnus and an Old Master at Purdue.

1996 **Henry T. Y. Yang,** *Chancellor, University of California, Santa Barbara.* Before assuming his current position with the University of California in 1994, Dr. Yang (Figures 10.6 and 10.7) served the Purdue faculty for 25 years with great distinction. In addition to an active teaching and research role in the School of Aeronautics and Astronautics, he was school head from 1980 to 84 and dean of the Schools of Engineering from 1984 to 1994. He has received numerous teaching awards from Purdue University, directed approximately 60 PhD and MS theses, and was the first Neil A. Armstrong Distinguished Professor of Aeronautics and Astronautics (1988-94). He is a member of the National Academy of Engineering, and is a fellow of the American Institute of Aeronautics and Astronautics and of the American Society of Engineering Education. He also has served on many Indiana, U. S. government, and industrial advisory committees.

Henry T. Y. Yang

1997 **John B. Hayhurst,** *BSAE '69, Vice President-The Americas, Boeing Commercial Airplane Group.* Mr. Hayhurst is currently responsible for the sale of Boeing airplanes and Boeing's business relationships with customers in North America and Latin America. This position builds on his experience in airplane sales, marketing, product development, and engineering. From 1994 to 1997, Mr. Hayhurst was vice president and general manager of the 747-X program and led the Boeing team responsible for planning the design, development, manufacture, and certification of larger, longer-range versions of the Boeing 747. Prior to that assignment, he was

John B. Hayhurst (BSAE '69).

involved in the development of other new and derivative Boeing airplanes. He has served as vice president of Boeing Commercial Airplane Group Marketing since 1987, playing a significant role in the launch of the Boeing 777. Mr. Hayhurst began his Boeing career in 1969 as a customer support engineer. From there, he was assigned to successive positions as a marketing analyst, financial planning supervisor, 747 marketing manager, 757 marketing manager, area sales director, and director for customer training and flight operations support. Mr. Hayhurst received the Distinguished Engineering Alumnus Award from the Purdue University Schools of Engineering in 1989.

2000 **David O. Swain,** *BSAE '64, DEA '93, OAE '99, Senior Vice President of Engineering and Technology for The Boeing Company.* Mr. Swain is president of Phantom Works and senior vice president of engineering and technology, both with The Boeing Company. As Phantom Works president, he oversees Boeing's research and development organization; as senior vice president, he is Boeing's chief engineer and a member of the Executive Council, which oversees all activities of the company. Born and raised in Lebanon, Indiana, Mr. Swain earned a Purdue bachelor of science degree in aeronautical engineering in 1964. He then began his career with McDonnell Douglas as an engineer on the Gemini spacecraft project. He subsequently was involved in the company's tactical missile programs. A range of execu-

David O. Swain (BSAE '64, DEA '93, OAE '99).

tive positions followed, including vice president/general manager of McDonnell Douglas Electronics Company and vice president for strategic business development for McDonnell Douglas Astronautics Company. As senior vice president and C-17 program manager for McDonnell Douglas Aerospace-West, he led the development and production of the first 10 C-17 Globemaster III aircraft, implementing a process-based management approach to help turn around the $36 billion C-17 development program. These efforts led to the Laurel Award from *Aviation Week* and *Space Technology* magazine.

2001 **Roy D. Bridges, Jr.,** *MSAE '66, DEA '98, OAE '99, Director, NASA Kennedy Space Center.* Mr. Bridges has distinguished himself in three careers. As a military officer, he retired from the United States Air Force with the rank of two-star major general. In a second career, he was an astronaut and logged 188 hours in space. Today, Mr. Bridges commands worldwide attention as director of NASA's John F. Kennedy Space Center. His responsibilities include managing all of NASA's facilities and activities at the center related to processing and launch of the space shuttle. A 1965 distinguished graduate of the U.S. Air Force Academy, Mr. Bridges received a master of science degree in

Roy D. Bridges, Jr. (MSAE '66, DEA '98, OAE '99).

astronautical and aeronautical engineering from Purdue in 1966. With the Air Force, Mr. Bridges served as a test pilot, instructor pilot, and special assistant to the deputy chief of staff for research, development, and acquisition at the Pentagon. He flew more than 226 combat missions in Vietnam. Selected as a NASA astronaut in 1980, he piloted the Spacelab-2 mission in 1985, which was the first to operate the Spacelab Instrument Pointing System. He was named director of NASA Kennedy Space Center in 1997, where he currently manages a team of 1,825 NASA employees and 10,000 contractor workers.

2003 **Charles Taylor,** *BSME '46, MSME '48, Professor Emeritus of Engineering Science at the University of Florida.* Mr. Taylor is a pioneering researcher in photoelasticity and holography and a gifted mentor who produced a new generation of mechanicists. Born in West Lafayette, Indiana, and raised in Monticello, Indiana, Mr. Taylor attended Purdue, earning his bachelor's degree in mechanical engineering in 1946 and his master's in engineering mechanics in 1948. He received his doctorate in theoretical and applied mechanics from the University of Illinois in 1953, and began his academic career there in 1954 as a professor of theoretical and applied mechanics. He remained at Illinois until 1981. He then became a professor of engineering science at Florida and retired there as professor emeritus in 1991. At both Illinois and Florida, Mr. Taylor was active with research in photoelasticity, holography, and experimental mechanics, for which he received several awards. His love of academic life apparently was contagious — almost all of his former PhD students are now professors

Charles Taylor (BSME '46, MSME '48).

at prestigious universities around the world, including Purdue. Mr. Taylor served as president of the Society of Experimental Mechanics, the Society of Engineering Science, and the American Academy of Mechanics, and was an associate editor for the American Society of Mechanical Engineers' journal. In 1979, Mr. Taylor was elected to the National Academy of Engineering for his pioneering developments in three-dimensional photoelasticity and in the use of laser and holography.

2004 **William O'Neil,** *BSAE'61, Chairman of the Space Exploration Committee of the International Astronautical Federation and is member of the International Academy of Astronautics.* Mr. O'Neil earned his bachelor's degree from Purdue in 1961 and his master's in aeronautical and astronautical engineering from the University of Southern California in 1967. He began his career at Boeing Airplane Company and moved to Lockheed Missiles and Space Company before joining the Jet Propulsion Laboratory in 1963. Assignments at JPL have included trajectory design and navigation for Surveyor, the first soft-landing lunar spacecraft. He was the navigation chief for Mariner Mars, the first United States spacecraft to orbit another planet, and for Viking, the first soft-landing craft on Mars. In addition, he was the science and mission design manager for Project Galileo during its development phase throughout the 1980s, and then became Galileo project manager shortly

William O'Neil (BSAE'61).

after its launch in 1989. That spacecraft, an orbiter and an entry probe, arrived at Jupiter in December 1995, becoming the first to penetrate an outer planet atmosphere and the first to orbit an outer planet in December 1997. Following Galileo's two-year primary mission, Mr. O'Neil was appointed chief technologist and then project manager for the Mars Exploration Program at the Jet Propulsion Laboratory. However, the project was postponed and he became manager of JPL's Systems Management Office until his retirement in 2001. He continues as chairman of the Space Exploration Committee of the International Astronautical Federation and is member of the International Academy of Astronautics. Mr. O'Neil has been honored with NASA's highest award, the Distinguished Service Medal, for his management of Project Galileo.

2005 **James D. Raisbeck,** *BSAE '61, Founder and CEO of Raisbeck Engineering Inc. and Raisbeck Commercial Air Group Inc.* Mr. Raisbeck has distinguished himself in aviation by combining a keen engineering ingenuity with a spirit of entrepreneurship that is unique in the modern-day aerospace industry. He is the founder and chief executive officer of Raisbeck Engineering Inc. and Raisbeck Commercial Air Group Inc., both in Seattle. The two companies focus on integrating advanced technology into existing aircraft in ways that increase their productivity and profitability. Born and raised in the Milwaukee area, Mr. Raisbeck earned his bachelor of science degree in aeronautical and aerospace engineering from Purdue in 1961. Soon after graduation, he went to work for the Boeing Airplane Company in Seattle as a research aerodynamicist, designing flaps. In 1969, Mr. Raisbeck left Boeing to become president, chairman and chief engineer for Robertson Aircraft Corporation, where he teamed up with other former Boeing engineers to design single- and twin-engine Pipers and Cessnas. Four years later, he founded Raisbeck Engineering, Inc., which has shaped several business turbine aircraft designs, including designing and producing the first supercritical wings to be used in general and commercial aviation. In 1996, he started Raisbeck Commercial Air Group, which specializes in designing noise-reduction systems for commercial aircraft needing to meet stiffer federal regu-

James D. Raisbeck (BSAE '61).

lations. Shortly after Sept. 11, 2001, the company began producing armored cockpit security systems that could be installed in existing commercial airliners. He established The Raisbeck Engineering Distinguished Professorship for Engineering and Technology Integration at Purdue in 2000.

2007   **Martin C. Jischke,** *President Emeritus, Purdue University.* Dr. Jischke received his bachelor's degree in physics from the Illinois Institute of Technology and his doctorate in aeronautics and astronautics from the Massachusetts Institute of Technology. He came to Purdue in August 2000 from Iowa State University, where he had served as its president for nine years. Before that, he was chancellor at the University of Missouri-Rolla and had been a faculty member, director, dean, and interim president at the University of Oklahoma. Under his leadership as President, the University is nearing completion of its Campaign for Purdue, which so far has brought in $1.64 billion in private

Martin C. Jischke

gifts. In carrying out the strategic plan, he oversaw the University's undertaking of more than 50 capital projects, including construction of 40 new buildings. Among those construction projects was the $350 million Discovery Park, Purdue's hub for interdisciplinary research that houses 10 primary centers focusing on everything from biosciences, the environment, and manufacturing to oncological sciences, cyberinfrastructure, and health care engineering. He earned numerous honors and distinctions during the course of his Purdue years. In 2006, he was appointed by President George W. Bush to the President's Council of Advisors on Science and Technology. He also received the Centennial Medallion of the American Society for Engineering Education.

Chapter 11

Re-Entry:~

Bell/Boeing
CV-22A Osprey

Reflecting on the Past — Looking to the Future

The goal of this volume has been to summarize the history of the Purdue University School of Aeronautics and Astronautics. This is a many-faceted story, as Purdue has grown to world prominence and leadership in aerospace education and research. Although Indiana has little aerospace industry, the Lafayette area has a long tradition exploring the frontiers of progress. Just a few miles southwest from the current Purdue campus, for example, Fort Ouiatenon was established as a French trading post with the Wea Indians in 1717. This was the first white settlement in Indiana, and Fort Ouiatenon became an important center for the Indian and white man to exchange respective technologies. The fact that interchanges between the two cultures were not always peaceful is also evident in local history, as the Battle of Tippecanoe occurred a few miles northeast on November 7, 1811. Here General William Henry Harrison, later to become the ninth president of the United States, fought an Indian confederation led by The Prophet, brother of the great Shawnee chief Tecumseh. This was the last large conflict between organized Indians and white men east of the Mississippi River, and was a milestone in bringing peaceful settlement to the Northwest Territory.

The Lafayette area also has a rich aerospace tradition. The first air delivery of U.S. mail originated by hot air balloon in Lafayette, Indiana, on August 17, 1859. The first airplanes came to the community on June 13, 1911, landing on the Purdue athletic field before 17,000 people. Purdue opened the first university-operated airport in 1934, and Amelia Earhart served on the Purdue faculty in the 1930s. Indeed, the aircraft for her final round-the-world trip was funded by the Purdue Research Foundation. The nation's space program has been intimately entwined with Purdue University, as the second U.S. citizen to fly in space (Gus Grissom) and the first (Neil

Armstrong) and last (Gene Cernan) men to step on the moon were Purdue engineering graduates. In all, 23 alumni have been selected for space travel, and more than one-third of all U.S. space crews have had at least one Purdue graduate. Two of these individuals (Mr. Grissom and Roger Chaffee) died in a flash fire in 1967, which destroyed their Apollo space capsule during a simulated moon launch. Their tragic deaths remind the world of the precious sacrifices needed to advance human progress.

Although the School of Aeronautics and Astronautics was formally established as a separate academic unit on July 1, 1945, aeronautical education and research at Purdue began much earlier. Aeronautical engineering courses were first offered by the School of Mechanical Engineering in 1921, and extensive World War II programs led to the first aeronautical engineering degrees awarded by the School of Mechanical and Aeronautical Engineering in August of 1943. Through December 2009, Purdue University has produced 6,848 BS degrees, more than 1,685 MS degrees, and 557 PhDs in aerospace engineering. It is believed that during the past decade, the school has graduated the most aerospace engineers in the country. These alumni have made many significant contributions to aerospace engineering and other technical and nontechnical fields, and their accomplishments have brought significant fame to their alma mater.

As the school reflects on its 65th anniversary in 2010, it is in an excellent position to strengthen its leadership in the aerospace arena. It has a dynamic faculty who are involved with a broad spectrum of basic research issues of national importance. These faculty are backed by an excellent support staff and well-equipped, modern laboratories and computational facilities. The school enjoys an enthusiastic and talented student body who are tremendously excited by the opportunity to further advance aerospace technology. Most of all, the school has a tradition of excellence and accomplishment, earned by prior generations of faculty and students, which encourages the current members to higher levels of achievement.

The challenge facing the school during its next 50 years will be to prepare students to provide the aerospace technology needed for the 21st century. Indeed, the real challenge will be to help government and industry determine future aerospace opportunities for our nation and the world. The end of the Cold War during the 1990s changed the apparent need for large conventional air forces, and public enthusiasm for space exploration has been dampened by other economic considerations. Thus, the aerospace industry has entered a period of changing missions and faces an uncertain future. History teaches us, however, that there will always be threats to national security, and that it behooves the country to remain vigilant and to anticipate what forms new aggression may take. Clearly, air power, including surveillance by aircraft and/or spacecraft, and a satellite-based communications network, will play an important role in achieving world peace. Moreover, a strong aerospace industry focused on leading-edge technologies would maintain the core of key scientists and engineers needed for a broad-based technical infrastructure which can quickly respond to both military and peacetime needs.

On the commercial side, the nation's airfleet is aging, and while research is being conducted to learn how to safely obtain the maximum benefit from these national assets, older aircraft will eventually need to be replaced with modern fleets of new air transports. In addition, future advances in supersonic flight provide the promise to form closer ties with Europe and the Pacific rim countries. These prospects, combined with continued developments in satellite communications, provide the opportunity for real economic growth, which will once again free the nation to pursue its innate curiosity about outer space.

Thus, the Purdue School of Aeronautics and Astronautics is facing the 21st century with confidence and optimism. It is proud of its accomplishments, and delights in those of its alumni. As it looks back on the past 65 years, the school looks forward to the future, and to playing an instrumental role in providing the world with undreamed-of opportunities for air and space travel. Truly, the Purdue University School of Aeronautics and Astronautics has contributed one small step in the advancement of aerospace technology.

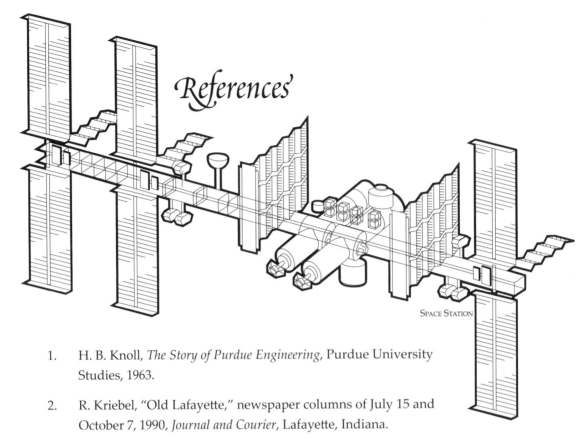

References

SPACE STATION

1.   H. B. Knoll, *The Story of Purdue Engineering*, Purdue University Studies, 1963.

2.   R. Kriebel, "Old Lafayette," newspaper columns of July 15 and October 7, 1990, *Journal and Courier*, Lafayette, Indiana.

3.   R. W. Topping, *A Century and Beyond: The History of Purdue University,* Purdue University Press, W. Lafayette, Indiana, 1988.

4   E. F. Bruhn, *A History of Aeronautical Education and Research at Purdue University For Period 1937–1950*, internal report, Purdue University, prepared during the 1966-67 academic year, printed September 1968.

5   E. F. Bruhn, *Analysis and Design of Aircraft Structures*, 1st edition, 1941. Subsequent revisions titled *Analysis & Design of Flight Vehicle Structures*, Tri-State Offset Company, 1965, revised 1973.

6.   "Purdue's First Aero Class - 1943," *The Purdue Alumnus*, Volume 57, Number 3, December 1969, pp. 6–12.

7.   P. Fountain, "Aviation Goes To College," *Flying*, 1945.

8.   L. H. Tally, "The Sky Was No Limit: Three Purdue Grads Who Quietly Made Aviation History," *The Purdue Alumnus*, October 1993, pp. 12–14.

9.   C.R. Hedden, "Head of the Class: Purdue," *Aviation Week and Space Technology*, August 18/25, 2008, p. 76.

APPENDIX A

BACHELOR OF SCIENCE DEGREES AWARDED

Note: This appendix includes bachelor's degrees granted by the combined School of Mechanical and Aeronautical Engineering prior to establishment of the independent School of Aeronautics on July 1, 1945. It also includes degrees granted by the Division of Engineering Mechanics, which became the Division of Engineering Sciences in 1954. That department merged with the School of Aeronautical Engineering to form the School of Aeronautical and Engineering Sciences in 1960, which was renamed the School of Aeronautics and Astronautics in 1973. Degrees are listed by academic program:

1. **BS in Aeronautical Engineering (1943–1966)** *or*
   BS in Aeronautics & Astronautics (1967–2009)   Pages 343-375

2. **BS in Air Transportation (1947–59)**   Pages 375-379

3. **BS in Engineering Sciences (1957–1976)**   Pages 380-383

BS IN AERONAUTICAL ENGINEERING (1943-1966) OR BS IN AERONAUTICS & ASTRONAUTICS (1967-2009)

August 1943
Allen, John Lee
Boswinkle, Robert Walter Jr.
Brink, Paul Louis
Brown, Sherwood Hulett
Dunn, James Robert
Essig, Robert Herman
Fleming, William Adam
Gaugh, William James
Harrison, John Cowels
Herrick, Robert Nicolls
Howland, Merville Chester
Jones, Lewis Posey
Kerr, Hans Irving
Lappi, Urho Oscar

Lowy, Stanley Howard
Ochiltree, David Warren
Pendley, Robert Elias
Rose, Leonard Murrell
Salisbury, Ralph Dewey Jr.
Schneiter, Leslie Earl
Trueblood, Ralph Beers
West, Franklin Edward Jr.
Zimney, Charles Marshall

October 1943
Smith, Walter Dixon

January 1944
Schaffer, Claude Edward

April 1944
Smith, David Bell

June 1944
Chia-Tseng, Chen
DuWaldt, Frank Arthur
Gault, Donald Eiker
Johnsen, Howard Henry
Loufek, John Elliott
Mori, Takeshi
Pliske, Daniel Robert
Williams, Arthur Llewellyn

September 1944
Gwinn, Robert Earl
Schroeder, Paul Martin

October 1944
Freed, Shervin
Lipes, Russell Magee Jr.
Neuss, Waldemar
Russell, Callen Patrick
Spalding, Carl William

February 1945
Connair, Sylvan Edward Jr.
Hagenmaier, Carl Frederick
Homsher, Paul Thurlow
Laskin, Nathan
McMurray, John Wightman Jr.
Palmer, George Marshall Jr.
Peine, John George
Skillen, Robert Caflisch

June 1945
Arnold, Mary Louise
Bonner, John Joseph
Carpenter, Charles Raymond Jr.
Dettwyler, Hans Rudolph
Fleek, David Neil Jr.
Flinn, Darius Smith
Foss, Richard Leander
Gillespie, Charles Austin
Given, Herbert Zanville
Gray, Robert Merle
Grey, Ralph Eugene Jr.
Haase, Richard Henry
Heile, Donald Henry
Hove, John Edward
Jack, Stephen
Jaeckel, George Oscar
McQuiston, Clyde Emmor
Reed, George Arthur
Roberts, Robert Earl
Schimmel, Robert Charles
Sowarby, Abram John Jr.

October 1945
Altenhaus, Julian LeRoy
Callner, Phillip David
Corden, Carlton Donald
Crawford, John Clements
Eger, Robert Carl
Goodman, Raymond Morris
Irgens, Rolf Nelson
Kahl, Fritz Otto
Ludwick, Richard Hand
Margolin, Louis
Mathias, John Grant
Montgomery, John Campbell
Sanders, John Lyell
Schmertzler, Alvin Louis
Smith, Joseph Patrick Jr.
Telle, George Robert
Treon, Stuart Leo

February 1946
Bateman, Robert Edwin
Black, Roger Lee
Catt, Lewis Richard

Cooper, Gearld Keith
Cox, Donald D
Crist, Lyle Martin
Dale, Alan Albert
Dufka, Rudolph Rank
Eichenberger, Daniel Robert
Folk, Robert Leroy
Galle, Kurt Robert
Gelder, Thomas Foster
Goebel, William Frederick
Gratza, John Richard
Harrison, Allyn W.
Hawkins, Jr., Ralph Donald
Holman, Jeremiah Alter
Howells, Frederick Whan
Johnson, Richard Snider
Kasler, Richard Eugene
Mix, John Martin
Nelson, Robert Leonard
Nickols, Frank Andre Jr.
Palmateer, Wallace Arthur
Parsons, Thomas Richard
Plush, Stanley Lawrence
Rasmussen, Eldon Earl
Schall, Harry Kermit
Snyder, Jack Richard
Thorman, Herman Carl
Wachold, George Raymond

June 1946
Andrade Q., Pablo Emilio
Brown, Robert Clinton II
Carman, Robert Russell
Carter, Roy Conway
Cohen, Jack Howard
Donaldson, Greta Cameron
Dudek, Edward Frank Jr.
Gatineau, Robert Jean
Gluhinich, Gabriel John
Hankins, Glenn Allen
Jansen, Emmert Tobias Jr.
Jenkins, Dale Gordon
Keller, Paul Frederick
Kerchelich, Martin Paul
Knight, Raymond Trevor
Lalvani, Shersing Jetmal
Lazarus, Irwin Philip
Ledyard, Arthur Harvey
McGuire, Elwood Franklin
MacReynolds, William Lamson
Marshall, Maurice Keith
Matisse, Albert Palmer
Matthews, Gordon Paul
Meise, Norman Russell
Parker, Edward Ingraham
Peterson, John Millard
Rozzell, George McAllaster Jr.
Schauble, John James
Spraker, Wilbur Allen Jr.
Sterling, James Berkel
Thielker, Armand Lahrmann
Weddington, Robert Lee
Yatsko, John

August 1946
Berryman, George William
Chiao, Cheng Chi
Hiatt, Melvin Albert

February 1947
Bedwell, Lowell Gerald
Cleaveland, Wendle Boyd
Downing, Noel Lester
Flatau, Abraham
Goldman, Jerome Marvin
Gruber, Joe Clarence
Holliday, Paul Trebel
Hunt, Raymond Sailors Jr.
Islinger, Joseph Stephan
Krivka, Alexander
Leake, Lewis Albert Jr.
Merkel, Eugene Clarence
Merrill, Ralph Lucas
Mills, Galen William
Nettesheim, Raphael Daniel
Perisho, Clarence Howard
Phillips, William Charles
Potter, Davis Ball
Ronan, Edward Thomas
Rose, Sam Edward Jr.
Sanker, William Curtis
Streicher, Albert Henry
Thomas, Richard Roy Jr.
Thompson, Bruce Lee
Thompson, William Dennison Jr.

June 1947
Arne, Vernon Lee
Bright, Loren Glen
Drake, Willis Kirk
Farnsworth, LeMoyne Elbert
Fisher, Philip Wayne
Ford, John Charles
Gleason, Robert Francis
Houston, Russell John
Kesilis, Stanley Paul
Levine, Sylvan Eugene
Nicholson, Robert Allan
North, Gilbert Bishop
North, Warren James
Oldenburg, Richard Leon
Roodhouse, Jr., Benjamin Ornan
Sawyer, Robert Edward
Scull, Wilfred Eugene
Strasser, Betty Mae
Sutton, Roland Clell
Wade, Merwin Gerald
Wallace, Floyd Douglas Jr.
Willer, Jack Edward
Wilson, Robert James

August 1947
Carley, Don
Golden, Norman Harold
Kamp, Armand
Nestel, John Ira
Slomski, Joseph Frank

Sofranko, Wilbur Carl
Williams, Murray Francis

February 1948
Adair, George Lysle
Brugge, Robert Morse
Burlew, Roy Francis
Cahill, John William
Collier, Thomas Clayton Jr.
Curdts, William Theodore III
Dickey, Myron Dwight
Dorn, John Louis
Faust, Jerome
Gibler, David Lee
Giles, Harry Leslie Jr.
Goldstone, Norman Jay
Goodrich, James Frederick
Hardy, Maurice Miller
Hedgepeth, John Mills
Huntington, Jr., Hugh
Jerome, Elwood L.
Johnston, George Stephen
Krug, James Lewis
Kutsch, Howard James
Landstrom, Walter Gordon
Mautner, Richard Curtis
Mershon, Jackson Lavene
Narigan, Harold William
Pryor, Harold Eugene
Reed, Verlin Dean
Roberts, Albert Osborne Jr.
Rosebrook, Theodore Lee
Rumple, Henry Irvin
Sagata, Juro
Sattler, Robert Irwin
Scalzo, Michael Angelo
Sines, Fred Emanuel
Sundquist, Alfred Emanuel
Vaught, John Manning
Wesbecher, James Joseph
Wilson, Richard Darrell

June 1948
Angelone, Joseph
Baldwin, Floyd Lee
Bassett, Denison Sisco
Billica, Farrett Marshall
Boothman, Irvine Scott
Brown, Arthur Edmands
Burgeson, Arthur Edmund Jr.
Calledare, William Charles Jr.
Campbell, George Aberdeen
Carter, David Lawrence
Collister, Richard Lawrence
Costilow, Eleanor, Louise
Cowdin, Paul Earl
Cox, Chester Thomas
Croker, Walter Davis
deBruyn, Herbert Gerard
Dennis, David Haselton
Dunlop, James Ringman
Garrett, Charles Gerald Jr.
Gibbs, Robert Lewis

Hahn, Paul Gilbert
Holzhauer, Earl John
Hostetler, William
Hummel, Kenneth Harold
Isaacson, Jerrold Anselm
Jackson, Charles Edward
Jundanian, Thomas
Junkin, William Pursell
Keller, Robert Homer
Kice, John Deming
Kier, Norman
Konkol, George
Koogler, Robert Earl
Lanham, Paul Edward
McComb, Harvey Godfrey Jr.
McDermott, Alexander Mitchell
Merriman, Roger Wayne
Michaels, Franklin
Miles, Joseph Howard
Morrison, Rogers Herbert
Ohrman, Clyde William
Overdeer, Robert Marvin
Peregrine, Moore William
Purdy, Glenn Gibson
Rathfon, Robert Edwin
Reed, Kermit Eugene
Reed, Marlin DeWitte
Robuck, Mead Kelsey
Rudd, Gordon William
Sandercock, Donald Miller
Schmidt, Franklin Robert Jr.
Schmitt, Alfred Frederick
Shaw, James Robert
Shewmaker, Bruce Patterson
Skinner, Ernest Harold
Slanker, Earl J.
Smith, Lester William
Snell, Gale Elwood
Steen, Clarence Herbert
Strathman, Arthur Edward Jr.
Streed, Rodney Warren
Terney, Norman Lee
Toomey, James Francis
Walters, Joseph Jarvis
Woodcock, Robert James
Younger, George Gene

August 1948
Akers, Marion Jesse
Bandelier, John William
Baumgartner, Cecil Alvin
DeMuesy, Thomas Gray
Griffin, Maurice Wayne
Griffith, John Hardy
Harvey, Julian Arthur
Hughes, Paul Richard
Hunter, Richard Earl
Lewis, Fran Barry
Peede, Loring Gregory
Thompson, Robert C.

February 1949
Abel, Albert Reed

Beckert, Richard Arthur
Bickford, Russell Harry
Bison, Walter E.
Brooks, Thurman Preston
Buccheler, George Ernest Jr.
Cook, Robert Beckman
Cooper, Herbert Hutchings Jr.
Drewe, William Frederick
Hartman, Jack Edward
Heaton, Thomas Rush
Klint, William Kenneth
Lancaster John Thomas
Larson, Kenneth Ernest
Lewis, Jack Malcolm
Luppi, Edward James
Marsh, Donald Leroy
Mayer, Alvin Henry
McDonald, Sherrill Rudd
Moeller, John David
Moore, John Kerwin
Parker, Harold Wayne
Rapp, John Edward
Ruhlman, Jon Randall
Stefan, Albert Joseph
Sutherland, Howard Lawrence
Thompson, Richard Lee
Trammell, Warren V.
VanDoren, Donaldson
Vannelli, Roger Bigum
Vautaw, Robert Lee
Voelz, Joseph Karl
Waltz, Donald McMillan
Weiler, Durand Edward
Wenger, Donald Ralph
Wheaton, Richard John
Wiemer, George Frederick
Wiener, Harry Emil
Young, Jacques Edward

June 1949
Adamson, Thomas Charles Jr.
Baird, Richard Balfe
Bauerband, Percival Robert
Blake, Donald Spencer
Blomquist, Richard Martin
Bock, Charles Cornelius Jr.
Bracke, Peter Paul
Bright, Charles William
Brower, Elayne Marian
Brown, James Dewey Jr.
Carnes, James Leland
Churchill, John Russell
Chute, Richard
Clampitt, William Dean
Clelland, Jack Desmond
Cuzner, Harry Frederick Jr.
Dabrowski, Edmond Joseph
Dale, Leland Howard
Dorsey, Edward Garland Jr.
Ellis, Donald Woodrow
Engerer, Edwin William Jr.
Graef, John Brewer
Hanson, Edwin Eugene

Hanson, Thomas Francis
Heassler, Reid Ashby
Hesse, Charles
Hunter, John Robert
Ikerd, Lynn Ray
Ingmire, Helen Virginia Haley
Ingmire, Robert Lewis
Johnson, Newton Dale Jr.
Jones, Jim Joe
Jones, Walter Paul
Kaleher, John Harold
John Robert Kennedy
Kincheloe, Iven Carl Jr.
Kinsey, Keith Charles
Knapp, Clyde Holman Jr.
Knuth, Eldon Luverne
Lacey, Thomas Robert
Lammering, Jack Lessing
Lanzer, Walter Herman
Lewis, Forrest Wills Jr.
Mahurin, Walker Melville
Mason, Robert Jr.
McCook, Thomas Joseph
McIntyre, Robert Geoge
Meech, Ella Mae
Miller, Fred Willard
Morrow, William James
Newill, Robert Brooks
Peticolas, Richard Phillip
Phipps, John McCormick
Place, William Morris
Powell, Harry Kilburn
Provart, Robert
Redenour, Joseph Dean
Reid, Charles Robert
Richey, John Mac
Rogers, Herbert Franklin
Ruhmann, Stanley Robert
Rupkey, Robert Hugh
Schatzman, Eckley Gaylor
Scheuer, James Carlyle
Schnakenburg, Edward Otto
Shank, Robert Eugene
Shearer Jerry Trego
Stoops, Ernest Samuel
Strodel, Donald Herbert
Tomlinson, Jerry Gene
Wallenhorst, Edwin Linus
Wiard, William Dwight
Young, Karl Henry
Young, Louis Charles

August 1949
Devlin, Howard Anthony
Devol, James Campbell
Dunn, Addison Gardner Jr.
Evers, James Donald
Hughes, Sherman William
Jones, Clifton Ellis
Lopp, Kenneth Paul
Maglietta, George Melvin
Martin, John Patrick
Rose, Dudley Bailey

Shaw, Reid Johns
Stalk, George
Watson, John George

February 1950
Adel, Robert Elliot
Alter, Robert Lewis
Axelrod, Melvin
Benson, Joe Arthur
Brown, Floyd Blaine Jr.
Collingworth, Jack Marvin
Colyer, Duard Browning Jr.
Fink, Russell Charles
Glass, John David
Gleeson, William Stoddard
Gordon, Henry Charles
Hall, Herman Lew
Hanson, Fred Alton
Herstine, Glen Leslie
Johnson, Kenneth Odell
Kartasuk, Raymond Heronim
Kietzman, Herman Robert Jr.
Klimczak, Anthony Joseph
Lazaro, Julio Jose
Leamon, John Ferris
Lemmon, Stanley Vaughn
Lippoldt, Otto Frederick
Martin, Reese Sargent
Matthews, Clyde Charles
McClanahan, Paul Glenn
Miller, Clemant Kennedy
Minger, Frederick Harvey
Mitchell, Deane Harold
Moffat, William Villeroy
Moose, Harvey Earl
Novak, Donald Hoyt
Patras, James Louis
Perless, Leonard Carl
Reeder, Robert Henry
Renbarger, Jack Lester
Rynearson, Glenn Orville
Schneider, Charles Joseph
Springer, Harold Leroy
Waterman, Hugh Everett
Wigley, Harold Willard

June 1950
Anderson, Thomas Edgar
Barfod, Frank
Bogan, Thomas Rex
Boll, Robert Thomas
Bosscher, James Peter
Brady, Robert Eldridge
Brodersen, Donale Hamley
Brunsman, George Edward
Burress, Joseph William
Cady, Joseph George
Cain, William Charles
Clark, Delbert Gene
Crawford, Charles Marion
Cutter, Lois Mae Niederloh
Davis, Dean Francis
Dunning, Leslie Leon

Dunton, William Harris
Frazier, Albert Aquila
Glidden, Fred Ray Jr.
Golden, Charles Carroll
Grabowski, Wallis Richard
Hale, William Richard
Harvey, Ray Wilson
Higgins, Charles Clifford
Holt, Robert Lee
Honsberger, Jacob Edward
Hunnicutt, Wayne Edward
James, Warren Kay
Jones, James Parsons
Jones, Robert Lewis
Kell, Robert John
Kennedy, Norman James
Lane, Richard Paul
Langley, John Arthur
Leakey, Bruce Hollinger
Lowes, Charles Robert
Marshall, John Thomas Jr.
Mayer, William Earl
McQueen, Kenneth Theodore
Miller, Garner Wakefield Jr.
Mushial, Francis Vincent
Nelsen, Carl Thorwald
Otto, Robert Cushman
Parrett, Donald James
Patrick, Richard Montgomery
Ramsey, Joseph Conner
Reed, Sally Ann
Rowan, Henry Ackley
Samsen, John Richard
Schnyer, Arnold Daniel
Sharp, Robert Lee
Shull, Oswald Warden
Smith, Alfred Keith
Snell, Rex Lee
Southland, Gerald Frederick
Swenson, Richard Frank
Talmage, Arthur Lee
Thelander, John Arden
Timmerman, Warren Harvey
Wait, Kenneth Albert
Ware, William Carter Jr.
Watson, Richard
Wells, Robert Francis
Whisman, Lynn Robert
Winkler, Richard Charles
Young, Richard Myers

August 1950
Beall, Charles Nelson
Childress, Glen Elmer
Georgas, John James
Gerhardt, Leslie Harold
Gibson, Carroll Benjamin Jr.
Greenberg, Arthur Bernard
Hartman, Francis William
Plouse, Henry
Rodgers, Robert Roy
Wright, Kenneth Frank

January 1951

Alles, Thomas William
Alman, Allen Edwin
Balian, Roxy Alice
Basil, Lloyd Ira
Burns, Marjory Joan
Downs, Thomas Bellwood
Ellington, Robert William
Felix, Leroy Dale
Hahn, Jack Ralph
Howell, Wayne Deloss
Jackson, David William
Klatt, Florine June
Larrimer, Walter Harrison Jr.
Marshall, Harold Malcolm
Mott, Harold Caven
Plafcan, Charles Joseph
Rea, Mead Melvin
Rose, Rudolph Frederick Jr.
Schepke, Thomas Leonard
Sedvert, Theodore Warren
Shah, Jayanti Shantilal
Spidell, Richard Henry
Stainback, Paul Calvin
Stammerjohn, Richard Max
Thelander, William Gordon
Zimmerman, Arthur Harold

June 1951

Anderson, William Ross
Bauermeister, Walter Karl
Beall, Richard William
Browne, William Albert
Burdick, George Standon
Burgess, Robert Kingsley
Burgos, Harry
Burns, Howard Paul Jr.
Cafarella, Frank Joseph
Carroll, Patrick Colin
Choquette, Raymond Emile
Clayton, Richard Max
Edstrom, Clarence Raymond
Ensley, William Tyrus
Ferreira, Paul Franklyn Jr.
Fortini, Anthony
French, William Arthur
Geye, Richard Paul
Hawk, George Wayne
Holmes, David Alvin
Kalb, Arno George
Kaplon, Ted Joseph
Kehres, James Robert
Kelly, Charles Peter Jr.
Kitchel, George Edward
Leissler, Leo Abbott
Lukenbill, Emery Don
MacFarlane, Bert Fraser Jr.
Mallett, William Ernest
Marty, John Lewis
Motsinger, Richard Newell
Muller, Charles Henry
Newgent, Jack
Olinger, Max Eugene

Perkins, Ronald Arthur
Perrone, Robert Gary
Person, Morris William
Plemel, James Eugene
Richter, Arthur
Rinearson, Richard Eugene
Robertson, John Milton
Rossetto, James Anthony
Schaper, Peter Wolgang
Scott, Ray Edward
Swanson, Andrew G.
VanderVelde, Wallace Ear
Volker, Charles Malcolm
Wernet, Robert Frederick

August 1951

Allen, Charles Gilpin
Blakeslee, Donald Jack
Bueker, Robert Arthur
Corbett, Robert Earl
Fehrs, William Jr.
Jessen, Allen Jackson
Lee, Robert Paul

January 1952

Corsette, Douglas Frank
Davies, Stephen Bresse
Endicott, Donald Lewis
Fisher, Raymond Alan
Henshaw, Marshall Darling Jr.
Kesling, Paul Jr.
McClure, George Wayne
Owens, Edward Gilbert
Peasley, Donald Lee
Rasey, John Allen
Veile, Wallace Wright
Zepf, Richard Joseph

June 1952

Begin, Robert Edwin
Bersinger, Everitt Vedder
Capasso, Vincent Nicholas Jr.
Corcoran, William Lawrence
Covert, Henry Earl Jr.
DeLalio, George Martin
Ernest, Alan Ward
Farrington, Ira Donald Jr.
Goranson, George Grant
Graham, Bruce Harlan
Hackman, Lloyd Eugene
Hartman, William Bruce
Jonas, Mary Jane
Kester, Ralph Eldon
Laymon, William Arthur
Levi, Oscar Arthur
Miller, Ralph Samuel
Milikan, David Lawerence
Miltonberger, Anna Georgene
Morgan, Homer Garth
Roof, Edwin Melon
Rupert, Richard Wayne
Sherrill, Robert Dee
Shuter, John Edward

Spak, Michael
Stanwood, Jay Wesley
Stiebling, Mary Lou
Tangler, Howard Paul
Umbreit, Joe Meloy
Vinson, Paul Webster
Wilmore, Dhalmas Otto
Young, Daniel Herbert III

August 1952

Cowdin, William Gros
Dodge, Eric Harvey
Goble, Gilbert James
Hanner, Oliver Mark
Kaszynski, Henry Joseph Jr.
Kester, Robert Eugene
McKaig, William Dewey
Nicholas, George James
Smith, Robert Everett

January 1953

Bowditch, David Nathaneil
Cowherd, Harold Jr.
Darling, Charles William
de Castongrene, Russell Orthomar Jr.
Fisher, Robert Erwin
Flores, John Rodney
Goldstein, Charles Max
Hamm, Donald Frederick
Hardesty, Edwin Leroy Jr.
Haykin, David Judson Jr.
Johnson, Lynn Danridge
Marcotte, Donald John
McAtee, Robert Francis
McLure, Richard Donald
Petty, Paul Eugene
Thorpe, Louis McCord
Wheasler, Robert Arthur

May 1953

Bigham, James Pollock Jr.
Booth, Richard Ross
Davis, Joseph William
Eklund, LaVerne Gene
Everts, Ronald, Jack
Fastiggi, James Joseph
Feeley, LeGrand Derby Jr.
Flinn, Evard H.
Fox, Joe Thomas
Freeland, Joseph Lynn
Garred, Mathew David Jr.
Heath, Webster Courtney
House, Richard Louis
Howley, Donald Louis
Huben, Clayton Antone
Irvine, Thomas McDonald
Jackson, Arthur Senior
Johnson, Robert Turner
Johnson, Sherwood Maurice
Kelly, Robert Clarke
Koziatek, Steve Joseph Jr.
Masters, Joseph Neil
Miller, Harvey Eugene

Myslak, Thaddeus Benedict
Neidhold, Carl Dudley
Probst, Harry Chester
Schultz, Robert Henry
Shaffer, Joseph Jr.
Small, J. William
Small, Robert Dave
Staples, Frederick Grant
Stash, William George
VanderHoven, John William
Weir, Samuel Gamble
Wisehart, William Richard

August 1953
Beck, James Kinter
Berry, Verne Edward
Bohl, Warren George
Schmetzer, William Montgomery
Shanahan, Billy Rex
Swanson, Daniel Carl

January 1954
Bellinfante, Robert Joseph
Conklin, Edmund Judson
Elliott, James Vincent
Lodge, Donald Waldo
Moore, Cecil Lesley
Reynolds, Warren Charles
Schellert, Jerry Lee
Schienberg, Marvin Bernard
Simmons, John Robert
Taylor, Norval Page
Thompson, George Edward

May 1954
Ahls, William Lloyd
Beers, Harold Simpson Jr.
Boehmer, Stanley Alfred
Brown, Robert Maurice
Carter, Mark Richard
Cingo, Ralph Paul
Clark, Thomas Sampson
Coates, Charles Robert Jr.
Da Costa, Augusto Vale
Decker, Vern LeRoy
Gronkiewicz, Thaddeus Robert
Howard, William Julius
Illium, Herbert Charles Jr.
Jawor, Cheslav Frank
Karrenberg, Hans Karl
Marquardt, Richard Eldon
Mascolo, Wilfred Delfino
Mertaugh, Lawerence Joseph
Nesse, Richard Bertan
Norton, Frederick Minter
Pasko, Donald Charles
Scott, Robert Bruce Jr.
Soberski, Floryan Ludwig
Steinmann, Herbert Henry
Truax, Phillip Park
Zapotowski, Bernard Patrick

August 1954
Clark, Harold Lee
Disser, William Charles
Russey, Robert Edmund
Smith, Dale Arlen
Thunstrom, Lennart Nils

January 1955
Armstrong, Neil Alden
Bancroft-Billings, Ralph
Blechschmidt, Carl Wilhelm Jr.
Drissell, Brian Arthur
Gardner, Donald Albert
Genens, Lyle Emil
Gugeler, Robert Charles
Hopf, Russell Lewis
Johnson, Ronald Paul
MacNicol, Robert Kenneth
Schlosser, Stanley Miles
Schneider, Paul William
Sievers, Gilbert Keith
Stoker, Gerald Lee
Wilson, Robert Allen

June 1955
Barthel, James Robert
Brant, William Warren
Brown, Robert Alden Warner
Burkhart, William Edwin
Caldwell, Walter Knowlton
Carnegis, Isidoros Angel
Cramer, William Kenneth
Davis, Duane Madison
Doyle, John Thomas
Fritz, Donald Clayton
Hardy, Halsey Dale
Hines, Lawrence David
Johnson, Jack
Kamien, Chaim Zelman
Kellam, John Miles Jr.
Landers, Robert Ashenfelter
Lang, Ronald Peter
McClaughry, Richard Chester
Naab, Kenneth Nicholas
Rodman, Ray Wilbur
Ross, Lee Edward
Sanderman, William F.
Sommer, Robert Wellington
Stephenson, David Skeen
Thomas, John Walter
Thompson, Donald Dwane
Toney, John Samuel
Ward, Donald Dale
Weakley, Thomas Lionel
Wernicke, Rene DuWayne
Wilke, Frederick Henry Jr.
Winstandley, William Carroll
Wooden, John Alfred
Zenor, Donald Dean

August 1955
Casey, John Carl
Higgins, Edward Kee

Hutchinson, Samuel Frederick
Irion, Clarence Eugene Jr.
O'Brien, Daniel Joseph
Parker, Richard Wilbur Jr.
Savage, Richard William
Thompson, Thomas Ross
Valor, Norman Harlev

January 1956
Ford, Dale Charles
Freund, William John
Halal, William Emitt
Hartman, Ernest, Lester
Henderson, Virgil Estle Jr.
Hodson, Charles Henry
Kraft, Eugene Charles
Pivirotto, Thomas James
Regnier, William Wayne

June 1956
Anderson, Curtis Barry
Andrews, John Franklin
Bolinger, Robert Paul
Bolster, Bradley Drake
Booth, George Calvin
Brandt, Robert LaVern
Bunyak, Ronald Steve
Cady, Edwin Cornell
Cappellari, James Oliver Jr.
Clingan, Bruce Edwin
Conley, Eugene Ogden
Creasey, Floyd Eugene
Deskins, Harold Eugene
Dezelan, Richard William
DonMoyer, Richard Golden
Everett, John Milton
Fava, Donald Gene
Fouts, William Bert
Hartke, Charles William
Hartke, Richard Henry
Hayden, William Donald
Heath, Gerry Clayton
Hofferth, Delbert Dean
James, Robert Nicholas
Johnson, Maynard Kent
Jones, Donald Lee
Khan, Ashraf Hasen
Kinneer, Billy Lee
Knapp, Robert Frederick
Krull, Nicholas Peter Jr.
Lindeman, Robert William
Long, William Robert
Mayerhofer, Robert Donald
Maynard, Gordon Earl
Miller, Franklin Elliott
Mittermaier, Norman Paul
Morgan, Bruce Joseph
Person, Hans Anton
Petersen, Richard Herman
Reed, Benjamin William Jr.
Ringgenberg, Robert Lee
Rittenhouse, Lewis Eugene
Schatz, William Jackman II

Swaim, Howard Lee
Swingle, Robert Lyle
Trenka, Andrew Richard
Tryon, Paul Franklyn
Unger, Charles Melvin
Wilson, Lionel George
Wolter, Willy August Christian
Woodhouse, Lucien Craig

August 1956
Butler, Ludlow S. Jr.
Chen, Franklin Yueh-Kun
Fagerstrom, Joseph Waynesworth
Gregor, John Robert
Hayes, Robert Tennyson
Marchand, Jack Austin
Obear, James Jackson
Seeley, Charles Thomas
Wilson, Dennis James

January 1957
Bagnoff, Donald
Byam, Richard Edward
Eng, James Charles
Fearing, George Neil
Ferman, Marty Allen
Forsmo, Dennes Phillip
Gaitatzes, George
Hassel, Robert Kenneth
Johnson, Charles Buddy
Johnstone, Edmund Hughes Jr.
Klippel, Gilbert Franklin
Marks, Lawrence Eugene
Mulholland, John Derral
Newgeon, Walter Ward
Noblitt, James Paul
Roland, George

June 1957
Arnett, Matthew Dean
Babcock, Charles Dwight Jr.
Blackiston, Harry Spencer Jr.
Bower, Richard Lee
Boyd, Delmar David Jr.
Brian, George Robert
Chaffee, Roger Bruce
Cleaver, Albert Leroy
Clingman, David Lee
Collins, Frederick John
Connor, Ronald Virgil
Crandell, Gloria Catherine
Dougherty, John Alexander Jr.
Drynan, Ronald Darren
Duhnke, Robert Emmet Jr.
Ellis, Stanley Aaron
Finch, Louis Ferdinand
Force, Charles Thomas
Fouts, William Brooke
Garmon, William Leron
Gibson, James Darrell
Glahe, Fred Rufus
Gotha, Fred
Gregory, Charles David

Hamaker, Robert Joseph
Harvey, Walter Douglas
Herdman, Howard Lee
Hutchins, John Frederick III
Janes, Chesterfield Howell Jr.
Jepson, William Donald
Johnson, Donald Wayne
Jurgovan, Roger Jacob
Kane, Milton Thomas
Kinsler, Caroline Alice
Larrison, John Lee
Leach, William Benjamin
Long, Basil Mallory
Lundahl, Duane Webster
Milton, Harold William Jr.
Morton, Donald Hope
Nencka, Walter David
Pendleton, Richard Wesley
Poel, Jerry Dean
Reinecke, William Gerald
Rencenberger, Robert William
Rittenhouse, James Freeman III
Rolland, Alan David
Setmeyer, Edward Frank
Seward, Arthur Lewis
Sharman, Douglas Thomas
Spokas, Romas Balys
Stewart, James Ferguson Jr.
Stouppe, David Edward
Swaim, Robert Lee
Swoger, William Frank
Szabo, Thomas Robert
Thompson, James Milton
Thompson, John David
Urick, Gilbert Hunt Jr.
Van Putte, Ronald Everett
Vinson, John Elden
Walburn, Clarence Gene
Ward, Gerald, George
Welzen, Joseph Andrew
Wennerstrom, Ernest Stanford
Wolf, John Thomas

August 1957
Beyer, Kenneth Emil
Cusack, George Bernard
Glancy, Jerry Leroy
Hensel, Thomas Edward
Murphy, John Michael
Study, Joe Allen
Wilhite, Jack Richard
Wynne, John Washington Jr.

January 1958
Beck, William Francis
Byrnes, John Marshall Jr.
Flee, Charles Richard
Gess, Lawrence Edward
Harvey, Philip Edward
Hoeks, Henry Jay
Horwitz, Gerald Allen
Lipp, Louie Jackson
Sandlin, Steven Monroe
Shackford, Joseph Edward

June 1958
Abbott, Anton Dwight
Abrams, Frank Joseph Jr.
Adams, Erwin Francis
Ambs, Phillip Henry
Auguston, Jon Roy
Baggett, Philip James
Bannon, Francis Patrick
Barkmann, Karl Emil
Bartholomees, George Hugh
Becea, Thomas Gregory
Black, John Allen
Bourne, William Roscoe Jr.
Burchfield, Charles Gilbert
Christensen, Carl Spencer
Clark, Edward Uhlan
Cook, Charles Ernest
Crawford, Charles Huelett
Cronin, Robert Joseph
Curry, Carlton Eugene
Daugherty, Clifford Bruce
De Wesse, James Henry
Doak, Don William
Engler, Walter Peter
Etheridge, James Dawson
Glaser, Frederick Carroll
Hoadley, John Stephen
Huber, Edward Gustave
Kerr, James Roy
Komechak, George Judy
Laird, William Frederick
Magee, David Robert James
Maldonado, Michael
Marlotte, Gary Lynn
McElvain, Robert James IV
Minick, Eugene Paul
Mondrzyk, Robert John
Moreland, Floyd Eldrige
Nelson, Wallace Emanuel Jr.
Norton, Allan Minter
Paris, Franklin Lane Jr.
Patterson, Charles Frank
Peters, Thomas
Pierson, Wayne Lineer
Pneuman, Gerald Warnick
Ramsey, Paul Elbert
Roper, Alan Thomas
Russell, Richard Allen
Ryals, Willard Glenn
Sherwood, Richard Bradley
Swingle, William Lloyd
Tabata, William Kenji
Tasch, Irene Eleanor
Thein, Maung Kyaw
Theodoroff, Richard Joseph
Usab, William James
Valenti, Joseph Francis
Wackrow, Richard Ronald
Weil, James Anthony
Wolpert, Richard Louis
Yeager, Walter Carl
Young, David George
Young, Kenneth Robert
Zalmanis, Andris

August 1958
Bennett, Thomas Sidwell
Binz, Ernest Fredrich
Blue, William Charles
Bluemenkranz, Jerald Martin
Bridwell, Gene Porter
Forsythe, Conrad Orville Jr.
Hays, Luther Marr
Johnson, Richard Allen
Kirk, Richard Carlyle Jr.
Kloth, Richard Emil
Lakamp, Larry Lee
Lamb, Edwin Dean
Rettich, Thomas Alden
Wilson, Lary Wade

January 1959
Chapin, Claire Edwin Jr.
Davidson, Roger Marion
Ditto, Joseph Carol
Durbin, William Hartley
Eversman, Walter
George, Homer Earl
Handly, Edward Clarence
Haynes, Norman Ray
Johnson, Duane Paul
Kellogg, Larry Gesley
Kennedy, James Clarence
LeFavour, Walter Gene
Lewis, William Frank
Mesec, Gilbert Ronald
Meyers, James Frank
Stabler, James Eugene
Tangedal, Norman Lloyd
Weller, Wallace LaRue
Whitacre, Walter Emmett
Wilson, Brian Lane

June 1959
Andrews, Paul Dwight Jr.
Atkinson, Philip Rudoloph
Bailey, Merlin Murray III
Bander, Thomas Joseph
Bauleth, Eloy Mario Peixoto
Bergbauer, Daniel Matthew
Bortz, Paul Issac
Brandewie, James August
Burch, James Dallas
Byrne, David Allen
Carlson, Richard Darrold
Cheng, Su-Ling
Cherry, Franklin Dee
Chichester, Frederick Donald
Christman, Charles Fredrick Jr.
Coffin, Clarkson Lee
Collins, Robert Lynndon Jr.
Cox, Walter, Stanton
Donelson, John Earl
Fankhauser, Earl William
Fletcher, Robert LaVerne
Foster, Kenneth Lloyd
Gates, Richard Merrill
Geiger, James Joseph

Geralde, Robert Edward
Giasolli, Mero Vincent
Goetz, Donald Howard
Graham, Charles Bernard
Hargraves, Charles Ray
Hillsamer, May Eugene
Hoch, Roy John Jr.
Homburg, Tracey George
Horsewood, Jerry Lee
Hull, David George
Jecmen, Dennis Michael
Kanouse, Kenneth Lee
Kaufman, James Harry
Kay, Harry Gustaf
Kelly, Paul Donald
King, Phillip Andrew
Kintzel, Craig Sempill
Kirkham, Frank Stevens
Laird, John Dale
Leech, Thomas Franklin
LeRoy, James Vernon
Long, James Blair
Lusty, Arthur Herbert Jr.
Magera, John Jr.
Mantz, Kenneth Lee
Marshall, Frederick Carl
McGowan, Kenneth Dale
McMahon, Richard O'Neil
McMillan, Jack Downes
Mead, Charles Thomas
Meyer, James Edward
Motzny, Kenneth Guy
Myles, Keith Hastings
Ng, David Ping-Kan
Notestein, John Edward
Pelton, Donald David
Plunkett, Dennis Allen
Powell, Richard Allen
Pritchard, Robert Emlyn
Reed, Vernon Lawrence
Reeder, John Albert
Rich, Daniel Thomas
Roth, Peter
Sallada, Robert Velmont
Schuning, Kenneth William
Schweikle, John David
Scott, Paul Witherspoon III
Shaffer, Richard Bradley
Smith, Joseph Dale
Smith, Malden Douglas
Smith, Thomas Mitchell
Stumph, John Stephen
Taylor, James Jacob
Taylor, Ronald Gene
Teuber, Roland Frederick
Toth, Richard
Vail, James Russell
Velligan, Frederic Andrew
Vertigan, Richard
Waara, Fredrick Marion
Wallace, Donald Wayne
Walters, Robert Charles
Woebkenberg, William Henry Jr.

Wolfe, Kenneth Eugene

August 1959
Castelluccio, Warren Kent
Cawood, Carl Vance
Dreher, Carl Edward Jr.
Fleener, Wendell Aaron
Gebert, Carl
Hepler, Leslie Jackson
Hilbert, John Elmer
Karleen, Ronald Edward
Kraus, David Michael Jr.
Ladd, Howard Vincent
Mason, Joseph David
Nering, Paul Dale Jr.
Rickard, William Dean
Simons, Laurence Edward
Skaret, David Jon
Tarnow, Herman Elmer
Voyls, Donald Walter
Vyzral, Francis John
Wontorek, Bernard Francis

January 1960
Allen, Gerald Ashley
Beardsley, Peter Paul
Bryan, Kenneth Eugene
Butler, Robert Lee
Cosgrove, John Patrick
Cox, Dean Maurice
Czumak, Frank Michael
Fromme, Joseph Archibald
Holder, William Glenn
Lipscomb, Robert Dabney Jr.
Mayes, Gary Earl
Mickelsen, Harvey Peter Jr.
Mitchell, Walker Hugh
Nichols, Roger Dale
Saaris, Gary Robert
Schmidt, Hans Walter
Schneider, Thomas Gordon
Schwab, Avon Hugh
Thompson, Harry Brownlee Jr.
Tobias, Marshall Allen

June 1960
Albers, Gerald Elmer
Broadwell, Ronald Earl
Carnes, Frederick Eugene
Cork, Max Joseph
Esterline, Richard Alan
Everett, Guy Bennett
Gaffey, Troy Michael
Gambaro, Ernest Umberto
Haws, Lawrence Burrows Jr.
Kummerer, Keith Robert
LeCount, Ronald Lane
Maguire, Michael Gene
Marut, Charles Martin
Meleason, Edward Thomas
Miertl, Donn Alan
Mitchell, Gary Edward
Moll, Richard Lee

Mucha, Thomas Jerome
Ozols, Modris
Pierick, Jerry Devoe
Roberts, Raymond Ralph
Strack, William Carl
Strickler, Robert Leslie
Stubbs, Daniel Franklyn
Tall, Wayne Allen
Ullrey, Paul William
Ward, Terry Leon
Wilke, Richard Otto
Willich, Wayne

August 1960
Boussom, Robert David
Freeman, Mark Leslie
Gerbasi, Martin Jon
Gleiter, John Peter
Hall, James Frederick
Igoe, John Edward Jr.
Ranes, Richard Lee
Rivir, Richard Bryam
Spicer, Robert Lee

January 1961
Barnacastle, Joseph Sidney
Crowder, Charles O'Rourke
DeMichieli, William Pietro
Friedman, Gerald Joseph
Healy, James Joseph
Hughes, Jerry Lee
James, Nolan Elliott
Janssen, Werner Carl
Kershner, David Harold
Kester, John Paul
Leedy, Larry Lee
Lohmueller, Bruno Ludwig
Mais, Ronald Weller
Medanic, Stanley
Miller, Terry Gene
Nangle, Donald Leo
Nomanson, Jack Harland
Novak, Peter Anthony
Randall, James Edward
Stutts, Harry Carey Jr.
Waltz, Frank Friedl
Whitman, Gerald Andrew
Weise, John Otto
Willman, Jerry Lee
Wright, John Elmer

June 1961
Alden, Ralph Edward
Alkire, Frederick Gene
Bailey, Jerry Lee
Bennett, Merritt Duane
Bergesen, John Eric
Bethel, Howard Emery
Bogemann, Lawrence Leo
Boisvert, Raymond Gerald
Bowers, Jeffrey Lytle
Brown, James Edward
Burgess, David Charles

Campbell, George Vincent Jr.
Davis, Lowell Keith
Ervin, John
Erwin, Lowell Ray
Frearson, David Eric
Hess, Edgar George
Huh, John Syenghee
Krasne, Robert John
Lahs, Robert Alan
Lott, Robert Alan
Lung, Joseph Orley
Miller, Bruce Alan
Murtaugh, Thomas Dennis
Naggs, Karle Frederick
Newbold, John William
Nichols, Jack Louis
O'Neil, William John
Quick, David Hugh
Radcliffe, Roger Lee
Raisbeck, James David
Ross, Thomas Harvey
Sanderson, Lawrence Everett
Schlaefi, John Louis
Schuerman, J. Allan
Simo, Miroslav Andrew
Starr, Stottler King
Staudt, Fredrick J.
Suits, Norman Lee
Suter, Edgar Thomas
Szwabowski, Edward Julius
Wade, Barton Scott
Walters, Larry Rae
Watson, Calvin Franklin Jr.
Werner, Fredrich Clemens
West, Roland Houston Albert
Whitehead, Kenton David

August 1961
Brandvold, William Clarence
Lamb, Charles Melford
LeRoy, John Earl
Tanimoto, Paul Mamoru

January 1962
Andrews, James Howard
Bell, Merlin Gene
Carle, Burton Gordon
Carlson, Burnette John
Fanning, Robert Clark
Feldman, Sol Mack
Grellmann, Hans-Werner
Lui, Oscar Ye-Ling
Scarl, Eugene
Schiffer, John Richard Jr.
Yoder, Lee Orden

June 1962
Clayton, Frederick Imel
Crouse, Gordon Lee
Halyard, Raymond James
Johnson, Gearold Robert Jr.
Karlakis, Theodore
Kring, Larry Albert

Moynihan, Philip Irwin
Reddall, Walter Frederic III
Ruterbories, Burghard Herman
Schroeder, Howard Lee
Smrz, John Phillip
Stultz, J. Warren
Traylor, Paul Keith
Wampler, Michael Thomas
Wells, Raymond Paul
Wong, Bradley Yew Hin
Zeps, Leopold Peter

August 1962
Chamberlain, Donald Robert
Coffman, James Bruce
Heavin, Myron Gene
Roberts, Franklin D.
Scribner, James Edward
Yamane, Daniel Fujio

January 1963
Barlock, Lawrence Andrew
Bertram, Lee Alfred
DeWitt, Charles Edwin
Greening, James Everette
Harmon, Timothy Jesse
Leonard, James Richard
Lorette, Michael Lawrence
Mattes, Richard Joseph
McHale, Charles Everett Jr.
Nadeau, Fred Russell
Shew, Chloral Lansing
Stoller, Edward David
Strack, Kenneth John
Todd, Myron Eugene
Unger, John Fred

June 1963
Bremer, Robert James
Brinkmann, Frederick Henry
Csaszar, Bruce Joseph
Eastwood, James Albert
Foggatt, Charles Edward
Krueger, Donald Gene
Lausch, Keith Myron
Lewin, Margaret Fern
Mattes, Robert Eymard
McGlothlin, Charles Edward
Murphy, Lana Gale
Neulieb, Robert Loren
Perham, David George
Raschka, James Gregg
Reed, Richard Phillip
Seto, Rodney King Mun
Todd, Marvin Lee
Weber, James Alec
White, Douglas Charles

August 1963
Loudenback, Leland Davis
McKay, Raymond Angus
Meier, John King
Wear, David Ervin

January 1964
Baither, Richard Allen
Corley, James Milton
Duvall, Arthur Julian
Evans, Paul Francis
Flowers, Ronald John
Guenther, Rolf Adolf
Gum, John Arthur
Hahn, Dennis Walter
Haven, Theodore Charles
Jones, Hamilton Michael
Kessler, William Clarkson
Kinney, Douglas Glenn
Lippincott, Robert Park
Moffett, Paul Warren
Parkinson, Thomas William
Stalnecker, Wayne Edward
Tuszynski, Stephen Walter
Wacker, George Frederick
Wahl, Roger Louis
Weber, Jack George
Whitacre, Horace Edward
Yancy, Gene
Zyskowski, Stanley John

June 1964
Badeau, Lydia Annin
Barr, Joseph Gerard
Baturevich, George Alan
Bennett, Franklin Odbert Jr.
Cannon, Thomas Calvin Jr.
Ciskowski, Thomas Michael
Deis, Frank Robert Jr.
Eckler, Gilbert Ross
Elkins, Ronald Gale
Furst, David Lowell
Harris, Edward Park III
Harris, Randall Lee
Head, Thomas Eugene
Hedegard, Alan Harald
Hesprich, Glen Vincent
Hilgenberg, Dennis Paul
Hines, Norman Eugene
Hinkle, Thomas Vernon
Jones, William Judson
Keenan, Frank James III
Kelly, Ronald Dean
Krieps, Richard Neal
Krueger, George Richard
Kuczynski, William Anthony
Kuhn, Russell Paul
Lenahan, Dean Thomas
Little, Daniel Raymond
Moore, Ronald Lee
Navarro, Richard Alan
Palmreuter, Edsel Charles
Ricks, James Michael II
Safranski, Stanley Gerard
Schuh, Delbert Jay II
Stanton, David Mac
Stiles, Thomas James
Swain, David Oscar
Sweet, Joel

Vetter, Donald Robert
Von Rosen, Thomas Alexander
Wilson, Robert Lee
Wuichet, Stephen Anthony

August 1964
Cardinal, James Joseph
Henkel, Charles Edward
Hesler, Lee James
Wasserman, Jacob Frederick

January 1965
Al-Essa, Khalid Sultan
Chen, James Tze-Chuang
Conley, Ralph Robert Jr.
DeKuyper, Robert Eduard E. H.
Dunville, Eric Gordon
Gill, Robert Allan
Knapp, Lloyd Stephen
Lanterman, Benny Joe
Laskody, Jerome Robert
Lehmuth, Lee Richard
Muhl, Charles Phillip Jr.
Oetzel, Rodger Philip
Poplawski, Ronald Eugene
Rimkunas, Sal
Schroedel, Scott Andrew
Seymour, John Christopher
Werhane, Randy Brian
Wilson, Frank William Jr.

June 1965
Arden, Robert William
Ashley, Stephen Ray
Banks, Roosevelt Maynard
Baughn, Terry V.
Boyd, Carl Ritter
Cole, Roger Ellison
Cox, Wayne Paul
Doyle, George Robert
Ely, Wayne Lee
Evans, Larry Gerald
Evans, Maynard Jackie
Fong, Ping Jr.
Gordon, John Bernard
Harrington, Douglas Edgar
Hibbard, George Lewis
Jeffrey, William David
Kamis, Donald Neal
Klopp, Robert Richard
Langdoc, William Albert
Laughlin, Michael John
Lewis, David Stanley
Markl, Robert Henry
Marley, Arley Cleveland III
Nemes, Albert George
Novick, Allen Stewart
Oeding, Robert George Jr.
Pater, Stephen Walter
Ramsey, Daniel Eugene
Reichmanis, Ilmars
Richardson, Louis Eugene
Russo, Robert Steven

Ruzzo, Tom E.
Schmidt, David Kelso
Scohy, James Richard
Skele, Peters
Smith, Virgil Kirkland III
Springman, Gillie Frederick
Stephenson, Randolph Wallace
Swift, Thomas Raymond
Thomas, Theodore James
Volkay, William John
Whitworth, Jon Alan
Wolber, Donald Harry
Zealey, Gilbreath Byron

August 1965
Beck, Wayne Thomas
Garrahan, Joseph Richard
Murach, Benjamin Vargas
Roelke, Donald Paul
Saal, Karl William Jr.
Schneider, Alfred
Setzler, Paul Elliott

January 1966
Avery, Malcolm Lynn
Brewer, Jerry Dean
Casagrand, Robert Ashley Jr.
Daugherty, Gerry Robert
Frick, William Charles
Gardner, James Robert
Helton, Michael Richard
Jonaitis, Kenneth Wayne
Means, James Le Roy
Neville, Douglas Lee
Patrick, Gerald John
Priem, Roger Lynn
Senchak, William Emil
Sosnay, Richard Gordon
Steer, Donald Nelson
Willen, Gary Scott

June 1966
Avers, Margaret Louise B.
Ayson, Edmund Dante
Baker, William Morse
Bowman, Ralph Dennis
Campbell, Donald Scott
Chandik, Thomas John
Colvert, Elston S.
Cook, Richard Drury
Crook, James Leslie
Decoste, Dennis Joseph
Dolson, Richard Charles
Fidler, Gary Lester
Getz, William Carl
Gruber, Larry Everett
Hogan, Brain Dennis
Hopping, Cecil Rex
Isaacs, Fred Wesley
Jacobson, Robert Arthur
Marshall, Gerald Charles
McKane, Gordon Thomas Jr.
Meek, David Duane

Miller, Albert Ray
Moebs, David Sylvester
Mort, Raymond William Jr.
Myer, Paul Henry
Okaro, Chukwuemeka James M.
Olzmann, Norbert Alfred
Readnour, Jack Lee
Reed, Ralph Herbert Jr.
Siegel, Alan Elliott
Smith, Darrell Franklin Jr.
Snyder, Howard Allen
Thomas, Robert John
Wantuch, Richard Alan
Winkeljohn, Douglas Marshall
Zellers, Jon Lance

August 1966
Best, Eric
Blodget, Elweyn C.
Gaudette, Theodore Louis
Held, Kenneth James
Miller, Kenneth Gregory
Slamkowski, Harry Leon
Southerland, Mark James

January 1967
Conner, William Russell
Conrad, James William
Dilmaghani, Jalal
Gross, Darrell Eugene
Harley, Charles Young
Harrington, Richard Michael
Lehmkuhler, Larry John
Mendenhall, Harry Keith
Mishler, Dennis Harold
Pekelsma, Nicholas John
Pflum, Ronald Joseph
Portinga, Peggy K.
Smith, Thomas Richard
Vitale, J. Gregory
Weaver, Leon B.
Wrucha, Jerry Andrew
Yarnell, Robert Stewart Jr.
Zutaut, Edward Carl

June 1967
Augusten, John Clifton
Barrett, Arthur Lee Jr.
Bristow, Dean Rowe
Deerhake, Allen Charles
Derrickson, Charles Michael
Director, Mark Nathaniel
Doane, Paul Michael
Drexler, James Howard
Evbuoma, Stanislaus Nosakhare
Flemming, Robert James Jr.
George, Barry Brian
Green, James Wilson
Heinmiller, Paul Jay
Hera, Ronald Wayne
Irace, William Richard
Jones, William Henry
Joyce, Douglas Allyn

Keller, Robert Michael
Kleeman, John Joseph
Klusman, Steven Arlen
Lockenour, Jerry Lee
Longstreet, Robert Alan
Marr, Clayton Lee
Nasal, Timothy Patrick
Rule, Timothy Theron
Sellberg, Ronald Paul
Sheets, Daniel Martin
Soukup, Stephen McKay
Stewart, William Becker
Street, Thomas Charles
Vairo, Vincent Ronald
Wasson, John Roger
Winter, Dean Carl
Young, Don Lee

August 1967
Labash, Richard Anthony
Reed, Claude Brinton

January 1968
Bauer, Aaron Otto II
Baughman, Wilfred Earl Jr.
Ekhaguere, Solomon
    Oghoereye
Hager, William Dennis
Howes, Robert Lloyd
Lackey, Robert Paul
Lee, Roy Robert
Malecha, Kenneth William
Phillips, Larry Lee
Romberg, Robert Athelstan
Thompson, Kenneth Reed Jr.
Van Allen, Richard Earl

June 1968
Anderson, Dennis Jay
Buechler, Ralph Lee
Butcher, David Neal
Byrne, Robert Michael
Cassano, James Sebastian
Dubbs, Roger Bruce Jr.
Dunville, Glen Richard
Epple, Dennis Norbert
Green, William Charles
Hall, Stephen Boyd
Hill, Steven Keith
Hodgson, Bruce William
Jacobs, Barry Jon
Kaminsky, James Adam
Kirkpatrick, Daniel William
Kmitta, Thomas Joseph
Koehler, Richard Evans
Krieg, Lenny Albert
Labus, Thomas Joseph
Mackora, Bryan Joseph
Mitchell, Stephen Craig
Myers, Richart Otto Jr.
Oakley, Robert Alan
Orazio, Fred Donald Jr.
Plank, Steven Kenneth

Prickett, James Michael
Readnour, Jon Stuart
Rush, Howard Michael
Schlegel, Mark Owen
Skorich, James Milan
Smoak, James Allen
Sprandel, Tom Gust
Thacker, Richard Leighton
Trice, Clifton Tony
Wheeler, John Ladd

August 1968
Riedel, Steven Charles
Shehadeh, Ahmad Mohamad
Vanderby, Ray Jr.

January 1969
Bofah, Lawrence Kete
Brown, Lionel Fleming
Cork, Thomas Ray
Dornan, James Michael
Harmening, Richard Daniel
Lerner, Paul Jerome
Nagy, Joseph Gabor
O'Brien, Donald Terrence
Richardson, Richard Jan
Shultz, Jerry Keith
Steinbarger, Jay Oliver
Strauss, Stanley Howard
Stuckas, Kenneth J.
Tolene, Gail Edwin
Warford, Michael Wayne
Willis, DeWitt William
Wilson, Gerald Grund
Zakem, Steven Bernard

June 1969
Alliston, Robert Earl
Amerman, David William
Andre, Wayne Allen
Arnold, John Edward Jr.
Barter, Stephen William
Bird, Thomas Arthur
Blanford, Robert Keith
Brown, James Hugh
Brown, Thomas George
Cieciwa, Gregory Alex
Corcoran, Richard Albert
Doran, Michael Edward
Duzak, Michael Charles
Fishback, Alan Curtis
Fossett, Wallace Lee
Fraley, David William
Gatewood, Stanley Dean
Givan, Max Eddy
Greenland, Richard Lee
Haak, Bogdan Michael
Hall, Dane Robert
Haverly, William Edmunds
Hellman, Michael Paul
Hemmig, Floyd Gene
Holbo, Lyle Edward
Hostetler, Stephen Charles

Huiatt, Keith Wilson Jr.
Ingle, Robert Lee
Johnson, Richard Westel
Joy, Gary Bremer
Kaemming, Thomas Alan
Karpyak, Steven Douglas
Kinney, Bruce Elliot
Kirkdorffer, Richard Allen
Kuhns, Randall Henry
Laue, Dennis Richard
Lawler, Roy Charles
Lewis, Donald George
Leyda, Bryan Douglas
Ma, Harry Hsiangpo
Maxwell, Thomas Lee
McKellip, Spencer Waring III
McKown, Philip Manly
Metz, Robert Arthur
Michel, William Barrett
Miller, James A.
Milne, Daniel Lawrence
Moran, Howard William
Nuss, Elmer Allen
Owara, Vernon Naoki
Pankiw, Henry Stanley
Pederson, Robert Carl
Russell, Richard Eugene
Samas, Robert
Scott, Gary Henry
Sgritta, Thomas Andrew Jr.
Shaw, Robert Linford
Smith, Jerry Allen
Smith, Stephen Kent
Spitz, Robert John
Stauffer, George Albert
Svendsen, Dick Gerald
Topai, Huba Zoltan
Townsend, Paul Jerome
Troyer, David Michael
Underwood, David William
Utterback, Robert Kirk
Volk, James Anthony
Walker, Neil Renz
Wertz, Faust Hissong

August 1969
Everly, Robert Edward III
Hayhurst, John Blake
Kraus, Fredrick Emil
Lu, Pei Chuan
Miller, Edward Max
Murphy, Kenneth Edward
Nienow, John Floyd
Stambolos, Angelo Anthony

January 1970
Bigler, William Boyd II
Blume, William Henry
Chipman, Philip Eugene
Deffenbaugh, Floyd Douglas
Dixon, Martin Clay
Dyer, Calvin Le Roy
Feige, Donald Alan

Gilson, Robert Linton
Hancock, Joseph Richard
Hannah, Darryl Eugene
Hiltebeitel, Edward William
Hughes, Glendon Edward
Humble, Charles Edward
Hummel, Terry Lee
Jones, William LeRoy Jr.
Kitterman, David Lynn
Landy, Michael Allen
Lui, Leon Yuen-Leung
Maners, Graham Stuart
Owczarek, Edward Anthony
Palmer, Robert Wayne
Smith, John William
Smith, Michael John
Stewart, John James
Tyler, Dale Keith
Vetters, Thomas Lynn
Williams, Dale Martin
Wilson, Garry Dale

June 1970
Anderson, Richard David
Armstrong, Richard Eldridge
Baughman, John Lewis
Biermann, John Alfred
Braciak, Stephen Henry
Braun, Robert Walter
Brown, Thomas Francis Jr.
Bullock, James Robert
Butler, Thomas Alva
Carmody, Lawrence Patrick
Croop, Harold Charles
Dalhart, Glenn Austin
Davis, Donald William Jr.
Di Bello, Michael Frank
Dominik, Thomas Gerald
Drouchner, Erich Emil
Duncan, Richard Russell
Dunn, Patrick Francis
Early, Dwight Holdridge III
Edwards, Joan Annette
Fozo, Steven Robert
Galler, James Franklyn Jr.
Galvas, Michael Robert
Gilsinger, Ronald Jay
Green, Robert Barry
Hackman, Barbara Jean
Haramoto, Cary
Herringshaw, Edward Dale
Horlacher, Walter Rawlins III
Kellen, Dennis Lee
Keller, David Charles
Kendall, Arthur Sherfey III
Kennedy, Michael Terrence
Kincheloe, Everet Vernon
Kirkley, Richard Alan
Krehbiel, John Stewart
Lamberty, Brett Dewain
Lane, Jan Michael
Large, Robert Alan
Leslie, Clark William

Livingston, Ray Walter
Lukavich, Gerald Lee
Luther, Ronald Elbert
Mastrangelo, Robert Joseph
McClure, Terrence John
Millstein, Leo
Niksch, Richard Alan
Nygaard, Erik Dan
Payne, Douglas Melville
Pierre, David Michael
Price, Howard James
Quarto, Alfred
Reeder, Walter Francis III
Reilly, William George
Roth, Larry Lee
Sanders, David Keith
Sellers, James Frederick
Silverthorn, James Taylor
Sinopoli, James Michael
Skinner, David Alan
Steinbarger, David Lynn
Steininger, David Anthony
Sterns, Gregory James
Stevens, Patrick Lee
Suppinger, Albert Vincent
Waibel, Randall Harold
Walters, Robert Michael
Williamson, James Paul
Wolff, Greg Lee
Wozniak, James Bartholomew
Yutmeyer, William Peter

August 1970
Barkley, John Michael
Burns, Charles Leroy
Dorozio, David Eugene
Kocyba, Thomas Edward
Misiuk, Stanley Joseph
Prymak, Thomas Nicholas
Smith, James Richard
Wells, Richard Michael

January 1971
Adams, Richard Roy
Alger, Jeffrey Lewis
Beaudet, Carl Alan
Carruthers, William Lawrence
Cors, John Thomas
Craig, Mark King
Forbes, Robert Charles Jr.
Foster, Thomas Francis
Fowler, Thomas Alan
Gerken, Dale James
Hall, Geoffry Thomas
Harrison, Walter Henry
Hermes, Robert Arnold
Host, David Allan
Huliba, David Allen
Hunthausen, Roger Joseph
Kielb, Robert Evans
Kirk, Allen Dewayne
La Manna, William Joseph
Lash, Ronald Joe

Mahoney, Melvin Merrill
McDowell, William John
McGinnis, Thomas Emmett Jr.
Michaud, Maurice George
Pegden, Claude Dennis
Reilly, Roger David
Roth, Ruth Nancy Vanerka
Salstrom, Roger Lee
Scheessele, David Walter
Sentman, Orville Lawrence
Springer, Steven Edward

June 1971
Anderson, Timothy Richard
Beecher, Alan Lee
Bennett, Michael Lewis
Bien, Jay Kent
Bowman, Bruce Kinser
Capin, John Richard
Capp, Frank William
Christiansen, Terry Gordon
Corso, Michael Joseph
Damrow, Ross Melvin
Diebold, Alfred John Jr.
Duffy, Raymond James
Ehlen, Clifton Wynn
Elble, Rodger Jacob Jr.
Erickson, Paul Arvid
Fleming, Williard Joseph Jr.
Frye, Donald William Jr.
Gardocki, Martin John
Gessley, Bruce Charles
Gilmore, Robert Thomas
Haase, David Scott
Harrison, Garry Lee
Haseltine, Raymond F.
Hayes, Jeffrey Mayer
Hendrickson, Jay Paul
Hildebrand, Daniel Hugh
Hiss, Glen Alan
Irvine, Colin Charles
Jarvis, Vernon Joseph
Kauffman, Lawrence M.
Kennedy, Peter
La Marca, James Edward
Lane, David Girdon
Link, Richard Alan
Little, Joseph Howard
Marler, Martin Leon
Maynard, David Paul
McConnell, David Scott
McHenry, William Irvin
McKay, Duane Leon
Monk, Donald Lee
Moppin, Kenneth Paul
Morgan, Terry Bruce
Obergfell, Richard Allen
Olsen, Albert D. Jr.
Owen, Mark Anthony
Patterson, David J. C.
Pearson, Terry Louis
Pouzar, Douglas Joseph
Reder, Thomas Joseph

Rhoades, Gene Jerald
Roberts, David Lee
Ruehr, William Carl
Russell, Robert Wayne
Saff, Charles Robert
Sakakini, John Steven
Schommer, John Michael
Simpkins, Earl Lewis Jr.
Skira, Charles Andrew
Smalley, Frederic Williams Jr.
Smith, Duke Alan
Smith, Wayne Raymond
Stecher, Mark Stephen
Steimel, John Charles
Stepro, Dennis Charles
Thome, William John
Thompson, Francis Marshall C.
Valrance, James Lumsden
Voight, William Melvin
Walker, Charles David
Wehrenberg, William Henry Jr.
Wojnar, Ronald Thomas
Wolf, Daniel Star

August 1971
Brunansky, Louis Stephan
Fagg, Robert Ervin
Gerber, James Curtis
Hess, Donald Ray
Julius, Edmund Paul
Lovick, Robert Gregg
Mavaddat, Afsaneh
McKinstray, James Merrill
Morris, Edward Andrew
Ranard, Douglas Rex
Vlchek, Frank Joseph
Wilson, Mark Kent

January 1972
Bauman, Kenneth Alvin
Boldt, Kenneth Roger
Branson, Roger Lee
Buell, Robert Kenyon
Cosner, Robert Raymond
Davis, Edwin Warren
Farris, Richard Lee
Green, Dennis John
Hall, Darryl Wayne
Harrell, John Mason
Higa, Earl Shigemasa
Ingwersen, Martin Lewis Jr.
Irvine, David Earl
Jewell, Gail Ann
Jones, Granville Paul
Jones, Lyle Evan
Jones, Robert Allen
Marquette, Michael Raymond
Meyer, Lloyd Jerome
Miller, Larry Wayne
Pumilia, Michael Peter
Ray, James Richard
Robinson, David Dean
Schwaiger, Stephen Keeler

Sims, Kendall Andre
Smuckler, Joseph Allan
Stanfield, David Michael
Turrelli, Robert Rudolph
Wright, Phillip Darrell

June 1972
Belloli, Anthony J.
Benda, Brian Joseph
Borrenpohl, Charles W.
Bowers, Douglas Lyle
Brock, Robert Dean
Bruce, Thomas William
Bruestle, Harry Rolf
Cantwell, William Stanley
Cervenka, James Allen
Clark, Richard Stewart
Combs, Richard Alan
Cook, Michael William
Davidzuk, Emile Joseph
Davignon, Richard H.
Dean, Dennis Ross
DeBaun, Gary Wayne
Drerup, Vincent Marion
Dunn, Gregory Alan
Eckstein, Donald Francis
Fielding, John Robert
Foley, Kenneth James
Gaskill, Richard Paul
Gengnagel, Robert Lynn
Gerber, Mark Alan
Gilkey, Samuel Charles
Gorrell, William Thaddeus
Graber, Patrick Matthew
Hahn, Stephen Louis
Hall, Dana LeRoy
Harries, Thomas Joseph
Harris, Terry Michael
Herrel, Randall Lee
Hill, Rikard Eugene
Hirko, George Steve
Hoilman, Kent Alan
Ives, Arthur Robert
Jahnke, Richard Earl
Keller, Raymond Lee
Kress, Jeffrey Allen
Krise, Steven Gerald
Kubiak, Andrew Mitchell
Kudlick, Dean Allan
Langmead, Larry Paul
Lovison, John Kenneth
Lowder, Harold Edward Jr.
MacFarlane, Larry James
Martello, Keith Wallace
Mattasits, Gary R.
McLaren, Bruce J.
Mills, Lawrence Duane
Moore, Carleton John
Morrison, Ronald Walter
Peters, William Lee
Richie, Joseph Simpson
Ritz, Mark Edward
Roth, Robert Howard

Schneider, Steven Joseph
Seum, Charles Stephan
Shafer, William James
Shoemaker, David Hal
Skinner, Thomas Lawrence Jr.
Stallbaum, Robin Ruth
Super, Peter Jude
Szmurlo, Charles Joseph Jr.
Vardaman, James Thomas
Vidmar, Dennis Wallace
Warres, Edward Clark
Webber, Alfred Wayne
Weber, Richard Jay
Willmot, Glenn Douglas
Wirth, Edward Paul
Yaeger, Larry Steven
Youngblood, Stanley B.

August 1972
Al-Rashid, Nasser Mohammad
Creighton, Mark Alan
Day, James Melvin
Espinosa, Angel Manuel
Hofferth, David Bruce
Laird, Donald Dwain
Rhue, George Truitt
Sims, David Lee
Taverni, Frederick Anthony
Travelbee, Charles Brian
Wantland, Edgar Wantland

December 1972
Biggs, Roger Duane
Bray, Philip Ervin
Crorey, William George Jr.
Day, Michael Lee
Des Forges, Daniel Thomas
Estes, Ronald Carl
Gailey, Thomas Glen
Gates, Frederick Michael
Heard, William Joseph
Huett, Gary James
Kasowski, Andrew Henry
Kern, John Philip
Krein, Steven Walter
Leech, James Nathan
Liebovich, Lou Robert
Manak, William Thomas
Manzoku, Roy Tamotsu
Moore, Terry Scott
Muller, Michael Anthony
Rogol, Charles Stephen
Schaffer, Van Anthony
Simmons, Michael Louis
Slone, Donald L.
Spencer, Steven Richard
Staab, George Hans
Tams, Robin Timothy
Warner, Stephen Roger
Webb, John Allen
Wolting, Duane Ernest
Yarrington, Allen Lee

May 1973
Bangert, Robert Lawrence Jr.
Brave, Mark Louis
Brown, Jeffrey Jonathan
Brown, Mark Neil
Brown, William Wesley Jr.
Burg, Kenneth Michael
Callaway, William Rodger
Crovesi, Marco Mario
Dye, Barry Cundiff
Epley, Lawrence Ernest
Esch, Fred Eyrol
Fleck, David Lee
Fratello, David John
Gilbreath, Gary Wayne
Glaros, Louis N.
Graf, Marta Louise
Gutknecht, James Edward
Harr, Calvin Dean
Harter, Peter Kent
Hays, Drew
Hench, Mark Anthony
Henderson, Thomas Best Gibson II
Johnson, John David
Kacur, Michael Brian
Kanning, James Lee
Kaul, Charles Edward
Kitagawa, Melvyn Sakae
Klein, James Timothy
Kuklinski, August Andrew
Kurth, Fredric Charles Jr.
Lake, Jerry Miles
Leischer, Dale Richard
Luckring, James Michael
Maine, Richard Edwin
McCaskey, Richard Michael
Mendenhall, Robert Morris
Miller, Ronnie Keith
O'Connell, Michael Roe
Patton, Donald Dennis
Peterson, Donald Earl
Pettee, Charles Wilson
Pohlar, Leonard Paul
Pouder, John Marsh
Rice, Thomas Stuart
Rodgers, Christopher Thomas
Samples, David Olin
Schumann, Douglas William
Shidler, Phillip Allen
Short, Gary Howard
Slama, John Thomas
Steinbeck, Andrew Raymond
Strauss, Richard Alan
Thompson, Jeffrey Allen
Thompson, Roger Gene
Vishnauski, Jon Michael
Warner, Dennis Eugene
Watland, Arnold Lee
Weiss, Arthur Harold
Welch, Gerald Michael
Wheadon, John Randolph Jr.
Zakrzewski, Thomas Joseph

August 1973
Jung, Jacky Yum-Kee
Muldary, Patrick Farrell
Scharnhorst, Richard Kent
Sims, Robert Louis

December 1973
Anderson, Wayne Ray
Arey, Stephen Heibert
Beal, George William
Edwards, Larry Edwin
Hoesly, Richard Lowe
Kraper, Kenneth Michael
McGuire, Danny Clay
Penny, Terry Robert
Rice, Janet Louise
Schuldt, Marc Alan
Shattuck, Paul Laurence
Sherman, Robert
Trapp, Michael Kelly
Weeks, Laurence Campbell

May 1974
Aho, Brian Matthew
Anderson, Mark William
Barkman, Danny Keith
Bastian, Thomas Wayne
Beebe, Craig Allyn
Bielski, Edward Mark
Bonk, Jon Craig
Brady, Daniel Lee
Creighton, Jo Anne
DeHaven, Leon Alan
Everly, Walter Keith
Ewing, Robert Daniel
Gambin, Wayne Peter
Gaudion, Thomas Joseph
Gran, Carl Stanley
Graviss, Kenton Joseph
Haueter, Thomas Edward
Hawk, John Dennis
Hirn, John Joseph
Jamrogiewicz, Roman Andrew
Koke, John Michael
Kroll, William Burton
Ledger, Alan Steven
Lord, Bruce James
Marks, Larry Erwin
Martin, Leonard W.
Matucheski, Joseph Victor
McCarty, Robert Lawrence
Miller, William Edward
Miner, Dennis Dale
Moy, Calvin Brian
Paul, Warren Kent
Powlen, Ronald J.
Prentice, Gerald Eugene
Renie, John Paul
Sesny, Paul Andrew
Smith, Doyne Edward
Somers, Dan Michael
Uffelman, Kenneth Eugene
Wade, David Earl

Walker, Bruce Keppen
Wannenwetsch, Greg David
Witte, Michael James
Yanner, Michael Lambert
Yeakley, Phillip Lynn
Zanton, David Francis

August 1974
Hannah, Michael Alan
Parker, Richard Lockhorn

December 1974
Foster, John Irving III
Fullman, Donald Grant
Giant, Stanley Charles
Johnson, Warren Herbert
Jones, Carl Michael
Merrell, John Spencer
Moore, Randi Caryl Manges
Moore, Robert Craig
Mulgrew, Michael Patrick
Northcraft, Stephen Andrew
Queen, Henry Caleb Jr.
Redmon, Danny Ray

May 1975
Abernathy, Charles Larry
Adams, Gerald James Jr.
Arcangeli, Gerald Thomas
Bellagamba, Laurence
Brown, Gregory William
Bushong, Gregory Brent
Campbell, Kevin Dean
Cornelius, Patrick Edward
Cupp, Terry Jo
Devitt, Daniel Frank
Douglass, David Alan
Doyle, Brian William
Eckerle, Wayne Allen
Evans, Robert Edward Ted
Greene, Timothy James
Haas, Bernard Charles Jr.
Halt, Gary Elmer
Hammiller, John Joseph
Haupt, Charles William Jr.
Hinsdale, Andrew Jeffrey
Hyman, Richard Alan
Keever, David Bruce
Kosloske, Douglas Lloyd
Kozdron, John Aloysius
Kress, Stephen Scott
Lafuse, Sharon Ann
Lenoch, Vlado
Lux, David Philip
Maikowski, Michael Francis
Meyer, Robert Ray Jr.
Morgan, David Llewellyn
Neville, Jeffrey William
Reed, Allan Thomas
Ridenour, Ronald Neal
Robinson, Michael Alan
Sass, Daniel Edwin
Schiffer, Albert Edward

Swan, Arthur Dan
Van Benschoten, John Ray
Wallen, Bruce Richard
Wetzel, Richard Bruce
Wible, Thomas Jeffrey
Wilkins, Bruce Gregory

August 1975
Stewart, Charles Raaen
Williams, Thomas Lee

December 1975
Adiwoso, Adi Rahman
Cheney, Daniel Roy
Corbett, Melvin C. III
Fajardo, Charlotte Helene
Jackson, Robert Lea
Jones, Ross Mitchell
Klutzke, Jerald Lee
Messerschmidt, Stephen J.
Porter, Tony Leevern
Scanlon, William Michael
Singleton, Gary Wayne
Strickler, John Peter
Sweeney, Mark Louis
Turelli, Thomas Walter

May 1976
Abreu, Michael Henry
Adams, Mary Ellen
Allen, Gerald Dale
Bartlett, Wayne Morris
Bates, Bradley Dale
Beecroft, Timothy James
Breit, Gregory Roy
Brennan, John Thomas
Bruner, Wesley Richard
Burdsall, Charles Wesley II
Chen, Ming Kay
Conkright, Gary William
Cowdin, David Harold III
Dest, Stephen Anthony
Gagnon, Michael Lee
Gorney, Jeffrey Lawrence
Grieger, Kenneth Allen
Harlan, Douglas Alan
Hixson, Kevin Eugene
Hollimon, Robert Alexander
Hyatt, Robert Kevin
Kary, Edmund George
Kline, Larry Robert
Konzak, James Paul
Lamparter, Jeffrey Alan
Lewitzke, Craig William
MacKechnie, Lewis Conrad
Moe, Alan Neal
Morgan, Thomas O'Neill
Oswald, William John
Paller, Donald Peter
Raymer, Daniel Paul
Rychnowski, Andrew Stefan
Secor, Randal Edward
Sibal, Stephen Donald

Slosson, David Evert
Smith, Sandra Lee
Striz, Alfred Gerhard
Thomas, Mark John
Thomas, Richard Glenn
Ullestad, Gary Stephen
Upton, Randall Vaughn
Yarbrough, Edwin Richard

August 1976
Zorich, Bruce Martin

December 1976
Brown, Paul Edward
Call, William Ray
Cass, Jane Carol
Clark, Stephen Craig
Francis, Daniel Wayne
Fugate, Jeffrey Earl
Gillespie, Donald Lester
Green, James Sidney
Lauria, Peter Joseph
Little, John Loren
Martin, James William
Matz, Dean George
Moos, Harold Eugene Jr.
Mundt, Mylen James Jr.
Overdorf, Roger Lee
Ripma, Edward Johannessen
Shutt, Dean Jerrold
Stephenson, Stephen Lee
Tilley, Donald Edgar
Tolbert, Ronald H. Jr.
Washburn, Karen Elizabeth

May 1977
Baker, Allen Dale
Barcelos, Celso Anthony
Benson, Scott William
Bohun, Michael Harry
Breese, Mark Leslie
Brinson, Michael Richard
Bucher, William Richard
Cahoon, William Timothy
Chim, Siu-Man Edwin
Coppock, Timothy Neal
Cox, Gregory John
Cross, Jeffrey Lynn
Davis, Craig Allen
Deckelbaum, Jeffrey David
Drake, Jason Andrew
Dunn, John Martin
Ehlers, Steven Michael
Eliasen, Rune Carl
Enyart, Thomas Alvin
Finnerty, Christopher Stanley
Forness, Stephen Douglas
Gerstenmaier, William Howard
Grover, Charles Howard
Haas, James Anson
Harker, Brian Gene
Hill, Seth Craig Whitman
Hogue, James Harvey

Hollenback, David Mark
Johanns, Michael Alan
Kelpe, David Walter
Khorrami, Ahmad Farid
Kress, Edward Walter Jr.
Letellier, Scott Pierson
Love, Michael Harry
McLaren, John Harold
Miller, Michael Barton
Mlcuch, Thomas Richard
Morrical, Daryl Gene
Morse, Dennis Frank
Mosolf, Roy Julius
Naville, Gary Lee
Palac, Donald Thomas
Reiber, Carl Edward
Riesner, Keith Alan
Roberts, Andrew Curtis
Robinson, Jane Ann
Schulenburg, Carl Robert
Schumann, Donald Edward
Skinner, David Lee
Snepp, David Karl
Soderland, Carl Ivan
Staehle, Robert Livingstone
Strickland, Gregory T.
Stropnik, Branko Louis
Weissinger, Glenn Emil
Wirt, Robert Orville Jr.
Wolfson, Craig Stanislaus

August 1977
Bosecker, Scott Edward
Fine, Sheryl Anne
Harrison, Pamela Jane Wetzel
Sprangers, Cornelius Anton

December 1977
Al-Mukhaizim, Saud Abdulla
Booher, Mark Edward
Gilbert, Michael Glenn
Laptas, Michael Henry
McCain, Charles Raymond
Waldron, James Scott

May 1978
Barger, Mary Elizabeth
Berkowitz, Lenard Franklin
Bogna, Albert John
Burgess, Mark Alan
Coker, Estelle M.
Dahl, Scott Alan
Dudley, Robert Baird
Duvall, Gilliam Elroy
Filler, Russell Edward
Fleming, Daniel Patrick
Flint, Joseph Walter
Greenwalt, Scott Lash
Harbaugh, Gregory Jordan
Hotz, Tony Francis
Jewett, Thomas Edward
Krupp, Michael Edward
Kubinak, Caryl Lee Johnson

Livingston, James W. Jr.
Lucas, David Edward
Marsden, James Myron Jr.
Masher, Kerry Dale
Matheney, Timothy James
Matson, Robert Elmer
Miller, David Paul
Nixon, Dalen Erle
Polansky, Mark Lewis
Power, Barry Dean
Rickershauser, Gary Gerard
Rinehart, Mark William
Robinson, Steven Paul
Scroggin, William Duane
Seffern, John Jay
Sheldon, Robert Gerard
Slomski, Joseph Francis Jr.
Smitheimer, Don Lawrence
Stanley, Keith Edward
Stogsdill, Steven Michael
Usab, William James
Vasta, Steven Keith
Whitcomb, Howard Clarkson III
White, Bonnie Lynn

August 1978
Brickman, Michael Alan
Hibbert, Harold Haggar
Howard, Jenny May
Poorman, Peter Charles

December 1978
Apter, Douglas Scott
Blaisdell, Debra Lee
Bland, Jeffrey Dylan
Clark, Ransom Parker
Davis, Darryl W.
Feliciano, Mario Antonio
Hunter, Robert William
Leewood, Alan Ross
Murphy, William Jacob
Oliphant, Robert Charles
Proulx, Roger Doria
Richards, William, Raymond
Scott, Alan James

May 1979
Arocha, Onex
Batina, John Terence
Bauer, Frank Henry
Beitel, Gregg Richard
Bogan, Kevin Scott
Burlison, Terrill Lynn
Crutchfield, Jeffrey James
Dawson, Robert Edward
Dougherty, Frances Carroll
Downs, Lawrence George Jr.
Fox, Steven Chabut
Gallotto, Anthony
Gamble, Ricky Alan
Ghergurovich, John
Hammerschmidt, Mark McCain
Heath, Bradley Eugene

Helbling, James Frederick
Hodges, Douglas Alan
Ingallinesi, Anthony Angelo
Kempski, Mark Henry
King, Jeffry Alan
Kladden, Jeffrey Lee
Krantz, Francis Richard
Law, Tracy Don
Marsden, Catherine Ethel
Martin, Bailey Eugene
Morrison, Stephen Choi
Mosher, Raymond Ellis
Nicosin, Pamela Sue
Ohrenberger, Kevin Thomas
Pinkowski, Stanley Michael
Rivera, Pedro L.
Rogers, Jon Paul
Shields, Randolph Conrad
Smallowitz, Jeffrey Mark
Stout, Anita
Szabo, Louis Steven
Thomson, R. Kent
Trout, Martin John
Vanek, James A.
Weber, Jeanine Marie
Weston, Glenn Peter
White, George Brian
Willis, Bruce Donald
Wingert, David Alan

August 1979
Dudley, David Welton
Kenworthy, Mark L.
McLaren, James David
Miner, Gary Allen
Wray, Rickey Lane

December 1979
Bacon, Barton Jon
Baldwin, Craig Wesley
Bonem, Peter Bryant
Bowyer, Joseph Alan
Dill, Brad J.
Dimitriou, Andrew
Harrington, John Henry III
McNichols, William Marsh
Modrey, Jon
Moore, Michael Walter
Pinella, David Francis
Pourmand, Mostafa
Rychlak, Michelle Elizabeth
Willhoite, Bryan Curtis
Wise, Stephen Marshall
Wright, David Stephen

May 1980
Allen, Charles Scott
Bain, Scott Avery
Bauer, John Linus
Baxter, Matthew Charles
Bentley, Eric Lee
Bernard, Chris Lynn
Blake, Robert Scott

Blank, Jeffry Joseph
Bonnice, William Fenton
Bowen, Steven Allen
Boylan, William Thomas
Bragger, John William
Bragger, Susan Jo Morgan
Breunlin, Richard Carl Jr.
Brown, Gary Lee
Brunner, Ann Kathleen Hanson
Carpenter, Michael Grant
Coccimiglio, Mark
Davis, Steven Mitchell
DeQuay, Laurence
Detzel, Michael Joseph
Dorste, Craig Thomas
Flora, Timothy Lee
Fulton, Cynthia Lynn
Gary, Wyndham F. Jr.
Halt, David Wayne
Herb, Robert Scott
Hess, Joseph Paul
Hoernlein, Stephen Lynn
Johnston, Mark Eaton
Lilley, Jay Stanley
Lynch, Lindley Jay
Maerki, Glenn Steven
Mithoefer, Jeffery Vernon
Molyneux, Harry Francis
Muigai, Jane Wengui
Murphy, Terrence Hyde
Nagai, Nelson Toshikazu
Neil, Robert Dyas
Noble, William R.
Orkiszewski, Charles Stanley
Oshaughnessy, Edward J.
Patton, Michael Earl
Raiford, Richard Renz
Richards, David Lee
Riley, Mark Alan
Saultz, James Edward
Saunders, Mark Perry
Sutton, Ronald Dale
Tierney, Paul Kevin
Trowbridge, Timothy L.
Velasco, Kirk Douglas
Vicroy, Dan Douglas
Wilson, Kenneth Alan
Zimpfer, Dennis Scott

August 1980
Enicks, Nicholas J.
Farris, Glenn Girvin
Harter, James Arthur
Papke, Curtis James
Riba, William Thomas
Weber, Margaret Rose

December 1980
Blomshield, Frederick Steele
Cain, Gary Lee
Flenar, Gregory Allen
Gernand, Joseph James
Grady, Joseph Edward
Hausfeld, Thomas Eugene

Kobee, Robert Matthew
Macocha, Michael Andrew
Murphy, Robert Alan
Nelson, David Thomas
Scholz, Edwin Feuer
Schroeder, David Bryan
Smith, David Randolph
Smolana, Kenneth Vitold
Tomasson, Eric
Utay, Arthur Wessel
Valle, Ronald Philip

May 1981
Alstott, Pamela Kay
Anderson, Edward L.
Barohn, Gary Alan
Beasley, Kevin Guy
Bennett, Brian Kendall
Boyd, Charles Neill
Cageao, Richard Philip
Clarke, Peter Paul
Coleman, Frederick Ray
Deur, John Mark
Dornseif, Mark Jonathan
Eggink, Roy Anthony
Farrell, Michael Kenneth
Feuerstein, Mark Gregory
Fletcher, Eric Ross
Foist, Brian Lee
Freeland, Gary Ray
Grove, Scott Allen
Guevara, Luis Alejandro
Hatfield, James Charles
Heuss, Matthew Allan
Hill, Donald Merton
Hoffschwelle, John E.
Jones, Michael Anthony
Jung, Joel Robert
Kishline, Scott Samuel
Klein, Michael Allen
Langhenry, Mark Thomas
Lanning, Mark Eric
Lee, Marshall Morgan
Leininger, Daniel Rentschler
Lewis, Mark Bradshaw
Lewis, Thomas Griffith
Marriott, Brent Christopher
McKinney, Leon Ellington Jr.
Merrill, Randall Ray
Miles, Mike W.
Moran, Maureen E.
Ohmit, Eric Eugene
Peters, George Patrick
Privoznik, Cynthia May
Rooney, William John Jr.
Sanger, Kenneth Bateson
Sawyer, Richard Steven
Scheele, Paul Matthias Jr.
Schwartz, Barbara Ann
Short, Randall George
Stephen, Eric Joe
Stukel, Stephen Peter
Tanner, Russel Roy

Tillman, Thomas Gregory
Townsend, Barbara Kay
Tritsch, Douglas Evan
Waszak, Martin Raymond
Webber, Paul David
Whiston, Stephen David
Wilcox, Peter Alan
Williams, Jeffrey Paul
Wroe, John Richard

August 1981
Cassady, Robert Joseph
Rabe, David Lee

December 1981
Arnold, Gary Edmond
Arrington, Cynthia Venita
Bauman, Douglas Gray
Bland, Geoffrey L.
Danley, Michael Lee
Donaldson, Steven Lee
Freeman, William Gregory
Galenski, Larry Alan
Gates, Thomas Scott
Heath, Bradley Jay
Jasper, Donald Wayne
Labee, Charles John
Newton, Christian David
Oritz-Ramirez, Rafael
Perez, Rigoberto
Resch, Stephen Alfred
Robinson, James Earl
Stephens, Kyle Richard
Willard, Thomas Jefferson

May 1982
Augustine, Thomas Wesley
Baker, Dennis Keith
Belcher, Bradley Duane
Bloemker, Eric Brian
Britt, James Kevin
Buethe, Scott Allen
Carlock, Daniel Edward Jr.
Cirioli, Adriano Alphonse
Coons, Clayton Amede
Cotta, Roy Basil
Davidson, John Burdyne Jr.
Davis, Francis Edward Jr.
Dremann, Christopher Charles
Dunkin, Daniel Scott
Easterbrook, Eric T.
Ebersole, Charles Douglas
Evans, Douglas J.
Fink, Carl J.
Geyer, Mark Stephen
Graft, Patrick Waldren
Gridley, James Dumont
Harris, Richard Alvin Jr.
Heilala, Mark James
Hoeferkamp, Richard Edward
Igli, David Andrew
Kelly, Scott Roger
Kenaga, Donald Wayne

Kinney, William Howard
Knight, Hilary Garth
Leach, Marisa Smith
Leban, Frank Anthony
Liston, Glenn Walter
Marstiller, John Winwood
McDonald, James Robert
McNulty, William David
Miller, Lezza Ann
Myers, Joe Gaither Jr.
Myers, Timothy F.
O'Connell, Kevin Bernard
Peterson, David William
Peterson, Douglas Gerald
Petri, David Arthur
Postma, Thomas Jay
Quinn, John Michael
Reed, Darren Keith
Reeve, Scott Richard
Sadorf, Kurt R.
Satterlee, Richard William
Schroeder, Douglas Scott
Sharp, Keith Wayne
Sharp, Tracy Alan
Smith, Philip John
Spearing, Scott Frederick
Stark, John Aldrich
Stringfield, Gregory James
Stucki, Jeffrey Alan
Sullivan, Craig Lawrence
Summerset, Twain Ki
Taylor, Norma Faye
Teets, Robert Bryan
Totah, Philip Edward
Tucker, Tenna Evelyn
Uhrig, John William Jr.
Walsh, John James
Weidmann, Daniel Joseph
Weygandt, Jonathan Birchard
White, Lawrence Russell
Wozniak, James Michael
Yost, Peter Walter
Young, Robert Allen
Ziegler, Ralph Conrad III

August 1982
Dooley, Steven Patrick
Ferguson, Paul Lawrence
Grady, James Michael
Olsen, Michael Erik
Whitehead, Douglas Scott
Wickliff, Jennifer Lynn

December 1982
Benich, Christopher Joseph
Champion, Jon Howard
Ellison, Joseph Fabian
Faas, Paul David
Fouts, Kenneth Allen III
Henry, Theron Joseph
Ledington, Hiram Matthew
Lehrich, Marc D.
Lindsley, Lance Jon

McKissack, Douglas Ross
Melton, Dennis Michael
Nolan, Brian Gregory
Norman, Timothy Lee
Smith, Gary James
Spurr, Patrick Michael
Swanson, Erik Robert
Swarts, Dale Frederick
Toner, Scott Leroy
Vance, William A. Jr.
Wehrhan, Tobias Guenther
Whiteman, Byron Gale

May 1983
Allen, Bruce Arthur
Allen, Michael D.
Baker, David Lee
Beeker, Gregory Leon
Berry, Dale Thomas
Boller, Benjamin Keith
Brentner, Kenneth Steven
Brokaw, Matthew Frank
Brown, Curtis Glen Jr.
Burkhart, Kevin Alan
Bustle, Arnold Silas
Cain, William James
Campbell, Michael Howard
Carnahan, Timothy Michael
Carter, Nelson David
Catalfano, Michael Salvatore
Chiodi, Bruce Meigs
Clark, Wesley Joseph
Conway, Timothy Brady
Dreessen, Michael Paul
Fargusson, Gary Ralph
Francisco, Bradley Jay
Fuglsang, Dennis Floyd
Gearhart, James Walter
Greenblatt, Jerry M.
Guest, Gregory Charles
Hawthorne, Kenneth John
Hodge, Kathleen Frances
Irving, George B.
Jennett, Lisa Ann
Johnson, Jon Carl
Jones, Frank Patrick
Koechle, Mark Edward
Lesieutre, Daniel Joseph
Little, Richard Frederick Jr.
Mannella, Marcus Gerard
Martin, Randall Eugene
McGrath, David Kentlow
McLaughlin, Timothy Paul
Murphy, James Thomas
Murray, Robert Francis
Nelson, Leslie Diane
Nottorf, Eric Walter
Oard, Ronald Joseph
Parry, Craig Owen
Pfeiler, Anna Catherine
Ray, Suvendoo, Kumar
Reuter, Abigail Leigh Hunter
Rogers, Steven Paul

Rudolph, Brian Norwood
Ruflin, Scott Joseph
Sanders, Michael James
Sharp, Patricia Ann Dolan
Shultz, James Edward
Smith, Jeffrey Thomas
Sorensen, Eric Paul
Stine, Scott Glenn
Stover, Paul Gene
Szarleta, Norbert Edward
Szarwark, Thomas Brian
Taylor, Bruce Lee
Thurman, Sam Wesley
Torgerson, William Ted
Van Atta, Joseph Patrick
Visich, Michael
Wade, Clark Tucker
Warns, Richard John
Wilson, Michael Grove
Wuthrich, Rex E.
Yates, David Eugene

August 1983
Bell, Larry D.
Colombero, Craig Stephen
Edwards, Gregory Charles
Grezeszak, Jeff Michael
Grue, Richard Allen
Kucek, John Patrick
Oritz, Joseph M.
Solberg, John B.

December 1983
Barron, Dean Alan
Beatty, Jeff Crawford
Brand, Christopher Brian
Carr, Steven Lee
Emmet, Brian Robert
Gamble, Kenneth Charles
Garman, Ronald Dean Jr.
Godfrey, Anne Michelle
Guilmette, Thomas Howard
Jennings, Nathaniel Hills
Kruse, William David
Ortman, Timothy Martin
Periard, Michael J.
Ryan, Gregory Gerard
Senty, Peter Daniel
Steeves, Guy William
Stone, Raymond Eugene
Stuve, David Alexander
Walker, Gregory Paul
Wirkkala, David Richard

May 1984
Aldrich, Daniel Ray
Alt, Terry Robert
Antreasian, Peter
Askin, Ron W.
Barta, Gregg Harris
Bastnagel, Philip Michael
Baxter, Michael Joseph
Baylor, Keith Ray

Behnke, Mark Everett
Bosworth, John T.
Breland, Jeffrey Alan
Conn, Andrew Dean
Curry, James Matson
Delli Santi, James William
Demidovich, Nicholas Michael
Devilbiss, Jonathan Frederick
Ebenreiter, Carol Lyn
Farrell, Mark Avery
Favour, Lee John
Flora, Mark Alan
Gallman, John Waldemar
Gaunce, Michael Thomas
Gibbons, Michael David
Goad, Jeffrey K.
Haas, Paul Allan
Hamke, Rolf Ehrhardt
Hemmig, David Brian
Hopping, Bradley Martin
Huie, John C.
Hunnell, Thomas David
Ingram, John Russell
Kastenholz, Charles Vincent
Keszei, Michael Lee
Koharko, Daniel Norman
Krafcik, James Thomas
Lavelle, Brian Christopher
Liebbe, Steven William
Lilley, Mark Scott
Lucas, Michael Anderson
Luebke, Kenneth Jay
McAdams, James Valen
McAllister, Robert Scot
McMinn, Howard Stephen
McSweeney, Kelly M.
Medley, John Abell III
Meyer, Kenneth Dale
Miller, Theron Jeffrey
Mitchell, Lloyd Kerry
Murray, Kelvin Lenior
Neuts, Catherine Jenny
Olsen, Douglas Carl
Orr, Mark Stephen
Pernicka, Henry John
Pollee, Dean Roy
Robarge, James Emerson Jr.
Schroeder, Jeffery Allyn
Shaw, Jon Joseph
Smith, Mark Dennis
Stevens, Mark Raymond
Stine, Todd Louis
Sturdevant, Dan Allen
Sutila, John Steve
Thorne, James Dana
Tsiao, Sunny
Turk, Mark Allen
Warren, Scott Andrew
Watson, Rodney Lamonte
Wendel, Thomas Robert
Wick, Darren Duane
Williams, Steven Lee
Wright, Troy Kevin

August 1984
Bigliano, Thomas Frank
Carpenter, Kent Alan
Levinson, Scott William
Morris, John Owen
Schliessmann, Thomas Lee
Soedel, Sven Michael

December 1984
Anderson, Mark Alan
Boner, Andrew Case
Brunn, Mark Evan
Carteaux, Mark Allen
Caruso, Peter Joseph
Clark, William Allen
Clifft, Gregory Paul
Cook, Joseph Robert
Fox, Jeannie M. Bergstresser
Green, Alan Huber
Groves, Jeffrey Edward
Helt, David Russell
Herron, Tadd Floyd
Kechkaylo, Karen Anne Jensen
Keen, Kevin Lawrence
McNeil, Michael Benton
Moore, Jeffrey R.
Nistler, Randolph Powell
Nunez, Samuel Leon
O'Donnell, Michael Scott
Panich, Daniel
Perkins, Jeffrey Ivon
Privette, James Mark
Quirk, Jane Mary
Rogers, Donald Howard
Rygaard, Christopher Andrew
Sallee, Vernon James
Seeman, Denise Marie
Sparks, Carson Wayne
Trese, Paul Francis
Vogelsang, Mary Frances
Zurschmiede, Eric W.

May 1985
Aberle, Victoria Elizabeth
Anderson, Brian Keith
Auer, Ronald Kevin
Augenstein, David Leroy
Baumgartner, Jay Alan
Berreth, Steven Patrick
Bohlmann, Jonathan D.
Bokash, David Eugene
Bough, Roger Marion
Boyce, Jeffrey John
Campbell, Robert E.
Carlson, Christopher James
Chan, Kent K.
Cieslak, Robert John
Clauss, Jon Paul
Crane, Steven Parker
Cunningham, Mark Alan
Davis, Gregg E.
Davis, Owen Kosloff
Denta, Brian John

Drajeske, Mark Howard
Fennessey, Richard Lee
Fields, James Richard
Fiorito, Lynn M.
Gates, Russell John
Grossman, Mary Alice
Halsmer, Dominic Michael
Hayes, Brian David
Heise, James Robert
Hobbs, Donald Charles
Ikenze, Chukwuemeka Michael
James, Paul Anthony
Joseph, Anthony Abraham
King, Jeffrey Scott
Koukoutsides, Demetrius John
Krozel, James Alan
Levine, Mark Scott
Lollock, Jeffrey Alan
Mark, Daniel F.
Norton, Thomas James
Noteboom, Robert James
Oeding, Brian Thomas
Palmby, Jeffrey Thomas
Patterson, Brian Paul
Pemberton, Nathan Kerby
Perry, Frank James
Petrus, Mary Michelle
Pettett, Thomas Gray
Polasek, Brian Basil
Queen, Jerry Joseph
Reed, Steven Andrew
Scheer, Steven Alan
Seketa, Paul Stephen
Sensmeier, Mark David
Smith, Joseph Mark
Smith, Karen Ann
Steeves, Kent Thurber
Strazzabosco, Donald George
Sulkoske, David Scott
Szobocsan, John Jude
Tatman, Randall Lee
Theriault, Ricky
Traster, John Ewart
Ulmer, John Philip Jr.
Vandenbrook, James David
VonKleeck, Brian Wade Jr.
Wanthal, Steven Paul
Wasikowski, Mark Edward
Wheeler, Gary Alan
Whitmore, Robert E. III
Wilson, Gregory Jon

August 1985
Alano, Christopher G.
Ammerman, Curtt Nelson
Pericak, Christopher

December 1985
Amaya, Mark Anthony
Atakar, Oktar Sevket
Atkinson, Daniel John
Bank, Roger Charles
Bates, Brent Lee

Bennis, Karim
Christ, John Alex
Conte, Peter J.
Crouch, Norman Charles Jr.
Delaney, Dawn Marie
Denney, David Edgar
Dick, Joseph A.
Edenborough, Kevin Lee
Ferraiolo, Nicholas Raymond
Graham, Andrew Dallas
Hale, Douglas John
Helmig, Laura M.
Herman, Jeffrey E.
Hubbard, Roy Bruce
Huck, James Robert
Humbert, Daniel Ramsay
Lee, Khueh Hock
Leigh, Jeffrey Daniel
Marsh, Steven Michael
McConville, Charles James
McNabb, Scott Lee
Messina, Michael Dean
Mohr, Bruce Joseph
Notestine, Kristopher Keith
O'Dell, Dale Alan
Padgett, Michael Leroy
Patton, William R.
Philippis, Michael Lee
Phillips, Michael Jerome
Rough, Todd William
Schmitz, Mary Catherine
Scott, Lynette Kay
Sleppy, Mark Alan
Smith, Ronald Steven
Tygert, Wayne S.
VanNorman, Eric Jon
Waldron, John Norman
Wernimont, Eric John

May 1986
Barrett, Gary Curtis
Bauer, Jeffrey Ervin
Beal, Alan Douglas
Beeman, David Enoch
Benavente, Javier Edgar
Bender, Jennifer Lynne
Bonich, James John
Boyle, Michael Alan
Brannock, Brian Kent
Burgess, Ronald Joseph
Campbell, Scott Douglas
Cenfetelli, Ronald Timothy Jr.
Christian, Julie Anna
Clements, Richard Louis
Coombs, David Charles
Cummings, Kenneth David
Danehy, Matt Ward
Dennerline, Rhett Rodney
Donnellan, Mary Kathleen
Ely, Todd Allan
Engelhardt, Douglas Brian
Ewart, Timothy Wayne
Fowler, Jerome Martin

Franz, Christopher Gustav
Fuhry, Douglas Paul
Gatewood, Melinda Kay
Goode, Randall Joseph
Gregory, Sharon Kaye
Grove, Jeffrey Louis
Guingrich, Duane Lee
Hertzberg, Steven Hugh
Hiday, Lisa Ann
Hiss, Steven Terry
Johnson, Jackie Kay
Johnson, Paul Allen
Jones, Neal B.
Joyce, James Andrew
Keen, Elliott David
Kershner, Kristina Karol
Kistler, Gordon Paul
Kor, Rina
Krizman, Kevin Joseph
Kurtz, Russell Dane
Layton, Jeffrey Boese
Leckinger, Robert Charles
Lee, Douglas III
Li, Sheng
Lipton, Mark Stephen
Makrides, Emilios Andreas
Martin, Robert Daniel
Miller, Gregory Alan
Miller, James Richard
Moes, Timothy Ray
Mohr, Ross Warren
Monell, Donald William
Monroe, Eric Kenneth
Nielander, James Ralph
Odom, Earl Boland
Papadakis, Philip Emmanuel
Paulsen, John Alexander
Pilla, Christopher Darius
Pinkerton, Pamela Joanne Wolosz
Pontecorvo, Paul Anthony
Powell, Keith Brian
Price, Vince Aaron
Ptacin, Richard Roy Jr.
Rein, Troy Jason
Riddle, Lisa Anne
Rogers, Derrick A.
Sanderson, John Scott
Schlossberg, Valerie E.
Schura, Michael Martin
Seal, Michael Damian III
Sharp, Robert Anthony
Sharps, Tawnya Joy
Sherrier, David Mathew
Smith, Barbara Lynn
Smith, Russell Edward
Speth, Joseph John Jr.
Sutherlin, Mark Earl
Thompson, Jeffrey Michael
Voigt, Dennis Matthew
Walker, Robert Brian
Wrobel, Michael Harold
Wyness, Gavin Alan

August 1986
Brouillard, Mark Raymond
Elbert, Edward Ray
Georgos, Mihail G.
Karnes, Jeffrey Alan
Sadenwater, Robert John Jr.
Schultz, Steven Robert

December 1986
Barkley, Blake Howard
Bercaw, Robert Hyer
Bishop, Brian Carnes
Booze, Bryan David
Bui, Yung The
Butcher, Michael Ronald
Clemons, Penny Jo
Cleveland, Christopher Dale
Demaise, Michael Paul
Gavin, John Richard
Gevers, Matthew Henry
Grandin, Christine Elizabeth
Gray, David Michael
Irmen, Greg Allen
Kelley, James Douglas
Khan, Imran Mohammed
Loppnow, Martin Lavern
Lopuszynski, Andrew Joseph
Martin, Richard Jay
Miller, Kenneth Lloyd
Neal, Bradford Alan
Olander, Douglas Michael
Pontious, Andrew Howard
Pope, Robert Warren
Prange, Gary John
Renna, James Paul
Royal, John Bernhardt
Runner, Bud William
Sayeedi, Naeem Rashid
Scherer, Todd Alan
Schwartz, Max K.
Sprunger, Leroy Clinton
Starts, Stephen Glenn
Strack, Gary A.
Swenson, Andrew Walker
Thornton, Rhonda Dawn Stanforth
Todd, Earl Michael
Vrahoretis, Robert Emam
Wade, Stephen Hall
Waggoner, Jeffrey Neil
White, Barry James

May 1987
Amstutz, Peter John
Armantrout, John Thomas
Armstrong, Gerald Lewis Jr.
Arrieta, Albert Joseph
Azzano, Christopher Paul
Bacon, Bruce Charles
Barker, Lisa Ann
Beutner, Thomas John
Blanford, Craig Scott
Brandt, Matthew Reed
Bruce, Gary W.

Bunce, Thomas James
Burke, Michael Allen
Burkle, Michael William
Cahill, Timothy Scott
Caralt, Fernando
Carney, Darrell Lynn
Chapel, Steven Edmon
Chonko, Darren John
Clokey, Craig Bedell
Collins, Kerry Scot
Collins, Robert Keith
Connolly, Paul Thomas
Crawford, David Bourke
Crawford, Larry Dean
Crescibene, Sam
Ebert, Daniel William
Eblen, Jeffrey Scott
Ewing, Curtis G. T. II
Fegelman, Todd M.
Fleming, Terence Franklin
Franklin, Eric Dean
Friedmann, David Scott
Gardner, William Jon
Garner, Gregory James
Garrett, Lorene Vayonna
Goeldner, Kevin Blake
Harker, Linda K.
Hart, Matthew Joseph
Held, Timothy James
Helvie, Lyle Andrew
Hoffman, Keith Frederick
Hughes, Holly Leigh
Isenhour, Walter Lee
Jackson, Wade Cecil
Jaggers, Mark Leon
Kotar, Steven John
Landman, Douglas Jay
Lehman, Stephen Edward
Lum, Stuart Anthony
Maxie, Melana Anne
McDowell, Kenneth Alan
Meijers, Paul Gerard
Michel, John Marc
Moore, Kelly Wayne
Mortenson, Carol Aileen
Mueller, Steven R.
Mullican, Andrew James
Novak, Michele Kristen
Oates, David John
Ohnesorge, David Reed
Otolski, Jeffrey Jerome
Passman, James Arthur
Patton, Jeffrey Helsley
Pekala, Robert Lance
Penczak, Frederick William
Penrod, Christopher Charles
Powell, Thomas David
Rappaport, Jonathan Richard
Rausch, Russ David
Reider, Thomas Allen
Remko, Steven Richard
Richard, Mark P.
Richards, Jeffrey Robert

Robinson, Brian Anthony
Roundy, Lance Michael
Rushinsky, John Joseph Jr.
Schipper, John Kenneth
Shade, Kurt Alan
Shikany, David Andrew
Snyder, Henry Todd
Springer, Kerry L.
St. Germain, Barry Warren
Stumpf, James Anthony
Suffoletta, Daniel Gerard
Sutton, Darryl Matthew
Taylor, James Theodore Jr.
Turnock, David Lewis
Van Campenhout, Adrian
Verhoff, Patrick Henry
White, Mark Harvey
Wickersham, David Keith
Wienholts, Erick Joseph
Williams, Andrew John
Wilson, Bruce Long
Winkler, Richard Charles II
Wright, David Andrew

August 1987
Bruner, Christopher William
Burgess, Jeffrey Scott
Curtis, Scott Andrew
Shidler, Geoffrey Lee

December 1987
Adams, Robert Charles Jr.
Albro, James Carter II
Ayer, Timothy Collins
Beckman, Arthur William
Biebesheimer, William Michael
Bovard, Robert Eugene
Breedlove, Andrew Philip
Bridge, Richard Louis
Brucker, Garrett Anthony
Coiduras, Jose M.
Dandridge, Derrick Minor
D'Angelo, John Louis
DeQuay, Roger
Duff, Jonathan
Edge, Randall Scott
Frye, Patrick Edward
Granger, Jeffrey David
Grubenhoff, Philip Michael
Horvath, Scott Mason
Jamieson, James Robert
Jentink, Thomas Neil
Johnson, William Scott
King, Brian Timothy
Kries, Jack Rudolph
Logan, Robert Francis Jr.
Mohamed, Zulkarnain Bin
Mooney, Joseph Keith
Moore, Michael Anthony
Nobbs, Steven George
Oppelt, John Benjamin
Slota, Kenneth John
Smith, Ronald Edward

Snyder, Thomas Stuart
Sprague, Michael Paul
Stave, Eric Alan
Steiger, Eric Stuart
Swihart, Matthew Scott
Verhoff, Patrick Henry
Vitale, Jeffrey Dunn
Waggoner, Brent Alan
Walsh, Brendan M.
Zenieh, Salah

May 1988
Ascough, Timothy Howard
Ault, Matthew Ned
Baird, Kenneth Allan
Barlow, Warren Phillip
Baxter, James Shannon
Benskin, Nancy Roxanne
Bowman, John David
Brown, Martha Celeste
Buening, Dennis John
Cappel, Murray Scott
Carey, Steven Phillip
Carpenter, Kenneth Allen
Clark, William Scott
Clemens, Eugene Franklin
Coak, Craig E.
Colby, Thomas David
Davis, Mark Allen
Davis, Paul Bradley
DeHoff, Eric Allen
Ditommaso, Darin Louis
Dwenger, Richard Dale
Esselman, Gregory Hubert
Fisher, Jeffrey P.
Forrest, David John
Gast, Mark Edward
Grove, David Michael
Haletky, Edward Lee
Hauser, Timothy Glenn
Hawes, Christopher Allan
Heeg, Jennifer
Hoffman, Randy W.
Hoops, John Robert
Jewell, Michael Alan
Johnson, Steven Andrew
Joyce, Brian David
Kennison, John Albert
Kleb, William Leonard
Kline, Thomas Anthony
Koontz, Steven Kent
Kresse, John Paul III
Lantz, William Robert
Leamy, Jeffrey Robert
Leete, Dennis Michael
Lung, Terence Tze-Ron
Maines, Brant Howard
Mancher, James Marc
Masden, Darrell Eugene
Mason, Steven David
McNew, Gregory Joel
Mekkes, Gregory Lee
Melonides, Stephen James

Moeller, Gerhard Walter
Molnar, Christopher James
Murphy, Patrick Kevin
Nelson, David C.
Nylec, Nicholas Theodore II
Oakeson, David Oscar
O'Neal, Patrick David
Popescu, Florin Codrut
Puig Suari, Jordi
Rallo, Rosemary Anne
Reilly, Thomas Patrick
Reznik, Carter Emil
Roland, Brian Gregory
Rowe, Pamela Sue
Sagarsee, Charles Kevin
Savin, Robin Francis
Scearce, Paul Taylor
Schultz, Joseph William
Scott, Robert Charles
Sents, Jeffrey Ryan
Sermersheim, Jeffrey Scott
Sherman, Robert
Siewiorek, Gregory Bernard
Simpson, Aaron Nelson
Stutts, Gary Wade
Thurman, Douglas Ray
Todd, Richard Emery
Trujillo, Steven M.
Tsouchlos, Timothy Andrew
Van Dam, Kurt David
Weber, Dominic Frank
Willhoff, John Weaver
Williams, Thomas Evan
Wu, K. Chauncey
Wullschleger, Donald Lee
Yorn, Kendall Luke

August 1988
Frahm, Tim Joel
Hindawi, Ahmad Thougan
Lightfoot, Samuel Jr.
Moore, Regina Marie
Sukandar, Inu Permana

December 1988
Beck, Lisa Anne
Bemtgen, Jean
Bovard, Karen Jean Severyns
Brown, Jerry Alan
Buchanan, Eric Scott
Cronin, James Michael
Dowdy, Angie Britta
Emerson, Thomas Duane
Feulner, Matthew Roger
Fisher, Larry Roger
Fong, James Gan
Freygang, William Gerard
Gant, Lance Todd
Gatewood, Barry N.
London, David Andrew
Keenan, Constance Lenore
Kenzler, Kurt Thomas
Kinsey, Dawn Daniel

Knight, Sean Christopher
Krueger, Hayden Anthony
Liba, Carl Jerome
McGruder, David Eugene
Meyer, Scott Allen
Mischel, Brian Charles
Myers, James Daniel
Myron, Patrick Robert
O'Brien, Timothy Richard
Oliver, Stephan Bruce
Porth, Donald Lester III
Powell, Craig Alan
Rees, Edward Charles
Sarks, Andrea Marie
Schatzka, Todd Michael
Scott, Kenneth James
Shultz, Jeffrey David
Webb, Tina Marie
Wolson, Brian John
Wynen, Monica Sue

May 1989
Alston, Brock Charles
Alt, Kevin Herman
Amos, Mark Harold
Asp, Darryl Alan
Barrett, Michael Joseph
Beers, Christina Lynn
Bravard, Bradley Scott
Brinzo, Brian Henry
Brost, Todd Michael
Burley, Russell William
Cooley, John William
Crussel, Roger Alan
De Wald, John Edward
Depaola, Kevin Joseph
Dickey, Walter Guynn
Fair, Matthew David
Focke, Jeffery Allen
Frietchen, Douglas Bradley
Garden, William John
Gausling, James Francis
Ghazali, Aminuddin Bin
Gose, Terry Wayne
Gould, Marston Jay
Gridley, Marvin C. III
Grigsby, Barry James
Harrison, James David Jr.
Hay, Stuart Scott
Heathcote, Robert Arnold
Henderson, Ronald Dean
Hucker, Scott Alan
Huss, Joe A.
Jablonski, Dean
Johnson, Thomas James
Kennelly, Edward Charles
Kerlin, Steven Kenneth
Koehlinger, Brian Lee
Kraft, Neal Douglas
Lages, Christopher Rowland
Langlois, Peter Andrew
Lee, John C. Y.
Lemoine, Patrick K.

Leong, Gary Norman
Longmeyer, Michael Henry
Mann, Ronald Carson Jr.
Martz, John Hughett
Mase, Robert Arthur
McGinnis, Mark Allen
Mesarch, Michael Andrew
Morris, Scott Richard
Muhamad, Sallehuddin
Navas Bernal, Jorge Ernesto
Newkirk, Joseph Troy
Nolcheff, Nick Adelbert
Nussel, Dale A.
Olsavsky, John George
Ortstadt, Tobin Carl
Paschal, Keith Barret
Peacock, Carolyn Frances
Phillips, Robbie Lee
Seiler, David Alan
Simcox, Matthew Robert
Simon, James Michael
Spencer, David Allen
Tyler, Joseph C.
Ventimiglia, Timothy Joseph
Walker, Charles David
Whittaker, Paula Jean
Williams, Richard Llewellyn
Yap, Federiko Louis Doornik
Zelenak, Maryann
Zimmerman, James William

August 1989
Eblen, David Michael
Hannah, Mark Adam
Jones, Patrick Michael
Mullen, Kyle David
Rakoczy, Michele Renay
Ryzewski, Jerome Leonard Jr.

December 1989
Beaudoin, John Louis
Bovaird, Raymond S.
Bowers, Lorie Jean
Conway, Scott Marshall
Crnarich, Daniel Edward
Day, Douglas Edward
Dippon, David Wayne
Doebling, Scott William
Fishel, Jeffery Lee
Hayashi, Mathew Sei
Hendrickson, Todd Richard
Hess, Amy Sue
Hook, David Ellis
Langford, James Eric
Mack, Jay Christopher
Mack, Thomas Lee
Matthew, Sally Anne
May, Raymond Bernard
McCurry, Craig Francis
Mountcastle, Mark Allen
Murphy, John Russell
Nelson, Chad Richard
Orrick, Alec Nicholas

Pace, Joseph Paul
Pawlik, Michael Edward
Piraino, Michael Charles
Pullins, Jeffrey Scott
Scheumann, Troy Duane
Swaby, Mark Lane
Teders, Rebecca Jean
Unsinger, David Allen
Van De Cotte, Michael Robert
Verde, Hector M.
Wilson, James Michael
Wood, Timothy Alan

May 1990
Ahler, Donna Marie
Allen, John David
Amos, Brian Paul
Anderson, D. Mark
Arnone, Stephen Peter
Bagadigng, Nolan Soriano
Bascom, Paul Stephen
Beard, Lisa Marie
Birnbaum, Richard Scott
Botos, Charles Douglas
Brown, Susan Jayne
Bullock, Steven Jerome
Burgett, Jonas Quintin
Burns, Robert Leo II
Busse, Lisa Marie
Buzon, Steven Moises
Cler, Daniel Lee
Conard, Henry Eugene Jr.
Cooper, Teresa Lynn
Cosky, Charissa, Jean
Dailey, Lyle Douglas
Dieter, Rebecca Janet
Dobosz, Christopher Allen
Doshi, Divyesh Nagindas
Elmore, David Gene Jr.
Emerson, John Major
Erdmann, Kenneth Leland
Erwin, Duane Scott
Flora, Spencer David
Fultz, Brian William
Garcia, John C.
Geenen, Robert Joseph
Gingiss, Anthony John
Hafterson, Katherine Leigh
Hatton, Michael Anthony
Hazlebeck, Thomas Evan
Herzner, Janine Elizabeth
Heuer, Stephan Eugean
Hill, Lisa Renee
Hines, Andrew Duncan
Hodgson, Robert Allan
Hooker, John Richard
Jaworski, Aloysius Andrew Jr.
Johnson, Kirk David
Kalke, Jerome Joseph Jr.
Kelso, Kristine Marie
Kennedy, Dale Michael
Kinsey, Gerald Lee
Korthauer, Ralph

Krause, Lisa Marie
Kuhn, Thomas Edwin
Kunack, Kelly Ann
Lamkin, Peter Joseph
Land, Christopher Kevin
Lax, Rodney Wayne
Lindauer, Patrick Alan
Lynch, Richard Jay
Mahaffey, Margaret Mary
Markland, Karen Lynn
Marriott, Arthur William IV
Mattox, Michael Joseph
Miller, Thomas Victor
Monzel, Christopher Robert
Morris, Mark Kirwan
Moulton, Bryan James
Nunez, Candida Lydia
Ondas, Michael Scott
Paige, Ryan Edward
Palomar, Linda Lorraine
Peterson, Craig Steven
Reid, Paul Alden
Remaly, Craig Michael
Ryan, Thomas Patrick
Schenk, Jennifer Lynn
Shirvinski, Frank Anthony
Smith, Gregory Lawrence
Snider, Timothy Owen
Spehar, Gregory David
Staats, Gary Scott
Tomek, William George
Tomlinson, Barbara Sue
Trisciani, Brenda Lee
Tucker, Hurd Charles
Turner, Simon
Vendrely, Timothy George
Wagner, Bertram William
Weeks, Craig Andrew
White, Kelli Ann
Wolf, Stacey Elizabeth
Woodard, Paul Robert
Yarger, Jill Michelle
Zuttarelli, Anthony Paul
Zydell, John Martin

August 1990
Barnes, Thomas Samuel
Flynn, Terrence Patrick
Groskreutz, Alan Ross
Petrin, Jeffrey Joseph
Ross, Brian Patrick
Screnci, Raymond

December 1990
Ash, Charles Thomas, III
Barrett, Tanya Marie
Bryant, Kevin Michael
Carpenter, Christopher Douglas
Clash, Daniel Robert
Cogar, Angela Renae
Cullather, Richard Ian
Dalton, David Robert

Darr, Terrance Brian
Desai, Shailen Dinubhai
Dickson, Paul Christopher
Fischer, Karen Elizabeth
Gale, Thomas Zell
Gilp, Brian Fredrick
Harris, Brock Alan
Harris, Drake Anthony
Hartshorn, Christian James
Kramer, Julie Ann
Kueber, Bret David
Kulam, Sheikna Lebbai
Lance, Christopher Alan
Lanman, David Marshall
Magee, Peter Jeffrey
Mains, Deanna Lynn Ivey
Mangel, Daniel Lee
Mann, James Bradley
Markopoulos, Peter
Merrill, Bradley Allen
Meyer, Scott Edwin
Mix, Charles Dequindre III
Montgomery, Robert Lee
Mosesson, Matthew Norman
Mowery, Jon Brett
Osborne, Bradley Alan
Phelps, Steven Lynn
Radecki, Matthew Paul
Raisch, John Lewis William
Reith, Timothy William
Richardson, Kevin Henry
Roberts, Todd William
Scott, Blake Francis
Seffernick, Neil J.
Shaffer, Kevin Scott
Shanmugasundaram, Ramakrishnan
Sriver, Todd Allen
Strysniewicz, Richard Edmund
Tomsits, Scott Robert
Tribble, Huerta Lee
Vanderwest, Scott Thomas
Wilson, Brett Stephen

May 1991
Ardalan, Mohammad Moez
Bailey, Gerry Scott
Baratta, Joseph Vincent
Barden, Brian Todd
Barry, David Michael
Bartick, Andy K.
Berry, Robert Everett
Blozy, Thomas Andrew
Boas, Mark Steven
Breiling, Kurtis Beamer
Brown, Jeffery Neil
Brown, Thomas William
Campbell, Bryan Thomas
Cass, Christopher Ryan
Cordero, Jorge Javier
Cosby, Ronald Maurice Jr.
Crispen, Rodney Earl
Davis, Emmett Jason
Day, Brian Carter

Dean, Loren Philip
Eaton, Ronald James
Feaster, Matthew O.
Gaines, James Barritt Jr.
Gasper, Joseph John Jr.
Gast, William Thomas
Gaynor, James Robert
Gnadinger, Philip Anthony
Gunawan, David James
Hayman, Greg Joseph
Heffern, Daniel Wayne
Hejl, Robert James
Howard, Stephen David
Hunt, Craig James
Hunt, Gordon Ashby
Kalupahana, Nandana Kamalasila
Katz, Jonathan William
Kemper, John Paul
Kendall, Orvin Dale
Kennedy, Stuart Brian
Laudeman, Robert Allen
Liese, Eric Arnold
Mai, Long K.
Mains, Benjamin Lynn
Marfil, Grace Diane
Maschino, Marc Laurent
McCool, Michael Lee
McCoy, Robert William
McNulty, Gregory Scott
Minor, Durand Edward
Moore, Stephen Scott
Mosey, Samantha Michele
Moy, Chi Kevin
Murray, Francis Todd
Naughton, Daniel Thomas
Nolan, Michael J.
Parent, Beth Ann
Patel, Moonish R.
Petri, Stephen Everett
Preuss, Julie Anne
Reed, Alyson Maxine
Reynolds, David Joseph
Rishko, Tara Lee
Saalwaechter, John Wesley
Sack, Elizabeth Elaine
Shannon, Brian Scott
Simpson, Philip Mason
Singer, Leonard Jr.
Smith, Bradley Roland
Spikings, David Kevin
Spott, Alan Joseph
Stephens, Brian Eugene
Strader, Larry Brian
Thomas, Paul Andrew
Till, Bradley Dean
Titzer, Christine Marie
Weid, Angela Renee
Weiss, Christopher Russell
Whitcraft, John Edward
White, Matthew Kendrick
Willis, Edwin Russell Jr.
Wilson, Roby Scott
Vaghnam, Nadeem Farid

July 1991
Capo, Armando
Fleming, Michael Patrick
Harrington, Brian David
Heflin, Edward Lee
Hinton, Steven
Miller, Christopher David
O'Donnell, Sean Robert
Runge, Thomas Andrew

December 1991
Alspach, David Allen
Arnold, Andrew
Beck, Rebecca Anne Dilling
Belcher, Bradley James
Cashman, Sean P.
Chlystun, Jeffrey J.
Cielaszyk, David Lee
Crocker, Alan Robert
David, Scott Allen
Dillon, Paul A.
Edwards, W. Glenn
Gerst, Mel Raymond
Glum, Martha Elizabeth
Gootee, Scott R.
Gries, Jason Wayne
Grindle, John Shannon
Holland, Julie Ann
Johnson, Michael Andrew
Kho, Cedric Ngo
Kleve, Robert Bruce
Knappenberger, Anjanette Sue
Lehman, Michael Lyn
Leong, Mark Stephen
McKesson, Kent Worth
Miller, Brent Klee
Moore, Robert Charles
Muller, Mark Brian
Myers, Andrew John
Noe, Christopher William
Nus, David Branden
Oesch, James Andrew
O'Keefe Karen L.
Pardieck, Matthew David
Peyton, James Logan II
Proffitt, Robert Jason
Rose, Glenn Charles
Salyer, Terry Ray
Scheitlin, David John
Schmitt, Steve Michael
Seale, Melanie Lynn
Selwa, Lynn Rose
Shepherd, Cari Lynn
Spoo, Mark Allen
Springer, David Gene
Stancliff, Stephen Bryan
Stokman, William Paul
Thomas, Michael Wayne
Vaughan, Tamara Jill
Walke, James Louis
Weismuller, Diane Lisa
Young, John Michael

May 1992
Abbott, Michael T.
Ahern, Gerald Robert
Alcenius, Timothy John
Arnold, Anthony Allen
Barnhart, Deron Warren
Barsun, Hans F.
Bissell, Christine Marie
Boley, William Christopher
Bolser, Stephen Todd
Borror, Sean L.
Bretthauer, Eric Richard
Bunnell, Robert Allen
Chambers, Robert Patrick
Coffey, Gregory David
Cole, Carolyn Erna Louise
Coombs, Cade Dale
Courtney, Scott D.
Davis, Bradley Alan
Doyle, Timothy
Eastwood, Brent James
Findley, Jeffrey Neil
Finkbeiner, Robert E. Jr.
Forster, Edwin Ewald
Frotton, Traecy Lynne
Gima, Kazuhiko
Graves, Eric Neil
Greenwald, R. Heather
Gruber, Jennifer Marie
Hacker, Gregory John
Halgren, Corey Thomas
Hall, Gregory Mark
Hall, Philip Godfrey
Heinimann, Markus Beat
Hilburger, Mark D.
Holtz, Michael James
Jankowski, James Walter
Kiger, Brian Vincent
Klug, John Charles
Klusman, Paul James
Kobza, John David
Koeske, Paul Philip
Kokal, Kevin John
Laczkowski, Douglas John
Lee, Shawn DeMoyne
Lighthill, Anthony Lynn
Lowery, Kirk Patrick
Mackey, Marc Elliot
Mangelsdorf, Mark Frederick
Marien, Ty Vincent
McDonald, Alan Lee
McGowan, Daniel James
McGuire, Joseph Byron
McLay, Courtlan Graham
Mechalas, John Peter
Meiss, Alan Richard
Menconi, Giore Anthony
Merrick, Kenneth Daniel
Mikulski, Daryl Wayne
Minniti, Robert Joseph III
Molony, John Michael
Myers, Stacey Rae
Napierkowski, Michael Edward

Nebuda, Sharon Elaine
Newton, Frances Kay
Peters, Brian Carl
Petersen, Timothy Wayne
Pongracic, Ivan
Queiser, Brian John
Rivas, Anna Maria
Salinas, Angel Geovanny
Sandys, Brian Peter
Schacht, Richard Alan
Scruggs, Leonard Andrew III
Shelby, J. Dugan
Spangler, Christopher Allan
Stattenfield, Kevin Mark
Struckel, Margaret Kathleen
Tamblyn, William Scott
Tartabini, Paul V.
Temores, Jesus B.
Wells, Paul Richard
Weltzer, Julie Ann
Williams, Pamela Jean
Wilson, Gregory Michael
Winkelman, James Richard
Zarobsky, Christopher John
Zezula, Chad Erik

July 1992
Boddy, Douglas Ernst
Bullifin, Kent James
Hall, Scott Monroe
Mount, Kimberly Louise

December 1992
Bond, Stacy Darren
Broviak, Bart Joseph
Chaney, Aaron Wendel
Chislaghi, Cathryn Jane
Croxall, Michelle Renee
Crump, Paul Anthony John
Erickson, Tiffany Joy
Filbern, Matthew Eric
Fisher, Michael William
Haven, Christine Elizabeth
Heitert, Amy Irene
Hilgefort, Kurt Alan
Hoover, William Frank III
Karver, Robert Walter
Kasun, Andrew James
Kegerreis, Kristopher Kahn
Mahan, Laura Jane
Moore, Daniel David
Nelson, Gregory Hill
Orlando, Peter John Jr.
Payne, Gregory Allan
Peters, Mark Eugene
Riffey, Todd Evan
Rogers, James Alan
Rzonca, Peter Joseph
Scott, Kelly Adam
Shanley, Robert Loren III
Stubbings, Natalie Beth
Tracy, Edward William Jr.
Whisman, Keith Alan

White, Jeffrey L.
Wood, Gregory Eugene
Yang, Wilson W.

May 1993
Andolz, Francisco Javier
Bates, Eric Joseph
Bayt, Robert Louis
Bennett, Eric Ransom Jr.
Bryant, Paul Joseph
Buswell, John Philip
Caperton, Edward Lynn
Carrasco, Jose Luis
Coley, Christopher Michael
Cooprider, Amy Lee
Crippen, Christopher M.
Debikey, E. Oliver
DeJoannis, Jeffrey Paul
Files, Bradley Steven
Fisher, Matthew Jeffrey
Gabor, Michael Joseph
Gillgrist, Robert Dallas
Gonzalez, Andres
Grubbs, Lana June
Guzman, Jose Javier
Hibler, John Paul
Hodge, Bertram Cort
Hoffman, Jeffrey Craig
Hoffstadt, Brett Alan
Hooper, James Anderson
Huston, William Allen
Kanuch, John Michael
Kline, Joseph Raymond
Lena, Michael Robert
Leung, Szu-Tao
Lin, Eric Y. J.
Loosemore, Robert Edward Jr.
Lugo, Jorge Neftali
May, Anthony Wayne
McAninch, Daniel Richard
Minnich, Randolph John
Moore, Brian Michael
Page, Jeffrey Stephen
Pearson, Erika Jane
Precup, Margaret Joy
Ringler, David Eric
Rutz, Mark William
Scheuring, Jason Nicholas
Seifert, Michael Scott
Semanik, David Edward
Smith, Troy Thomas
Summers, Darrell Lee
Sun, Peter Chao-Tsu
Szolwinski, Matthew Paul
Versteeg, Stephen Keith
Vonderwell, Daniel Joseph
Weisskopp, Andrew Mark
Werthman, William Lance
Wintring, Clifford Raymond
Wright, Stephen Christopher
Wu, Hsin-Cheng
Wyler, Derek Robert
Yuhas, Paul Jerome
Zakaria, Zaidi Bin

August 1993
Foster, Amy Elizabeth
Goebel, Ross Clair
Griffin, Jeffrey Franklin
Hare, Angela Marie
Nelson, Erik Lee
Schaeffer, Charles Edward
Vendramin, William Anthony

December 1993
Andrews, Jerry Wayne II
Brown, John Stewart, IV
Brumbaugh, Kathleen Marie
Brumfield, Alonzo Cartez
Bucci, Gregory Steven
Burbage, Brett Allen
Caravella, Joseph Robert
Dunlap, Tony Lavern
Fetty, Monica Rae
Gabert, Shawn Christopher
Godoy, Carolina
Goerges, Stephanie Lynn
Hartman, Michelle Lynne
Hensley, Daniel Paul
Marsh, Glenn MacArthur
Miller, Jeffrey Robert
Morris, Tammy Christine
Murphy, Patrick Keefe
Nichols, Todd A.
Pancake, Trent Alan
Pesut, Martin Russell
Rabideau, Kurt Michael
Rock, Kenneth Mark
Roth, Brent Ellison
Russell, James Long
Schroeder, Christine Ann
Slijepcevic, Stevan
Stecker, Scott David
Thompson, Troy Brian
Tyrcha, Jeffrey Michael
Varney, Bruce Edward
Walters, Leon Thomas
White, Sheri Neelam
Wiechart, Kent David
Wong, Keewah Gary
Zappe, Stephen Michael

May 1994
Adams, Douglas Scott
Bowman, Jason Christopher
Bremer, Donald Harold
Cooney, James Scott
Cox, Seth Adam
De Jesus, Rafael Omar
Doell, Christopher William
Douglas, Tamara Ann Nahrwold
Downing, David Michael
Emore, Gordon Lantz
Felten, Nancy Ann
Griese, Kimberly Ann
Hall, Michael Jerone
Hill, Scott Michael
Huffman, Richard Edwin Jr.

Hurst, Stephen Neil
Jasanis, Patrick Edward
Kakoczki, Steven Richard
Kam, Claude K.
Kauffman, John Christopher
Linder, Benjamin Charles
Lintereur, Louis Joseph
Lutterloh, Emily Clare
Martens, Kristina Leigh
Massa, David Jon
Mattick, Carolyn Sue
McCracken, Robert W.
Montes, Exor Luis
Morris, Richard Lindsay
Munro, Scott Edward
Murray, Ian Fisher
O'Herron, Daniel Paul
Payne, Elizabeth Anne
Pecson, Derek McDougall
Persinger, Michael Scott
Piatt, Gregory Alan
Prien, Jamison Richard
Prigge, Eric Allen
Rump, Kurt Matthew
Schlabach, Michael Alan
Schwartz, Robert Aaron
Stanard, James Mark
Strait, Carl Allen
Taylor, Michael Anthony
Tenaglia, Robert Anthony
Weaver, Marc Christopher
Wiley, Mark Anthony
Williams, Laura Ann
Williams, Todd Gordon
Witte, Gerhard Rudolf
Yeiter, Tobey Lee
Zink, Jonathan Todd

August 1994
Amrozowicz, Christie Marie
Chou, James Chin-Chia
Forester, Aaron Michael
Lushin, Phillip Andrew
Motiwalla, Darius Bomi
Robson, Teri Lynn

December 1994
Archambeau, Bruce Edmond
Bell, David Dale
Bigonesse, Rejean Alain
Borkowski, Jonathan Edward
Bubik, Jason Robert
Coussens, Michael Joseph
Dobson, Christopher Garrett
Doty, Kevin Andrew
D'Souza, Shariff Rosario
Emmert, Eric Alan
Holtzclaw, John David
Howland, Eddie Joseph
Kannal, Lance Eric
Khadder, George Assad
Langster, Travis Boyce

Marso, Robert Scott
Natividad, Roderick King
Oldenberg, Chadwick James
Pesek, Joseph Frank, II
Ruble, Tricia Ann
Sacharoff, Adam Kevin
Schneider, Julie A.
Sherman, David John
Sundaram, Kay M.
Szwajkowski, Darren Walter
Tani, Irwin Karman
Yue, Wei Thoo

May 1995
Ahmad, Salman
Anthes, Virginia Louise
Baker, Joseph Alan
Bougher, Jeffrey Alan
Casler, Daniel Paul
Comninos, Mark
Conlu, James A.
Cude, Jason Christopher
Devarajan, Giridhar
Eisele, David Andrew
Feeney, John Timothy
Gallagher, Eric Randle
Gates, Matthew David
Hendrickson, Heath Schuyler
Herman, William Charles
Houghton, Martin Bradley
Koselke, Anthony Heath
Krautheim, Michael Stephen
Kulas, Lisa Louise
Loh, Gregory William
Lore, Vito
Lucas, Amy Jo
Lynge, James Stephen
Miller, Kenneth Alan
Morrison, Daniel K.
Nuss, Jason Samuel
Ostlund, Brian Bevan
Peterson, Cara Lynn
Pham, Tuan Le
Roth, Gregory Lee
Shafer, Patrick Michael
Sichmeller, Scott Thaddeus
Steffen, Phillip Jay
Stephenson, Mark Thomas
Tack, Jeremy Charles
Torgerson, Shad David
Verville, Justin Martin
Wright, Jonathon Michael
Young, James Owen

August 1995
Kerns, Christopher M.
Lii, Neal Yi-Sheng
Morehead, Michael Reed

December 1995
Alikhan, Asif
Amann, Justin Carl
Anthony, Steven Ross

Ayala, Artagnan
Collins, Stephanie Jo
Davids, Eric John
Denzer, Brandon Scott
Erausquin, Richard German Jr.
Huycke, Jason Douglas
Majumdar, Neha Senajit
McDonald, David Bruce
Meiring, Eric Matthew
Peeples, John Churchill Humphrey
Piliang, Eddy Filanda
Polete, Sean Paul
Porter, Bradley Scott
Randolph, Christian Allen
Reeb, Amy Beth
Romeus, Loren Paul
Rumple, John Robert Jr.
Ryan, Thomas Francis
Searle, Joseph Mathias
Speckman, Keith David
Strazzulla, Frank Sebastian
Strong, Cynthia Anne
Sukandar, Ian Purbaya
Vandenboom, Michael Richard
Wininger, Clinton Michael

May 1996
Blomquist, Robert Joseph Jr.
Buttacoli, Michael Andrew
Caskey, Brent Allen
Christopher, Jennifer Ann
De Souza, Michael Justin
Dimopoulos, Apostolos Tolis
Dougherty, Chad Brian
Dumit, Jennifer Elizabeth
Eckstein, Michael Richard
Funk, Daniel Robert
Gardner, Kyle James
Gerber, Joseph David
Golden, Patrick John
Grasmeyer, Joel Mark
Harger, Jeffrey Scott
Henning, Kristine Kay
Hermansen, Jennifer Kay
Hudson, Julie Ann
Jacobs, Jason Jay
Jarvis, David Alan
Jones, Brian Robert
Joray, Brent William
Knauer, Chad Erick
Lawson, Charity Wynn H.
Lewis, Patrick Vincent
Lintereur, Beau Vincent
Metrocavage, Kevin Michael
Michlitsch, Jeffrey Michael
Miller, Matthew Christopher
Norris, Stephen Richard
Parsons, Kevin Kirk
Pitts, Torino Ray
Planey, Michael Charles
Puperi, Daniel Scott
Rippere, David Michael
Rogers, Stephen Kyle

Rosidi, Irfan
Schoonover, Philip Lee
Scipta, Ellen Lynn
Smith, Jennifer Marie
Snyder, Michael Joseph
Sorg, Brian Allen
Spence, Melanie Lynn
Wright, Jennifer Helene

August 1996
Sturm, Timothy Edward
Viehe, Darin Scott

December 1996
Bleil, John Jerome
Butcher, Michael Alan
Chin, See Loong
Crane, Douglas Michael
De Namur, Jordan Lesley
Driscoll, Nicole I.
Funk, John Eric
Hamlin, Louise Ann
Kawazoe, Michelle Keiko
Ong, Wee Gin
Owems, Jom Anthony
Ramirez, Penelope Gisella
Rasmussen, Amy Lynn
Ready, Paul J.
Ross, Tamaira Emily
Satoh, Hironori Neal
Van Gelder, Klaasjan Hendrik P.

May 1997
Bodony, Daniel Joseph
Bonfiglio, Eugene Peter
Crowley, Jennifer Kristen
Cruse, Khali Senuwae
Ehardt, Danielle Evelyn
Foti, Anthony Matthew
Freese, Rebecca Anne
Getten, Brian Robert
Giesler, Maureen Dianne
Gotschall, Patrick Owen
Haan, Robert Joseph
Hartigan, Daniel Jerome
Hasebe, Ryoichi Sergio
Hatcher, Jeremy Todd
Hayes, Heather Leigh
Jameson, John Ross III
Kiser, Russell Herbert
Marchand, Belinda G.
Mathews, Douglas William
Myers, Jason Neil
Pilcher, Katherine Mary
Reese, Steven Donald
Schwenk, Brian Kenneth
Smiczek, Daniel Troy
Starr, Matthew Jasen
Van Maasdam, Peter John
Walker, Angela Carmiela
White, Warren William III
Williamson, Craig Russell

August 1997
Kuka, Brandon James

December 1997
Flynn, Brian Edward
Gault, Matthew James
Ho, Loon David Shi
Kelly, Michael Everett
Lee, Mathew
Leech, James Robert
Miynarski, Philip Richard
Moretti, John Andrew
Rennick, Michael John
Ruzicka, Stephani Dawn
Strange, Nathan J.
Warren, Clinton John
Williams, Edwin Adams
Zimmermann, Jeramiah David

May 1998
Anderson, Jason Paul
Benish, Kerrie K.
Bowes, Angela Lynn
Cohea, Heather Jean
Derby, Lea Adrianna
Grobe, Shawn Alan
Hay, Kimberly Sue
Leistner, Elizabeth Rose
Liberti, Michael Andrew
Mahler, Cynthia Louise
Martin, Joel Andrew
Mc Daniel, Zeke Perry
Mc Kinley, David Patrick
Ramirez, Alfredo Enrique
Robertson, Brandan Richard
Schoenherr, M. Scott
Schneider, Elizabeth Alice
Shafer, Bart Marcus
Sharp, Barton Miles
Snell, Steven Christopher
Waninger, Sean Ryan
Webb, Scott Edward
Whited, James Christopher
Yargosz, Stephen Michael

August 1998
Awan, Kashif Qayyum
Meyer, Molly Kristen
Schlomer, Ryan A.

December 1998
Baumann, Ethan
Bae, Kyoung Duck
Bevis, Dwayne
Dobbins, Abigail
Frolik, Steven
Harris, Daniel
Harris, Laura
Louks, Daniel
Melchior, Michael
Mutchler, Bryan
Niksch, Matthew
Price, Daniel
Rentas, Raul

May 1999
Bockmiller, Daniel
Brilliant, Lisa
Burianek, Jeffrey
Cano, Helene
Champaigne, David
Cook, Andrea
Dressel, Kevin
Harmon, Michael
Hemann, Justin
Indermuehle, Kyle
Javorsek, Daniel
Johnson, David
Labuda, R. Gregory
Lee, Byung-Chan
Meyer, Eric
Miller, Daniel
Pomart, Cyril
Porras Alonso, German
Powers, Audrey
Ruiz, Jose
Sablotny, Timothy
Schenk, Peter
Schneider, Phillip
Strzyz, Joseph
Valentini, Luiz
Weinstock, Vladimir
Wright, Christopher

August 1999
Gifford, Bradley
Matlik, John

December 1999
Austin, Benjamin
Braeckel, Kurt
Davis, Robert
McKowen, Richard
Morgan, Alex
Okutsu, Masataka
Pilacik, Aimee
Schmidt-Lange, Thomas
Shurtleff, Andrew

May 2000
Asano, Hirokazu
Billingsley, Birk
Birkhauser, Emily
Bixby, Geoffrey
Blodget, Chad
Buennagel, Janet
Chew, Su-Lit
Curtiss, Ben
Debban, Theresa
Frecker, Tricia
Helms, William
Hoverman, Thomas
Kayir, Tulin
Key, Nicole
Knott, Mark
Koenigs, Michael
Kowalkowski, Matthew
Leong, Pooi-Fong

Liechty, Derek
Lim, Kok Hong
Matsumura, Shin
Moore, Jim
Panzenhagen, Kristin
Radocaj, Daniel
Remson, Andrew
Scheffler, Josh
Schultz, Jamie
Sloniger, Beth
St. John, Clinton
Sweet, Carmen
Wellner, Nicholas
Wulf, Steven

August 2000
Abdul Aziz, Hafizal Nor Bin
Hasebe, Keita
Kuo, Anthony C.
Prock, Brian Christopher
Wang, Mahammad

December 2000
Basiletti, Matthew Peter
Beech, Ryan Casey
Bhutta, Oneeb Ahsun
Bryant, Terrance Austin
Church, Jeremy Wayne
Curtis, Christopher Cary
Fitzpatrick, Bridget Maureen
Hout, Keith Russell
Martin, Eric Thomas
Mok, Jong Soo
Mustafi, Shouvanik
Patneau, Brian Wesley
Peters, Andrew Alan
Plank, Wendy Ann
Russell, Shawn Arlin
Spicer, Marc Tristan
Strutzenberg, Tyson Wade
VanMeter, Michael Gregory
Wilson, Nicholas Michael

May 2001
Abel, Brandon
Ackerman, Matthew
Blanton, Mark
Bliss, Brian
Burton, Kacie
Cardinal, Benjamin
Chen, Kuan-Hua Joseph
Cummings, Devin
Czapla, Nicholas
Delion, Ann (Dransfield)
Dempsey, Patrick
Fitzpatrick, Shannon
Garber, Heather
Garrison, Loren
Grasmick, Jake
Hanna, Stephen
Heidelberger, Chris
Irvine, Adam

Kennedy, Patrick
Kirchner, Casey
Klein, Debra
Lew, Phoi-Tack
Magdziarz, Adam
McCormick, Jeffrey
McElroy, Sean
Namwong, Bodin
Nokes, Michelle
Nour, Wael Aboul Haggag
Pavlicek, John
Peters, Christopher
Rodrian, Jeffrey
Saadah, Nicholas
Spreadbury, Sherri
Stein, William
Tuinstra, Tanya
VanY, Stephanie
Wong, Wiyan

August 2001
Hamilton, Donald Craig
Resat Koray

December 2001
Barnett, Brian Kent
Butt, Adam
Dahya, Kevin Raju
Dankanich, John Walter
Darr, Brian Wayne
Davis, Jeremy Paul
DeVoto, Pamela Maureen
Edwards, Jonathan Michael
Jungquist, Michael Joseph
Keune, John Nelson
Kuhn, Julia Darcey
Kuruvilla, Santosh Joseph
Landau, Damon Frederick
Matthews, Jaret Brice
McKenzie, Michael John
Roark, Tamara Sue
Spencer, Alec Thomas
Squibb, Daniel Robert
Wiratama, Andy

May 2002
Agrawal, Pooja
Allen, Amanda
Andersen, Randy
Beaver, Angela
Benner, Robert
Bertuccelli, Luca
Brooks, Evan
Castro, John
Chiew, Lee
Covarrubias, Gina
Ebetino, Ricardo
Eggink, Hunter
Fleck, Alex
Founds, Timothy
Gean, Matthew
Goetz, Giles
Gowan, John

Guthrie, Sarah
Hasebe, Yoske
Helfrich, Timothy
Hemler, Jeremy
Hiraoka, Daisuke
Huber, Nathaniel
Kakoi, Masaki
Kobza, Lisa
Lambeth, Allison
Lynch, Curtis
Mehta, Jatin
Merchant, John
Miller, Michael Raymond
Naessens, Lauren
Newenhouse, Amanda
Niemczura, Jaroslaw
Pan, Fong Loon
Peterman, Jonathan
Precoda, Paul
Saliers, Meredith
Sangngampal, Nattaporn
Schreiner, Michael
Simmons, Aric
Srogi, Ryan
Tan, Chong Sing
Tennant, Robert
Villamarin, Jesus
Wilson, Rayjan

August 2002
Aguayo, Eduardo
Allen, Jamie
Burnside, Christopher
Gowan, John
Mah, Jonathan
Merchant, Munir

December 2002
Bischoff, Gregory
Brower, Paul
Chan, Ke-Winn
Chen, Jit-Tat
Dudley, Rohan
Fosness, Thomas
Gates, Kristin
Glaser, Mellisa
Grehan, Heather
Hawkins, John
Hrach, Michael
Lash, Ryan
Loffing, David
Lovera, Javier
Martin, Thomas
Northcutt, Brett
Nugent, Nicholas
Perotti, Michael
Phang, Yin Fee
Sanders, Tim
Sipe, Colin
Spidale, Anthony
Toombs, Rory

May 2003
Adkins, Matthew
Bash, Stephen
Bohnert, Alex
Bouton, Matthew
Briggs, Eric
Brophy, Daniel
Brown, Laurel
Chapoy, Enrique
Crook, Douglas
Dafler, David
Djibo, Louise-Olivia
Eichel, Brenda
Fisher, Charles David
Fredlake, Joshua
Gedmark, John William
Ghesquiere, Joseph
Gordon, Cristina
Grimes, Jessica
Gromski, Jason
Habrel, Christopher
Harber, Joseph
Hidayat, Jimmy
Hoyle, Victoria
Hoyt, James Edwin Jr.
Kacvinsky, Rebbecca
Kolbert, Whitney
Kong, Jin Yong
Kost, Valerie Janis
Kuipers, Fred Mark
Ladd, Adam Franklin
Lamberson, Steven E. Jr.
Lee, See-Chen
Lucas, Jason Peter
Mane, Muharreh
Marando, Michael
Martinez, Samantha
Mathur, Ravishankar
Miller, Kevin Joel
Murphy, Aaron
Newsome, Elizabeth
Northam, Carly
Ooi, Chieh Min
Osier, Geoffrey
Owens, Brandon
Page, David
Pawley, Heather
Pieri, Gina Lee
Pinson, Robin
Pinyerd, James
Reitenour, Brian
Scholz, Raymond
Sherrick, Joseph
Shockling, Michael
Skillen, Michael
Steber, Stephanie
Stout, Matthew
Tanner, Travis
Taylor, Jordon
Trigg, Daniel
Vaughan, Emily
Wade, Kevin Lloyd

August 2003
Barr, Charles M.
Darkes, Kevin D.
Miller, Quinn
Skoog, Jonah
Troester, Jeffrey
Watson, James A. III
Yu, Yen Ching

December 2003
Ahn, Jung Hyun
Akhter, Ali
Allakki, Sandeep
Bagg, Matthew
Bailey, Justin
Bender, Daniel
Bies, Christopher
Bird, Scott
Brzezinski, Jennifer
Byron, Jennifer
Chesko, Brian
Chakraborty, Daniel
Davendralingam, Navindran
Downes, Michael
Falardeau, Joel
Granum, Geoffrey
Haase, Kelby
Hile, Grant
Hronchek, Brian
Hutter, Jared
Jos, Cyril
Jura, Melanie
Ko, John
Komives, Jeffrey
Leong, Kai Hui
Light, Theodore B.
Lisan, Darwin
Lo, Gerald
McCormick, Patrick
Mousseau, Douglas
Padmanabhan, Arun
Palmer, James Joseph
Rashiduzzaman, Mir
Ricca, Anthony
Rist, Amber
Robbins, Brent
Seah, Heng Keong Kelvin
Sigari, Cyrus
Sufana, Michael
Sydnor, Andrea
Tan, Alethea
Tchilian, Emil
Uhle, Albert
Umberger, Sarah
Watson, Jennifer

May 2004
Abdullah, Maizakiah
Bahnsen, Allison
Barua, Debanik
Berger, David
Blaske, Steven
Blattner, Eric

Bradford, Anthony
Branson, Matthew
Browning, Robert
Capdevila, Lucia
Clark, Stephen
Daly, Kevin
Dunn, Heather
Faust, Andrew
Fisher, Christopher
Fitting, Devin
Goedtel, David
Grebow, Daniel
Grosse, William
Hankins, Franklin
Heckler, Gregory
Holden, Douglas
Hsieh, Kelli
Huebsch, Louis
Ianniello, Alessandro
Janes, Leigh
Jocher, Glenn
Kacmar, Andrew
Kalb, Brady
Karnes, Rebecca
Khoo, Teng
Kobin, Noah
Krukowski, Christopher
Ladisch, Nikolaus
Lambert, Steven
Maier, John
Malashock, Rachel
Manning, Robert
Mazur, Marina
McMillan, Wade
Miller, Christopher
Mseis, George
Myer, Andrew
Nakaima, Daniel
Nelson, Patrick
Pahn, William
Paunicka, Joseph
Pender, Adam
Phillips, Benjamin
Pramann, Brian
Prentkowski, Mara
Redman, Bryan
Rutovic, Mihailo
Schoening, Brian
Shew, John
Silosky, Melanie
Spuzzillo, Michael
Tan, Eng Kee Ian
Thaivasigamony, Kapila
Toleman, Benjamin
Ulrich, Christopher
Ventre, Brian
Whitley, Ryan

August 2004
Browning, Robert
Mitchell, Jacob
Thompson, Bradley
Tucker, Justin

December 2004
Atsuta, Masaaki
Bruno, Nickolas
Capria, Matthew
Cavanaugh, Edward
Cocks, Rachele
Cook, Alexander
Croisetiere, Louis-Olivier
Damhauser, Nancy
Davis, Milton
Deitemeyer, Adam
Denton, Jamie
Dierdorf, Jeffrey
Eagon, Eric
Ernst, Matthew
Fedorczyk, Douglas
Gonzalez, Miguel
Hart, Robert
Hayes, Patrick
Irwin, Ryan
Jackson, Whitney
Jacobs, Robert
Joshi, Manasi
Karashin, William
Kirk, Kenneth
Krakover, Jamie
Lynch, Rebecca
McCaffery, Daniel
McGuigan, Theresa
Nalin, Joel
Navarro, Robert
Nerderman, Jay
Ptak-Danchak, Chad
Reitz, Adam
Robinson, Jeffrey
Rowe, Courtney
Schuff, Reuben
Setar, Nicholas
Shearer, Jonathan
Siler, Joshua
Smith, Kyle
VanderPyl, Timothy
Vos, Nathan
Weaver, Charles
Wong, May Ee

May 2005
Adams, Brian
Amback, Jon
Apostol, John
Badger, Kevin
Bogenberger, Brienne
Boopalan, Avanthi
Bradley, Kathryn
Browning, Peterson
Budd, Eric
Burke, Caley
Carpenter, Michael
Cenci, Nicholas
Chernish, Brian
Clark, Kevin
Clark, Randall
Cornelis, Kristine

Cunha, Christopher
Davidson, Gregory
Davis, Joseph
Doan, Melissa
Eash, Brendan
Evans, Meredith
Feuerborn, Steven
French, Mark
Friel, Jason
Gard, Leo
Gramm, Paul
Gray, Stas
Grupido, Christopher
Gustafson, Eric
Guttromson, Jayleen
Haddin, Jeffrey
Halla, Daniel
Halstead, Steven
Hargraves, Jason
Harrigan, Matthew
Hauser, Aaron
Heinemann, Matthew
Hoff, Philip
Hossain, Asif
Houlton, Brendan
Husein, Ezdehar
Jaron, Jacqueline
Karni, Etan
Kesling, Jared
Kloster, Kevin
Klutzke, Douglas
Kottlowski, Aaron
LaMaster, Christopher
Long, Angela
Long, Hafid
May, Ross
McCoy, Colleen
McCurdy, Justin
McKinnis, John
Metzger, Jeri
Mondino, Kathleen
Mook, Joshua
Mostrog, Todd
Mrozek, Kimberly
Naylor, Christian
Needler, Eamonn
Niziolek, Paul
Parkison, Benjamin
Pedersen, Stacie
Pollock, George
Rainbolt, Colleen
Renstrom, Brian
Reyzer, Charles
Rodkey, Samuel
Roesch, James
Shaw, Thomas
Sippel, Aaron
Smith, Austin
Spindler, Phillip
Stalbaum, John
Stangle, Charles
Statler, Christopher
Stefanczyk, Damian

Szamborski, Tim
Taylor, Joseph
Trkulja, Lazo
Vahle, Mark
Verbeke, Matthew
Voo, Justin
Walker, Scott
White, Nicholas
Wilhelm, Tyler
Wolf, Eric
Wright, Raymond
Wyman, Leah
Yochum, Robert
Yoke, Jeffrey
Zeszotek, Michelle

August 2005
Kallimani, James
Rodenbeck, Andrew
Tang, Jason

December 2005
Allensworth, Timothy
Anderson, Robert
Baker, Nicholas
Caldwell, Michael
Chiu, Jimmy
Christensen, Patrick
Conlin, Matthew
Crosson, Bradley
Droppers, Lloyd
Ewing, Joseph
Fernandez, Nina
Figueroa, Cristina
Foertsch, Matthew
Fosler, Jeffrey
Franklin, Kevin
Galotta, Edward
Golbov, Conrad
Grant, Michael
Gray, Brian
Hahn, Jeeyeon
Hartwell, Melissa
Hedberg, Thomas
Henning, Gregory
Hollingsworth, Kevin
Jennings, Thomas
Kobyra, James
MacDermott, Robert
McDaniel, Kyle
Mehta, Neel
Mills, Anne
Moessner, Michael
Morales, Dalia
Mourad, Ramy
Mullins, Maria
Rosenthal, Lisa
Rybarczyk, Gretchen
Sanders, Anthony
SreeRaman, Sachin
Tan, Lionel
Tellefsen, Rolf
Vartak, Nihar

Weisbrod, Sara
Wills, James

May 2006
Abbott, Ryan Michael
Alanis, Miguel Angel
Alkari, Ameya
Beirne, Stephen Bailey
Berger, Richard
Blank, Ryan Andrew
Boettcher, Philipp Andreas
Boucher, Paul Albert
Boyd, Jeffrey Alan
Brower, Laura Beth
Calderwood, Erin Elizabeth
Cashbaugh, Jasmine Lacy
Chachor, Nick Mitko
Chander, Arun Ram
Chepko, Ariane Brooke
Christensen, Patrick Michael
Coduti, Lee
Coffery, Adam Michael
Collins, John Quinn
Dale, Rebecca Marie
Darraugh, Megan Elizabeth
Daugherty, William Anthony
Davia, Sara Louise
Davis, Chad Richard
Duckett, Brian Christopher
Dumas, Michael Ryan
Fairbanks, James Earl
Febbraro, Pete James
Fink, Robert Henry
Fles, Christopher Alan
Gillies, Daniel Eric
Gordon, Dawn Perry
Hahn, Seth Derek
Hicks, Kimberly Chalmers
Holley, Eric Howard
Humble, David Joseph
Hurwitz, Bradley Jonathan
Ichikawa, Chiaki
Kester, Christopher William
Kim, Woo-Jae
Kite, Timothy Matthew
Kotegawa, tatsuya
Ksander, Jeffrey Scott
Lashkari, Dheer Vijay
Leonard, Eric John
Link, Douglas Waldemar
Malda, Jonathan Joseph
Maurer, Andrew Ludwig
    (BS in Applied Physics &
    Aero)
Mayne, Aaron Arthur
Miller, Daniel Franklin
Naramore, Adam Frank
Nash, Michael William
Nordstrom, Lauren Christina
Olmstead, Jason Michael
Parent, Jillian Rose
Patel, Hiren B.
Patel, Lakme

Pattee, Niccole Wheaton
Poonawala, Husein Imtiaz
Poppelreiter, Heidi Marie
Poulin, Christine Elizabeth
Radtke, James Thomas
Rohde, Justin T.
Rush, Charles Marshall III
Schreiner, Christopher Alexander
Schulz, Ryan David
Scott, Benjamin Walter
Settineri, Gaetano Luca
Shaw, Zade Latif
Sherer, Bridgt Ann
Smith, Clayton A.
Smith, Matthew Craig
Sparks, Timothy Dale
Strohecker, Seth Michael
Talbert-Goldstein, Benjamin Alex
Tan, Lionel Zhen Yu
Thomson, Ian
Uffelman, Daniel Patrick
Wang, Zheng
Weaver, Mark Graham
Wedde, Brandon Harry
Weise, Sarah Elizabeth
Willis, Garret M.
Wilson, David Louis
Wilson, Gregory Scott
Wolfe, Elizabeth Ann
Yamashita, Yusaku

August 2006
Jennings, Thomas

December 2006
Ali, Hadi Walid
Andrews, Nicholas
Batema, Dane
Bittner, Craig
Braun, Jonathan
Brawner, Ashley
Capps, Drew
Dennis, Matthew
Fitzgerald, Cynthia
Fredericks, William
Fromm, Jonathan
Glim, Beau
Gordon, Ashley
Harvey, Matthew
Hatem, Miles
Helderman, David
Herbertz, Norman
Jones, Jesse
Kim, Ki-Bom
Kneitz, Henry
Koch, Mark
Lossman, Matthew
Martin, Andrew
Miller, April
Moeller, Jacob
Mulligan, Ryan
Murphy, Christopher
Negilski, Matthew

Patel, Ravi
Pearcy, Samantha
Roman Asyn, Patriia
Ruic, Ashley
Schinder, Aaron
Schleucher, Kyle
Selby, Christopher
Sochinski, Nicholas
Tapee, John
Trafton, Tara
Trager, Joseph
Tran, Tung
Trivedi, Pinak
White, Stephanie

May 2007
Akaydin, Kirk
Anderson, Brandon
Aungst, Robert
Boruch, Slawomir
Boyer, Brian
Bush, Christopher Dane
Cotter, Kevin Richard
Cottle, Andrew Eugene
Cunningham, Mark Shelby
Darsono, Sumitero
Dauby, Boyce Heron
Drodofsky, Matthew James
Duncan, Sean Avery
Fay, Keith Anthony
Fox, Matthew Nicholas
Frey, Catherine Fenner
Gentz, Joel Christopher
Gohn, Nicholas Ros
Goppert, James Michael
Gray, Mattehw David
Guyon, Mathew Jordan
Haack, Lindsay Marie
Hagenbush, Charles Robert
Hancock, Richard C.
Hanssens, Elisabeth Anne
Hassan, Syed Afzaal
Higdon, Richrd Keith
Horst, John Mattie
Huang, Xing
Hutchings, Rick Thomas
Jamison, Benjamin Charles
Kassab, Stephen Raymond
Kim, Seung-Il
Kovach, Andrew Francis
Kubiak,Jonathan Matthew
Lewis, Matthew Ivan
Lobdell, Trent Mathew
Maguire, Daniel Patrick
Mazzarella, Adrian Nicolas
McNutt, Steven Lee
Md Ishak, Nizam Haris
Moonjelly, Paul Varghese
Odle, Jared Andrew
Oechsner, Stefan
Palumbo, Michael Joseph
Pane, Nathan Edward
Poliskie, Thomas Joseph

Pothala, Daniel
Rabczak, Paulina Anna
Raje, Akshay Prakash
Ricchio, Fredrick Joseph
Richter, Matthew Robert
Rodewald, Joshua Vaughn
Roth, Brian Christopher
Ryan, Kyle Patrick
Schmitt, Matthew Robert
Soyfoo, Adeel Mohummud
Stinson, David George
Studtman, Jeffery Kyle
Thornton, John P.
Tippmann, Jeffery Dwayne
Trivedi, Pinak M.
Vemulapalli, Kautilya
Vergara, Juan Carlos
Voss, Bethany Carol
Wagenbach, Phillip F.
Wampler, Brandon Loy
Wheeler, Justin Michael
Wilcox, Crystal Ann
Woock, Sean Ryan
Wooten, Breanne Kay
Wypyszynski, Aaron John
Zaubi, Alexander John

August 2007
Abdul, Mohammad
Dalton,Derek
Handa, Manish
Risley, Nicole
Scott, Ryan

December 2007
Altchuler,Joshua
Andrei, Alexandru
Ballard, Brett
Barker, John
Bianco, Michael
Byford, Dorothy
Candee, Elizabeth
Carlen, Christopher
Datta,Neelam
Estes, Alexandria
Fallon, Joseph
Ford, Colin
Gagnon, David
Garba, Jummie
George, Christopher
Hamlin, Adam
Harp, Nicholas
Hendricks, Lynn
Huang, Xin Zhao
James,Mark
Kowalkowski, Michael
Krieger, Andrew
Kumar, Atul
Kwan, Kevin
Kwon,Kibum
Lee, Brian
Lee, Julim
McRorie, Johnson

Meyer, Andrew
Nagappan,Nivas
Nam, Sangtae
Nishiie, Takayuki
Olikara, Zubin
Ooi, Abel
Pugh, Michael
Reinebold, Sara
Robinson, Lucas
Rogge, Courtney
Romanotto, Matthew
Shoemaker, Benjamin
Tassan, Sara
Valentine, Michael
Wallintin, Gail
Watt, Jacob
Yip, Alvin

May 2008
Acker, Brian
Bearman, James
Beasley, John
Beckett, Christopher
Berger, Andrew
Bluestone, Stephen
Breitnegross, Scott
Briden, Amanda
Brinker, Andrew
Bryan, Nicole
Bryson, Dean
Budzinski, Brian
Chaney, Albert
Childers, David
Conrad, Matthew
Darby, Jason
Dias, Joshua
Doyle, Jessica
Ferris, Bradley
Fosler, Matthew
Gershkoff, Brian
Guo, Kuo
Guzik, Allen
Harkness, Elizabeth
Heacock, Daniel
Henrich, Joseph
Henricks, Lance
Hinton, Richard
Horney, Patrick
Imel, Paul
Izzo, Steven
Kanehara, Junichi
Kayser, Matthew
Krupski, Peter
Lattibeaudiere, Dana-Dianne
Lynam, Alfred
Markee, Laura
Mason, Joshua
Mckenna, Seann
Mehmedagic, Amir
Milmoe, Ryan
Mody, Pritesh
Norris, David
Noth, Kyle

Olsten, Jonathan
Paladino, Robert
Rosin, Jamie
Schoenbauer, Jessica
Schwing, Alan
Shoemaker, Sarah
Shurn, Stephan
Smith, Aaron
Spoonire, Ross
Strauss, Christopher
Teixeira, Vincent
Tucker, Camrand
Wahl, Elisabeth
Waite, Adam
Waltke, William
White, Brandon
Wilcox, Nicole
Wilde, Dennis
Woods, Alexander
Yaple, Danielle
Zott, Jayme

August 2008
Chua, Daniel
Darlage, Colby
Lobo, Kevin
Mitchell, Kathryn
Sutton, Christopher
Washington, Brandon

December 2008
Adams, Theodore
Balta, Jerald
Bhise, Sean
Bociaga, Michael
Breeden, Daniel
Brite, Kyle
Catania, Phillip
Chan, Shu Sum Sumkie
Creighton, Ford
Cunningham, Bret
Deal, Matthew
Donahue, Kyle
Downard, Brian
Foor, Andrew
Grider, Nicholas
Guzman, Oscar
Haas, Dustin
Hanrahan, Christopher
Heath, Christopher
Horan, Thomas
Jackson, Jennifer
Kamaruddin, Ahmad Faizal
Kane, Molly
Kelsey, Nathan
Kim, Hayne
Kirkegaard, Jonathan
Lash, Ryan
Leahy, Patrick
Lorenzana, Timothy
Ma, Timothy
Manship, Timothy
Mitchell, Kurt

Morris, Stephanie
Okon, Shira
Pfeil, Mark
Predis, David
Price, Andrew
Procelli, Michael
Rhodes, Justin
Robinson, Aaron
Salazar, Gerardo
Seagle, Jonathan
Simerly, Stephanie
Stewart, Karrie
Stuart, Jeffrey
Stump, Jennifer
Trouw, Ruan
Vazquez, Nicholas
Walker, Michael
Wirth, Jason
Wolf, Erik

May 2009
Aitchison, John R.
Akgulian, Kara M.
Appel, Bradley C.
Ashok, Akshay
Azmi, Muhammad F.
Barney, Diane J.
Barrett, Donald A.
Blockton, Zarinah C.
Brown, Levi A.
Burkett, Jeffery A.
Cofer, Anthony G.
Coffey, Michael T.
Coughlin, Jonathon A.
Damon, Andrew R.
Day, Stephen A.
Dixon, John M.
Elmshaeuser, Joshua L.
Fakhry, Adham A.
Freeman, Gregory M.
Fulford, Marques A.
Gedrimas, Jacob G.
Glover, Mark J.
Grimes, Andrew G.

Guiles, Mark A.
Gurtowski, Nicholas L.
Halsmer, Thaddaeus I.
Ho, Motohide
Huff, Victoria C.
Huffman, Gregory D.
Kalaria, Poorvi C.
Kim, Mintae
Knowlton, Jeffrey M.
Kolencherry, Nithin J.
Lange, Nixon G.
Le Mond, Korey A.
Leffel, Kelly R.
Lehto, Ryan M.
Lincoln, Kevin M.
Loveless, Bryan L.
Lukasak, Joshua S.
Maire, Romain P.
Marrinan, Patrick M.
Mc Kay, Caitlyn A.
Mc Peake, Michael
Mizener, Andrew R.
Muller, Trenten L.
Nelson, Ryan J.
Palmer, Tara A.
Ramaiah, Sanjeev Kumar
Rebold, Timothy W.
Scheid, Jared B.
Skare, Steven E.
Smith, Eric J.
Tanvir, Saad
Troy, Christine L.
Vandenburg, Jessica E.
Vandendriessche, Andrew M.
Waletzko, Brittany M.
Wang, Richard Y.
Weber, William T.
Westerman, Solomon K.
Whiteman, Alexander T.
Yang, Jack
Zander, Michael E.
Zaseck, Christopher R.

August 2009
Knowlton, Jeffrey
Wiraatmaja, Dodiet

December 2009
Alban, Cory
Ang, Wei Phing
Bartholomay, Meghan
Bodey, Skyler
Buchanan, Christopher
Butler, Austin
Chargualaf, Jay
Christopher, Michael
Cooney, Patrick
Duquette, Timothy
Erson, Brian
Hadsell, Karen
Hong, Ji Hye
Huseman, Jonathan
Klein, Gregory
Kuhlman, Kevin
Lee, Seung Min
Lee, Shian
Leppanen, Curtis
Markoff, Joseph
Meier, Jon
Mikolaj, Elizabeth
Miller, Jonathan
Nguyen, Tuan
Pryzgoda, Nicholas
Sharkey, Matthew
Steck, Kevin
Stern, John
Stohler, James
Stover, Matthew
Thomas, Stephanie
Valenta, Richard
Young, Kyle

BS IN AIR TRANSPORTATION (1947-1959)

February 1947
Bryan, Emory Burns
Burkholder, Howard Carroll
Chestnutt, Robert Walker
Herbster, Constantine Luther
Ritchie, Harry Russell Jr.
Sogge, Richard Chadwick
Todd, Robert Stewart

June 1947
Armstrong, Robert Emerson Jr.
Bischoff, Richard Olaf
Bishop, William Jones

Bridges, Richard Warren
Cole, Luther Heilhecker
Corbell, Louis Manen
Dreyer, Fred Marcus
Garlic, William Loren
Hake, Murray
Harper, Joseph Doyle
Harpster, Willis Dale
Hinderks, Wardell E.
Laughery, Robert Bruce
LeMaster, Jean Richard
LeRoy, Charles Hayden Jr.
Linn, Wendell Alton

Litle, Robert Forgie Jr.
McKinstray, Joseph Lewis
Meyer, Frederick Eugene
Miller, Paul William
Murphy, Lewis Jr.
Newlin, Robert Jack
Pedersen, Norman Elwood
Petro, Gloe Edwin
Phillips, Richard Doran
Rodgers, David Hedden
Samuel, Arthur Howe
Sanderlin, Lloyd Winfree
Shama, Habbib Rex

Thompson, Wayne Bascom
Trask, James Edward
VanGundia, William Lee
Wanless, Francis Marion
Williamson, Marion Roy Jr.

August 1947
Baumbartner, Leslie Arthur Jr.
Dace, Fred Walton
Neal, Richard Lee
Osborn, DeVerle Ross
Primeau, Louis Joseph
Sprowl, Arthur Vernon

February 1948
Andersen, Richard Jay
Anderson, Byron Lesley
Anderson, Willis Haldyn
Becker, Richard Anthony
Bee, Thomas George
Bosworth, Everett S.
Boyum, Ambrose David
Brady, Louis Philip
Bratt, Albert Verner Jr.
Brown, Charles Eldridge
Burwell, Richard Eugene
Canning, Glenn Donald
Chambers, Eugene Wesley
Chell, George Warren
Costin, Thomas C.
Fearn, Dale Hughes
Forshee, William Alva Jr.
Herb, Dona Jane
Hook, Fred Joseph
Jantsch, Albert Ferdinand
Koerner, William Frederick Jr.
Kohl, John Franklin
Matthews, Harry Hargan
McKelvey, Collins Mead
Nellis, Donald Joseph
Oyler, James Ross
Rench, John Junior
Scal, Richard Copeley
Scott, James Nelson
Seeger, William Francis
Stewart, Robert Lacey
Stoune, John Austin
Studabaker, Hugh Donald
Taylor, Martin Watson
Urban, John Henry
Wiley, Thomas Allen Jr.

June 1948
Allison, Irving Maloy
Anderson, Earl George
Asbury, Richard Trask
Benedict, Thomas Milton
Bennett, Donald Cecil
Bettis, Loyd Malcom
Biechteler, Carl Ernest
Bigelow, Eugene Ashley
Blaauw, Harold Arthur
Boyd, Alvin Landren

Bristol, Harry Haulenbeck Jr.
Brown, Oliver Floyd
Christena, George Howard
Clay, Thomas Charles
Cline, Maurice Dwane
Cooley, Kenneth Charles
Cyr, Roderick Joseph
Dimond, Horace Everett
Dreisbach, Thomas Swope
Fenton, Thomas Aquinas
Fricke, Lawrence Paul
Frisina, Fran Richard
Garrett, Wendell Grey
Gelwicks, Johnston Ellsworth
Gilbert, Robert Beach
Haher, John James
Hanson, Emory Scott
Harry, James Bolitho Jr.
Harvey, Robert Parker
Hawthorne, William Carl Jr.
Hoskins, John Wallis
Lehman, Roger Williams
Maass, William George
Martinson, Donald William
Miller, Robert Earl
Norris, Paul Wesley
Parrish, Henry Grady Jr.
Pierce, Edward Roland
Pittman, Robert Jackson
Reineck, Raymond E.
Rood, Noel Francis
Scace, Charles Edward Jr.
Schliesser, Phillip Duane
Sprinkle, William Carroll
Stanley, John Kevin
Staton, Willis Howard
Stone, Robert Mac
Trebilcock, James Harry
Vaughan, John Wesley
Vining, Car Louis
Wagner, Philip Emerson
Watson, James Broadus
Whitlock, Wayne
Woodall, George William
Youngren, Harold Lincoln

August 1948
Amberson, Robert Charles
Cooning, Leon Joseph Jr.
Cosler, Robert Douglas
Graber, John Franklin
Lawton, Gerald Benson
McCorkle, John Walter III
McGinley, Andrew Dodds Jr.
Miller, Julius David
New, William Neil
O'Mahoney, Robert Michael
Quebberman, Lawrence Adam
Singh, Prithipal

February 1949
Agnew, Robert Lyle
Anderson, George Arthur Jr.

Boretsky, Clarence Louis
Burgess, Seth Edward
Carlson, Emory Rolland
Cotton, Floyd E. Jr.
Crone, Dale Merle
Czuczko, Michael Nicholas
Donovan, Robert Lewis
Edwards, William Correll Jr.
Evans, William Buck
Fitzsimmons, John Wayne
Hamilton, Ralph William
Hamlin, Ross Eldon
Kroh, Charles Freeman
Leary, Raymond Abbott
Lehman, Ralph Stanley
Loveless, Harry Edward
Moehle, Ivan Lloyd
Mullen, William Robert
Olson, Marvin Edward
Perry, Robert Ackerman
Peters, William Joseph
Reed, Robert Johnson
Seitz, Leo Raymond
Shepherd, Noble Nye Jr.
Snyder, Elemuel Max
Stecker, Richard Henry
Sticka, Virgil Louis
Stockdell, Victor Bruce
Strasburg, Roger Earl
Sutter, Robert Adam
Thomas, Arthur Daniel
Thomas, William Kerr
Townsley, James Robert
Ulrich, John Joseph
Walker, Roger Wellington
Ward, George J. Jr.
Webster, Grove
Winkelhake, Donald Charles
Yates, Robert Warren

June 1949
Acker, William James
Angrick, Harry Francis
Biggs, Richard Merle
Boldt, Walter Cyril
Boyle, Thomas Doctor
Cavell, Matthew Stuart
Delahanty, Ruth Elaine
Duerkop, Robert William
Eggers, Richard Frederick
Elliott, Carl Edward
Epstein, Monroe Perry
Findley, John Nolf
Franklin, Claude Truman
Fredericks, Jack Hudson
Hantzsch, Ralph Eugene
Hart, John William
Hellstein, Theodore Joseph Jr.
Howard, William Shultz
Joyce, Joseph James
Karolich, John Herbert
Kennie, Frederick John
Koehler, George John

Kurtz, Harrison David
Kurtz, Howard Edward
Mathena, Edward Hanna
May, Charles Dudley
McMahon, Kevin Francis
McMurry, Robert Dean
Mikelsoh, Dwane Gail
Minton, Joseph Paul
Mullin, David James
Norwood, Richard Case
Proctor, Roy Vance
Roth, William Ralph
Saul, Warren Elmo
Sexton, Burton Hathaway
Sisto, Oronzo Rosario
Strack, George Wellman
Sweeney, Edwin Francis Sweeney
Taylor, Harvey Joseph
Tropp, Henry Seymour
VonGoey, Henry Richard
White, Roy Jr.

August 1949
DeTamble, Richard James
Ferrell, William Keith
Foster, Dudley Woodbridge
Goff, Robert Arthur
Hurt, Doris Mildred
Kuehn, Richard Leonard
Lefevre, Ralph Lee
Maas, Louis Marshall
Mason, Donald Thomas
Pratt, Charles Winship
Ritchie, Edwin Sims
Schnedler, Albert August
Toth, Charles Jr.

February 1950
Caldwell, Bruce Draper
Cearing, Jack Russell
Coppage, John Fielding
Davis, Leroy Gene
Denny, William Howard
Doty, William James
Doyle, Robley Eugene
Gilbert, Ralph Louis
Hagen, Rex Bernard
Hanson, Jean Vandervort
Hinlicky, George Joseph
Joyce, Dale Hugh
Kusak, Nick
Layne, William J.
Luckas, Edmund
Margindale, John Henry Jr.
McGee, John Joseph
Nielsen, Rolf
Odell, Gordon Wendell
Oetting, William Frank
Parry, Addison Julius
Porter, Robert Leslie
Sides, Jack Fremont
Smith, Albert William
Stoughton, Chancey Ellsworth

Thayer, Allen Birney
Ward, Richard Carleton
Yarber, Charles Joseph

June 1950
Beck, George Eugene
Canfield, Jacqueline Jane
Carahan, Grove Cleland
Cassady, Edward Weeks Jr.
Castro, Ricardo Jaime Jr.
Ciambrone, John Pat
DenUyl, Dean
Foster, Harry Lawton Jr.
Frederick, Kenneth Robert
Gilbert, George Taylor
Herr, Lloyd Edward
Hiatt, James LeRoy
Hogsett, Richard Hadden
Jankowski, Raymond Augustine
Kovach, Cornelius Anthony
Lewiecki, Stanley Francis
Little, Hobart Meredith
Mager, George LeRoy
Matney, Noel Madison
McKay, Desco Esthal
Randall, William Coahran
Riffe, Charles Ward
Ring, Robert Elden
Scalf, Richard Leon
Schmitt, Stephen Peter
Shapen, George Theodore
Smith, Douglas Roddy
Tharp, Donald Eugene
Walker, Howard Curtis
Weiss, Marvin
Young, William Benjamin

August 1950
Blankenship, Wallace Preston Jr.
Colvard, Harold Louden
Elliott, Franklin Wisener
Freeman, Richard Dean
Mack, Philip George Jr.
Moore, Ernest Lee Roy
Selman, Clifford Gene
Washington, Thomas Lackland

January 1951
Bell, Gene Edward
Berghorn, William Charles
Carter, Albert LeRoy
Contompasis, James Charles
Covington, William Schau
Deluca, Joseph Alfred
Droom, David Bernhard
Gredy, John Henry
Hodges, Harold Lewis
Howard, Robert Arthur
Malmfeldt, Gordon Elmer
McPartland, Hugh Joseph
Peralme, Austin
Reichenbach, Calvin Hugh
Snyder, Robert J.

VanGorp, Peter Harve
Wright, Jason Carl

June 1951
Adams, Berkley Ernest
Bare, Earle Henry Jr.
Baumgardner, Kemit Paul
Bellish, Arthur Daniel
Britton, Samuel Lew
Carswell, John Davidson Jr.
Chaddock, Frank H. Jr.
Cochran, Thomas Walter
Cole, Merrill Lee
Cunningham, Richard Leon
Fass, Franklin Roderick Fielding
Fistori, Philip Arthur Jr.
Hallum, Augustus Felton
Hollingsworth, Rita May Rickard
Huffington, James Everett
Kiesel, Robert John
Koogler, Floyd Franklin Jr.
McBride, Elizabeth Ellen
Melloncamp, Walter Martin
Olson, Jack Taylor
Pavelka, Jerry Frank
Ravinet, Ernest
Reinecke, Herold Hess
Ricke, William Henry
Robinson, Samuel Martin
Rodriguez-Torres, William F.
Sekadlo, Roger George
Selenko, John
Simonin, LaVerne Delos
Sorenson, Robert William
Sowinski, Chester Edward
Spargur, William Burge
Umpleby, Arthur Norman
Vovolka, John Jr.
Waddell, Raymond Wallace
Wagner, Edward James
Wallace, Andrew Laird
Weigand, Frederick Arthur
Welker, Russell Gene
Welton, Ralph Ervin Jr.
Whiteway, Roger Earle

August 1951
Elliott, Dennis Clinton
Gentry, Robert Richard
Holmes, Kenneth Allen
Jordan, Kenneth Harlin
Krupsaw, Manuel Stanley
Smith, Grover Emerson
Stidham, John Walter
Whitlock, Robert Lee

January 1952
Andresen, Robert Henry
Athanasulis, Paul William
Bajwa, Mohammed Aslam
Bloch, Robert August
Brandt, Kenneth Herbert
Butler, Ludlow S. Jr.

Dora, Maurice Reed
Hunter, James Ross
Irwin, Russell Gene
Kelly, Graham Malone
Krawiec, Roman John
Kucaba, James Frederick
McCall, David Scott
McCarthy, Walter Bertrum
Norem, James Edwin
Norris, Marshall David
Samsen, Robert Eugene
Tate, Ralph Jr.
Thomson, Jack Richard
Watt, John Edward

June 1952
Beenaus, Albert, Raymond
Bloomstein, Elias
Brewster, Vernon Harcourt Jr.
Carmichael, Keith Lee
Carroll, Richard Martin
Clingenpeel, William Arnold
Clones, Nicholas John
Crane, James Winslow
Cunningham, John James
Dillingham, Wallace James
Eiler, T. Richard Jr.
Falta, Jerome Francis
Hilderbrand, Richard Eugene
Jack, Charles E. Jr.
Jones, Robert Warren
Lange, Harry Ronald
Leonard, Walter William
Marshall, John Edward
McElheny, Ralph Arthur
Monfort, Elias Riggs III
Moses, William Franklin Jr.
Neuman, Melvin Eugene
Ormesher, David Thomas
Paul, Richmond Walker Jr.
Pearson, Richard Gustave
Suite, Max Huston

August 1952
Ashley, Jackie Lee
Bohls, Allen Howard
Carter, Robert Charles
Eckman, Hanford Louis
Jordan, Jack Reid
Tuesburg, William Claude

January 1953
Brockman, Charles Edward
Fleming, Raymond Thomas Jr.
Harms, Jack Wert
Hiernaux, Richard N. Jr.
Raab, Ronald Buxton
Rich, J. M. Sanford
Scott, Thomas Lindsay
Sobas, Leonard, John
Stanley, Russell Curtis Jr.
Stephens, Carl William

May 1953
Able, Warren Frederick
Bean, Neil Tyler
Bushong, Brent
Buxton, John Alan
Campbell, Robert Edward
Eggerman, Wesley Vernon Jr.
Ferringer, Lawrence Henry
Fisher, Richard Byron
Gilson, Arthur Scott III
Gudeman, Alan Keith
Hertwig, Lee Joseph
Johnson, Charles William Jr.
Johnson, David Lee
Kennedy, William George Jr.
Lewis, Harold Over
Myers, Martin Lee
Pedjoe, John Paul
Samuels, Oren DuBois Jr.
Schroeder, Richard Leonard
Smith, Edward Everett Jr.
Smith, Robert Nicholas
Taylor, Nancy Ann Kiebler
Townley, Lloyd Moreland
Tuchek, John Bruce
Vehling, David Read
Wertenberger, James Wendell
Wiggins, Arthur Brenton
Winks, Donald Keith

August 1953
Gyorgyi, John Carl
Holmes, William Harrison
Kravet, Robert David
Meade, Jack Sommers
Pesaturo, Emil William Jr.
Silvernail, David Hodgman
Strohl, Thomas Merrill

January 1954
Crouch, Claybourne Addison
Ditzler, Worden Lodge
Dosmann, James Arnold
Grahn, Robert Allen
Ingersoll, Norwin Paul
Ishimoto, Norman Masao
Kozumma, Roger Tadashi
Raymond, Murray Roscoe Jr.
Smedley, William Eugene
Smith, David McKinley
Stuart, Edgar George
Stugart, Herbert Kurt
Weichbrodt, Wilbert Carl
Zergiebel, Charles Locke

May 1954
Buchwald, Graeme
Corrigan, Alvin Raymond
Edwards, Ronald Lee
Fender, Charles Marion
Ferrell, Billy Gene
Finch, Kenneth Eugene
Freeland, Edward Russell

Harmon, Robert Eugene
Jakimcius, Zigmant Aleksandras
Klinker, Robert Stephen
Kozminski, Donald Frank
Krisciunas, John Paul
Larsh, Robert Keith
Morris, William Franklin
Newkirk, Irvin Richard
Rich, John Lewis
Sell, John William
Soughers, Allen Perry
Walters, Gary Lee
Winkler, Maurice Frederick Jr.

August 1954
Balter, Wesley Dean
Dotterer, Richard Travis
Johnston, Gregory Roger
Ramsdell, Richard Gardner
Roberts, James Elwood
Waite, Lawerence Albert Jr.
Zwick, Stephen Louis

January 1955
Alford, Joanne
Brauer, Walter Henry
Carlson, David Alton
Gochenour, Jack Williard
Grassan, George Donald
Green, Stanley Jack
Miller, Jack Lee
Pate, Roger Calvin
Scales, Charles William
Tillotson, James Allen Jr.

June 1955
Baas, Michael Robert
Bauerle, John William
Beaven, Frank Newell Jr.
Beaver, David Lee
Clegg, David Howard
Cross, Jerry Lee
Dills, Teddy Eugene
Dudley, Richard Owen
Ginn, Robert William
Hruskovich, Roger John
Johnston, Lowell Eaton
Kauchak, William Earl
Kester, Walter William Jr.
Knepper, Robert Rulo Jr.
Lawrence, Robert Perry
McGarry, Donald Patrick
McLean, Robert Burton
McNally, Edward Russell Jr.
Savery, William Richard
Schnabel, Robert Harold
Schubert, James Joseph
Steffey, Richard Gerald
Workinger, Thomas Gene

January 1956
Britt, Frank Allen
Darby, Ralph Edwin Jr.

Jones, Robert Peter
Kapsalis, Thomas
Monger, Robert Gerald
Sandoval, Jaime Peredo
Reid, Arend Harris

June 1956
Ackor, William Russell Jr.
Austin, Roy Frank
Balis, James Daniel
Beckman, Norbert Joseph
Benjamin, David Walsh
Burritt, John Christopher
Castell, John Stanley
Cho Chun, Benjamin Kwai
Cook, Augustus, Dillon
Denney, Deryl Louis
Fait, William Jr.
Fowler, George Stephen
Geib, Richard Gail
Greiner, Calvin John
Groover, Stanley Lee
Henn, Robert Lee
Hounshell, William Harrison
Kapsalis, William
Klein, Austin David
Knollmueller, Joe William
Lowe, Thomas Howard
MacGregor, John Lindsay
Maloney, Richard Arden
Markl, Arthur Anthony C.
Marunde, Glen Edward
Mishler, David Lee
Mishler, Roger Gene
Nason, George Horace
Peck, Victor Orland
Peters, Donald Charles
Preucil, Ronald George
Robinson, C. B. II
Rutkowski, Robert Louis
Schultz, William Clyde
Sells, Robert Richard
Stabler, Allen Wayne
Titus, Thomas Richard
VanSickle, Ramon Loren
Woo, Chi Hyong
Zimmerman, Robert Earl

August 1956
Hadley, Richard Martin
Harper, James Loren
Hunt, James Lewis
Matin, David Ray
Musgrave, Charles David
Warnel, Charles Eugene

January 1957
Bolles, William Robert II
Burns, Robert Leo
Campbell, Robert Loy
Cleaver, Harold Thomas
English, Walter Myron Jr.
Jecha, Ronald Rudolph

Johnston, Edwin Jr.
Kelly, John Rodman
Loeffler, Edward Herman
Lohss, Wilbur Frederick
McDaniel, Billy Blaine
Mills, Orton Leroy
Newman, Max Edward

June 1957
Bailey, Charles Edwin Jr.
Bidwell, John Cushing
Bigham, Robert Terrence
Brown, George McCullough
Church, James Matthew
Clark Marvin Donald
Denneny, James Ade Jr.
Dillon, Donald Dale
Drummond, James Albert
Eberhardt, David Howard
Eodice, Peter Phillip
Etchison, Jack Pruitt
Fleming, Donald DeHass
Gardner, Eugene George
Green, William Clayton Jr.
Hagberg, Charles Paul
Hardwicke, Roger Mikhail Smith
Hepperleln, Harry Michael III
Hesterman, Dale Carl
Laden, Thomas Ray
Lane, Richard Eugene
McCabe, Barbara Joan
Meyers, Thomas Miller
Mohardt, John Harrison
Morton, James Douglas Jr.
Myles, Bruce Larsen
Roberts, Rufus Weber
Southerland, James Francis
Stroud, Robert Earl
Voras, Albert Henry
Wares, Dennis Allen
Wasson, John Nichols
Webster, Donald Spencer
Will, Daniel William
Zutavern Jerry Bruff

August 1957
Baffer, Norman Benjamin
Barnes, David Earle
Fisher, Thomas Allan
Harmeyer, Harry William
Humphrey, Joseph Dell
Jones, Alan Stanley
Lee, Raymond Dale
Polsky, Robert Harvey
Schwartzkopf, Lynn Arthur
Wert, Donald Ellsworth

January 1958
Beckerich, James Victor
Blagg, Harold Dean
Dunbar, Willard Parker Jr.
Gray, Harold, Michael
Hooper, Richard Keith

Lester, Richard Lee
Loomis, Dean Allen
Marshall, Philip Ray
Milligan, Ronald Dean
Pahmeier, Max Clarence
Reed, Philip B.
Riggs, William Crowder
Siefers, Jerry Lee
Siegel, Thomas Edgar
Simon, David William
Stoudt, Frank Larry
Sunkes, James Allen
Talbott, Robert Paul
Yauch, John Alden
Zeller, Mathias Thomas Jr.

June 1958
Cave, Joseph T.
Craig, Barry Somerville
Deming, James Edwin
Duesler, Robert Laingen
Dunham, Warren George
Eaton, John William
Griswold, Charles Joe Jr.
Heller, James Kirby
John, Lowell George
Kowal, William Thomas
Larkin, Lawrence Bernard Jr.
Linder, Robert Joseph Jr.
Morrissey, Douglas Michael
Shook, William Hanley
Weber, Thomas William
Williams, Ralph Whitson
Wilson, Nathaniel Jr.
Young, Delmar Dirk

August 1958
Hindmarch, John Canfield
Monahan, Robert Edward
Puzey, Robert Leavitt
Willen, James Merrill

January 1959
Farell, David Allen
Lincoln, Raynard Chalcroft Jr.
Logue, William Ralph
Miller, John Charles
Wilken, Herbert Joel

May 1959
Gasper, Ralph Louis

BS in Engineering Sciences (1957-76)

June 1957
Bernard, Michael Charles
Brown, Robert Earl
Commons, Theodore Louis
Fries, Richard Edward
Greiling, David Scott
Hering, Bruce Richard
Hesch, Dale Herbert
Hinchman, John Ralph
Jezik, Robert James
Lescinsky, Frank William
Offhaus, Richard Charles Jr.
Ottlinger, John Alexander
Prozan, Robert Joel
Winje, Robert Andrew

January 1958
Rauch, George Bradford Jr.

June 1958
Bender, Henry Edwin Jr.
Black, Charles Emory III
Clark, Alfred
Fleig, Albert Joseph Jr.
Gallagher, Alan Charles
Gibboney, James Gustin
Hosack, Grant Austin
Noble, Thomas Franklin
Steele, Jack Ronald
Swanson, Alan Arthur
Taylor, Douglas Manning
Vestal, Marvin Leon
Weeks, James Allan

August 1958
Ottaway, Robert William

January 1959
Ritchey, James Franklin
Williams, Stewart John

May 1959
Beckwith, Neil Evan
Bernstein, Edward Linde
Bloom, Joseph Morris
Brinsley, James Rothwell
Cooper, Leonard Yale
Dalzell, Donald Richard
Davidson, Allen Roger Jr.
Davidson, Michael Loughran
Dunlap, Ronald William
Fowler, Burl DeWitt
Fredrickson, Fritz Arne
Graff, Karl Frederick
Kennedy, William Stoddert
Lahs, William Robert Jr.
Lynch, Thomas Henry Jr.
McGovern, Joseph Daniel Jr.

McLean, James Alfred
Miller, Charles Eugene Jr.
Morgan, George Everett
Peale, Stanton Jerrold
Porter, James Colegrove
Scanlan, John William
Scheuneman, John Herman
Schwartz, Nate Jay
Shirley, Leland Kimble
Shoup, Gordon Scott
Silverman, Edwin Barry
Smith, Robert Earl Jr.
Thebault, John Robert
Tults, Peter Priit
Willingham, Donald Edward

June 1960
Allen, Robert Thomas
Blair, John Milton
Bratkovich, Richard Nicholas
Burwell, William Howard
Carrel, Donald Clark
Carson, William David
Fellers, Jon Ian
Fetz, Bruce Henry
Hauck, Clifford Albert
Huffman, Ronald Ray
Isenburg, Jerry Eugene
Janus, John Peter
Jensen, Douglas Warner
Johnson, Roy Arnold
Kingsley, Stuart Charles
Lucas, Jack Miller
Mathias, Richard Alan
Meier, Lewis III
Mitchell, Richaed Rowe
Niemeier, Byron Matthew
O'Brien, Robert Andrew Jr.
Palmer, James Thomas
Powers, Edwin Lowell
Rehm, Ronald George
Severance, Alan Wayne
Sherwood, Bruce Arne
Spade, Jerry Lee
Sturhahn, Robert Joseph
Summers, John Oliver
Webster, David Charles
Whitney, Myron Edward Jr.
Willett, Thomas Joseph

August 1960
Dubren, Sherwin
Mearns, Richard Lee

January 1961
Niemann, Arthur Frederick Jr.

June 1961
Bergstedt, Lowell Charles
Brandon, Nancy Lou
Coulter, Lawrence Joseph
DeLaCroix, Robert Frank
Doolen, Gary Dean
Gerardo, Dominic Francis
Grot, Richard Arnold
Hawk, James Gregor
Hill, Donald James
Hofferberth, James Edward
Jason, Richard Harry
Linnerud, Harold Jacob
Manlief, Scott Kieth
Marrone, James Irving
Miller, Merlin Glenn
Nordgren, Brett Marcus
Olson, David Gardels
Podney, William Nicholas
Puterbaugh, David Ward
Ransburg, David Preston
Rollins, Roger William
Rosfeld, Howard Bernard
Shanabarger, Mickey Rock
Siepker, Frank Groves
Stein, Arland Thomas
Stiller, Thomas Michael
Wallace, Ronald Kent
Williams, Richard R.
Yung, Benjamin
Ziemer, Charles Omar

August 1961
Lang, George Gordon
Prezbindowski, David Leon
Remde, Richard Hugo
Tomlinson, Robert William
Vadeboncoeur, James Ronald
Welch, James Douglas

January 1962
Ashbaugh, Noel Edward
Crozier, Alan Keith
Dilger, Joseoh William
Gordon, Curtis Lee
Hufford, Arthur Allen
MacGregor, Ronald John
Schmutzler, David Lee
Strange, Thomas Ivan
Theil, Charles Conrad Jr.

June 1962
Bechtel, Jon Harold
Clark, Ronald Ray
Delutis, Thomas Gregory
Drobish, Robert Gary
DuChateau, Paul Christian
Euler, James Alfred

Gersting, John Marshall Jr.
Harasty, Gerald Andrew
Jones, Michael Paul
Lau, Joseph Po-Keung
Lindsey, Thomas Harold
Lotz, Robert
Munson, Bruce Roy
Ring, Robert Clinton
Rogovein, Lawrence Jacob
Sandifer, Jerry Brown
Schneider, Bill Edwin
Schwiesow, Ronald Lee
Thimlar, Merlin Eugene
Thomas, John Howard
Truitt, Robert Blair
Wachter, George Thomas
Weber, Frank Robert
Wisehart, Kenneth Martin
Zung, Lawrence Bei-yu

August 1962
Hendey, Edwin Van
Jacoby, Jimmy Kay
McKay, Clair Albert
McKinley, Edward Lynn

January 1963
Bauer, John Joseph
Booth, Thomas Cob
Craven, Kenneth Allen
Eichler, Thomas Vaughan
Hines, Kenneth Wayne
Hugus, Jack William

June 1963
Bailey, David Arthur
Blum, Joseph William
Coy, Brent Eugene
Criswell, Michael Leland
Curtis, Howard Duane
Dreblow, Glenn Richard
Ehardt, John III
Gardner, Richard Allen
Graham, Thomas Hawkins
Green, Ivan Dale
Hahn, Edwin William
Hendicks, Charles Leroy
Jackson, Thomas Diack
Jensen, Ronald Roger
Joyce, James Michael
Kalan, James Edward
Klingsporn, Harold Allen
Landeck, Bruce Warren
Marbach, Terry Blaine
Pike, Herbert Edward Jr.
Ratcliff, Roger Allen
Schmitendorf, William Ernest
Skurchak, John Wallace
Spencer, Richard Frank
Stall, Richard John
Tucker, Stephen Bern
Wire, Charles Eugene

August 1963
Flack, James Robert
Johnson, David Harley
Scaramozzino, Phillip Joseph
Weaver, Charles Thomas

January 1964
Ford, Charles Thomas
Pickett, Frank Joe
Rains, Charles Porter
Sun, Pershing Bit-Sing
Wolsko, Thomas David

June 1964
Bache, Thomas Carter III
Beard, Richard Vernon
Beaty, Lolitia Frances
Brand, Timothy James
Brasseale, Kent Alan
Claus, William Davenport Jr.
Cramer, Bruce Armstrong
Doll, Theodore John
Dunn, William Elson
Eisele, John Andrew
Fein, Kenneth Joel
Haas, Mary Elizabeth
Habegger, Loren James
Hardin, Jay Charles
Harris, Jack Raymond
Heller, George Doscher
Hellman, Karl Harry
Hostetler, Robert Dalen
Hunsicker, James Alan
Kooker, Douglas Edward
Leet, Warren Edward
Liporace, Louis Anthony
Miller, John Thomas
Mlynarczyk, Francis Alexander Jr.
Paddock, Robert Maxfield
Pritchett, Michael Farlow
Salpietro, James Daniel
Schlueter, Roger Selig
Schwalm, Edward Albert
Schwartzkopf, Paul Kenneth
Skridulis, James Charles
Stevens, Leroy George
Vasicek, Daniel Joseph
Vogt, Joseph Paul
Wiley, John Richard
Wong, Felix Shek-ho

August 1964
Reed, Robert Alan

January 1965
Dane, Thomas William
Dasher, Randall Eugene
Neuzil, Peter Anthony
Panesar, Jarnail Singh
Teich, Leonard Max

June 1965
Bani-Asadi, Mohammad Hossein
Begeman, Paul Charles
Bissey, Stanley Ellis
Briscoe, William Frederick R. Jr.
Bruno, Richard Ormsby
Chapman, Richard Bruce
Connan, Jerald Mark
Darrah, Gillespie Blaine III
Deleget, James Leslie
Frey, William Hugh
Herderhorst, Robert Philip
Keiffer, Larry Lee
Kerber, Ronald Lee
Knapp, James Ronald
La Duc, John Thomas
Lindeman, Arthur John
Long, Parley Carper
McVoy, Charles Felter III
Montgomery, Harold Arthur
Mueller, David Walter
Negele, John William
Niksch, Gerald Louis
Oates, William Robert
Moghadam, Hossein Parsapour
Silvers, John Byron
Staiger, Peter Juergen
Tollinger, Jeffry Monroe
Waid, Donald Eugene II
Wiley, Robert Joe
Yoshimura, Allan Ken

August 1965
Miller, Malcolm Robert
Pfister, Carl John Jr.

January 1966
Gittemeier, Joseph Donald
Libster, Harold Lew
Lundberg, Gordon Keith
Myers, George Michael
Osburn, Carlton Morris
Reid, Robert William Jr.
Selzer, Kenneth Edwin
Starr, John Edward

June 1966
Carter, David John
Compani-Tabrizi, Behrooz
Cox, Arthur Morris
Craig, John Robinson
Davis, Kenneth Warren
Halkyard, John Edwin
Hoffman, James Cecil
Kafadar, Charles Bell
Leedom, Charles Melvin
McDonald, Michael Joseph
Mohler, Charles William
Reynolds, Peter Tee
Tonkin, Russell Joseph
Warren, Joe Edward

August 1966
Loehr, Richard William

January 1967
Avramidis, Stellios
Breakfield, Charles Dickerson
Elkins, George Henry
Luo, Siong Siu Tan
Mecredy, Robert Clark
Shields, David Warren

June 1967
Bailey, Robert Warren Jr.
Bird, John Albert
Bradshaw, Ross Edward
Burns, Lawrence Edward
Carroll, John Raymond
Carson, Charles Converse
Gault, James Richard
Gootee, David Paul
Hinzman, Harry LeRoy Jr.
Hollingsworth, Stephen D.
Kelso, David Matthews
Kidder, Ralph Eugene
McColgin, William Carey
Miller, Stephen Michael
Neidig, Stephen Lowell
Singleton, Harold Ross
Whicker, Donald

June 1968
Berger, Ronald Bruce
Blum, Dan
Carnahan, James Vernon
Curtis, Blaine Lawrence
Dye, Edward Carroll
Emigh, Leonard Craig
Fleischer, Lawrence Stephen
Hilkene, John Bradley
Loehlinger,Mervin William
Larsen, Ronald LeRoy
Laymon, Stephen Alan
Luck, Larry Ben
Maslyar, George Andrew III
McElwee, Jerry Wayne
Mellott, Forrest Dixon
Michaelsen, Roger Allan
Richards, David Duffield
Richeson, Kim Ervin
Ridberg, Michael David
Salo, Donald Roy
Shuppert, James Harland
Smith, Clifton Barton
Srnka, Leonard James
Tramm, Tom Robert
Waid, Phillip Alan
Woodworth, William M. O.

January 1969
Martin, Frank Jeffrey

June 1969
Aument, John Robert

Bever, Robert Charles
Blessing,William Douglas
Bole, Robert Earl II
Campbell, Don Merrit
De Lafosse, Peter Henry
Drinkut, Samuel Arthur
Ehardt, Thomas
Geyer, Howard Karl
Hehner, Andrea Allyne
Kaser, Michael Dee
Kayser, Kenneth William
Lam, Paul Chi-King
Lau, Wally Po-Wah
Riley, James Ward Jr.
Siefken, John Richard
Sittner, Eric Henry
Steury, Ronald Lee
Wagner, Stephen Everett

August 1969
Nagey, David Augustus
Sze, Nien Dak

June 1970
Armstrong, Michael Spaine
Brzezinski, Thomas Anthony
Davis, Wayne Joseph
Diehl, Roger Earl
Faller, Frederick Brice Jr.
Feil, Peter James
Fuller, James Leslie
Hedayat, Gholamali
Lenglade, Charles Richard Jr.
Mahan, John Edward
Malinowski, Terry John
McCann, Larry Dean
Meyer, Thomas George
Nelson, John Robert
Nippert, James William
Orban, David John
Richards, Riger Keith Jr.
Schoonover, Kenneth
Waite, Lynn Lewis
Whipple, Christopher George
Yue, William Wai

August 1970
Fillnow, Douglas Lee
Kahl, Michael Robert
Melrose, Richard Stockton
Neitzke, John William

January 1971
Garver, William Robert

June 1971
Altobellis, Richard M.
Champlin, Richard Eugene
Day, Douglas Bernard
Evans, Gregory Herbert
Farrell, Thomas Domonic
Faulkner, William Henry
Gilligan, John Gerard

Glaser, Robert Harry
Keller, Stuart Robert
McDonald, Thomas Austin
McWhirter, Wilford Jr.
Merriman, David Berry
Montgomery, Stephen Tedford
Norris, John Anthony
Roski, Richard Arthur
Shirley, Michael Alan
Sprandel, James Kimberly
Toth, Dennis Dale
Walker, William Francis
Watson, Steven carl
Wells, James Randolph
Yates, George Thomas

August 1971
Benson, Carl Douglas
Newsom, Donald Edward

January 1972
Murdock, Frank Kenniston Jr.
Sterner, Robert Charles

June 1972
Bare, James Edward
Brown, Richard Lee
Chivington, Steven Paul
Davis, Edward Libretta
Gable, Bruce Kent
Gartland, Eugene Charles
Grot, Arnold Stephen
Gyolai, James Joseph
Kaltwasser, Wayne Aurelius
Kelley, John Wilbur
Kervin, James Ernest
Khoo, John Man-Fong
Kopscick, George Michael
Krieger, William Fredrick
Lantz, Gregory David
McKim, Michael Lee
McKinnis, David Ray
Ongman, John Will
Reyzer, Kendall Charles
Rosquette, Hector Julian
Runge, Steven Michael
Vondersaar, Frank Joseph
Wesley, David Evan

December 1972
Espinosa, William Michael
Gurthrie, Gregory Robert
Kinsey, Brian Douglas
Madden Michael John

May 1973
Giloy, Douglas Edwin
Kaupish, Paul Max
Pitz, William John
Sand, Mark Evan
Schlicher, George Carlton
Uland, Michael James
Vandenberg, Denise M. Onchak
Workman, Stephen Keith

August 1973
Burns, Robert Donald III
Massena, Roy Phillip

December 1973
Cox, Jack Cecil
Paulson, Stephen Robert

May 1974
Baldwin, Mark Lewis
Bender, Edward Joseph
Bratkovich, Alan Wayne
Douglas, Robert Lewis
Lamberson, Steven Edward
Lawrence, Stephen Richard
Lawrence, Timothy Charles
Macal, Charles Martin
McFadden, James Robert
Minton, Michael Olan
Parmelee, William Gordon
Reeves, Thomas Wilmeth
Richards, Mark Prescott
Wolf, Eric John

December 1974
Graff, Dale Lawrence
Herald, Michael James
Jaap, Joseph Bradford
Sugawara, Yoshio

May 1975
Atkisson, Craig Fenton
Greenburg, Alan Morris
Julow, James Raymond
Kimpel, Charles Henry
Kuehne, Bruce Edward
Lathan, Donald Marce
La Vassaur, Gregory Joseph
Longwell, Richard Lee
Powell, George Douglas

August 1975
Howe, David Julin

December 1975
Knaebel, Michael Lee
Voss, Janice Elaine

May 1976
Amacher, Charles Patrick
Berry, Charles Brent
Crago, Steven Charles
Doligalski, Thomas Lawrence
Guernsey, Carl Scott
Hefner, Rick Dale
Tyra, Gerard Richard
Witten Mark Alan

August 1976
Gregory, Charles Ziegler Jr.
Riechers, John Thomas

APPENDIX B

MASTER OF SCIENCE DEGREES AWARDED & THESIS TITLES

Note: This appendix includes a list of master's degrees and thesis titles following formation of the independent School of Aeronautics on July 1, 1945. The first MS degrees offered by the new school were awarded in 1947. Graduates are listed below by the following degree titles:

1. **MS in Aeronautical Engineering (1947–1968)** *or*   Pages 386-394
   MS in Aeronautics and Astronautics (1968-2009)

2. **MS in Astronautics (1964–1968)**   Page 395
   This special degree program is described in Chapter 5.

3. **MS in Engineering (1967–2009)**   Pages 396-399
   Per Purdue policy, the MS in Engineering (MSE) degree is awarded to MS candidates whose BS degree is not in the same technical area as their graduate degree. While the School of Aeronautics and Astronautics has independent records of MSE degrees awarded after 1967, University records prior to this date do not distinguish MSE degrees by individual engineering school. Thus, it has not been possible to determine which MSE degrees prior to 1967 were granted to AAE students. A similar problem occurs with the master's degrees granted by the Engineering Sciences program. Those degrees are classified as MS in Engineering in University records, and are also grouped by the Registrar together with all MSE degrees (e.g. ME, CE, EE, etc.). Again, it has not been possible to determine a definitive list of MS graduates from the Engineering Sciences program. Engineering Science graduates after 1967 are, however, included below under the MSE category, although we cannot distinguish between AAE and Engineering Science students. We regret that we are unable to recognize the MS in Engineering (AAE or Engineering Science) graduates prior to 1967 in the following compilation.

4.   **MS Thesis Titles (1938–2009)**   Pages 400-424

Master's candidates who completed the thesis option are listed along with their thesis title and advisor. This list begins in 1938 with theses conducted under the Aero option of ME before the creation of the independent School of Aeronautics in 1945, and runs through degrees awarded by December 2009. Unfortunately, we do not have records of MS theses granted by the Division of Engineering Sciences before it merged with the School of Aeronautical Engineering in 1960.

MS IN AERONAUTICAL ENGINEERING (1947-68) OR MS IN AERONAUTICS & ASTRONAUTICS (1968-2009)

February 1947
Ayres, Langdon Ford
Richardson, Robert Lyle

June 1947
Boyer, Luther John
Boyer, Rodney Leonard
Brown, Sherwood Hulett
Corden, Carlton Donald
Head, Robert Eugene
Hedrick, William Sutter
Leake, Lewis Albert
Marshall, Edmund Valentine
Nelson, Robert Leonard
Rohert, Robert Edward
Rosenberg, Reinhardt Mathias

August 1947
Khan, Ahtesham Ali
McMurray, John Wightman Jr.
Povalski, James Albert
Vanderbilt, Vern Corwin
Zimney, Charles Marshall

February 1948
Spraker, Wilbur Allen Jr.

June 1948
Carpenter, Charles R. Jr.
Duncan, Richard L.
Nicholson, Robert Allan
O'Brien, Denis Henry Jr.
Scull, Wilfred Eugene
Wade, Merwin G.

August 1948
Gatineau, Robert Jean
Robinson, Robert Frank

February 1949
Keailis, Stanley Paul
Misener, Walter Stuart

June 1949
Carter, David Lawrence
Chapman, Rowe Jr.

Costilow, Eleanor Louise
Hsu, Hua Fang
Kordes, Eldon Eugene
Kovacik, Victor Paul
Maxey, Horace Hagood
Roberts, Robert Earl
Rodean, Howard Carroll
Savage, William Frederick
Sellers, John Paul Jr.
Skinner, Earnest Harold
Snell, Gale Elwood

August 1949
Akers, Marion Jessee
Bennett, Robert H. Jr.
Campbell, George Aberdeen
Fowler, John R.
Jackson, Charles Edward
Kuhlman, William Herman
Michaels, Franklin
Schmitt, Alfred Frederick
Thompson, Robert C.

February 1950
Klockzien, Vernon George
McComb, Harvey G. Jr.
McDermott, Alexander Mitchell
Sapowith, Alan David
Streed, Rodney Warren

June 1950
Chickering, John B.
Fellows, Walter S. Jr.
Heaton, Thomas Rush
Jones, Walter Paul
Knuth, Eldon Luverne
May, James Edward
Penn, William W. Jr.
Robinson, James B. III
Toomey, James Francis

August 1950
Hahn, William Clarence

January 1951
Chan, Kwok Keung
Herstine, Glen Leslie

June 1951
Harvey, Ray Wilson
Holbrook, Charles Tyler
Minger, Frederick Harvey
Novak, Donald Hoyt

January 1952
Margolis, Paul K.
Marshall, Harold Malcolm
Patrick, Richard Montgomery

June 1952
Allen Charles Gilpin
Hildebrandt, James Edwin
Kell, Robert John
McGinness, William Thornton
Saavedra, Joaquin Antonio

August 1952
Greenberg, Arthur Bernard
Holt, Robert Lee
Lyon, Herbert Arthur

January 1953
Oder, Harold S.

May 1953
Fortini, Anthony
Hromas, Leslie Alan
McCollough, Chester Ellsworth Jr.

August 1953
Bowditch, David Nathanial
Schaper, Peter Wolfgang

January 1954
Altiere, Fred Joseph
Capasso, Vincent Nicholas Jr.
D'Arcy, Edward James
Edstrom, Clarence Raymond
Stuber, Harold Britton

May 1954
Bailey, Cecil DeWitt
Barnes, William Henry
Madayag, Angel Feratero

Miller, Clement Kennedy
Schiele, Joe Scott
Wheasler, Robert Arthur

August 1954
Afifi, Hoda Mohamed
Kentzer, Czeslaw Pawel

January 1955
Yu, Ying-Nien

June 1955
Layton, James Preston
Truax, Philip Park

August 1955
Cingo, Ralph Paul
Morgan, Homer Garth

June 1956
Barthel, James Robert
Cirineo, Godofredo de Venecia Jr.
Drissell, Brian Arthur
Hayes, James Eugene
Sherrill, Robert Dee
Williams, Frances Marion

August 1956
Da Costa, Augusto Vale

January 1957
Johnston, Jack
Thompson, Thomas Ross

June 1957
Cappellari, James Oliver
Decker, Vern Leroy
Higgins, Daniel Leo
Hodson, Charles Henry
Oetting, Robert Benfield
Swanson, Andrew G.

August 1957
Bosscher, James Peter
Mertaugh, Lawrence Joseph Jr.
Shastri, Rajaram Mangalore

January 1958
Chow, Chuen-Yen
Norman, Wendell Smith

June 1958
Balaraman, Krishnaswami
Gearan, William Keaveny
Hill, William Watson Jr.
Ohira, Hiroichi
Parker, Richard Wilbur Jr.
Sedlak, Raynold Jacob
Seward, Arthur Lewis

January 1959
Gibson, James Darrell
Kalish, Eva Renate

Swaim, Robert Lee
Zak, Adam Richard

May 1959
Blackiston, Harry Spencer Jr.
Clingman, David Lee
Cook, Charles Ernest
Eddington, Robert Barnes
Herdman, Howard Lee
Murphy, John Michael
Pneuman, Gerald Warnick

July 1959
Paris, Franklin Lane Jr.

January 1960
Norman, Renny Stuart

June 1960
Bander, Thomas Joseph
Smith, Phillip Ronald
Swartz, Ronald Jacob

August 1960
Dezelan, Richard William

January 1961
Middendorf, Heinz Dieter

June 1961
Cork, Max Joseph
Gambaro, Ernest Umberto
Jecmen, Dennis Michael
Mickelsen, Harvey Peter Jr.
Sykes, Donald Kunkel

August 1961
Beveridge, Robert Bruce
Butler, Blaine Raymond Jr.
Diran, Oetarjo
Fahrner, Richard Lee

January 1962
Chichester, Frederick Donald
Czumak, Frank Michael
Gundling, David Roland

June 1962
Baker, Harold Lee
Cheng, Su Ling
Dorr, Robert Eugene
Graetch, Joseph Ernest
Kelly, Paul Donald
Lee, William Leung
Neal, Cecil Leon Jr.
Strickler, Robert Leslie
Yeung, Kin-Wai

June 1963
Beregesen, John Eric
Fanning, Robert Clark
Stutts, Harry Carey Jr.
Winder, Stephen William

Wu, Thomas Kong

August 1963
Grellmann, Hans Werner
Gromek, John Martin

January 1964
Lui, Oscar Ye-Ling
Nelson, Harlan Frederick
Spicer, Robert Lee

June 1964
Wong, Ka–Pang

August 1964
Bertram, Lee Alfred

January 1965
Buckley, Bruce
Cazier, Frank William Jr.
Lapine, Lucien
Yaros, Steven Francis

June 1965
Evans, Paul Francis
Guenther, Rolf Adolf
Haas, Mary Elizabeth
Head, Thomas Eugene
Hua, Hsi-Chun
Kessler, William Clarkson
Lippincott, Robert Park
Lydick, Larry Niles
Seto, Rodney King Mun
Sweet, Joel
Weber, James Alec

August 1965
Burk, Robert Clyde
Eckler, Gilbert Ross
Nahra, John Elias

January 1966
Cannon, Thomas Calvin
Flowers, Ronald John
Hesprich, Glen Vincent
Kissinger, Robert Dohn
Nuyts, Jacques Paul

June 1966
Capriotti, Larry Allan
Hesler, Lee James
Palmreuter, Edsel Charles
Pater, Stephen Walter
Venkataraman, Nellore S.

August 1966
Lewis, David Stanley
Markl, Robert Henry
Papailiou, Demosthenes D.

January 1967
Baughn, Terry V.
Brosseau, Jon Andrew

Daugherty, Gerry Robert
Doyle, George Robert Jr.
Ehmsen, Ronald John
Novick, Allen Stewart
Oeding, Robert George Jr.

June 1967
Evans, Larry Gerald
Roos, Rudolf

August 1967
Mazuel, Gilles Pierre

January 1968
Becker, Douglas James
Dilmaghani, Jalal

June 1968
Brown, Vernon Eugene

August 1968
Rule, Timothy Theron

June 1968
Bristow, Dean Rowe
Conner, William Russell
Egan, John Joseph III
Jones, William Henry
Joyce, Douglas Allyn
Keller, Robert Michael

August 1968
Bachelor, Robert Ross
Brown, Charles F. Jr.
Challe, Philippe Rene A.
Doane, Paul Michael
Green, James Wilson
Jonaitis, Kenneth Wayne
Miller, Alan Michael
Rule, Timothy Theron
Simica, Charles A.
Stewart, William Becker
Whicker, Donald

January 1969
Baer, Craig Alfred
Bettcher, James Robert
Burke, Charles Cecil
Covey, Richard Oswalt
Crimmel, William Wear
Curtiss, Walter Dallas
Eller, Thomas Julian
Howes, Robert Lloyd
Lutton, Paul Howard
May, Gary Allen
Morgan, Felix Evan
Oderman, Dale Barton
Rose, Eugene Arnold III
Schenk, Donald Edward
Tedor, John Barry
Thomas, Edwin Arthur
Towt, Howard Carnes

June 1969
Bauknight, Gerald Conrad
Bredvik, Gordan D.
Karpyak, Steven Douglas
Mac Bain, James Carter
Van Allen, Richard Earl

August 1969
Jacomein, William M.
Rizzetta, Donald P.
Schlegel, Mark Owen
Washburn, Donald Charles

January 1970
Bottomly, Roc
Deaver, Maurice Ardo Jr.
De Zonia, John Michael
Diehl, Ronald Lee
Franchi, Joseph L Jr.
Gardner, Guy Spence
Golart, Craig Stephen
Havrilla, Robert John
Head, Roger Carlos
Herklotz, Robert L.
Hodges, Terry Boyd
Killeen, Joseph Michael
Kirkpatrick, Daniel William
Kirkpatrick, Robert Jamison Jr.
Kumar, Ravindra
Lee, Charles William
Lisowski, Ronald James
Marvel, William Michael
Moore, Lynn Howell
Olafson, Frederick K.
Terhune, James Allen
Tucker, Bartow Charles
Walker, Robert Anthony

June 1970
Barter, Stephen William
Cherry, Clyde S. Jr.
Lewis, Donald George
Shanbhag, Ramesh Krishna
Stewart, Kirk Douglas
Walker, Neil Renz

August 1970
Eagles, Donald Earl
Laue, Dennis Richard
Lutterbie, Thomas Paul
Meredith, Joseph W. Jr.
Townsend, Paul Jerome
Trout, Kermit Edward Jr.

January 1971
Acurso, Jeffrey Louis
Allain, Richard S. Jr.
Baker, Richard Allen
Batuski, David John
Berta, Steven Joseph
Blume, William Henry
Carlson, Douglas Martin
Collins, Patrick Alan

Craigie, Ronald Patrick
Deffenbaugh, Floyd Douglas
Forkois, Jerald Lynn
Gilles, Gregory Lee
Habelt, William Walter
Hannah, Darryl Eugene
Harmon, Chester B.
Huttenlocker, Donald John
Irace, William Richard
Irwin, Stuart Edward
Johnson, Craig Lione
Kinnan, Timothy Alan
Kirkpatrick, Douglas H.
Knapp, Lloyd Stephen
Martinson, John Hamilton
Murrow, Richard Craig
Patel, Phiroze Dinshaw
Rossetti, Paul
Silverthorn, James Taylor
Vesel, Fred Henry
Weiland, Lewis Seaman

June 1971
Kim, Hang Wook
Lui, Leon Yeun-Leung
Smith, Stephen Kent

August 1971
Cooper, James Russell
Croop, Harold Charles
Dunn, Patrick Francis
Vyzral, Francis John
Yutmeyer, William Peter

January 1972
Autry, Larry Dale
Barngrover, Gary Clark
Cassano, Ronald
Chowdhury, Somen
Davis, Walter S. III
Hanus, Gary Joseph
Joseph, Daryl James
McKenzie, Mark Almon
Meixner, Robert Hay Jr.
Murchison, David C.
Payton, Gary Eugene
Sandstorm, James
Schunk, Jack Phillip
Seltzer, Robert L.
Waibel, Randall Harold

June 1972
Atkins, Jerome Anthony
Best, Eric Nicolas
Cosner, Raymond Robert
Dudley, William Craig
Gilmore, Robert Thomas
Hall, Darryl Wayne
Hildebrand, Daniel Hugh
Lamanna, William Joseph
Parameshwaran, Subramaniam
Skinner, David Alan
Voight, William Melvin II

Witt, Richard Michael

August 1972
Roberts, Philip Arnold

December 1972
Ferraioli, Richard A.
Harrison, Garry Lee
Hayes, Jeffery Mayer
Johnson, Curtis Dean
Mayward, Richard M.
Mehrotra, Sunil
Porter, James Howard
Wagner, Roger James
Ward, Morris Ardell Jr.

May 1973
Bruestle, Harry Rolf
Irvine, David Earl
Leddy, Michael Kevin
Wojnaroski, John W.
Wolf, Daniel Star
Yarrington, Alan Lee

August 1973
Balliett, Timothy Dean
Pegden, Claude Dennis
Staab, George Hans
Willen, Thomas Bernard

December 1973
Clark, Richard Stewart
Glaros, Louis N.
Manak, William Thomas
Welch, Gerald Michael

May 1974
Boyle, James Martin
Day, Michael L.
Fraser, Robert Stuart
Hoover, Alan David
Hudson, John Lester
Huett, Gary James
Lahiri, Subir
Luckring, James Michael
Marinella, Stephen John
Minto, David Walker
Moudry, James Arthur
Seder, Kenneth Edward
Smith, John William
Vishnauski, Jon Michael

August 1974
Drochner, Erich Emil
Han, An-Dong
Harter, Peter Kent
Liebovich, Lou Robert
Long, Richard Wayne

December 1974
Bastian, Thomas Wayne
Crovesi, Marco Mario
Gran, Carl Stanley

Paul, Warren Kent
Weeks, Laurence Campbell

May 1975
Fazio, Salvatore Jr.
Jamrogiewicz, Roman A.
McKean, Kenneth
Newsome, Richard W. Jr.
Peterson, Donald Earl
Uffelman, Kenneth Eugene
Wallingford, Stephen H.
Zanton, David Francis

August 1975
Baig, Mirza Irfan
Davis, Donald W. Jr.
Gloria, John Alfred

December 1975
Baughman, John Lewis
Chen, Chun-Chien
Julius, Edmund Paul Jr.
Ledger, Alan Steven
Miller, Ronnie Keith
Robinson, Michael Alan
Vahadji, Mohammad-Reza

May 1976
Bellagamba, Laurence
Eckerle, Wayne Allen
Fajardo, Charlotte H.
Fullman, Donald Grant
Hass, Bernard Charles J.
Hinsdale, Andrew Jeffrey
Keever, David Bruce
Larison, Clifton Dennis
McCorry, Daniel C. Jr.
Meteer, Philip Capaldi
Neville, Jeffery William
Okeefe, William Shawn
Raymer, Daniel Paul
Sibal, Stephen Donald
Siefke, Stanley Paul
Stewart, Charles Raaen
Stults, John Collier Jr.
Sweeney, Mark Louis
Wilkins, Bruce Gregory
Williams, Thomas Lee

August 1976
Abernathy, Charles Larry
Howard, Richard Moore
MacDonald, Richard E.

December 1976
Aralu, Boniface B.
Campbell, Kevin Dean
Evans, Robert Edward Ted
Morgan, David Llewellyn
Stritz, Alfred Gerhard
Wetzel, Richard Bruce

May 1977
Arcangeli, Gerald Thomas
Beecroft, Timothy James
Doligalski, Thomas L.
Kline, Larry Robert
Nguyen, Thanh Dieu
Thomas, Richard Glenn
Ullestad, Gary Stephen
Webster, James LeRoy

August 1977
Allen, Gerald Dale

May 1978
Chim, Siu-Man Edwin
Domon, Kozo
Ehlers, Steven Michael
Hollenback, David Mark
Reiber, Carl Edward
Snepp, David Karl
Tan, Shi Min
Washburn, Karen E.
Yarbrough, Edwin Richard

August 1978
Dahl, Scot Alan
Jones, Brian Watson
Matz, Dean George

December 1978
Cox Gregory John
Greenawalt, Scott Lash
Haas, James Anson
Kustura, John Brinsley
Polansky, Mark Lewis
Reymond, Michael Arthur
Rois-Mendez Armando Francisco

May 1979
Leung, Pui Kin
O'Brien, James Michael
Slomski, Joseph Francis Jr.

August 1979
Burgess, Mark Alan
Chin, Hsiang
Morse, Dennis Frank
Pollock, Peter Brian
Robinson, Jane Ann
Vogelsberg, Clifton Krell

December 1979
Chang, Yuan-Bin
Takezaki, Juro

May 1980
Bager, Mary Elizabeth
Bauer, Frank Henry
Davis, Roger Edward
Leewood, Alan Ross
Scurria, Norman Virgil Jr.

August 1980
Seffern, John Jay

December 1980
Batina, John Terence
Moore, Carleton John
Prasad, Shivshankar Narasimha

May 1981
Smith, Crawford Frederic III
Sundar. R. M.
Webb, Thomas Philip

August 1981
Hess, Joseph Paul
Weber, Margaret Rose

December 1981
Hu, Anren
Lilley, Jay Stanley
Parke, Curtis James

May 1982
Chen, Alexander Te
Harter, James Arthur
McKinney, Leon Ellington Jr.
Parpia, Ijaz Husain

August 1982
Blomshield, Frederick Steele
Foist, Brian Lee
Grady, Joseph Edward
Guevara, Luis Alejandro

December 1982
Bacon, Barton Jon
Deur, John Mark
Gilbert, Michael Glenn
Necib, Brahim
Rechak, Said
Stukel, Stephen Peter
Yucuis, William Allen

May 1983
Bourne, Simon Marcus
Gates, Thomas Scott
Heath, Bradley Jay
Reeve, Scott Richard
Sanger, Kenneth Bateson
Sawyer, Richard Steven
Steffens, Brucw Charles

August 1983
Allen, Charles Scott
Perez, Rigoberto
Tritsch, Douglas Evan

December 1983
Cassady, Robert Joseph
Dremann, Christopher Charles
Garrett, Daniel Ramon
Mignery, Lezza Ann Miller
Stucki, Jeffery Alan

May 1984
Boyd, Charles Neill
Dornself, Mark Jonathan
Geyer, Mark Stephen
Igli, David Andrew
Marstiller, John Winwood
Sefcik, Thomas John
Waszak, Martin Raymond
Williams, Jeffery Paul
Zhou, Shu Gong

August 1984
Kelly, Scott Roger
Olsen, Michael Erik
Parry, Craig Owen
Pinella, David Francis
Summerset, Twain Ki
Visich, Michael

December 1984
Alsip, Matt Myron
Buethe, Scott Allen
Kucek, John Patrick
McGrath, David Kentlow
McKissack, Douglas Ross
Privoznik, Cynthia May
Ray, Suvendoo Kumar

May 1985
Berry, Dale Thomas
Berry, Howard Maurice
Davidson, John Burdyne Jr.
Fuglsang, Dennis Floyd
Stone, Raymond Eugene
Walsh, Michael Kevin

August 1985
Bastnagel, Philip Michael
Miller, Theron Jeffrey
Spencer, David Bradley

December 1985
Jih, Chan-Jiun
Ledington, Alan H.
McAdams, James Valen
Morrison, Adrian Scott
Rizzi, Stephen Anthony
Soedel, Sven Michael
Sutila, John Steve
Tsiao, Sunny
Wendel, Thomas Robert

May 1986
Delli Santi James William
Gallman, John Waldemar
Gibbons, Michael David
Johns, John B.
Lollock, Jeffrey Alan

August 1986
Barron, Dean Alan
Gerrish, Harold Paul Jr.
Hotz, Anthony Francis

December 1986
Batcho, Paul Francis
Berreth, Steven Patrick
Brockman, Mark L.
Drew, Myres Nelson
Drouin, Donald Valmont Jr.
Halsmer, Dominic Michael
Hsieh, Chen
Kurtz, Russell Dane
Lum, Gregory Tsun-Fai
Norman, Timothy Lee
Pernicka, Henry John
Sallee, Vernon James

May 1987
Baxter, Michael Joseph
Bohlmann, Jonathan D.
Bokash, David Eugene
Clafin, Scott Evan
Cox, Dennis Eugene
Mavromatis, Theofanis
Seltzer, Robert Michael
Vance, Samuel Matthew

August 1987
Clark, Christopher Martin
Davidson, Donald Carl Jr.
Harris, Jeffrey Eric
Jones, Stanley Eugene
Levine, Mark Scott

December 1987
Augenstein, David LeRoy
Bachelder, Edward Nelson
Gaunce, Michael Thomas
Hamke, Rolf Ehrhardt
Heinstein, Martin Wilhelm
Jiang, Yi-Tsann
Kistler, Gordan Paul
Larrimore, Scott Charles
Mohr, Ross Warren
Murdock, Kelly Daniel
Shoults, Gregg Allan
Walker, Kevin Frank
Wanthal, Steven Paul

May 1988
Benavente, Javier Edgar
Ely, Todd Allan
Krozel, James Alan
Layton, Jeffrey Boese
Lee, Khueh Hock
Marsh, Steven Michael
Monroe, Eric Kenneth
Schierman, John David
Seal, Michael Damian II
Sherrier, David Mathew
Thompson, Jeffrey Michael
Townsend, Barbara K.
Tucker, Tenna Evelyn
Wetlesen, David Christian
Williamson, Robert Zane

August 1988
Silk, Anthony Bernard
Witkowski, David Paul
Woods, Jessica Amanda

December 1988
Bauer, Jeffrey Ervin
Cahill, Timothy Scott
Chen, Gang
Chu, Gou-Don
Gertz, Diana Louise, Leeper
Goeldner, Kevin Blake
O'Toole, Stephen James
Pitman, Frank Albert III
Probert, Andrew Arthur
Roundy, Lance Michael
Rushinsky, John Joseph Jr.
Schlossberg, Valerie, E.

May 1989
Chang, Chhchen
Coiduras, Jose M.
Damra, Fayez Mohammed
Koper, Judith Lee, Kennedy
Robinson, Brian Anthony
Wang, Shaupoh

August 1989
Drajeske, Mark Howard
Jentink, Thomas Neil
Lee, Elizabeth Mae
Ratzer, David Leo
Rausch, Russ David
Thompson, David Dean

December 1989
Ayer, Timothy Collins
Butler, Anthony Edward
Drury, Steven Charles
Dubke, Junathan Paul
Grove, David Michael
Henkener, Julie Aline
Kim, Hee Jun
Lian, David Yin Wei
Probert, Todd Charles
Woods, Martha Ann
Yorn, Kendall L.

May 1990
Armstrong, Gerald Lewis Jr.
Bemtgen, Jean
Blatt, Paul Andrew
Cooley, John William
Kleb, William Leonard
LeBan, Frank Anthony
Leong, Gary Norman
Martin, Michael Tavis
O'Neal, Patrick David
Puig Suari, Jordi
Scott, Robert Charles
Schultz, Jeffrey David
Snaufer, Mark Jeffrey
Todd, Richard Emery

August 1990
Danial, Albert Naguib
Dwenger, Richard Dale
Phillips, Ross Alan
Yap, Federiko Louis Doornik
Zenieh, Salah

December 1990
Ammerman, Curtt Nelson
Breda, David Andrew
Burley, Russell William
Frietchen, Douglas Bradley
Hucker, Scott Alan
MacLean, Roderick J.
Morris, Scott Richard
Schroeder, Jeffery Allyn
Wehmeier, Eric John

May 1991
Demir, Ismail
Doebling, Scott William
Enrico Santana, Cesar Alesandro H.
Friedmann, David Scott
Harrison, James David Jr.
Jacobs, William Blake
Longmeyer, Michael Henry
Mase, Robert Arthur
Mesarch, Michael Andrew
Moody, Thomas Bernard
Ross, Hugh Anthony

July 1991
Bell, Julia Lea
Hayashi, Mathew Sei
Scheumann, Troy Duane

December 1991
Allen, John David
Bullock, Steven Jerome
Doshi, Divyesh Nagindas
Erwin, Duane Scott
Saunder, Terrence John
Spencer, David Allen
Tao, Xuefeng
Wernimont, Eric John

May 1992
Allen, Carl Lee
Eudaric, Alain Marie
Kunack, Kelly Ann
Meyer, Scott Edwin
Paige, Derek Andrew
Rishko, Tara Lee
Wang, Hongli
Woodard, Paul Robert

July 1992
Bosler, Patrick Joseph
Brown, Susan Jayne
Eidson, Robert Carr
Hill, Lisa Renee
Hooker, John Richard
Lance, Christopher Alan

Weeks, Craig Andrew
Wilson, Brett Stephen

December 1992
Baratech, Manuel Jose
Berry, Robert Everett
Bowman, Erik Charles
Gunawan, David James
Hilbing, James Henry
Martinez Correas, Jose Luis
Moen, Michael Jon
Prado, Richard Adolfo
Turner, Simon

May 1993
Ankireddi, Seshasayee S.
Burke, Michael Joseph
Campbell, Bryan Thomas
Chambers, Robert Patrick
Cosky, Charissa Jean
Dean, Loren Philip
Doman, David B.
Elder, Robin Christy
Forsyth, Eric Nielsen
Gasper, Joseph John Jr.
Holland, Julie Ann
Katz, Jonathan William
Mains, Benjamin Lynn
Mains, Denna Lynn
McCoy, Robert William
Petropoulos, Anastassios Evangelos
Peyton, James Logan II
Spikings, David Kevin
Vitale, Jeffrey Dunn
Vukits, Thomas Joseph

August 1993
Hejl, Robert James
Hoffenberg, Robert
Laing, Peter
Patel, Moonish R.
Sims, Jon Andrew
Vaidya, Rajesh Suresh

December 1993
Buhler, Kimberley
Campbell, Eric Thomas
Courtney, Scott D.
Detert, Bruce Raymond
Diamant, Bryce Leylan
Doerfler, Mark Thomas
Mamner, Marvine Paula
Jones, Mark Alan
Leeks, Tamara J. Vaughan
Quinn, Brian John
Romanoski, Mark David
Stephens, Brian Eugene
Wilson, Roby Scott

May 1994
Beck, Rebecca Anne Dilling
Harrington, Brian David
Heinimann, Markus Beat

Kiger, Brian Vincent
Lauer, Mary Elizabeth
Mechalas, John Peter
Muller, Mark Brian
Sack, Elizabeth Elaine
Salyer, Terry Ray
Stokman, William Paul
Tragesser, Stephen Gregory
Zezula, Chad Erik

August 1994
Arendt, Cory Peter
Forster, Edwin Ewald
Keeter, Timothy Mark
Klug, John Charles
McGuire, Joseph Byron
Meiss, Alan Richard
O'Donnell, Sean Robert
Spangler, Christopher Allan
Yen, Chih-Chieh
Young, Steven Souglas

December 1994
Alcenius, Timothy John
Barden, Brian Todd
Coste, Keith
Miller, Jeffery Robert
Shannon, Brian Scott
Stubbings, Natalie Beth
Stuerman, Michael Todd
Temores, Jesus B.
Vonderwell, Daniel Joseph

May 1995
Berge, Sten Egil
Guzman, Jose Javier
Haven, Christine Elizabeth
Leung, Szutao
Seibert, Tomasz Ramy
Shastri, Ramachandra Parameshwar
Slijepcevic, Stevan
Wu, Hsin-Cheng

August 1995
Bayt, Robert Louis
Chou, James Chin-Chia
Fuh, Yiin-Kuen
Lena, Michael Robert
Love, James Glenn
Maley, Scott
Scheuring, Jason Nicholas
Stone, David Lamar
Summers, Darrell Lee
Szolwinski, Matthew Paul
Wiedman, Heiko Wilheim

December 1995
Burns, Steven Patrick
Heidenthal, Todd
Hodge, Kathleen Frances
Lengyel, Attila Jozsef
McVeigh, Pamela Alison
Pecson, Derek McDougall

Rutz, Mark William
Weaver, Marc Christopher

May 1996
Adams, Douglas Scott
Law, Ah-Hock
Miller, Jeffrey Robert
Neal, Bradford Alan
Peters, Mark Eugene
Powell, Shepard
Williams, Laura Ann

August 1996
Caravella, Joseph Robert
Ganapathy, Harish
Monsen, Quinn Boyd
Murray, Ian Fisher
Staugler, Andrew James

December 1996
Bucci, Gregory Steven
Emore, Gordon Lantz
Gandolfo, Bruno Maurice
Kannal, Lance Eric
Morris, Richard Lindsay
Munro, Scott Edward
Pham, Tuan Le
Rump, Kurt Matthew
Zink, Jonathan Todd

May 1997
Bunnell, Robert Allen
Cooney, James Scott
D'Amato, Fernando Javier
D'Souza, Shariff Rosario
Gates, Matthew David
Khadder, George Assad
Koutsavdis, Evangelos
    Konstantinos
Kuchnicki, Stephen Nicholas
Kuo, Yu-Hung
Roth, Gregory Lee
Setiawan, Emanual Rizal
Steffen, Phillip Jay
Szwajkowski, Darren Walter
Tinfina, Joseph Michael
Xi, Yueping

August 1997
Henry, Mark Edward
Sexton, Darren Gordon
Torgerson, Shad David

December 1997
Fang, Feng Jr.
Golden, Patrick
He, Min

May 1998
Batista-Rodriguez, Alicia
Bougher, Jeffrey Alan
Fanjoy, David William
Kim, Byoung Do

Nuss, Jason Samual
Ross, Tamaira Emily
Weimerskirch, Todd Duan
Witte, Gerhard Rudolf

August 1998
Baker, Joseph Alan
Kim, Jinsuck
Norris, Stephen Richard

December 1998
Erausquin, Richard German Jr.
Green, Stephen John
Hasebe, Ryoichi Sergio
Hwang, Yoola
Jones, Brian Robert
Palayathan, Ashootosh S.
Patanella, Alejandro Javier
Schlabach, Michael Alan
Tsai, Jia-Lin
Vandenboom, Michael Richard

May 1999
Huang, Hung-Chu

August 1999
Bonfiglio, Egene Peter
Bowman, Jason Christopher
Funk, John Eric
Mailhe, Laurie Marie
Narayanan, Shriram
Yoon, Suk Goo

December 1999
Adams, Robert Charles
Bodony, Daniel Joseph
Chandra, Budi Wijaya
Williams, Edwin Adams
Fernandez-Garcia, David

May 2000
Cho, Jeong-Min
Guille, Marianne
Hahn, Darren
Jacobson, Steven
Lachendro, Nathan
Lim, Sun-Wook
Srinivasan, Balaji
Strange, Nathan

August 2000
Frolik, Steven Andrew

December 2000
Bae, Kyoung Duck
Cho, Sung-Man
Cook, Andrea Marie
Haradanahalli, Murthy N.
Heaton, Andrew
Lee, Andrew
Marchand, Belinda
Alonso, German Porras
Yeh, Che-Ping

May 2001
Henderson, Joseph
Matlik, John
McKowen, Richard
Paek, Cholwon

August 2001
Beutien, Trisha
Bevis, Dwayne

December 2001
Anderson, Jason
Lee, Jungmin
Lim, Kok Hong
Miller, Daniel
Okutsu, Masataka
Radocaj, Daniel
Schmidt-Lange, Thomas
Skoch, Craig
Valentini, Luiz
Weinstock, Vladimir

May 2002
Brodrecht, David
Debban, Theresa
Helms, William
Hoverman, Thomas
Kowalkowski, Matthew
Krautheim, Michael
Peters, Christopher
Prater, Daniel
Russell, Shawn

August 2002
Austin, Benjamin
Kayir, Tulin
Martin, Eric
Prock, Brian
St. John, Clinton

December 2002
Gregory, James
Kothandaraman, Govindarajan
Swanson, Erick
VanMeter, Michael

May 2003
Chen, Kuan-Hua Joseph
Fitzpatrick, Shannon
Koenigs, Michael
Norman, Timothy
Rodrian, Jeffrey
Solmaz, Selim

August 2003
Dankanich, John Walter
Krishnan, Arvind
Landau, Damon F.
Page, David Andrew

December 2003
Ananthula, Vikram
Czapla, Nicholas

Garrison, Loren
Gnanamanickam, Ebenezer
Hasebe, Yoske
Herescu, Alexandru
Matsumura, Shin
Matsutomi, Yu
Merchant, Munir
Mok, Jong Soo
Sangngampal, Nattaporn
Stein, William

May 2004
Abdullah, Ermira
Aguayo, Eduardo
Benner, Robert
Bult, Jeffrey
Child, David
Corpening, Jeremy
Edwards, Jonathan
Gean, Matthew
Hiraoka, Daisuke
Keune, John
Lew, Phoi-tack
Mehta, Jatin
Moraines, Christophe
Niemczura, Jaroslaw
Northcutt, Brett
Precoda, Paul
Toombs, Rory
Tullos, James

August 2004
Chiew, Lee
D'Alto, Luis
Gates, Kristin
Loffing, David
Martin, Thomas
Melchior, Michael
Nugent, Nicholas
Subrmaniyan, Arun
VanY, Stephanie

December 2004
Chibana, Micah
Hrach, Michael
Osier, Geoffry
Pan, Fongloon
Parlee, Allison
Walters, Leon

May 2005
Eggink, Hunter
Harmon, Michael
Kong, Jin Yong
Main, Benjamin
McGilley, Joseph
Miller, Kevin
Newsome, Elizabeth
Pender, Adam
Tsohas, John
Wiratama, Andy

August 2005
Abdullah, Maizakiah
Gordon, Cristina
Harber, Joseph
Ladd, Adam
Lamberson, Steven Jr.
Lee, See-Chen
Mane, Muharrem
Nankani, Kamlesh
Nehrbass, Jonathan
Ooi, Chieh
Robarge, Tyler
Travis, Tanner,
Yu, Yen Ching

December 2005
Bies, Christopher
Briggs, Eric
Butt, Adam
Choi, Yoonsugn
Djibo, Louise-Olivia
Eichel, Brenda
Gonzalez, Pol-Enrico
Grosse, William
Habrel, Christopher
Hidayat, Jimmy
Iqbal, Liaquat
Jos, Cyril
Kacmar, Andrew
Kakoi, Masaki
Yepes, Javier Lovera
Maier, John
Padmanabhan, Arun
Skillen, Michael
Ian, En Kee

May 2006
Adeisa, Mobolaji
Ahn, Jung-Hyun
Chan, Ke-Winn
Chen, Jit-tat
DeHoyos, Amado
Deitemeyer, Adam
Fritsch, Geraldine
Gonzalez, Miguel
Grebow, Daniel
Grupido, Christopher
Hsieh, Kelli
Ianniello, Alessandro
Kalb, Brady
McCaffery, Daniel
Miller, Christopher
Moyle, Nicholas
Mseis, George
Park, Hwun
Phillips, Benjamin
Revenaugh, Joshua
Ulrich, Christopher
Ventre, Brian

August 2006
Delgado, Jorge
Falardeau, Joel

Joshi, Manasi
MacDonald, Megan
Manning, Robert

December 2006
Hauser, Aaron
Clark, Randall
Garner, James
Heckler, Gregory
Khoo, Teng Thuan
Nakaima, Daniel
Ozimek, Martin
Tenny, David
Wright, Raymond

May 2007
Boopalan, Avanthi
Bruno, Nickolas
Evans, Meredith
Grinham, Matthew
Karni, Etan
Kloster, Kevin
Lim, Sungyoun
Pollock, George
Rodkey, Samuel
Sindiy, Oleg
Torres, Anwar
Wheeler, Emily

August 2007
Byron, Jennifer
Droppers, Lloyd
Henning, Gregory
Jaron, Jacqueline
Schuff, Reuben
Wilson, Gregory

December 2007
Alkari, Ameya
Ballard, Christopher
Brophy, Daniel
Ewing, Joseph
Link, David
Onken, Jeffrey
Patel, Hiren
Davendralingam, Navindran

May 2008
Antaran, Albert
Aungst, Robert
Braun, Jonathan
Cashbaugh, Jasmine
Chigullapalli, Sruti
Hahn, Jeeyeon
Harriss, Ernest
Hatem, Miles
Huang, Shih-Hsuan
Huang, Xing
Kotegawa, Tatsuya
Kovach, Andrew
Ladisch, Nikolaus
Linke, Kevin
Otterstatter, Matthew

Puga, Alejandro
Sengstacken, Aaron
Tichy, Jason
Voo, Justin
Weise, Sarah

August 2008
Ali, Hadi
Berger, Richard
Chiu, Jimmy
Gordon, Dawn
Hannon, Michael
Irrgang, Lucia
Lossman, Matthew
Trebs, Adam

December 2008
Bush, Christopher
Coduti, Lee
Conway, Matthew
Hassan, Syed
Kapilavai, Sravan Dherraj
Kube-McDowell, Matthew
Leonard, Eric
Moonjelly, Paul
Pothala, Daniel
Selby, Christopher
Shearer, Jonathan
Tate, Nathan
Uffelman, Daniel
Vavrina, Matthew
Zhao, Jia

May 2009
Ayyaswamy, Venkattraman
Blank, Ryan
Hiu, Chii-Jyh
Jones, Jesse
Kwan, Kevin
Md Ishak, Nizam
Ohlsson, Linnea
Palewicz, Alexis
Phillips, Timothy
Ritchey, Andrew
Sandroni, Alex
Skube, Seth
Walsh, Kelly

August 2009
Ahn, Benjamin
Aligawesa, Alinda
Browning, Rebecca
Casper, Katya
Hamel, Peter
Helderman, David
Kowalkowski, Michael
Londner, Edward
Schmidt, Amanda
Tapee, John
Wypszynski, Aaron

December 2009
Guo, Kuo

Horst, John
Kim, Seung-il
Lin, Zengyue
Liu, Weiyi
Lynam, Alfred
Needham, Galen
Palumbo, Michael
Pomeroy, Brian
Roth, Brian
Tyagi, Ankit
Vemulapalli, Kautilya
Wheaton, Bradley
Wilcox, Nicole

MS IN ASTRONAUTICS (1964-1968)

June 1964
Barrett, Francis Llewellyn
Butterfield, Douglas Holman
Cox, Joseph James Jr.
Eckelkarnp, Vincent Clemence
Elfers, John Edwin
Hegstrom, Roger Joel
Horton, Henry Turnage
Meier, Thomas Charles
Mordan, Charles Raymond
Sherrill, Gerald Hardee
Wells, Norman Earl

January 1965
Bedarf, Richard Allan
Besch, Lawrence Edward
Budinoff, Jerold Edels·
Butler, Jerome Joseph III
Gruters, Guy Dennis
Kurz, Edward Stephen
Lodge, Robert Alrred
Long, James Edward
Matthes, Gary William
Smith, Gary Mark
Sue, James Edward

January 1966
Aicale, Ronald Rocco J.
Barton, Richard Jr.
Blaha, John Elmer
Bridges, Roy Dubard Jr.
Edwards, James Craig
Gados, Ronald George
Grosvenor, Williard
Klein, James Raymond
Kruczynski, Leonard Richard

Lawrence, Stephen Hard
Newendorp, James Vernon
Sabin, Marc Leslie
Soulek, James Walter
Thompson, Howard Clarence
Wood, James Michael

August 1966
Cooper, William Dean

January 1967
Boyd, Stanley Eugene
Carlson, Randal Davis
Casper, John Howard
Conrad, John Cosley Jr.
Dominiak, Stanley Walter Jr.
Gourley, Laurent Lee
Johnson, Howard Conwell Jr.
Leydorf, William Francis Jr.
Milberg, Raymond Fredrick
Nielsen, Reese Robert
Sarff, Charles Michael
Simmons, Michael Wayne
Smith, Lee Thomas
Walker, Robert Allan
Willett, David Anthony
Womack, Carl Lavan

June 1967
Betts, John Thomas
Shimomura, Ryoji

August 1967
Burns, James Richard

January 1968
Bauer, Christian Andreas
Bradley, Ronald Gay
Burnett, Paul Cather
Carleton, Roger Eugene
Femrite, Ralph Byron
Fuller, John Howland Jr.
Giles, Michael Neil
Hartley, Gerald Gordan
Hendrickson, Wylie Craig
Pechek, Phillip John
Selke, Robert Karl
Shriver, Loren James

August 1968
Thoden, Richard Walter

MS In Engineering (1967-2009)

January 1967
Nelson, Donald Lewis
Osburn, Carlton Morris
Reid, Robert William Jr.
Warren, Jerald William

June 1967
Chen, Kuo-Hwei
Clemmer, Lewis Eugene
Green, Charles Shaw Jr.
Zoeller, Michael Allen

August 1967
Chiu, Palmer Bang-Ming

January 1968
Aloisio, Charles Joseph
Becker, Douglas James
Carter, David John
Chao, Chien-Ming
Hunt, Richard Myron
Milotich, Francis John
Mlakar, Paul Francis

June 1968
Bevilaqua, Paul Michael
Carson, Charles Converse
Cook, Richard Drury
Gault, James Richard
Kidwell, Harlan J. Jr.
Luo, Siong Siu Tan
McColgin, William Carey
Neidig, Stephen Lowell
Petrolino, Joseph A. Jr.
Riddhagni, Prabaddh
Trent, Ronald

August 1968
Whicker, Donald

January 1969
Hager, William Dennis
Jamieson, John Jr.
Pope, Rhall Edward
Reed, Claude Brinton

June 1969
Oelke, Robert Jan
Riggers, Wilton Elmer
Shankar, Nellur S.

August 1969
Bond, Richard Joseph
Gray, Donald Dwight
Hsia, Margaret Huey-Yun

January 1970
Chang, Dow-Shing

Duan, Jen-Reen
Fung, Siu Bong
Means, Richard Terry
Pandalai, Raghavan P.
Picologlou, Basil F.

June 1970
Barron, Max Reeves
Emigh, Leonard Craig
Kayser, Kenneth William
Paquette, Gerald Raymond

June 1971
Davis, Wayne Joseph
Feil, Peter James
Keramidas, George A.
Mahan, John Edward
Mutyala, Bramaji Rao
Nippert, James William
Shirley, Michael Alan

August 1971
Allen, Michael Duane
Dalhart, Glenn Austin

January 1972
DeEskinazi, Jozef
McCann, Larry Dean
Smittasiri, Sakda
Tuan, Jen Lee

June 1972
Bauman, James Michael
Chen, Fang-Pai
Kaye, Michael Alan
Merriman, David Berry
Montgomery, Stephen T.
Papastavridis, John G.
Saleh, Salah Eddin
Toth, Dennis Dale

August 1972
Nawrocki, Paul Edward
Schmiesing, James F.

December 1972
Horvath, John Andrew Jr.
Wells, James Randolph

May 1973
Cermak, Robert John
Faulkner, William Henry
Kelly, John Wilbur
Mac Farlane, Larry J.
Newson, Donald Edward
Warden, David Edward

August 1973
Burns, Robert D. III
Elliott, Bruce H.
Kung, Joseph Hsiang-I
Massena, Roy Phillip

December 1973
Cox, Jack C.
Gloria, John Alfred
Heard, William Joseph
Pitz, William John

May 1974
Bratkovich, Alan Wayne
Elliott, Bruce H.
Giloy, Douglas Edwin
Newman, Richard L.
Sung, Shung Hsing
Tapia, Patricio Castillo

December 1974
Baldwin, Mark Lewis
Jaap, Joseph B.
Lamberson, Steven Edward
Lawrence, Timothy C.
Lawrence, Stephen R.
Thurman, Weir Peake
Wolf, Eric John

May 1975
Douglas, Robert Lewis
Fink, Raymond Keith
Guthrie, Gregory Robert
Imada, Wesley Tatsumi
Kuehne, Bruce Edward
Reeves, Thomas Wilmeth
Sprandel, James Kimberly
Triplett, Mark Bond

August 1975
Cohen, Allan Robert
Morehead, Robert

December 1975
Greenburg, Allan Morris
Less, Michael Clement

May 1976
Ayangade, Peter A.

August 1976
Richardson, Nancy May

December 1976
Sung, Shung Hsing

May 1977
Chung. Yun-Shung

Corbett, Michael Wade
Riechers, John Thomas
Thomas, Richard Glenn

August 1977
Hefner, Rick Dale

December 1977
Crago, Steven Charles
Rychnowski, Andrew S.

August 1978
Hsu, John-Shyong

December 1978
Allen, James Jardine Jr.
Gregory, Charles Z. Jr.

May 1980
Butz, Larry Albert
Fork, William Earl Jr.
Miller, Christopher John

August 1980
Kelly, Kent Calvin
Robert, Michel
Watzin, James Gerard

December 1980
Dimitriadis, Steve

May 1981
Foley, Robert Charles
Kullegowda, Hogegowda
Pollack, Douglas Arthur
Voit, Peter Michael

August 1981
Asteriadis, Athanasios

December 1981
Nack, Kevin Karl

May 1982
Tse, Frank Chee Wei

August 1982
Frederick, Robert Alvin Jr.

December 1982
Steele, William Gene

May 1983
Rottier, Stephen Richard

December 1983
Cordova, Carlos, A.
Ryan, Rosemary Jean

May 1984
Davisson, Richard Ray
Matson, John Charles

August 1984
Kenaga, Donald Wayne

December 1984
Anderson, Mark Ronald
Khawam, Maurice Anthony

May 1985
Estes, Donald Edward
Lesieutre, Daniel Joseph
McComb, Thomas Harvey
Peak, Edward John

May 1986
Hall, Louis Duvivier III
VanSice, James Carter

December 1986
Moster, Gregory Edward
Villemagne, Christian De

May 1987
Norris, Gregory Allen
Thomas, James Sanford Jr.

August 1987
Pitts, Dwight Adair

May 1988
Okada, Kenji
Evans, James E.

December 1988
Behrens, Ricky D.

May 1989
Baumgartner, Paul David
Budde, Patrick William
Sorge, Marlon Edmond

August 1989
Bryant, Jennings Ray
Chen, Shi-Yew
Chilcot, Kimberly Jane

May 1990
Yalch, Thomas John

August 1990
Williams, Steven Neal

December 1990
Black, Robert Edward III
Chen, Yen-Meng
Kott, Trevor Real

May 1991
Neussl, Michael Alfred

July 1991
Minster, Todd Alan
Smith, Monty J.

December 1991
Kirschner, William A.
Melville, Reid Barlow

May 1992
Grindey, Gregory John
Lange, Christopher Mark
Ong, Cheng Huat
Ventura, Mark Christopher

December 1992
Buckley, Michael Dean
Fishburn, Matthew Brian
Spillman, David Carl

May 1993
Aucoin, Brent Alan
Moukawsher, Elias James
Steinbach, Timothy Mark
Thompson, John Elliot
Vermillion, Brian Arthur

August 1993
Lai, Chang-Ching

December 1993
Schreiter, Daniel John

May 1994
Connor, Douglas Charles
Mann, James Bradley
Mihelic, Joseph Edward

August 1994
Hsieh, Der-Ning

December 1994
Kimura, Tsunekazu
May, Scot Terrell

May 1995
Barker, Jeffery Harold
Bramantya, Eka
Moses, Michael Patrick
Varney, Bruce Edward
Wade, Thomas David

August 1995
Beaumont, Matthew Hall

December 1995
Bilodeau, Brian Allen
Dunlap, Tony Lavern

May 1996
Baker, Joseph Thomas
Randall, Laura Anne
Carmel, Mark Stephen
Goetz, Andreas Richard
Rifani, Andreas Ismael

August 1996
Viassolo, Daniel Edgardo

December 1996

May 1997
Butt, Mark Eugene

August 1997
Carpenter, Michael James
Pakalapaati, Rajeev Tirumala

December 1997

May 1998
Mello, Charles William
Rahman, Naveed

August 1998
Kompella, Sridhar
Reimann, William James

December 1998

May 1999
Sanchez, Paul Kevin
Roeder, Blayne Alan
Ninan, Lal
Winz, Werner Andreas
Hua, Yuan

August 1999
Chang, I-Ling
Gupta, Anurag
Sharma, Anindya

December 1999
Li, Chunsu
Arrieta, Hernan Victor
Sakaue, Hirotaka
Tieche, Christopher Raymond
Xi, Jun
Zaini, Raafat Mahmoud

May 2000
Bartha, Bence
Bouboulis, Melvin
Bundechanan, Thra
Duke, David
Lindqvist, Jens
Pamu, Gautham

August 2000
Long, Matthew Robert
Bhatia, Deepak
McInnes, Allan Ian
Ong, David

December 2000
Kim, Jeesoo
Partouche, Ashers
So, Jin Wook
Epps, Jonathan
Kim, Young-Jae
Rufer, Shann
Yarnot, Vincent
Roberts, Dyan Renee

May 2001
Kwon, Hyuk-Bong

August 2001
Perez, Eddie
Rangarajan, Balaji

December 2001
Bouras, Constantinos
Fischer, Brian
Lasic, Victor
Shimo, Masayoshi

May 2002
McConaghy, Thomas
Palmer, Robert
Bodily, Brandon
Bulathsinghala, Ivor
Garcia, Daniel
Hatfield, Charles
Joshua, Raymond

August 2002
Aratama, Shigeki
Lamb, Gregory

December 2002
Canino, James

May 2003
Dyer, Donald
Tu, Min-Cheng
Garman, Karl
Roth, Brian
Ward, Mark

August 2003
Sisco, James C.

December 2003
Ali, Syed
Lana, Carlos
Richardson, Renith
Shrotri, Kshitij
Thabet, Atef

May 2004
Kluzek, Celine
Vetter, Chad
Wright, Charles
Langenbacher, Erik
Portillo, Juan
Yam, Chit

August 2004
Pearson, Nicholas
Qian, Haiyang
Kumari, Shyama

December 2004
Nguyen, Anh-Thu
Rubel, Ken

Smith, Justin A.
Dumong, Xavier
Kibbey, Timothy
Liang, Liang
Quintana, Juan
Simmons, Aric

May 2005
Genov, Dentcho
Patterson, Christopher
Talbert, Brian
Deo, Amitabh
Smith, Justin J.

August 2005
Rausch, Raoul

December 2005
Borg, Matthew
Smajlovic, Dino
Thicksten, Zachary
Ge, Yun
Genco, Filippo
Genco, Giacinto
Jo, Daeseong
Merrill, Marriner
Spohn, Jason

May 2006
McDonald, Seth
Agarwal, Avinash
Churchfield, Matthew
Hartnett, Randall
Mathis, David
Pan, Yi
Sardeshmukh, Swanand
Smith, Tracey
Sudo, Machiko
Wagner, Christoph
Zakharov, Sergey

August 2006
Oliver, Anthony
Saheba, Ruchir
Smith, Randolph
Wennerberg, Jason

December 2006
Balogun, Olaniyi
Haberlen, Philip
Han, En-Pei
Joseph, Gregory
Juliano, Thomas
Nightingale, Jay
Schrik, Daniel

May 2007
Ohlaug, Loryn
Siehling, John
Sturridge, Melissa
Childress, Karla
Fry, Donald
MacMillan, Michael
Uzmann, Joseph

August 2007
Gujarathi, Amit
Pinheiro, Jacob

December 2007
Kloeden, Richard
Segura, Rodrigo
Case, Erle
Dambach, Erik
Dufraisse, Gabriel
Nierling, Jonathan

May 2008
Cheng, Annie
Gangestad, Joseph
Kumar, Anant
Werner, Vincent
Eckstein, James
Galitzine, Cyril
Moore, James Whitley

August 2008
Moss, James
Salek, Robert
Park, Ki Sun
Pensky, Yury
Samson, Daniel

December 2008
Brown, Todd
Ghose, Shayani

May 2009
Johnson, Kay (Kenneth)
Cummings, Cornelius
Erickson, Brian
Han, Seung Yeob
Kotecha, Pankit
Lin, Wei-Nan
Rupakula, Aparna
Schlueter, Andrew
Tatineni, Prashant

August 2009
Dikshit, Prakash
Garibaldi, Oscar
Leng, Yujun
Petrov, Eugene
Sharma, Bhisham

December 2009
Ambalavanan, Manikkavasagan
Chauhan, Digvijay
Dustin, Joshua
Fu, Yuqiang
O'Hara, Loral
Roa, Mario
Sgro, Titus

MASTER'S THESIS TITLES (1938-2009)

1938

Emmons, Paul Clyde, "Developments of Apparatus for the Determination of Propeller Characteristics of a Light Airplane in Flight," (Professor K. D. Wood).

Goksel, Mehmet Ridvan, "A Survey of Specifications and Estimates of Characteristics of Military Airplane of 1937."

1939

Baals, Donald Delbert, "The Design and Calibration of the 1939 Purdue Wind Tunnel Balances," (Professors K. D. Wood and J. Liston).

Wilkins, Dale Emmett, "Supercharging of Air-Cooled Aircraft Engines of Low Power Output," (Professor J. Liston).

1940

Shepard, William Bowers, "Comparison of Bending and Compression Tests on Stiffened Sheets," (Professor K. D. Wood).

Wickersham, Robert Owen, "The Design, Construction and Testing of a Plywood Covered Full Cantilever Aircraft Wing."

1941

Liu, Nai-Chien, "Wind Tunnel Test of $^1/_2$- Scale-Model of the XP-1 Airplane," (Professor K. D. Wood and J. Liston).

Yui, Sing-Fuh, "Tests of Thin-Walled Plastic Cylinders to Determine the Shape of the Allowable Stress Interaction Curves Under Combined Compression, Bending and Torsional Loads," (Professor E. F. Bruhn).

1942

Anderson, Seth Bernard, "The Use of a Pitot Tube Integrating Rake to Determine Propeller Thrust Coefficient," (Professor K. D. Wood).

Brown, Clinton Eugene, "The Development of Apparatus and Technique for the Measurement of Drag In Towed Flight," (Professor K. D. Wood).

Dickinson, Richard, "Ultimate Stress Interaction Curves in Combined Compression, Bending, and Torsion on Thin-Walled Circular Cylinders," (Professor E. F. Bruhn).

Formanek, Edward Bernard, "Combined Stresses in Thin-Walled Plywood Structures," (Professor E. F. Bruhn).

Woodward, Walter Ray, "The Development of Torque and Thrust Meters for Measurement of Propeller Efficiencies in Flight," (Professor K. D. Wood).

1943

Johnson, Alan Dillistin, "Redesign and Calibration of a 30" X 48," Throat Wind Tunnel," (Professors K. D. Wood and R. G. Binder).

1944

Stitz, Erwin Otto, "A Photoelastic Study of Stresses in Spur Gear Teeth," (Professor C. S. Cutshall).

1946

Wang, Shou-Wu, "Materials for Three Dimensional Photoelasticity," (Professor R. G. Sturm).

1947

Ayres, Langdon Ford, "Design of Turbosupercharger Hot Gas Test Equipment and Performance Testing of Type B Turbine," (Professor J. Liston).

Boyer, Rodney L., "The Effect of Design Variables Upon the Characteristics of Burners for the J-23 (I-40 Type) Turbo-jet Aircraft Engine," (Professor J. Liston).

Brown, Sherwood H., "Tip Effect and the Performance of Low Aspect Ratio Wings," (Professor J. R. Weske).

Clark, John Wood, "Stress Analysis of a Railway Freight Car Truck Side Frame."

Corden, Carlton Donald, "The Static Longitudinal Stability of a Flexible Airplane," (Professor R. M. Rosenberg.

Head, Robert E., "Preliminary Design for A Six-Component Balance for a Wind-Tunnel."

Hedrick, William S., "Tip Effect and the Performance of Low Aspect Ratio Wings," (Professor J. R. Weske).

Kahn, Ahtesham Ali, "A Synoptic Survey of the Theory and Operation of Ejectors," (Professor J. W. McBride).

McMurray, John W., Jr., "A Comparative Study of the Performance Characteristics of Model Jet Burners," (Professor J. Liston).

Nelson, Robert Leonard, "Performance Characteristics of a Merz 17.6 Test Engine," (Professor J. Liston).

O'Brien, Hubbert L., "Method and Techniques for Calibrating the SR-4 Strain Gage Under Hydrostatic Pressure," (Professor R. G. Sturm).

Povalski, James Albert, "A Study of Internal Combustion Engine Pressures with a Cathode-Ray Oscillograph," (Professor P. E. Stanley).

Richardson, Robert L., "Design of Turbosupercharger Compressor Test Equipment and Performance Testing of Type B Compressor," (Professor J. Liston).

Wetterstrom, Edwin, "Discontinuity Stresses in Pressure Vessels," (Professor R. G. Sturm).

1948
Anderson, Reidar L., "A Unification of Some Theories of Elastic Strength," (Professor R. G. Sturm).

Carpenter, Charles Raymond, "Comparative Analysis of Actual and Analytical Operating Variables of a Single Cylinder Engine," (Professor J. Liston).

Cunny, Robert W., "Process Control for Cold Worked Stainless Steel," (Professor L. A. Kessley).

Gatineau, Robert Jean, "Gyroscopic Vibrations of Propellers," (Professor T. J. Herrick).

Hesse, Walter J., "The Design and Instrumentation of the Rocket Test Facility at Purdue University," (Professor M. J. Zucrow).

Horner, John T., "Tensor Analysis of Strain in Elasticity and Hydronamics," (Professor R. G. Sturm).

Kessley, Lester A., "Process Control for Cold Worked Stainless Steel," (Professors C. W. Breese and R. G. Sturm).

McNeilly, Vance H., "Force Analysis of Freight Car Draft Gears," (Professor R. G. Sturm).

Nicholson, Robert Allen, "On the Influence of Flow Through a Spanwise Slot on Drag of an N.A.C.A. 0018 Wing," (Professor J. W. McBride).

O'Brien, Denis H., Jr., "A Survey of Supersonic Diffusers Applicable to Ramjets," (Professor M. J. Zucrow).

Petrick, Ernest Nicholas, "The Design and Development of Equipment for the Static Testing of an Aviation Gas Turbine Engine," (Professor M. J. Zucrow).

Robinson, Robert Frank, "The Effect of Shear on a Pitot," (Professor J. W. McBride).

Scull, Wilfred Eugene, "Correlation of Actual and Analytic Cylinder Phenomena," (Professor J. Liston).

Sellers, John Paul, Jr., "The Problem of Escape from the Earth and the Stepped-Rocket," (Professor M. J. Zucrow).

Spraker, Wilbur Allen, Jr., "On the Influence of Flow Through a Spanwise Slot on Lift of a 0018 Wing," (Professor J. W. McBride).

1949
Akers, Marion Jesse, "The Effects of Sound Energy on Liquid Atomization," (Professor J. Liston).

Chapman, Rowe, Jr., "Shear Flow About a Pitot Tube," (Professor M. E. Shanks).

Costilow, Eleanor Louse, "Loss Trends in an Infinite Array of Airfoils," (Professor M. E. Shanks).

Daane, Robert A., "Thermal Stresses in Thick Walled Hollow Cylinders," (Professor N. Little).

Evans, Jack, "An Investigation of Waterproofing, SR_4 Strain Gages," (Professor R. G. Sturm).

Jackson, Charles Edward, "The Flutter of a Uniform Cantilever Wing with Tip Weight by the Response Function Method," (Professors H. Serbin and T. J. Herrick).

Kordes, Eldon Eugene, "First-order Vibration of Propellers," (Professor T. J. Herrick).

Marsh, Herbert Warren, "Metallurgical Examination of a Pressure Vessel," (Professors R. G. Sturm and G. M. Enos).

McGillam, Clare Duane, "Some Vibration Characteristics of Steel Roller Chain," (Professor R. G. Sturm).

Michaels, Franklin, "Measurement of Liquid Rocket Propellant Flow Rate," (Professor M. J. Zucrow).

Misener, Walter Stuart, "A General Method of Gas Turbine Analysis."

Peugh, Darrel E., "A Method for Investigation of Interior Ballistic Forces in Guns," (Professor R. G. Sturm).

Roberts, Robert Earl, "A Survey of the Literature on the Statistical Theory of Turbulence," (Professor M. E. Shanks).

Savage, William F., "Range of Jet-Propelled Aircraft."

Schmitt, Alfred Frederick, "A Survey of Theories of Plasticity as Applied to Structural Analysis," (Professor E. F. Bruhn).

1950

Balling, N. R., "Correction Factor Method of Thermodynamic Cycle Analysis," (Professor M. J. Zucrow).

Carlson, Robert Lee, "Applications of Finite Differences to Stability Problems," (Professor N. Little).

Cox, Paul B., "Dynamics of Electric Generators."

Hazen, Thamon Edson, "Utilization of Whole Vegetal Stalks Bonded with Adhesives for Building Boards and Structural Panels," (Professors H. J. Barre, R. G. Sturm and R. L. Whistler).

Helms, Harold Eugene, "Size Effect on Endurance Limit Stress for SAE 4140 Steel," (Professor R. G. Sturm).

Holden, Gifford Merrill, Jr., "Analytic Study of Semi-empirical Boundary Layer Solutions," (Professor M. E. Shanks).

Isaacson, Jerrold Anselm, "Photoelastic Check of Theoretical Stresses in a Curved Beam," (Professors N. Little and E. O. Stitz).

Johnston, George S., "Fatigue Strength Under Combined Stress," (Professor R. G. Sturm).

Knuth, Eldon Laverne, "The Introduction of Fluids into a Moving Gas Stream Through Parallel Disks as Related to Film Cooling," (Professor M. J. Zucrow).

Lewiecki, Edward M., "Determination of the Effect of Pressure on the SR_4 Strain Gage," (Professor R. G. Sturm).

McComb, Harvey, G., Jr., "Symmetric Local Face Buckling of Curved Sandwich Panels," (Professors B. Klein, T. J. Herrick and Hsu Lo).

McDermott, Alexander, M., "The Application of the Response Function to the Calculation of Lateral Control Effectiveness for Elastic Airplanes," (Professor H. Serbin).

Penn, William W., Jr., "The Effects of Compressibility on the Lift Wings," (Professor H. Serbin).

Sapowith, Alan David, "Pitching Moment of Bodies of Revolution in Subsonic Flow."

Saunders, K. D., "Thermal Stresses in the Uncased Formation Hole," (Professor N. Little).

Steed, Rodney Warren, "An Experimental Investigation of the Local Buckling Strength of Curved Sandwich Panels Loaded in Axial Compression," (Professors B. Klein and T. J. Herrick).

Toomey, James F., "Experimental Investigation and Development of a High Compression Combustion Chamber," (Professor V. C. Vanderbilt).

1951

Dykhuizen, Marvin G., "Strength Analysis of an Escalator Frame," (Professor J. L. Waling).

Gray, Robert Merle, "Stress Concentrations in Tank Car Shells Near Anchor Plate Welds," (Professors C. W. Lawton and E. O. Stitz).

Harvey, Ray Wilson, Jr., "Analysis, Design and Construction of a Free Piston Gas Generator," (Professors V. C. Vanderbilt and J. Liston).

Herstine, Glen Leslie, "Experimental Investigation of a High Compression Engine," (Professors W. B. Barnes and J. Liston).

Minger, Frederick H., "The Construction and Testing of a Subsonic Air Thermocouple Probe Calibration Rig," (Professor J. Liston).

Novak, Donald Hoyt, "Calculation of Steady Two-Dimensional Supersonic Flow Near the Nose of Blunt Bodies," (Professor M. E. Shanks).

Radcliffe, Byron M., "I. Investigation of Post-yield Strain Gages at High Strains, and II. Response of SR-4 A-5 Strain Gages at Elevated Temperatures," (Professor E. O. Stitz).

Salmassy, Omar Kayyam, "An Investigation of the Properties of Sintered Iron," (Professor E. C. Thoma).

Weydert, John C., "A Study of Stresses in Thick-Walled Elliptical Pressure Tubes," (Professor E. O. Stitz).

Swick, Howard A., "Effect of Forging Impact Pressures Upon Several Mechanical Properties of SAE 4340 Steel," (Professors R. G. Sturm and E. O. Stitz).

1952
Corsette, Douglas Frank, "An Analysis of Equipment Utilization and Ground Time Controlling Factors in Airport Operation," (Professor W. J. Richardson).

Douglas, Robert Alden, "The Design of a Hydraulic Test Apparatus for Metal Sheet," (Professor W. E. Cooper).

Greenberg, Arthur Bernard, "The Stability and Flow of Liquid Films Injected Into An Air Duct Through Spaced Parallel Disks in the Two-Dimensional and Three-Dimensional Cases," (Professor M. J. Zucrow).

Hamilton, William Wilbur, "A Photoelastic Study of Shrink Fits," (Professors C. W. Lawton and E. O. Stitz).

Hassemer, Lt. Col. David W., "Utilization of the Photo-Grid Process in Preparation of Training Aids for Engineering Mechanics Causes," (Professor C. S. Cutshall).

Kell, Robert J., "Rolling Effectiveness and Aileron Reversal of Low Aspect-Ratio Rectangular Wings at Supersonic Speeds," (Professor H. Serbin).

Lambert, John Wallace, "Changes of Material Properties Due to Straightening Operations," (Professor R. G. Sturm).

Margolis, Paul K., "Experimental and Theoretical Study of the Lift Distribution of Flexible Straight Wings," (Professor Hsu Lo).

Marshall, Harold M., "The Effect of Coriolis Acceleration on the Bending Vibration of a Rotating Beam," (Professor Hsu Lo).

Nester, John H., "An Investigation of the Development of Stresses Induced by Arc Welding," (Professor N. Little).

Oder, Harold S., "Experimental Investigation of the Ultimate Energy Absorption Characteristics of Simple Structural Units," (Professor E. F. Bruhn).

Patrick, Richard M., "Supersonic Flow Around Bodies of Revolution," (Professor H. Serbin).

Smith, William Douglas, "Development and Evaluation of a Small Free-Piston Gasifier," (Professor V. C. Vanderbilt).

1953
Fortini, Anthony, "Development and Evaluation of a Free-Piston Gasifier," (Professor V. C. Vanderbilt).

Hartsaw, W. O., "Applications of Photoelasticity as Visual Aids for Basic Instruction in Mechanics of Materials and Machine Design," (Professor C. S. Cutshall).

Hromas, Leslie A., "Problems in Diabatic Flow," (Professor H. M. DeGroff).

Kiper, A. Muhlis, "Recovery Factors in Heat Transfer at High Velocities — Their Interpretation and Correlation," (Professor D. Y. Touloukian).

Koziuk, Frank Stanley, "The Instrumentation of a Shock Tube," (Professors A. C. Todd and J. Cage).

Pyke, Donald Leaming, "Stress Analysis of Pad-eyes Having Welded Reinforcing Plates," (Professor J. L. Waling).

Schaper, Peter W., "The Design and Construction of a Shock Tube," (Professor H. M. DeGroff).

Zebrak, Walter A., Jr., "Methods in Determining the Air Passenger Potential of a Community," (Professor M. M. Erselcuk).

1954

Kentzer, Czeslaw P., "Calibration of a Shock Tube," (Professor H. M. DeGroff).

Rubinstein, Joseph, "A Comparison of Performance on Three Space-Perceptual Tasks," (Professor L. M. Baker).

Timm, Edwin Marx, "A Study of Static and Dynamic Yield Stress of Rotating Band Materials," (Professors P. F. Chenea and J. L. Waling).

1955

Houstrup, John Peter, "A Universal State in Three-Dimensional Photoelasticity," (Professor E. O. Stitz).

Layton, James Preston, "Heat Transfer in Oscillating Flow," (Professor M. J. Zucrow).

Wainwright, William L., "The Design of a Compression Impinger," (Professor J. L. Waling).

1957

Bosscher, James P., "An Investigation of Some Dynamic Properties of Silicone Rubber Sponge," (Professors R. J. H. Bollard and P. E. Stanley).

Cappellari, James Oliver, Jr., "Approximate Solutions for Trajectories of Minimum Time for Rocket-Powered Aircraft," (Professor A. Miele).

Decker, Vern L., "The Synthesis of an Electronic Analog Computer Simulation of a Twin Spool Turbo-Prop Engine," (Professor P. E. Stanley).

Higgins, Daniel Leo, "The Study of an Integral Panel System Under Various Loading Conditions," (Professor J. Bollard).

Oetting, Robert S., "Temperature Measurement in Aircraft Power Plants," (Professor J. Liston).

Shastri, Rajaram Mangalore, "Preliminary Selection and Design of Regenerators," (Professor P. S. Lykoudis).

Thompson, Thomas R., "On the Hot-Wire Anemometer at Very Low Speeds," (Professors DeGroff and P. S. Lykoudis).

1958

Hill, William W., Jr., "Study of Calibration Methods for Wind Measuring Equipment," (Professor L. Hromas).

Parker, Richard Wilbur, Jr., "Instrumentation of a Hypersonic Wind Tunnel," (Professor P. E. Stanley).

1959

Paris, Franklin Lane, Jr., "Parametric Study of the Pressurized Argon Electric Arc and Related Electrode Heat Transfer," (Professor G. M. Palmer).

Pneuman, Gerald Warnick, "Magneto-fluid-mechanics of a Viscous, Electrically Conducting Fluid Contained Within Two Finite, Concentric, Rotating Cylinders in the Presence of a Magnetic Field," (Professor P. S. Lykoudis).

Swaim, Robert Lee, "The Effects of Time-varying Parameters on the Aeroelastic Stability of A Supersonic Airfoil," (Professor Hsu Lo).

1961

Kelly, Paul Donald, "Applications of Molecular Theories to Aerodynamics," (Professor R. Goulard).

1962
Yeung, Kin-Wai, "The Viscoelastic Analysis of an Elliptical-Holed Plate Subjected to Stresses at Infinity," (Professor G. Lianis).

1963
Fanning, Robert Clark, "A Study of Free Molecule Flow," (Professor W. A. Gustafson).

Grellman, Hans W., "Experimental Study of a Rotating Electric Arc in a Strong Magnetic Field," (Professor G. M. Palmer).

Stutts, Harry Carey, "Application of DC-Board Technique to Plasticity Problems," (Professor N. Inoue).

1964
Barrett, Francis L., "Trajectory Optimization Using the Adjoint Method," (Professor S. J. Citron).

Bertram, Lee A., "Viscoelastic Thermomechanical Response to Infinitesimal Cyclic Load on Finite Static Load," (Professor G. Lianis).

Butterfield, Douglas H., "The Effects of Continuous Low Thrust in the Meridian Plane on Satellite Orbits," (Professor Hsu Lo).

Cox, Joseph James, Jr., "Transfer From an Elliptical Orbit to a Coplanar, Co-focus, Non-intersecting Circular Orbit by Two or More Impulses."

Eckelkamp, Vincent C. J., "A Multiple Current Dipole Model of the Human Heart," (Professor P. E. Stanley).

Elfer, John E., "Spectroscopic Analysis to Determine the Electron Concentration in the Wake of a Re-entry Body as Influenced by Seedi Implants of Aluminum in the Model and Tested in a Plasma Tunnel," (Professor G. M. Palmer).

Horton, Henry Turnage, "Lateral Control of a Terminal Guidance System for Lunar Landing," (Professor S. J. Citron).

Meier, Thomas Charles, "An Analytic Solution for the Entry of Vehicles Into Planetary Atmospheres," (Professor S. J. Citron).

Mordan, Charles Raymond, "Experimental Evaluation of a Low Speed Flight and Landing Characteristics of Hypersonic Aircraft," (Professor G. M. Palmer).

Nelson, Harlan Frederick, "Absorption of Solar Energy in an Atmospheric Layer," (Professor R. Goulard).

Sherrill, Gerald Hardee, "Longitudinal Control of a Terminal Guidance System for Lunar Landing," (Professor S. J. Citron).

Wells, Norman Earl, "An Analysis of the Method of Gradients for the Optimization of Flight Paths," (Professor S. J. Citron).

1967
Mazuel, Gilles Pierre, "Review and Extension of Inversion Techniques in Radiation Transfer," (Professor R. J. Goulard).

1968
Riddhagni, Prabaddh, "An Experimental Investigation of an Incompressible Turbulent Wake Behind a Sphere," (Professor P. S. Lykoudis).

1971
Dunn, Patrick F., "Magneto-Fluid-Mechanic Natural and Forced Heat Transfer from Horizontal Hot-Film Probes," (Professor P. S. Lykoudis).

1973
Warden, David Edward, "A Dynamic Model for Evaluating Man's Impact on Global Climate."

1974
Tapia, Patricio C., "The Ill-Posed Problem of System Input Identifications," (Professor V. Vemuri).

1978
Domon, Kozo, "Analysis of Contact Force and Dynamic Response of Beams and Plates Subjected to Impact of an Elastic Sphere," (Professor C. T. Sun).

Jones, Brian, "Development of a Fixed-Base Flight Simulator on an Interactive Minicomputer- Graphics System," (Professor D. K. Schmidt).

Matz, Dean, "Modeling Techniques and Preliminary Wind Tunnel Testing of an Eight-Bladed Propeller with Non-Swept and Swept Blades," (Professor J. P. Sullivan).

1979

Pollock, Peter, "Development of Instrumentation for a Laser Doppler Velocimeter and Its Use in Analyzing the Flow Field of an Eight Bladed Propeller," (Professor J. P. Sullivan).

Takezaki, Juro, "The Photoelastic Analysis of Mixed-Mode Fracture Mechanics," (Professor J. F. Doyle).

1980

Barger, Mary Elizabeth, "The Effect of Erosive Burning on Temperature Sensitivity in Composite Solid Propellants," (Professor J. R. Osborn).

Butz, Larry, "Turbulent Flow in S-Shaped Ducts," (Professor J. P. Sullivan).

Scurria, Norman, "Three Dimensional Photoelastic Analysis of the Shrink Fit Problem by Stress Freezing," (Professor J. F. Doyle).

Seffern, John Jay, "Modeling Techniques and Preliminary Testing of an Eight-Bladed Prop-Fan With and Without Proplets," (Professor J. P. Sullivan).

1981

Voit, Peter Michael, "Improving Strength in Composite Laminates Through Use of Hybrid Composites," (Professor C. T. Sun).

Sundar, R. M., "Performance of Wind Turbines in a Turbulent Atmosphere," (Professor J. P. Sullivan).

Lilley, Jay Stanley, "The Design and Operation of a Servo-Controlled Solid Propellant Strand Window Bomb," (Professor J. Osborn).

Nack, Kevin Karl, "A Study of the Aerodynamic Breakup of Molten Aluminum Particles in Two-Phase Flow," (Professor J. Osborn).

1982

Harter, James A., "Fatigue Crack Growth Characteristics of Embedded Flaws in Plate and Lug Type Fastener Holes," (Professor A. F. Grandt, Jr.).

Blomshield, Frederick S., "Pressure Oscillation Analysis of a Research Dump Combustor Ramjet Experiencing Combustion Instability," (Professor J. R. Osborn).

Foist, Brian Lee, "Aeroelastic Stability of Aircraft With Advance Composite Wings," (Professor T. A. Weisshaar).

Frederick, Robert A., Jr., "A Photographic Study of Effect of Acceleration on Composite Propellant Combustion," (Professor J. R. Osborn).

Grady, Joseph, "Use of Cracked Beam Finite Elements to Estimate Stress Intensity Factors," (Professor C. T. Sun).

Bacon, Barton, "A Modern Approach to Pilot/Vehicle Analysis and the Neal-Smith Criteria," (Professor D. K. Schmidt).

Deur, John, "A Study of Non-steady Combustion on Heterogeneous Propellants," (Professor R. Glick).

Gilbert, Michael, "Dynamic Modeling and Active Control of Aeroelastic Aircraft," (Professor D. K. Schmidt).

Steele, William, "Aeroelastic Stability of a Forward Swept Wing Aircraft with Wing Tip Stores," (Professor T. A. Weisshaar).

Stukel, Stephen, "Aeroelastic Tailoring for Passive Flutter Suppression," (Professor T. A. Weisshaar).

Yucuis, William, "Computer Simulation of a Multiaxis Air-to-Air Tracking Task Using the Optimal Pilot Control Model," (Professor D. K. Schmidt).

1983

Bourne, Simon, "Bi-Partitioned Estimation and Its Application to Pilot Control Strategy Identification in the VTOL Tracking Task," (Professor D. Andrisani, II).

Heath, Bradley J., "Stress Intensity Factors for Coalescing and Single Corner Flaws Along a Hole Bore in a Plate," (Professor A. F. Grandt, Jr.).

Gates, Thomas, "Acoustic Emission Behavior in 3D Carbon/Carbon Composites," (Professor C. T. Sun).

Perez, Rigoberto, "Initiation, Growth and Coalescence of Fatigue Cracks," (Professor A. F. Grandt, Jr.).

Tritsch, Douglas, "Prediction of Fatigue Crack Lives and Shapes," (Professor A. F. Grandt, Jr.).

Cassady, Joe, "Measurement of the Arc Chamber Pressure Distribution in Magnetoplasmadynamic Thruster," (Professor J. R. Osborn).

Mignery, Lezza (Miller), "Stitching in Laminated Composites," (Professor C. T. Sun).

Ryan, Rosemary, "The Effect of Aeroelastic Tailoring in the Passive Control of Flutter and Divergence of Aircraft Wings," (Professor T. A. Weisshaar).

Stucki, Jeffrey Alan, "Impact Effects on SMC-R50 Composite," (Professor C. T. Sun).

1984

Boyd, Charles, "Airfoil Design by Numerical Boundary Layer Optimization," (Professor J. P. Sullivan).

Geyer, Mark, "Solid Propellant Flame Temperature Measurements," (Professor J. R. Osborn).

Marstiller, John W., "Effects of Viscosity and Modes on Transonic Aerodynamic and Aeroelastic Characteristics of Wings," (Professor H. T. Yang).

Matson, John, "Combustion Response Measurements of Solid Rocket Propellant by a Microwave Technique," (Professor J. R. Osborn).

Waszak, Martin, "Analysis of Flexible Aircraft Longitudinal Dynamics," (Professor D. K. Schmidt).

Zhou, Shu-gong, "Failure of Quasi-Isotropic Composite Materials Under Off-Axis Loading," (Professor C. T. Sun).

Kelly, Scott, "Characterization of Mechanical Properties of a Fiberglas Composite," (Professor C. T. Sun).

Kenaga, Donald W., "Elastic-Plastic Material Characterization and Fatigue Crack Growth in Unidirectional Metal-Matrix Composites," (Professors J. F. Doyle and C. T. Sun).

Pinella, David, "Finite Element Analysis of Cracks on Orthotropic Elastic-Plastic Materials," (Professor J. F. Doyle).

Anderson, Mark R., "Closed Loop, Pilot/Vehicle Analysis of Approach and Landing," (Professor D. K. Schmidt).

Buethe, Scott, "A Study of Optimal State-Rate Feedback Explicit Model-Following in its Application to V/STOL Aircraft," (Professor D. Andrisani, II).

McGrath, David K., "Propellant Solid Phase Temperature Measurements," (Professor J. R. Osborn).

Ray, Suvendoo, "Three Dimensional Crack Closure Measurements in Polycarbonate," (Professor A. F. Grandt, Jr.).

1985

Berry, Dale, "Feedback Control, Large Amplitude Vibration, and Postbuckling of Space Lattice Beams Using Simplified Finite Element Models," (Professor H. T. Yang).

Berry, Howard, "Viscous Effects on Transonic Airfoil Stability and Response," (Professor H. T. Yang).

Davidson, John, "Flight Control Synthesis for Flexible Aircraft Using Eigenspace Assignment," (Professor D. K. Schmidt).

Fuglsang, Dennis, "Nonisentropic Unsteady Transonic Small Disturbance Theory," (Professor M. H. Williams).

Lesieutre, Daniel J., "The Theorem Performance of Counter Rotating Propeller Systems," (Professor J. P. Sullivan).

Miller, Theron, "Control Effectiveness of a Composite Swept Back Wing," (Professor T. A. Weisshaar).

Spencer, David, "The Gravitational Influence of a Fourth Body on Periodic Halo Orbits," (Professor K. Howell).

Jih, Chan-Jiun, "The Evaluation of Stress Intensity Factors by Finite Element Method," (Professor C. T. Sun).

Morrison, Adrian Scott, "Development of Software for the Control of Spectrum Loading Fatigue Tests," (Professor A. F. Grandt, Jr.).

McComb, Thomas Harvey, "Interaction and Coalescence of Fatigue Cracks at a Hole in a Plate," (Professor A. F. Grandt, Jr.).

Rizzi, Stephen, "Fatigue Crack Growth in Boron-Aluminum Metal Matrix Composites," (Professor C. T. Sun and J. F. Doyle).

Wendel, Thomas, "Closed-Loop, Pilot/Vehicle Analysis of Approach and Landing of Pitch-Rate Command Aircraft," (Professor D. K. Schmidt).

1986
Gallman, John, "A Computational Transonic Flutter Boundary Tracking Procedure," (H. T. Yang).

Gibbons, Michael, "Nonisentropic Unsteady Three Dimensional Transonic Small Disturbance Potential Theory," (Professor M. H. Williams).

Barron, Dean, "LDV Setup in an Annular Combustion Model," (Professor J. P. Sullivan).

Berreth, Steven P., "A New End Tab Design for Off-Axis Tension Tests of Composite Materials," (Professor C. T. Sun).

Pernicka, Henry, "The Numerical Determination of Lissajous Orbits in the Circular Restricted Three-Body Problem," (Professor K. C. Howell).

Sallee, Vernon J., "A Study of Integrated Structure and Control Design," (Professor T. A. Weisshaar).

Villemagne, Christian de, "Model and Controller Reduction: A New Approach Using Oblique Projections and Canonical Interactions," (Professor R. E. Skelton).

1987
Baxter, Michael, "Some Considerations of Actuator Dynamics in the Control of a Flexible Beam," (Professor K. C. Howell).

Bohlmann, Jonathan, "The Static Aeroelasticity of a Composite Oblique Wing," (Professor T. A. Weisshaar).

Cox, Dennis E., "Effect of Adhesive on Impact Damage Using a Rubber Capped Impactor," (Professor C. T. Sun).

Mavromatis, Theofanis M., "Target Tracking of Maneuvering Aircraft's and Tracker Parameter Optimization," (Professor D. Andrisani, II).

Norris, Gregory, "Selection of Non-Ideal Noisy Actuators and Sensors in the Control of Linear Systems," (Professor R. E. Skelton).

Levine, Mark, "Body-Conforming Grids for General Unsteady Airfoil Motion," (Professor M. H. Williams).

Gaunce, Michael T., "A Study of Temperature Sensitivity and Variable Pressure Exponents in AP/HTPB Composite Propellants," (Professor J. R. Osborn).

Hamke, Rolf, "A Theoretical Study of the Temperature Sensitivity of a Solid Rocket Motor," (Professor J. R. Osborn).

Heinstein, Martin, "Adaptive 2D Finite Element Simulation of Metal Forming Processes," (Professor H. T. Yang).

Kistler, Gordon, "Fatigue Crack Growth and Retardation Behavior of Aluminum-Lithium Alloy 2090-T8E41 and Ternary Al-Li-Zr," (Professor A. F. Grandt, Jr.).

Mohr, Ross W., "Effects of Mach Number and Sweep Variation of Transonic Aeroelastic Forces and Flutter Characteristics," (Professor H. T. Yang).

Murdock, Kelly D., "The Application of Subarcsecond Star Tracker Technology to Interplanetary Navigation," (Professor K. C. Howell).

Shoults, Gregg A., "Filtering Strategies for Non-Gravitational Accelerations with Application to the Global Positioning System," (Professor K. C. Howell).

Walker, Kevin F., "The Effect on Fatigue Crack Growth Under Spectrum Loading of An Imposed Placard, 'G', Limit," (Professor A. F. Grandt, Jr.).

Wanthal, Steven, "Formulation of Three Simple Triangular Plane Stress Anisotropic Finite Elements for a Microcomputer," (Professor H. T. Yang).

1988

Ely, Todd Alan, "Orbit Determination of a Spacecraft with Undetermined Control Inputs," (Professor K. C. Howell).

Krozel, James, "Search Problems in Mission Planning and Navigation of Autonomous Aircraft," (Professor D. Andrisani, II).

Lee, Alex Khueh Hock, "A Computational Investigation of Propeller/Wing Integration," (Professor J. P. Sullivan).

Marsh, Steven M., "Sun-Synchronous Trajectory Design Using Consecutive Lunar Gravity Assists," (Professor K. C. Howell).

Okada, Kenji, "Sensitivity Reducing Controller Design Method for Uncertain Parameter Systems," (Professor R. E. Skelton).

Schierman, John, "Tracking Fixed Wing Aircraft and Helicopters Using Attitude and Rotor Angle Information," (Professor D. Andrisani, II).

Seal, Michael D., II, "An Experimental Study of Swirling Flows as Applied to Annular Combustors," (Professor J. P. Sullivan).

Townsend, Barbara, "The Evaluation of Three Control Laws for the CH-47 Helicopter," (Professor D. K. Schmidt).

Williamson, Robert Z., "Microcomputer-Based Flight Simulation," (Professor D. K. Schmidt).

Silk, Anthony B., "Modeling Human Perception and Estimation of Kinematic Responses During Aircraft Landing," (Professor D. K. Schmidt).

Witkowski, David P., "Experimental Investigation of Propeller-Wing Interaction," (Professor J. P. Sullivan).

Woods, Jessica A., "Dynamic Aeroelastic Stability of a Generic X-Wing Aircraft Configuration With Design Parameter Variations," (Professor T. A. Weisshaar).

Bauer, Jeffrey E., "Identification of Equivalent Short Period Dynamics of the X-29A," (Professor D. Andrisani, II).

Behrens, Ricky D., "An Experimental Investigation of Wind Tunnel Quality," (Professor J. P. Sullivan).

Chen, Gang, "Nonlinear Finite Element Analysis of Plates and Shells of Composite Laminates," (Professor C. T. Sun).

Schlossberg, Valerie E., "Earth-Mars Trajectories Using a Lunar Gravity Assist," (Professor K. C. Howell).

1989

Chang, Chihchen, "Random Vibration of Uncertain Beam Structures With Geometrical Nonlinearity," Professor H. T. Yang).

Damra, Fayez M., "Optimal Covariance Control for Linear Continuous and Discrete Time Systems," (Professor R. E. Skelton).

Robinson, Brian Anthony, "Aeroelastic Analysis of Wings Using the Euler Equations With A Deforming Mesh," (Professor H. T. Yang).

Wang, Shaupoh, "Adaptive 2D Finite Element Method for Analysis of Phase Transformation," (Professor H. T. Yang).

Chen, Shi-Yew, "Applications of Three Dimensional Fracture Mechanics to Micro-Indentation Testing and Scribing," (Professor T. Farris).

Drajeske, Mark, "An Experimental Investigation of Cooperative Control Synthesis and a Lateral Position Tracking Task," (Professor D. K. Schmidt).

Jentink, Thomas N., "Formulation of Boundary Conditions for the Multigrid Acceleration of the Euler and Navier Stokes Equations," (Professor W. Usab).

Rausch, Russ David, "Euler Flutter Analysis of Airfoils Using Unstructured Dynamic Meshes," (Professor H. T. Yang).

Ayer, Timothy, "Airfoil Wake Models for Unsteady Full Potential Applications," (Professor M. H. Williams).

Drury, Steven Charles, "A Study of Probabilistic Durability Analysis," (Professor A. F. Grandt, Jr.).

Henkener, Julie Aline, "The Fatigue Crack Growth Behavior of an Experimental Ternary AL-2.6LI-0.09ZR Alloy," (Professor A. F. Grandt, Jr.).

1990
Armstrong, Gerald L., Jr., "An Analysis of the Instantaneous Performance Characteristics of Nozzleless Solid Rocket Motors," (Professor J. R. Osborn).

Blatt, Paul Andrew, "Evaluation of Fatigue Crack Initiation Behavior of an Experimental Ternary Aluminum-Lithium Alloy," (Professor A. F. Grandt, Jr.).

Kleb, William Leonard, "Temporal-Adaptive Euler/Navier-Stokes Algorithm for Unsteady Analysis of Airfoils Using Triangular Meshes," (Professor M. H. Williams).

Martin, Michael T., "Power Flow Analysis of Wave Propagation in Trusses," (Professor J. F. Doyle).

O'Neal, Patrick D., "Design of a Roll Control System for a Smart Projectile Using Loop Transfer Recovery," (Professor D. Andrisani, II).

Scott, Robert Charles, "Control of Flutter Using Adaptive Materials," (Professor T. A. Weisshaar).

Danial, Albert Naguib, "A Parallel Lanczos Algorithm for the Solution of the General Eigenvalue Problem," (Professor J. F. Doyle).

Dwenger, Richard Dale, "Laser Doppler Velocimeter Measurements and Laser Sheet Imaging in An Annular Combustor Model," (Professor J. P. Sullivan).

Williams, Steven Neal, "Automated Design of Multiple Encounter Gravity-Assist Trajectories," (Professor J. M. Longuski).

Chen, Yen-Meng, "Effect of Hydrostatic Pressure on Surface Roughness of Pinch-Off Phenomenon," (Professor T. N. Farris).

Hucker, Scott, "A Boundary Element Approach to Thermoelastic Crack Problems," (Professor T. N. Farris).

MacLean, Roderick J., "The Flowfield Around a STOVL Aircraft Model in Ground Effect," (Professor J. P. Sullivan).

Morris, Scott R., "The Characterization of Damage Due to Out-of-Plane Loading in the AS4/PEEK Composite," (Professor C. T. Sun).

Schroeder, Jeffery A., "Design, Analysis, and Flight Test of Pilot Displays for Hovering Aircraft," (Professor D. Andrisani, II).

1991
Harrison, James D., Jr., "Implicit Multigrid Acceleration of the Euler and Navier Stokes Equations," (Professor W. J. Usab).

Bell, Julia Lea, "The Impact of Solar Radiation Pressure on Sun-Earth L1 Libration Point Orbits," (Professor K. C. Howell).

Minster, Todd A., "Prediffuser-Combustor Model Studies Under Conditions of Water Ingestion in Engines," (Professor S.N.B. Murthy).

Scheumann, Troy D., "A Numerical and Experimental Investigation of the Effects of Notches on Fatigue Behavior," (Professor A. F. Grandt, Jr.).

Smith, Monty J., "Structural Optimization Using Passive and Active Control," (Professor R. E. Skelton).

Melville, Reid, "A Non-Linear Vortex-Lattice Model for Leading-Edge Separation On a Prop-Fan Blade," (Professor M. H. Williams).

Saunder, Terrence John, "Fatigue Crack Growth From Coldworked Eccentrically Located Fastener Holes in 7075-T651 Aluminum," (Professor A. F. Grandt, Jr.).

Spencer, David Allen, "Multiple Lunar Encounter Trajectory Design Using a Multi-Conic Approach," (Professor K. C. Howell).

Tao, Xuefeng, "The Generic Aircraft Tracker: Theory and Computer Implementation," (Professor D. Andrisani, II).

1992

Eudaric, Alain Marie, "Ellipsoid Methods for Multiobjective Control With Quadratic Performance Measures," (Professor M. Rotea).

Paige, Derek Andrew, "Anisotropic Composite Panel Flutter Suppression Using Adaptive Materials," (Professor T. A. Weisshaar).

Woodard, Paul Robert, "Grid Quality Assessment for Unstructured Triangular and Tetrahedral Meshes and Validation of an Upwind Implicit Euler Solution Algorithm," (Professor H. T. Yang).

Bosler, Patrick J., "The Design and Control of a Digital Controller for a Laboratory Helicopter Model," (Professor R. E. Skelton).

Hill, Lisa R., "A Spectral Boundary Element Method for Transient Heat Conduction," (Professor T. N. Farris).

Hooker, John, "Spatial and Temporal Adaptive Procedures for the Unsteady Aerodynamic Analysis of Airfoils Using Unstructured Meshes," (Professor M. H. Williams).

Weeks, Craig Andrew, "Design and Characterization of Multi-Core Composite Laminates," (Professor C. T. Sun).

Moen, Michael Jon, "Design, Testing and Analysis of a High-Speed, Time-Resolved Non-Intrusive Skin Friction Sensor System," (Professor S. Schneider).

Turner, Simon, "A Feasibility Study Regarding the Addition of a Fifth Control to Rotorcraft In-Flight Simulator," (Professor D. Andrisani, II).

1993

Campbell, Bryan T., "Temperature Sensitive Fluorescent Paints for Aerodynamics Applications," (Professor J. P. Sullivan).

Dean, Loren P., "Methods for Resolving Visual Ambiguities During Optical Target Tracking," (Professor D. Andrisani, II).

Doman, David, "An Experimental Study of Pilot-Vehicle Dynamics Using a Tilt Rotor Flying Machine," (Professor D. Andrisani, II).

Forsyth, Eric, "Initiation and Growth of Small Fatigue Cracks at a Notch," (Professor A. F. Grandt, Jr.).

Mains, Deanna Lynn, "Transfer Trajectories From Earth Parking Orbits to L1 Halo Orbits," (Professor K. C. Howell).

Moukawsher, Elias James, "Fatigue Life and Residual Strength of Panels With Multiple Site Damage," (Professor A. F. Grandt, Jr.).

Vukits, Thomas Joseph, "Low and High Speed STOVL Configurations in Ground Effect," (Professor J. P. Sullivan).

Hejl, Robert James, "On Solid Rocket Motor Grain Burnback Analysis Using Adaptive Grid Techniques," (Professor S. Heister).

Hoffenberg, Robert, "Filtered Particle Scattering: Laser Velocimetry Using an Iodine Filter," (Professor J. P. Sullivan).

Laing, Peter, "Gas Turbine Prediffuser — Combustor Performance During Operation With Air-Water Mixture," (Professor S.N.B. Murthy).

Patel, Moonish R., "Automated Design of Delta-V Gravity-Assist Trajectories for Solar System Exploration," (Professor J. M. Longuski).

Vaidya, Rajesh S., "On Predicting Elastic Moduli and Plastic Flow in Composite Materials Using Micromechanics," (Professor C. T. Sun).

Buhler, Kimberley, "A Study of Fatigue Crack Growth of Fuselage Panels Containing Multiple Site Damage," (Professor A. F. Grandt, Jr.).

Detert, Bruce, "The Design and Testing of a High Reynolds Number Vortex Tube," (Professor J. P. Sullivan).

Doerfler, Mark T., "Determination of Stress Intensity Factors for Cracks at Deep Notches," (Professor A. F. Grandt, Jr.).

Jones, Mark Alan, "A Feasibility Study of the Use of Axisymmetric Thrust Vectoring for Enhanced In-Flight Simulation," (Professor D. Andrisani, II).

Leeks, Tamara Jill, "Optimal Design of Partial-Plate Piezoelectric Actuators for Maximum Panel Deflection," (Professor T. A. Weisshaar).

Quinn, Brian J., "Comparative Evaluation of Failure Analysis Methods for Composite Laminates," (Professor C. T. Sun).

Stephens, Brian E., "Dynamic Loading of Cracked Panels," (Professor J. F. Doyle).

Wilson, Roby S., "A Design Tool for Constructing Multiple Lunar Swingby Trajectories," (Professor K. C. Howell).

1994

Heinimann, Markus Beat, "Numerical Determination of Stress Intensity Factors for Cracks at Deep Notches," (Professor A. F. Grandt, Jr.)

Harrington, Brian David, "Failure Initiation and Ultimate Strength of Composite Laminates Containing A Center Elliptical Hole," (Professor C. T. Sun).

Mann, James Bradley, "CBN Grinding of M2 Steel: Manufacturing Aspects and the Performance of Ground Surfaces," (Professor T. N. Farris).

Muller, Mark Brian, "Transonic Drag Reduction Through the Use of Induced Strain Actuators to Form an Adaptive Airfoil," (Professor T. A. Weisshaar).

Salyer, Terry Ray, "Quantitative Noise Reduction Measurements of A Multiple-Source Schlieren System," (Professor S. D. Collicott).

Stokman, William Paul, "Design and Characterization of Fiber/Metal Laminate Composite Materials," (Professor C. T. Sun).

Zezula, Chad Erik, "The Influence of Initial Inhomogeneities on Notch Fatigue of 7050-T7451 Aluminum Plate," (Professor A. F. Grandt, Jr.).

Sack, Elizabeth Elaine, "Evaluation of Finite Element of Piezoelectric Materials With Holes and Cracks," (Professor T. N. Farris).

Forster, Edwin, "Flutter Control of Wing Boxes Using Piezoelectric Actuators," (Professor H.T. Yang).

Keeter, Timothy M., "Station-Keeping Strategies for Libration Point Orbits: Target Point and Floquet Mode Approaches," (Professor K. C. Howell).

McGuire, Joseph Byron, "Fluid Dynamic Perturbations Using Laser Induced Breakdown," (Professor S. D. Collicott).

O'Donnell, Sean Robert, "Low Cycle Plasticity of IM7/8320 Thermoplastic Composite," (Professor C. T. Sun).

Young, Steven Douglas, "Particle Image Velocimetry Applied to Flow in the Wake of a Flat Plate," (Professor J. P. Sullivan).

Alcenius, Timothy John, "Development of Square Nozzles for High-Speed, Low-Disturbance Wind Tunnels," (Professor S. P. Schneider).

Kimura, Tsunekazu, "A Fault-Tolerant Controller Design Using Alternating Projection Techniques," (Professor R. Skelton).

Spangler, Christopher, "Nonlinear Modeling of Jet Atomization in the Wind-Induced Regime," (Professor S. Heister).

Stuerman, Michael, "Surface Thermal Loading Imposed by a Sonic or Supersonic Impinging Jet," (Professor N. L. Messersmith).

1995

Bayt, Robert Louis, "Prediction of End-Cap Effects on Equilibrium Helium Bubbles in the Gravity Probe-B Spacecraft," (Professor S. H. Collicott).

Lena, Michael Robert, "Repair and Reinforcement of Cracked Aluminum Plates With Adhesively Bonded Composite Patches," (Professor C. T. Sun).

Love, James Glenn, "Scalability of Mechanical and Thermal Loads in Underexpanded Jet Impact," (Professor S. N. B. Murthy).

Scheuring, Jason Nicholas, "An Evaluation of Aging Aircraft Materials," (Professor A. F. Grandt).

Szolwinski, Matthew Paul, "Mechanics of Fretting Fatigue Crack Formation," (Professor T. N. Farris).

Varney, Bruce, "Mechanics of Roller Straightening," (Professor T.N. Farris).

Bilodeau, Brian, "Application of the Spectral Element Method to Interior Noise Problems," (Professor J.F. Doyle).

Burns, Steven Patrick, "Fluorescent Pressure Sensitive Paints for Aerodynamic Rotating Machinery," (Professor J.P. Sullivan).

Dunlap, Tony, "Technique for Experimental Grinding Ratio Measurement," (Professor T.N. Farris).

McVeigh, Pamela, "Finite Element Analysis of Fretting Fatigue Stresses," (Professor T.N. Farris).

Rutz, Mark, "Effect of Transverse Acoustic Oscillation on the Behavior of a Liquid Jet," (Professor S.D. Heister).

1996

Adams, Douglas, "Efficient Finite Element Modeling of Thin-Walled Structures with Constrained Viscoelastic Layer Damping," (Professor C.T. Sun).

Peters, Mark, "Development of a Light Unmanned Aircraft for the Determination of Flying Qualities Requirements," (Professor T.A. Weisshaar).

Randall, Laura, "Receptivity Experiments on a Hemispherical Nose at Mach 4," (Professor S. P. Schneider).

Rifani, Andreas, "Investigation of 3-D Fatigue Crack Growth Behavior in a Double-T Cross Section," (Professor A.F. Grandt, Jr.).

Caravella Jr., Joseph R., "Experimental Investigation of Combustion in a Radial Flow Hybrid Rocket Engine," (Professor S.D. Heister).

Murray, Ian, "Modeling Acoustically Induced Oscillations of Droplets," (Professor S.D. Heister).

Ganapathy, Harish, "Modeling of Plate/Fastener Contact: Application to Fretting Fatigue," (Professor T.N. Farris).

Viassolo, Daniel E., "Implementation of Digital Controllers," (Professor M.A. Rotea).

Bucci, Gregory Steven, "Modeled High Lift Wake Flows at High Reynolds Number," (Professor J.P. Sullivan).

Emore, Gordon L., "Computational Modeling of Geometric and Material Nonlinearities with Applications to Impact Dynamics," (Professor H.D. Espinosa).

Kannal, Lance, "Spectral Super-Elements for Wave Propagation," (Professor J.F. Doyle).

Munro, Scott Edward, "Effects of Elevated Driver Tube Temperature on the Extent of Quiet Flow in the Purdue Ludwieg Tube," (Professor S.P. Schneider).

Rump, Kurt, "Modeling the Effect of Unsteady Chamber Conditions on Atomization Processes," (Professor S.D. Heister).

Zink, Jonathan, "Mechanical Design and Control of Flexible Joint Robotic Manipulators," (Professor C.T. Sun).

1997

D'Amato, Fernando J., "Control of Uncertain Lightly Damped Systems," (Professor M.A. Rotea).

Gates, Matthew, "A Crack Gage Approach to Monitoring Flaw Growth Potential in Aircraft," (Professor A.F. Grandt, Jr.).

Koutsavdis, Evangelos, "An Investigation of Kirchhoff's Methodologies for Computational Aeroacoustics," (Professor A.S. Lyrintzis and Professor G.A. Blaisdell).

Henry, Mark Edward, "An Experimental Investigation of a Cavitating Slot Orifice," (Professor S.H. Collicott).

Pakalapaati, Rajeev, "Simulation of Heat Treatment of Steels: Experimental Verification of Distortion Due to Martensitic Transformations," (Professor T.N. Farris).

Sexton, Darren G., "A Comparison of the Fatigue Damage Resistance and Residual Strength of 2024-T3 and 2424-T3 Panels Containing Multiple Site Damage," (Professor A.F. Grandt, Jr.).

Torgerson, Shad, "Pressure Sensitive Paints for Aerodynamic Applications," (Professor J.P. Sullivan).

Golden, Patrick John, "A Comparison of Fatigue Crack Initiation at Holes in 2024-T3 and 2524-T3 Aluminum Alloys," (Professor A.F. Grandt, Jr.).

1998

Bougher, Jeffrey, "Part Deflection in Precision Turning of Hardened Steels," (Professor T.N. Farris).

Fanjoy, David, "Using the Genetic Algorithm for Multidisciplinary Helicopter Rotor Airfoil Design: Shape and Topology Approaches," (Professor W.A. Crossley).

Nuss, Jason, "The Use of Solar Sails in the Circular Restricted Problem of Three Bodies," (Professor K.C. Howell).

Ross, Tamaira Emily, "Designing for Minimum Cost: A Method to Assess Commercial Aircraft Technologies," (Professor W.A. Crossley).

Witte, Gerhard R., "Experimental Investigation of a 40% thick Half-Span Boundary Layer Control Wing," (Professor J.P. Sullivan).

Kompella, Sridhar, "Techniques for Rapid Characterization of Grinding Wheel-Workpiece Combinations," (Professor T.N. Farris).

Norris, Stephen, "Longitudinal Equilibrium Solutions for a Towed Aircraft," (Professor D. Andrisani II).

Reimann, William J., "Three Dimensional Form Generation During Infeed Centerless Grinding," (Professor T.N. Farris).

Erausquin, Richard, "Cryogenic Temperature- and Pressure-Sensitive Fluorescent Paints," (Professor J.P. Sullivan).

Green, Stephen, "Fatigue Crack Initiation, Growth, and Coalescence at Notches in High Strength Steel," (Professor A.F. Grandt, Jr.).

Hasebe, Ryoichi Sergio, "Performance of Sandwich Structures with Composite Reinforced Core," (Professor C.T. Sun).

Jones, Brian R., "Aerodynamic and Aeroacoustic Optimization of Rotor Airfoils via Parallel Genetic Algorithm," (Professor W.A. Crossley and A.S. Lyrintzis).

Patanella, Alejandro, "A Novel Experimental Technique for Dynamic Friction Studies," (Professor H.D. Espinosa).

Tsai, Jia-Lin, "Dynamic Delamination Crack Propagation in Polymeric Composites," (Professor C.T. Sun).

VandenBoom, Michael, "A Thermostructural Analysis of a Lined Composite Rocket Nozzle," (Professor S.D. Heister).

1999

Hua, Yuan, "Effective Crack Growth Model and Pseudo Three Dimensional Analysis of Composite Laminates," (Professor C.T. Sun).

Ninan, Lal, "High Strain Rate Characterization of off-Axis Composites Using Split Hopkinson Pressure Bar," (Professor C.T. Sun).

Roeder, Blayne Alan, "Impact of Thermally Confined Alumina/Aluminum Laminates: Experiments and Modeling," (Professor C.T. Sun).

Sanchez, Paul Kevin, "Rigorous Investigation of Cavitation in a 2-D Slot Orifice," (Professor S.H. Collicott).

Bonfiglio, Eugene Peter, "Automated Design of Gravity-Assist and Aerogravity-Assist Trajectories," (Professor J.M. Longuski).

Bowman, Jason C., "The Effects of Low-Observable Requirements on Tanker/Transport Aircraft," (Professor T.A. Weisshaar).

Chang, I-Ling, "Domain Switching Effect on Fracture and Fatigue Behavior of Piezoelectric Materials," (Professor C.T. Sun).

Funk, John, "Ignition Delay Analysis of Non-Toxic Hypergolic Miscible Fuels," (Professor S.D. Heister).

Mailhe, Laurie M., "Design of a High Thrust/Low Thrust Orbital Transfer Vehicle Considering Van Allen Radiation Belts," (Professor S.D. Heister).

Yoon, Suk Goo, "Simulation of the Nonlinear Dynamics of Charged Liquids Using Boundary Element Methods," (Professor S.D. Heister).

Adams, Robert C., Jr., "Force Identification on General Structures," (Professor J.F. Doyle).

Arrieta, Hernan Victor, "Dynamic Testing of Advanced Materials at High and Low Temperatures," (Professor H. Espinosa).

Bodony, Daniel Joseph, "Turbulence Model Computations of an Axial Vortex," (Professor G.A. Blaisdell).

Chandra, Budi Wijaya, "Experimental Investigations of the Unsteadiness in a Cavitating Slot Orifice," (Professor S.H. Collicott).

Li, Chunsu, "Analysis of Cracks in Thin Layers," (Professor C.T. Sun).

Sakaue, Hirotaka, "Porous Pressure Sensitive Paints for Aerodynamic Applications," (Professor J.P. Sullivan).

Tieche, Christopher, "Three-Dimensional Finite Element Analysis of Finite Width Contact," (Professor T.N. Farris).

Williams, Edwin A., "Satellite Constellation Design Methods," (Professor W.A. Crossley).

2000

Bartha, Bence, "Performance of Hard Turned Surfaces," (Professor T.N. Farris).

Cho, Jeongmin, "Modeling and Lowering Thermal Residual Stresses in Bonded Composite Repairs of Metallic Aircraft Structure," (Professor C.T. Sun).

Duke, David K., "Induced Drag Reduction Using Aeroelastic Tailoring and Adaptive Control Surfaces," (Professor T.A. Weisshaar).

Guille, Marianne, "Luminescent Paints Measurement Systems for Aerodynamics Applications," (Professor J.P. Sullivan).

Hahn, Darren James, "Development of Mechanically Fastened Joint Specimens to Rank Aluminum Alloy Fatigue Performance," (Professor A.F. Grandt, Jr.).

Lachendro, Nathan, "Flight Testing of Pressure Sensitive Paint Using a Phase Based Laser Scanning System," (Professor J.P. Sullivan).

Lindqvist, Jens Richard, "Aerodynamic Losses in a Strut and Duct Flow with Lift," (Professor S.H. Collicott).

Frolik, Steven A., "Hypergolic Liquid Fuels for Use with Rocket Grade Hydrogen Peroxide," (Professor J.J. Rusek).

Long, Matthew R., "Characterization of Substrate Geometries for the Catalytic Decomposition of Hydrogen Peroxide," (Professor J.J. Rusek).

McInnes, Allan Ian Stuart, "Strategies for Solar Sail Mission Design in the Circular Restricted Three-Body Problem," (Professor K.C. Howell).

Ong, David, "Investigation of Cavitation in Circular and Slot Orifices with a Step," (Professor S.H. Collicott).

Cho, Sung Man, "A Sub-Domain Inverse Method for Dynamic Crack Propagation Problems (Professor J.F. Doyle).

Cook, Andrea, "Genetic Algorithm Approaches to Optimizing the Location and Number of Smart Actuators on an Aircraft Wing," (Professor W.A. Crossley).

Haradanahalli, Murthy, "Modeling of Fretting Fatigue and Life Prediction in Blade/Disk Contacts," (Professor T.N. Farris).

Heaton, Andrew Floyd, "A Systematic Method for Gravity-Assist Mission Design," (Professor J.M. Longuski).

Marchand, Belinda G., "Temporary Satellite Capture of Short-Period Jupiter Family Comets from the Perspective of Dynamical Systems," (Professor K.C. Howell).

Rufer, Shann J., "Development of Burst-Diaphragm and Hot-Wire Apparatus for Use in the Mach-6 Purdue Quiet-Flow Ludwieg Tube," (Professor S.P. Schneider).

2001

Henderson, Joseph A., "Formal Design Space Evaluation and Optimization for Innovative Aeroelastic Concepts," (Professor Terrence Weisshaar).

Matlik, John, "Motivation, Design, and Calibration of a High Frequency Fretting Fatigue Load Frame," (Professor T.N. Farris).

Beutien, Trisha R., "A Ceramic-Manganese Oxide System for the Decomposition of High Concentration Hydrogen Peroxide," (Professor S.D. Heister).

Pérez-Ruberté, Eddie, "Elasto-Plastic Finite Element Analysis of Contacts with Applications to Fretting Fatigue," (Professor T.N. Farris)

Rangarajan, Balaji, "Characterization of Crack Growth in Elastic-Plastic Materials," (Professor C.T. Sun).

Anderson, Jason P., "Earth-to-Lissajous Transfers and Recovery Trajectories Using an Ephemeris Model," (Professor K.C. Howell).

Lim, Kok-Hong, "Anodic Studies of the Aluminum-Hydrogen Peroxide Semi-Fuel Cell," (Professor John J.J. Rusek).

Miller, Daniel O., "Electrocatalysis in Hydrogen Peroxide Fuel Cells," (Professor S.D. Heister).

Radocaj, Daniel J., "Experimental Characterization of a Simple Gas Turbine Engine Sump Geometry," (Professor S.H. Collicott).

Shimo, Masayoshi, "Modeling of Oil Film Behavior on a Seal Runner and a Sump Wall," (Professor S.D. Heister).

Skoch, Craig R., "Final Assembly and Initial Testing of the Purdue Mach-6 Quiet Flow Ludwieg Tube," (Professor S.P. Schneider).

Valentini, Luiz F.R., "Aft-Fuselage Modification of the Gulfstream IV," (Professor J.P. Sullivan).

Weinstock, Vladimir D., "Modeling of Drop Trajectories and Drop/Wall Interactions in Gas Turbine Engine Sumps," (Professor S.D. Heister).

2002

Bodily, Brandon H., "Mechanical Properties of Single-Walled Carbon Nanotubes by Modeling as Pin-Jointed Trusses," (Prof. C.T. Sun).

Brodrecht, David John, "Parametric Studies of an Aluminum-Hydrogen Peroxide Semi-Fuel Cell," (Professor S.D. Heister).

Bulathsinghala, Ivor J., "Trajectory Control Through Thrust Vector Rotation," (Professor S.D. Heister).

Garcia, Daniel Benjamin, "Fractographic Analysis of Fatigue Failures with Fretting Induced Damage," (Professor A.F. Grandt)

Hoverman, Thomas J., "Performance of New Air-Assist/Air-Blast Atomizer for Large Liquid Mass Flows," (Professor S.H. Collicott).

Kowalkowski, Matthew Kenneth, "Trajectory Control for Agile Landmines," (Professor S.D. Heister).

Krautheim, Michael Stephen, "Downstream Measurements of a High Reynolds Number Horseshoe Vortex," (Professor S.H. Collicott).

Palmer, Robert K., "Development and Testing of Non-Toxic Hypergolic Miscible Fuels," (Professor Steve Heister).

Aratama, Shigeki, "The Static and Fatigue Characteristics of a Notched Hybrid Composite Laminate," (Professor C.T. Sun).

Austin Jr., Benjamin L., "Characterization of Pintle Engine Performance for Nontoxic Hypergolic Bipropellants," (Professor S.D. Heister)

Lamb, Gregory John, "Fatigue Performance of an Advanced Fastener Hole Cold Working Method," (Professor A.F. Grandt).

Martin, Eric Thomas, "Multiobjective Optimization Using Genetic Algorithms with Applications for Aircraft Design," (Professor W.A. Crossley).

Prock, Brian C., "Energy Based Design for Morphing Aircraft Wings," (Professor Terrance Weisshaar).

Remson, Andrew E., "The Non-Catalytic Ignition of High-Performance Monopropellants," (Professor S.D. Heister).

Canino, James, "Characterization of The Turbulent Windage in an Annulus and its Incorporation into Two Sump Design Codes," (Professor S.D. Heister).

Gregory, James W., "Unsteady Pressure Measurements in a Turbocharger Compressor Using Porous Pressure-Sensitive Paint," (Professor J.P. Sullivan).

Kothandaraman, Govindarajan, "Parameter Estimation of the G-12 Parachute," (Professor M.A. Rotea).

Swanson, Erick O., "Mean Flow Measurements and Cone Flow Visualization at Mach 6," (Professor S.P. Schneider).

VanMeter, Michael G., "Experimental Investigation of Hybrid Rocket Combustion Instabilities," (Professor S.D. Heister).

2003

Fitzpatrick, Shannon Lee, "Study of Hydrogen Peroxide Low-Gravity Control for Propellant Management Devices," (Professor S.H. Collicott).

Garman, Karl Edwin, "Design and Testing of a Portable Flight Test Instrumentation and Data Acquisition System for Aircraft," (Professor D. Andrisani)

Koenigs, Michael Richard, "Numerical Methods for Modeling Gas Turbine Combustors," (Professor W.A. Crossley).

Rodrian, Jeffrey Eugene, "Incorporating Real World Engineering Experiences into Aeronautical and Astronautical Engineering Education," (Professor A.F. Grandt, Jr.).

Roth, Brian Douglas, "Aircraft Sizing with Morphing as an Independent Variable: Motivation, Strategies, and Investigations," (Professor W.A. Crossley).

Sisco, James C., "Autoignition of Kerosene by Decomposed Hydrogen Peroxide in a Dump Combustor Configuration," (Professor W.E. Anderson).

Ali, Syed Tabrez, "Heterogeneous Decomposition of Hydrogen Peroxide and Its Stability Margin," (Professor W.E. Anderson).

Ananthula, Vikram, "Computations of an Axial Vortex Using Turbulence Models and Investigation of Rotation Corrections," (Professor G.A. Blaisdell).

Armstrong, Kimberly, "Fundamentals of the Mass Ejection Propulsion System for Mobile Landmines," (Professor S.D. Heister)

Garrison, Loren Armstrong, "Jet Noise Models for Forced Mixer Noise Predictions," (Professor A.S. Lyrintzis).

Gnanamanickam, Ebenezer P., "Piezoelectric Ceramics for High Temperature Aerospace Applications," (Professor J.P. Sullivan).

Lana, Carlos, "Characterization of Color Printers Using Robust Parameter Estimation," (Professor M.A. Rotea).

Matsumura, Shin, "Streamwise-Vortex Instability and Hypersonic Boundary-Layer Transition on the Hyper-2000," (Professor S.P. Schneider).

Matsutomi, Yu, "Impulse Measurement and Modeling of Cyclic Pulse Detonation Engines," (Professor S.D. Heister).

Mok, Jong Soo, "Investigation of Simultaneous Vaporization and Decomposition of Hydrogen Peroxide Using Multiple Jets in Crossflow," (Professor W.E. Anderson).

Stein, William B., "A Theoretical Basis for Electrokinetic Propulsion," (Professor J.J. Rusek and Professor S.D. Heister).

2004

Child, David Rex, "Experimental Validation of Mode I Stress Intensity Factors for The Single-Cracked Pin-Loaded Lug," (Professor A.F. Grandt, Jr.).

Corpening, Jeremy, "Experiments On Combustion of Advanced Hybrid Rocket Fuels Using Hydrogen Peroxide Oxidizer," (Professor S.D. Heister).

Edwards, Jonathan Michael, "Impulse Measurements of Momentum Exchange Propulsion for Mobility Applications," (Professor S.D. Heister).

Gean, Matthew, "Finite Element Analysis of the Mechanics of Blade Disk Contacts," (Professor H. Kim).

Keune, John, "Development of a Hail Ice Impact Model and the Dynamic Compressive Strength Properties of Ice," (Professor H. Kim).

Lew, Phoi-Tack, "Effects of Inflow Forcing On Jet Noise Using 3-D Large Eddy Simulation," (Professor A.S. Lyrintzis).

Niemczura, Jaroslaw, "Evaluation of Positive Displacement Pump Types for Small Scale, Scroll Driven Propellant Feed Systems," (Professor S.D. Heister).

Wright, Charles, "Investigating Correlations Between Reynolds Averaged Flow Fields and Noise for Forced Mixed Jets," (Professor A.S. Lyrintzis).

Chiew, Lee, "Stress-Intensity Factors Solutions for Countersunk Holes Under Pin-Loading," (Professor A.F. Grandt, Jr.).

D'Alto, Luis, "Incremental Quadratic Stability," (Professor M.J. Corless).

Kumari, Shyama, "A Statistical Account of the Effect of Arbitrary Profile Variation on Fretting Fatigue Behavior of the Contact Surface of a Dovetail Joint," (Professor T.N. Farris).

Martin, Thomas N. III, "Infrared Spectroscopy for the Measurement of In-Situ Deposition of Thermally Stressed Hydrocarbon Fuels," (Professor W.E. Anderson).

Nugent, Nicholas J., "Control and Data Acquisition System for Rocket Combustor Experimentation," (Professor W.E. Anderson).

Pearson, Nicholas S., "Vaporization and Decomposition of Hydrogen Peroxide Droplets," (Professor W.E. Anderson).

Qian, Haiyang, "Improved Double-Strap Joint Design," (Professor C.T. Sun).

Subramaniyan, Arunkarthi, "Effect of Nanoclay on Compressive Behavior of Polymeric Composites," (Professor C.T. Sun).

Hrach, Michael A., "Investigation of Oil Drainage from the Bearing Chamber of a Gas Turbine Engine," (Professor S.D. Heister).

Kibbey, Timothy Paul, "Impinging Jets for Application in High-Mach Aircraft Thermal Management," (Professor S.D. Heister).

Liang, Liang, "Numerical Simulation of GPS Code Tracking Loops," (Professor J.L. Garrison).

Pan, Fongloon, "The Use of Surface Integral Methods in Computational Jet Aeroacoustics," (Professor A.S. Lyrintzis).

Quintana, Juan Antonio, "A Global Search Method for Damage Detection in General Structures," (Professor J.F. Doyle).

Rubel, Ken, "Thermal Decomposition of an Advanced Monopropellant," (Professor S.D. Heister).

Smith, Justin A., "Qualification Testing and Fluorescent Imaging of Biobased Aviation Deicer Fluids," (Professor J.P. Sullivan).

Walters, Leon T., "Experimental Study of Diffusing S-Ducts," (Professor J.P. Sullivan).

2005

Deo, Amitabh, "Effect of Shape and Thickness on Properties of Short Fiber Composites," (Professor C.T. Sun).

Genov, Dentcho Angelov, "Electromagnetic Confinement of Nuclear Plasma: Propulsion Perspectives," (Professor I.A. Hrbud)

Harmon, Michael, "Investigation of Combustible Liners for Altitude Compensation in Liquid Rocket Nozzles," (Professor S.D. Heister).

Main, Benjamin, "Evaluation of Continuing Damage From a Cracked Hole with a Failed Ligament," (Professor A.F. Grandt, Jr.).

Miller, Kevin, "Experimental Study of Longitudinal Instabilities in a Single Element Rocket Combustor," (Professor W.E. Anderson).

Patterson, Christopher, "Representations of Invariant Manifolds for Applications in System-to-System Transfer Design," (Professor K.C. Howell).

Tsohas, John, "Altitude Compensation Using Combustible Nozzle Liners," (Professor S.D. Heister).

Lamberson, Steven E., Jr., "Composite Laminate Optimization Techniques Applied to Load-Bearing Circuit Board Design," (Professor W.A. Crossley).

Lee, See-Chen, "Investigation of Transient Ionospheric Perturbations Observed in the GPS Signal," (Professor James L. Garrison).

Mane, Muharrem, "Allocation of Variable Resources and Aircraft Design Using Multidisciplinary Optimization for Systems of Systems," (Professor W.A. Crossley).

Nankani, Kamlesh, "Optimization Approaches for Morphing Airfoils Using Drag and Strain Energy as Objectives," (Professor W.A. Crossley).

Nehrbass, Jonathan, "Drag Benefits of Formation Flight and Morphing on Transatlantic Point-to-Point Commercial Aircraft Service," (Professor W.A. Crossley).

Rausch, Raoul, "Earth to Halo Orbit Transfer Trajectories," (Professor K.C. Howell).

Robarge, Tyler W., "Laminar Boundary-Layer Instabilities on Hypersonic Cones: Computations for Benchmark Experiments," (Professor S.P. Schneider).

Tanner, Travis N., "Applications of Crack Tip Opening Angle for Ductile Fracture Analysis," (Professor C.T. Sun).

Yu, Yen Ching, "Liquid Film Cooling in a Model Combustor Using Swirl Injection," (Professor W.E. Anderson).

Bies, Christopher, "Analysis of an Advanced Aero-Engine Bearing Chamber," (Professor S.D. Heister).

Briggs, Eric, "Comparative Study of Pyrolytic Coking of JP-8 and JP-10 on Inconel and Stainless Steel Surfaces," (Professor W.E. Anderson).

Butt, Adam, "Acoustic Inertial Confinement Fusion: Potential Applications to Space Power and Propulsion," (Professor Ivana Hrbud).

Djibo, Louise-Olivia K., "Natural Conformations: a Study of the Shape of Structures in Equilibrium," (Professor J.F. Doyle).

Eichel, Brenda, "Model-Based Compression of GPS Ephemeris," (Professor J.L. Garrison).

Ge, Yun, "Radial Inertia Effects on Dynamic Response of Extra-Soft Materials in Split Hopkinson Pressure Bar Experiments," (Professor W. Chen).

Habrel, Christopher, "Addressing Uncertainty in Aircraft Sizing During Conceptual Design," (Professor W.A. Crossley).

Jos, Cyril, "Direct-Connect Testing of a Rocket Based Combined Cycle Engine," (Professor W.E. Anderson).

Kakoi, Masaki, "Transfers Between the Earth-Moon and Sun-Earth Systems Using Manifolds and Transit Orbits," (Professor K.C. Howell).

Lovera Yepes, Javier A., "Algorithms for Aircraft Intent Inference and Trajectory Prediction," (Professor M.A. Rotea).

Merrill, Marriner H., "Development of a Simple and Inexpensive Method for Manufacturing High Quality Nanoparticle/Polymer Nanocomposites," (Professor C.T. Sun).

Skillen, Michael, "Developing Response Surface Based Wing Weight Equations for Conceptual Morphing Aircraft," (Professor W.A. Crossley)

Smajlovic, Dino, "Obtaining Ocean Roughness Statistics from Reflected GPS Signals," (Professor J.L. Garrison).

Spohn, Jason, "The Design, Build, and Testing of an Infrared Aim Point Biasing Structure," (Professor H. Kim).

2006

Chen, Jit-Tat, "Use of a PID Control Structure To Design an Autopilot For an Autonomous Aircraft," (Professor D. Andrisani).

Churchfield, Matthew, "Numerical Computations of a Wingtip Vortex in The Near Field," (Professor G.A. Blaisdell).

Deitemeyer, Adam, "Three-Dimensional Characteristics of the Plastic Zone and Necking in the Crack Tip Region," (Professor C.T. Sun).

Grebow, Daniel, "Generating Orbits in The Circular Restricted Three-Body Problem with Application To Lunar South Pole Coverage" (Professor K.C. Howell).

Grupido, Christopher, "Development of a Split Hopkinson Pressure Bar Method for Intermediate Strain Rate Experimental Testing," (Professor W. Chen).

Hsieh, Kelli, "Study of Fatigue Behavior of Embedded Copper in Multifunctional Composite Materials," (Professor H. Kim).

McDonald, Seth, "Design Methodology for Airships of Non-Axisymmetric Cross-Section Geometry," (Professor H. Kim).

Miller, Christopher, "High Altitude Airship Simulation and Control," (Professor J.P. Sullivan).

Moyle, Nicholas, "Experimental Determination of the Mode I Stress Intensity Factor for a Corner Cracked Lug using a Marker Banding Technique," (Professor A.F. Grandt).

Mseis, George, "Mechanical Characterization of an Electrodeposited Nanocrystalline Copper," (Professor T.N. Farris).

Pan, Yi, "The Upper Limit of Constant Strain Rate in a SHPB Experiment for a Linearly Elastic Material," (Professor W. Chen).

Park, Hwun, "Resistance of Adhesively Bonded Composite Lap Joints to Damage by Transverse Ice Impact," (Professor H. Kim).

Sardeshmukh, Swanand, "Performance Analysis of Coupled and Segregated CFD Methods," (Professor C. Merkle).

Ventre, Brian, "Open-Loop Tracking of an Occulting GNSS Signal," (Professor J.L. Garrison).

Delgado, Jorge, "Design and Validation of a Test Facility for Coaxial Radio Frequency Plasma Thrusters," (Professor I.A. Hrbud).

Joshi, Manasi S., "Effect of Temperature on Mode I Fracture Toughness of Certain Polymeric Foams," (Professor R.B. Pipes).

MacDonald, Megan, "On the Nonlinear Dynamic Response of Plain Orifice Atomizers/Injectors," (Professor S.D. Heister).

Manning, Robert, "Gas Bubble Stability in a Sphere Layer," (Professor S.H. Collicott).

Oliver, Anthony, "Evaluation of Two-Dimensional High-Speed Turbulent Boundary Layer and Shock-Boundary Layer Interaction Computations with the OVERFLOW Code," (Professor G.A. Blaisdell and A.S. Lyrintzis).

Saheba, Ruchir, "Real-Time Thermal Observer for Electric Machines," (Professor M.A. Rotea).

Smith, Randolph, "Computational Modeling of High Frequency Combustion Instability in a Single-Element Liquid Rocket Engine," (Professor C. Merkle).

Wennerberg, Jason, "An Experiment to Study the Effects of Aspect Ratio on Rocket Thrust Chamber Cooling Channel Performance," (Professor W.E. Anderson).

Haberlen, Philip, "Supercritical Fuel Film Cooling in a RP-1/Gox Staged Combustion Rocket," (Professor W.E. Anderson).

Heckler, Gregory, "Implementation and Testing of an Unaided Method for the Acquisition of Weak GPS C/A Code Signals," (Professor J.L. Garrison)

Juliano, Thomas J., "Nozzle Modifications for High-Reynolds-Number Quiet Flow in The Boeing/AFOSR Mach-6 Quiet Tunnel," (Professor S.P. Schneider).

Khoo, Teng Thuan, "Effect of Bondline Thickness on Mixed Mode Fracture of Adhesive Joints," (Professor H. Kim).

Nakaima, Daniel Kenithi, "Ductile Fracture of Polycarbonate," (Professor C.T. Sun).

Nightingale, Jay Michael, "Experimental Correlation Between Intrinsic Material Properties and the Failure of Composite Laminates Under Ice Impact," (Professor H. Kim).

Ozimek, Martin, "A Low Thrust Transfer Strategy to Earth-Moon Collinear Libration Point Orbits," (Professor K.C. Howell).

2007
Childress, Karla, "Supersonic Analysis of Plume Characteristics for Asymmetric Nozzles," (Professor C. Merkle).

Grinham, Matthew James, "Parametric Study of Damage Containment Features in Integral Structures for Optimum Crack Retardation," (Professor A.F. Grandt, Jr.).

Karni, Etan, "Experimental Characterization of Opposed Oscillating Wings for a Hovering MAV," (Professor J.P. Sullivan).

Rodkey, Samuel Cole, "Dynamics of Bearing Chamber Air and Oil Flows in Gas Turbine Engines," (Professor S.D. Heister).

Sindiy, Oleg V., "A System-of-Systems Framework for Improved Decision Support in Space Exploration," (Professor D.A. DeLaurentis).

Byron, Jennifer, "Maneuver Costs to Transfers Between Halo and Halo-Like Orbits in the Circular Restricted Three-Body Problem" (Professor K.C. Howell).

Droppers, Lloyd, "Study of Heat Transfer in a Gaseous Hydrogen Liquid Oxygen Multi-Element Combustor," (Professor W.E. Anderson).

Gujarathi, Amit, "Analysis of an Axisymmetric Rocket Based Combined Cycle Engine," (Professor C. Merkle).

Jaron, Jacqueline, "Static Two-Phase Solutions in Laterally-Compressed Circular Cylinders in Zero Gravity," (Professor S.H. Collicott).

Pinheiro, Jacob R., "Aerodynamic Characteristics and Dynamic Modeling of a Lenticular Airship," (Professor J.P. Sullivan).

Schuff, Reuben, "Experimental Investigation of Asymmetric Heating in a High Aspect Ratio Cooling Channel," (Professor W.E. Anderson).

Wilson, Gregory S., "Improving Fatigue Crack Growth Performance of Structural Components Through Consideration of Residual Stress in Design and Manufacture," (Professor A.F. Grandt, Jr.).

Dambach, Erik M., "Development of a Hybrid Rocket Engine for Evaluation of Altitude Compensation Using Ablative Nozzle Liners," (Professor S.D. Heister).

Davendralingam, Navindran, "Modeling of Flexible Structures," (Professor J.F. Doyle).

Kloeden, Richard, "Modeling Crack Propagation in Modern Stiffened Fuselage Structures," (Professor A.F. Grandt, Jr.).

Segura, Rodrigo, "Oscillations in a Forward-Facing Cavity Measured Using Laser-Differential Interferometry in a Hypersonic Quiet Tunnel," (Professor S.P. Schneider).

2008
Braun, Jonathan, "Zero Gravity Two-Phase Stability Solutions of Droplets in a Bent Circular Cylinder," (Professor S.H. Collicott).

Cashbaugh, Jasmine, "An Agent-Base Modeling Approach to a Study of the Effects of Network Logic and a New Aircraft Design on an Air Force Supply Network Located in the Middle East," (Professor D.A. DeLaurentis).

Chigullapalli, Sruti, "Application of High-Order Numerical Schemes for the Boltzmann Transport Equations to Non-Equilibrium Gas Flows," (Professor A. Alexeenko).

Hahn, Jeeyeon, "A Rule-based Conflict Resolution Algorithm for the Next Generation Air Transportation System (NextGen)," (Professor I. Hwang).

Otterstater, Matthew, "Design of an Altitude Simulation Facility for Testing Ablative Nozzle Liners," (Professor S.D. Heister).

Chiu, Jimmy, "A Simulink Model for Vehicle Rollover Prediction and Prevention," (Professor M.J. Corless).

Gordon, Dawn, "Transfers to Earth-Moon L2 Halo Orbits Using Lunar Proximity and Invariant Manifolds," (Professor K.C. Howell).

Hannon, Michael, "Evaluation of Diffuser Modifications for the Boeing/AFOSR Mach-6 Quiet Tunnel," (Professor S.P. Schneider).

Irrgang, Lucia, "Investigation of Transfer Trajectories to and from the Equilateral Libration Points L4 and L5 in the Earth-Moon System," (Professor K.C. Howell).

Lossman, Matthew, "Design and Analysis of Lenticular and Conventional Airships," (Professor J.P. Sullivan).

Moss, James, "Development of Tools for Experimental Analysis of Erosive Burning in/Near Slots in Solid Rocket Motors," (Professor S.D. Heister).

Park, Ki Sun, "Modeling Dense Sprays Produced by Pressure-Swirl Atomizer," (Professor S.D. Heister).

Samson, Daniel, "Micro-Sphere Adhesive System for Quick Repair," (Professor R.B. Pipes and Professor C.T. Sun).

Trebs, Adam, "Biannular Airbreathing Nozzle Rig Facility Development," (Professor S.D. Heister).

Brown, Todd Steven, "Multi-Body Mission Design in the Saturnian System with Emphasis on Enceladus Accessibility," (Professor K.C. Howell).

Conway, Matthew, "A Computational Investigation of Flow Through an Axisymmetric Supersonic Inlet," (Professor G.A. Blaisdell).

Kube-McDowell, Matthew, "Parametric Study of the Generation of Shocks in Near-Critical Turbofan Nozzles," (Professor A.S. Lyrintzis and Professor G.A. Blaisdell).

Moonjelly, Paul, "Design and Implementation of a Purely Electromagnetic Attitude Control System for a Nano-Satellite," (Professor M.J. Corless & Professor D.L. Filmer – co-chairs).

Selby, Christopher, "High Altitude Airship Station Keeping and Launch Model Development Using Output from Numerical Weather Prediction Models," (Professor J.P. Sullivan).

Tate, Nathan John, "Experimental Determination of Mode I Stress Intensity Factors for Elevated Temperature Fretting Fatigue of a Nickel-based Superalloy," (Professor T.N. Farris).

Vavrina, Matthew, "A Hybrid Genetic Algorithm Approach to Global Low-Thrust Trajectory Optimization," (Professor K.C. Howell).

2009

Ayyaswamy, Venkattraman, "Simulation of Low-Density Gas Droplet Supersonic Flows Expanding into Vacuum," (Professor A. Alexeenko).

Jones, Jesse, "Wind Tunnel Testing of a Supersonic Cruise Nozzle in Subsonic Ejector Configuration," (Professor J.P. Sullivan).

Kotecha, Pankit, "Tube Network and Dynamic Sectorization of Airspace for Next Generation Air Transportation System (NextGen)," (Professor I. Hwang).

Md Ishak, Nizam Haris, "Inlet Total Pressure Distortion Effects on Transonic Fan Performance," (Professor N. Key).

Sandroni, Alexander, "Plume and Performance Measurements on a Plug Nozzle for Supersonic Business Jet Applications," (Professor S.D. Heister).

Schlueter, Andrew, "Effects of Z-pinning on Mode I Dynamic Delamination of Woven Composite Laminate," (Professor W. Chen).

Skube, Seth, "Large Strain Compression Testing of Ductile Materials at Quasi-static and Dynamic Strain Rates," (Professor W. Chen).

Ahn, Benjamin, "Forced Excitation of Swirl Injectors Using a Hydro-Mechanical Pulsator," (Professor S. Heister).

Aligawesa, Alinda, "The Detection and Localization of Traffic Congestion for Highway Traffic Systems Using Hybrid Estimation Techniques," (Professor I. Hwang).

Casper, Katya M., "Hypersonic Wind-Tunnel Measurements of Boundary Layer Pressure Fluctuations," (Professor S. Schneider).

Dikshit, Prakash, "Development of an Airport Noise Model Suitable for Fleet-Level Studies," (Professor W. Crossley).

Garibaldi, Oscar D., "Unmanned Aerial Platform for Atmospheric Flux Measurements," (Professor J. P. Sullivan).

Helderman, David, "Measurement and Analysis in A Subscale Rocket Combustor," (Professor W. Anderson).

Kowalkowski, Michael A., "Flame Sensing in Pulse Detonation Combustors Using Diodes, Ion Probes, and High Speed Video," (Professor S. Heister).

Londner, Edward, "The Use of an Inertial Measurement Unit in the Flight Testing of a Small, Remote-Piloted Air Vehicle," (Professor J. P. Sullivan).

Tapee, John, "Experimental Aerodynamics Analysis of a Plug Nozzle for Supersonic Business Jet Application," (Professor J. P. Sullivan).

Wypyszynski, Aaron, "Flight Testing of Small Remotely Piloted Aircraft for System Identification," (Professor J. P. Sullivan).

Kim, Seung-il, "Damage Detection of Sandwich Structure Using Time Domain Reflectometry," (Professor R. B. Pipes).

Needham, Galen J.R., "Small Crack Initiation and Formation: An Investigation into Equivalent Initial Flaw Size and Parallel Offset Surface Crack Interactions," (Professor A. F. Grandt).

O'Hara, Loral, "Investigations Into An Unsteady Heat Release Model Applied to an Unstable Model Rocket," (Professor W. Anderson).

Palumbo, Michael J., "Predicting the Onset of Thermoacoustic Oscillations in Supercritical Fluids," (Professor S. Heister).

Tyagi, Ankit, "Investigating Long-Range Aircraft Staging for Environmental, Economic and Travel Time Impacts," (Professor W. Crossley).

Wheaton, Bradley, "Roughness-Induced Instability in a Laminar Boundary Layer at Mach 6," (Professor S. Schneider).

Appendix C

PhD Thesis Titles

Note: PhD theses indicated by (*) were granted by the Division of Engineering Mechanics, which became the Division of Engineering Sciences in 1954. That department merged with the School of Aeronautical Engineering in 1960 to form the School of Aeronautical and Engineering Sciences. After this date, the asterisk is omitted, and no distinction is made between PhDs granted by the combined school. Since all PhDs require a thesis, this list also summarizes all PhDs awarded by the school.

1949

* Wang, Shou-Wu, "A New Method for Calculating the Cohesive Energy and the Compressibility of Metals, Applied to the Sodium Crystal," (Professors R. G. Sturm and H. M. James).

1950

Duncan, Richard Levi, "Analytical Investigation of the Performance of Gas Turbines Employing Air-Cooled Turbine Blades," (Professor M. J. Zucrow).

1951

* Ata, Abdel-Kerim Mohamed, "Hyper-Elastic Pure Bending Treated by True-Stress True-Strain Relationship," (Professor E. O. Stitz).

* Cooper, William Eugene, "Determination of Principal Elastic Strains," (Professor J. O. Hancock).

Hesse, Walter John, "Analysis of Turbojet Thrust in Flight," (Professor M. J. Zucrow).

* Wetterstrom, Edwin, "An Analysis of the Graver Cylindroid Pressure Vessel," (Professor R. G. Sturm).

1952

* Avery, James Paul, "A Numerical Procedure for Stress Determination in Certain Plain Stress Problems," (Professor N. Little).

* Smith, Lester William, "An Investigation of Discontinuity Stresses in Pressure Vessels by Means of Three Dimensional Photoelasticity," (Professor E. O. Stitz).

1953

* Belsheim, Robert O., "Delayed-Yield Time Effect in Mild Steel Under Sinusoidally-Varying Axial Loads," (Professor J. O. Hancock).

* Kececioglu, Dimitri B., "A Determination of the Plastico-Dynamic Physical Properties of Machinable Metals by Metal Cutting," (Professor J. O. Hancock).

Schmitt, Alfred Frederick, "A Study of the Dynamic Ultimate Energy Absorption Characteristics of Structural Elements," (Professor E. F. Bruhn).

1954

Bollard, Richard J. H., "Effect of Secondary Stresses on the Dynamic Characteristics of Aircraft Structures," (Professor H. Lo).

* McNeilly, Vance Hill, "The Dynamic Stability of a Large Forging Press," (Professor N. Little).

* Neff, Richard Clark, "Separation of Stresses in Photoelasticity by Means of Poisson's Integral Formula," (Professors N. Little and E. O. Stitz).

1955

* Cooper, Gerald Keith, "Prediction of Fracture in the Deep Drawing of Thin Sheet Metal," (Professor N. Little).

* Gray, Robert M., "The Elastic Sphere Under Dynamic and Impact Loads," (Professor A. C. Eringin).

1956

* Douglas, Robert Alden, "Some Thermal Stress Problems in Cylinders," (Professor E. A. Trabant).

Greenberg, Arthur B., "The Mechanics of Flow on a Vertical Surface," (Professor M. J. Zucrow).

* Lambert, John Wallace, "Fluid Flow in a Non-Rigid Tube," (Professor A. E. Eringen).

Novak, Donald Hoyt, "On Detached Shock Waves," (Professor M. E. Shanks).

1957

* Samuels, J. Clifton, "On Stochastic Linear Systems," (Professor A. C. Eringen).

Hromas, Leslie A., "Unsteady Couette Flow," (Professor H. M. DeGroff).

* Lenzen, Kenneth H., (Professor R. G. Sturm).

1958

Goulard, Madeline, "Torsion With Warping Restraint of Thin-Walled Tubes," (Professor H. Lo).

* Hallman, Theodore Morgan, "Combined Forced and Free Convection in a Vertical Tube," (Professor R. J. Grosh).

Kentzer, Czeslaw Paul, "Subsonic Compressible Flows With Circulation by a Method of Correspondence for Cascades and Isolated Airfoils," (Professor M. E. Shanks).

* Woolridge, Charles B., "The Effect of Heater Surface Grain Size on Nucleate Pool Boiling," (Professor R. J. Grosh).

Goulard, R., "The Role of Catalytic Recombination Rates in Stagnation Heat Transfer at Hypersonic Speeds," (Professor DeGroff).

1959

Modi, Vadilal J., "On the Vibrations of Heated Cylinders and Plates," (Professor H. Lo).

* Richardson, B. L., "Some Problems in Horizontal Two-Phase Two-Component Flow," (Professor E. A. Trabant).

1960

Barthel, James Robert, "The Magneto Fluid Dynamics Problems of Stokes Flow and of Hypersonic Flow Around a Cone," (Professor P. S. Lykoudis).

1961

Cappellari, James Oliver, Jr., "The Effect of the Geomagnetic Field on the Orbit of a Charged Satellite," (Professor H. Lo).

Dunkin, John W., "On the Propagation of Waves in an Electromagnetic Elastic Solid," (Professor A. C. Eringen).

Norman, Wendell S., "Lateral Stability of a Vehicle Equipped With a Hydraulic Power Steering System," (Professor H. M. DeGroff).

Zak, Adam Richard, "Buckling by Thermal and External Compressive Loads of Thin Cylindrical Shells Filled With an Elastic Core," (Professor A. J. H. Bollard).

1962
Koh, Severino L., "On the Foundations of Non-Linear Thermo-Viscoelasticity," (Professor A. C. Eringen).

Li, Frank Kuang-Hua, "The Thermal Elasto-Plastic Analysis for Shells of Revolution," (Professor H. Lo).

Seward, Arthur Lewis, "Studies on Thermal Stresses in Elastic, Plastic and Viscoelastic Materials Subjected to Transient Temperatures," (Professor G. Lianis).

Soong, Tsu-teh, "On the Dynamics of a Disordered Linear Chain," (Professor J. L. Bogdanoff).

Valanis, K. C., "Studies in Stress Analysis of Viscoelastic Solids Under Non-Steady Temperature Gravitational and Inertial Loads," (Professor G. Lianis).

Bailey, Cecil, "Vibration and Buckling at Thermally Stressed Plates of Trapezoidal Planform," (Professors H. Lo and Bollard)

1963
Elnan, Odin R. S., "A Feasibility Study of Interception of a High Speed Target by a Beam Rider Missile," (Professor H. Lo).

Jordan, N. F., "On the Static Nonlinear Theory of Electromagnetic Thermo-Elastic Solids," (Professor A. C. Eringen).

1964
Bennett, A. G., "On the Effect of the Sun on the Stability of the Lagrange Points of the Earth and Moon — A Study of Linear Differential Systems with Periodic Coefficients," (Professor H. Polard).

Bober, William, "The Generalized Boltmann Equation and Its Application to the Shock Structure Problem," (Professor S. S. Shu).

Glauz, William Donald, "Aerodynamic Forces on Oscillating Cones and Ogive Bodies in Supersonic Flow," (Professor S. S. Shu).

Jones, Alan Lytton, "Wave Propagation in Optical Fibers," (Professor S. J. Citron).

Jones, John Paul, "The Propagation of Elastic Waves in a Porous, Saturated, Elastic Solid," (Professor A. C. Eringen).

Naghdi, Amir Khosrow, "Stress Distribution in a Circular Cylindrical Shell With a Circular Cutout," (Professor. A. C. Eringen).

Pneuman, G. W., "Hydromagnetic Waves in a Current Carrying Plasma Column With Non-Uniform Mass Density," (Professor P. S. Lykoudis).

Sweet, Arnold Lawrence, "A Stochastic Model for Predicting the Subsidence of Granular Media," (Professor J. L. Bogdanoff).

Yu, Chia-ping, "Magneto-Atmospheric Waves in a Horizontally Stratified Conducting Medium," (Professor P. S. Lykoudis).

Dixon, Roy Conrad, "A Dynamical Theory of Polar Elastic Dielectrics," (Professor A. C. Eringen).

1965
Allen, Robert Thomas, "Theory of the Thermal Accommodation Coefficient," (Professor P. Feuer).

Alspaugh, Dale William, "Development of a Method for the Analysis of the Nonlinear Behavior of Thin Shells of Revolution," (Professor J. E. Goldberg).

Amba-Rao, Chintakindi Lalita, "Fourier Transforms in the Analysis of Thick Plates With an Application to Birefringent Coatings," (Professor M. M. Stanisic).

Bernard, Michael Charles, "Buckling of a Column With Random Initial Displacements," (Professor J. L. Bogdanoff).

Butler, Blaine R., Jr., "The Stuttering Problem Considered from an Automatic Control Point of View," (Professor P. E. Stanley).

DeHoff, Paul Henry, Jr., "Theoretical and Experimental Investigation of Finite Linear Viscoelasticity," (Professor G. Lianis).

Gaston, Charles Arden, "The Singular Case in Trajectory Optimization," (Professor H. Lo).

Hosack, Grant Austin, "On Kinetic Theory of Fully Ionized Gases," (Professor S. S. Shu).

Ingram, John D., "Continuum Theory of Chemically Reacting Media," (Professor A. C. Eringen).

Lear, William M., "On the Use of Ultrastable Oscillators and a Kalman Filter to Calibrate the Earth's Gravitational Field," (Professor J. E. Gibson).

McNitt, Richard Paul, "Irreversible Thermodynamics and Continuum Mechanics in the Presence of Electro-Magnetic Field," (Professor M. M. Stanisic).

Murty, Sistla Sri R., "Two-Dimensional Radiative Transfer in Attached Shock Layer," (Professor R. Goulard and G. M. Palmer).

Sidwell, Kenneth Ward, "The Electromagnetic Basis of the Radiative Transfer Theory," (Professor J. C. Samuels).

1966
Brouillette, Eugene C., "Experimental and Theoretical Analysis of Magneto-Fluid-Mechanic Channel Flow," (Professor P. S. Lykoudis).

Daugherty, Jack Donald, "A Study of the Damping of the Tonks-Dattner Plasma Wave Resonances," (Professor H. DeGroff).

Grot, Richard A., "Relativistic Continuum Mechanics," (Professor A. C. Eringen).

Howsmon, Alan Johnston, "Dynamics Scattering of Particles by Perfect Crystals," (Professor P. Feuer).

Kiel, Roger Eugene," Collisionless Plasma Flow Fields," (Professor W. A. Gustafson).

Mucha, Thomas J., "A Feasibility Study of Pilot Controlled Terminal Phase Rendezvous," (Professors H. Lo and S. J. Citron).

Ramirez, Gilbert Aguirre, "A Relativistic Theory for Viscoelastic Solids," (Professor G. Lianis).

Thomas, John Howard, "A Theoretical Investigation of Magnetohydrodynamic Turbulence," (Professor M. M. Stanisic).

Beil, Robert Junior, "Experimental Investigation of Magnetic Coupling in Dielectric Solids," (Professor A. C. Eringen).

Parfitt, Vaughn Rupert, "Reflection of Plane Waves From the Flat Boundary of a Micropolar Elastic Halfspace," (Professor A. C. Eringen).

Palmer, James Thomas, "Sufficient Conditions for Almost Sure Lyapunov Stability for a Class of Linear Systems," (Professor F. Kozin).

1967
Chapin, Claire Edwin, "Nonequilibrium Radiation and Ionization in Shock Waves," (Professor R. Goulard).

Goldberg, William, "An Experimental Investigation of Nonlinear Isothermal Viscoelasticity," (Professor G. Lianis).

Hugus, Jack William, "An Investigation of the Technique of Statistical Linearization," (Professor F. Kozin).

Li, Nelson Chin-Shui, "On the Theory of the Plasma Resonance Plane Probe," (Professor W. A. Gustafson).

MacGregor, Ronald John, "A Quantitative Statement of the Generator Theory of Nerve Cell Function," (Professor P. S. Lykoudis).

Olson, Carl Adelbert, "Solar Loop Prominences a Theoretical Model," (Professor P. S. Lykoudis).

Pike, Herbert Edward, Jr., "Control of a Distributed Parameter System: the Slab Reheating Furnace," (Professor S. J. Citron).

Uherka, Kenneth L., "A Model for the Evolution of Spiral Galaxies in the Presence of a Magnetic Field," (Professor P. S. Lykoudis).

1968

Groves, Richard Newland, Jr., "A Direct Interaction Approximation of Magnetohydrodynamic Turbulence," (Professor M. M. Stanisic).

Hua, Hsichun M., "Heat Transfer From a Constant Temperature Circular Cylinder in Cross-Flow and Turbulence Measurements in an MFM Channel," (Professor P. S. Lykoudis).

McGuirt, Charles William, "Studies of Thermomechanical Constitutive Equations for Nonlinear Viscoelastic Media," (Professor G. Lianis).

Nelson, Harlan Frederick, "Structure of Shock Waves with Nonequilibrium Radiation and Ionization," (Professor R. Goulard).

Schmitendorf, William Ernest, "On the Riccati Transformation Technique and the Conjugate Point Condition for Optimal Control Problems and Differential Games," (Professor S. J. Citron).

Vogt, Ernest Doyle, "The Perturbation Effects of a Magnetic Dipole On a Charged Satellite," (Professor H. Pollard).

Wiley, Jack Cleveland, "Nonlinear Behavior of Elastic Systems Subjected to Follower Forces," (Professor J. Genin).

1969

Coy, Brent Eugene, "A Theoretical Investigation of Temperature Dispersion in Magnetohydronamic Turbulence," (Professor M. M. Stanisic).

Gardner, Richard Alan, "Magneto-Fluid Mechanic Pipe Flow in a Transverse Magnetic Field With and Without Heat Transfer," (Professor P. L. Lykoudis).

Graetch, Joseph Ernest, "The Electron Density and Electrostatic Potential Field of an Ionospheric Satellite," (Professor W. A. Gustafson).

Hahn, Edwin William, "Minimum Weight Elastic and Plastic Design of Beams and Simple Structures," (Professors S. J. Citron and D. W. Alspaugh).

Hardin, Jay Charles, "A Stochastic Model of Turbulent Channel Flow," (Professor A. L. Sweet).

Ho, Patrick Yu-sing, "Dynamics of Bag-Type Breakup of Droplets in Various Flow Fields," (Professor R. F. Hoglund).

Huffman, Ronald Ray, "The Dynamical Behavior of an Extensible Cable in a Uniform Flow Field — An Investigation of the Towed Vehicle Problem," (Professor J. Genin).

Jenkins, Rhonald Milburn, "A Unified Treatment of Metal Oxide Particle Growth Histories in Rocket Chambers and Nozzles," (Professor R. F. Hoglund).

Lou, Alex Yih-Chung, "Viscoelastic Characterization of a Nonlinear, Fibrous Composite Material," (Professor R. Schapery).

Nahra, John E., "The Balance Function and the Riccati Second-Vibration Optimization Technique," (Professors H. Lo and D. W. Alspaugh).

Regain, Joseph Francis, "Two-Dimensional Analysis of Transonic Gas-Particle Flow Fields in Axisymmetric Nozzles," (Professor R. Hoglund).

Simcox, Craig Dennis, "Theoretical Investigations of Acoustic Turbulent Interactions with Applications to Free-Jet Spreading," (Professor R. Hoglund).

Sweet, Joel, "Dynamic Stability of Rotating Systems," (Professor J. Genin).

Weber, James Alec, "Dispersion Relations for a Collisionless Plasma In An Applied Magnetic Field," (Professor W. A. Gustafson).

1970

Aloisio, Charles Joseph, Jr., "The Application of a Nonlinear Viscoelastic Theory to the Characterization of an ABS Plastic," (Professor R. A. Schapery).

Betts, John Thomas, "An Approximation Technique for Determining the Optimal Control of a Distributed Parameter System," (Professor S. J. Citron).

Cannon, Thomas Calvin, Jr., "A Three Dimensional Study of Towed Cable Dynamics," (Professor J. Genin).

Curtis, Howard Duane, "A Study of Relativistic Constitutive Equations of Continua," (Professor G. Lianis).

Humphreys, Robert Ples, "Design of Maximum Thrust Plug Nozzle for Fixed Inlet Geometry," (Professor J. D. Hoffman).

Katari, Rama Murthy, "Convective Instability of Time-Dependent Flows by the Method of Energy," (Professor M. M. Stanisic).

Luo, Siong Siu Tan, "Discrete Time Series Synthesis of Randomly Excited Structural System Response," (Professor W. M. Gersch).

Mitchell, Richard Rowe, "Necessary and Sufficient Conditions for Sample Stability of Second Order Stochastic Differential Equations," (Professor F. Kozin).

Reid, Robert William, Jr., "On Methods of Calculating Noninferior Performance Index Vectors," (Professor S. J. Citron).

Roos, Rudolf, "Peristaltic Transport in Physiological Systems and the Ureter in Particular," (Professor P. S. Lykoudis).

Shalaby, El-Said El-Sayed, "Relationship of Casting Parameters and Cutting Performance of Cast Cutting Tools — Their Compatibility With Wrought Tools."

Sharpe, David Robert, "An Investigation of the Fitting of Mixed Autoregressive-Moving Average Models to Second Order Stationary Time Series Data," (Professor W. Gersch).

Thiel, Charles Conrad, Jr., "A Method on the Response of Structures to Moving Mass Distributions," (Professor M. M. Stanisic).

Venkataraman, Nellor S., "Introduction and Alfven Wave Drag on Long Cylindrical Satellites," (Professor W. A. Gustafson).

Vogt, Joseph Paul, "Uniaxial Wave Propagation in Nonlinear Viscoelastic Materials," (Professor R. A. Schapery).

Winder, Stephen William, "Estimation for Differential Equation Parameters in the Presence of Additive Noise," (Professor W. M. Gersch).

Wu, Thomas Kong, "Optimal Control of Spacecraft Rendezvous," (Professors M. Goulard and Hsu Lo).

1971

Kissinger, Robert Dohn, "Numerical Solution of a Viscous Compressible Swirling Flow Through a Delaval Nozzle," (Professor A. Ranger).

Luo, Fang-Fu, "Non-Linear Oscillations of a Liquid Excited by Periodic Vertical Forces Under Weak or Zero Gravity Conditions," (Professor S. S. Shu).

Neulieb, Robert Loren, "Relativistic Effects of Nonuniform Motion on the Propagation of an Electromagnetic Wave," (Professor F. J. Marshall).

Seltzer, James Edward, "A Modified Specular Point Theory for Radar Backscatter," (Professor S. S. Shu).

Wang, Jon Yi, "Mathematical Analysis of Optical Remote Sensing of a Thermal System," (Professor R. Goulard).

1972

Booth, Thomas Cobb, "Higher-Order Effects Associated With Steady Streaming About A Vibrating Sphere," (Professor F. J. Marshall).

Chen, Cheng, Jen, "Velocity Oscillations in Solar Place Regions,"(Professor P. S. Lykoudis).

Chen, Kup-hwei, "Dynamical Response of Cylindrical Shells Due to Moving Excitations," (Professor M. M. Stanisic).

Chiu, Palmer Bang-Ming, "A Proposed Technique for Discretization of Boundary Conditions in Time-Dependent Viscous, Compressible Flows," (Professor C. P. Kentzer).

Euler, James Alfred, "Surface Wave Interaction Between An Incompressible Viscous Electrically Conducting Fluid and a Turbulent Inviscid Non-Conducting Fluid,"(Professor M. M. Stanisic).

Ferreira, Sergio Magalhaes Martins, "Forced Convection Condensation of Vapor Flowing Around a Circular Cylinder in Presence of Inert Gas, Gravitational Field and Superheating," (Professor C. P. Kentzer).

Haars, Neil W., "Forces and Torques on Satellites Moving Through Rarefied Plasmas," (Professor W. A. Gustafson).

Helfritch, Dennis J., "The Utilization of a Two-Temperature Gas-Particle Suspension in MHD Power Generation," (Professor W. A. Gustafson).

Hendricks, Charles L., "Optimization of Discrete Systems Containing Random Final Time and Delays with Applications to Production-Inventory Problems," (Professor A. J. Koivo and S. J. Citron).

Kuo, Ying Ming, "A Comparison of Theoretical and Experimental Investigations of Transcapillary Exchanges," (Professor W. A. Gustafson and J. J. Friedman).

MacBain, James C., "Response of Materially Damped Timoshenko Beams Considering Damped Flexible Supports," (Professor J. Genin).

McFarland, Alvin L., "Intermittent Positive Control of Air Traffic in a Horizontal Plane," (Professor D. W. Alspaugh).

Novick, Allen S., "Fluid Dynamics of Meteor Explosives in a Heterogeneous Atmosphere," (Professor C. P. Kentzer).

Picologlou, Basil J. P., "Two Problems of Fluid Mechanics of Physiological Systems: I. Biorheological Aspects of Intestinal Motility; II. Radial Migration of Dilute Suspensions in Poiseuille Flow," (Professor P. S. Lykoudis).

Pope, Rhall Edward, "Design of Stability Augmentation Systems for Decoupling Aircraft Responses," (Professor J. Modrey).

Radwan, Hatem R., "Nonlinear Dynamics of Thin Elastic Shells," (Professor J. Genin).

Schmidt, David K., "Optimal Multiple Aircraft Control For Terminal Area Approach," (Professor R. L. Swaim).

Thurneck, William J., Jr., "Branched Trajectory Optimization by the Method of Steepest Descent," (Professor R. L. Swaim).

Wang, Johnson Chong-Tsun, "Applications of Wierner-Hermite Expansion to the Theories of Turbulence," (Professor S. S. Shu).

Zmuda, Joseph Francis, Jr., "Experimental Investigation of the Multiple Integral Form of the Constitutive Equation for a Nonlinear Viscoelastic Material," (Professor G. Lianis).

Zoeller, Michael A., "The Stability of Time-Dependent Flows of Conducting Boussinesq Fluids by the Method of Energy," (Professor M. M. Stanisic).

1973

Bevilaqua, Paul Michael, "Intermittency, the Entrainment Problem," (Professor P. S. Lykoudis).

Carnahan, James Vernon, "An Analysis of Pedestrian Flow: A Simulation Model for an Airport Pier Arm," (Professor J. L. Bogdanoff).

Cunningham, Thomas Browning, "The Design of a Pilot Augmented Landing Approach Control System," (Professor R. L. Swaim).

Huang, Shiao-Nan, "Minimum Weight Designs of Sandwich Structures," (Professor D. W. Alspaugh).

Kayser, Kenneth William, "A New Method for Estimation of Response in Complex Systems," (Professor J. L. Bogdanoff).

Kim, Hang Wook, "A Study of Free Vibrations of Thin Elastic Shells Subjected to Initial Loads Using Finite Elements," (Professor H. T. Yang).

Kumar, Ravindra, "Emission and Reflection from Healthy and Stressed Natural Targets with Computer Analysis of Spectroradiometric and Mustipsectral Scanner Data," (Professor L. F. Silva).

Whicker, Donald, "Applications of Relativistic Continuum Mechanics to the Interaction of Electromagnetic Waves with Moving Media," (Professor G. Lianis).

Ailor, William Henry III, "Shape and Thickness Effects on Lift Developed by a High Speed Ground Transportation Vehicle Using Ram Air for Support," (Professor W.R. Eberle).

1974

Cheng, Nai-Chung, "Electromechanically Coupled Vibrations and Surface Waves in Piezoelectric Media with Overlays," (Professor C. T. Sun).

Deffenbaugh, Floyd Douglas, "The Aerodynamics Forces on Bodies of Revolution in Separated Flow and the Unsteady Cross Flow Analogy," (Professor F. J. Marshall).

Holaday, Burton, "Minimum-Time Rescue Trajectories Between Spacecraft In Circular Orbits," (Professor R. L. Swaim).

Shafey, Naiim A., "Nonlinear Wave Propagation in Heterogeneous Anisotropic Plates," (Professor C. T. Sun).

Wagner, Roger, "A Finite Element Displacement Approach for the Elasto-Plastic Analysis of Thin Cylindrical Shells," (Professor H. T. Yang))

Chao, Chien-Ming, "Flame Diagnostics By Infrared Inversion Techniques," (Professor R. Goulard).

Cook, Richard D., "A Mathematical Theory of Viscous Fluids With Microstructure and Microtemperature," (Professor S. L. Koh).

Dunn, Patrick Francis, "The Rheological Characterization and Experimental Determination of Biological Material Properties," (Professor P. S. Lykoudis).

Zakem, Steven B., "A Physical and Mathematical Model for the Flow of Suspensions," (Professor P. S. Lykoudis).

1975

Geyer, Howard Karl, "A Contribution To The Stability of A Conducting Boussinesq Fluid With Heat and Mass Transfer," (Professor M. M. Stanisic).

McMasters, John H., "The Optimization of Low-Speed Flying Devices by Geometric Programming," (Professor G. M. Palmer and R. Goulard).

Porter, Milton B. Jr., "Effects of Stability Augmentation on the Gust Response of A STOL Aircraft During a Curved Manual Approach," (Professor R. L. Swaim).

Shashaani, Gholam Reza, "Propagation of Supersonic Rotor Sound Radiations in Jet Engine Inlet Ducts," (Professor P. G. Vaidya).

Chen, Fang-Pai, "Optical Measurement of Jet Engine Exhaust Pollutant Levels," (Professor R. Goulard).

Kelley, John Wilbur, "A Complete Nonconforming Convergent Displacement Formulated Singular Finite Element for Linear Elastic Static and Dynamic Crack Problems," (Professor C. T. Sun).

Montgomery, Stephen T., "An Upper Boundary Turbulent Heat Transport Across a Layer of Electrically Conducting Fluid Under a Magnetic Constraint," (Professor M. M. Stanisic).

1976

DeEskinazi, Jozef, "A Finite Element Model for the Stress Analysis of Pneumatic Tires in Contact with a Flat Surface," (Professor H. T. Yang).

Horvath, John, "A Third-Order Curvilinear Finite Element Utilizing Displacement Derivative Degrees of Freedom and Partitioning," (Professor H. T. Yang).

Patrick, Rayford P., "Magneto Fluid Mechanic Turbulence," (Professor P. S. Lykoudis).

Ymada, Shoji, "A Theoretical and Experimental Study of Strength and Its Distribution in Fiber-Reinforced Composites," (Professor D. W. Alspaugh).

Silverthorn, James Taylor, "Manual Control Displays for a Four Dimensional Approach," (Professor R. L. Swaim).

Papastavridis, John George, "Frames and Coordinates in General Relativity and Their Applications to Mechanical and Electromagnetic Phenomena," (Professor G. Lianis).

Roberts, Philip Arnold, "Effects Of Control Laws And Relaxed Static Stability On Vertical Ride Quality of Flexible Aircraft," (Professor R. L. Swaim).

Wolf, Daniel Star, "Some Problems of Interfacial Cracks," (Professor C. T. Sun).

Best, Eric N., "Null Space and Phase Resetting Curves for the Hodgkin-Huxley Equations," (Professor A. T. Winfree).

Reed, Claude B., "An Investigation of Shear Turbulence In The Presence of Magnetic Fields," (Professor P. S. Lykoudis).

Kunoo, Kazuo, "Minimum Weight Design of Cylindrical Shell with Multiple Stiffener Sizes Under Buckling Constraints," (Professor H. T. Yang).

Staiger, Peter J., "Generation of Surface and Internal Waves in a Viscous Electrically Conducting Density Stratified Fluid by Surface Pressure Fluctuations," (Professor M. M. Stanisic).

Sung, Shung-Hsing, "Finite Element Flutter Analysis of Flat Panels in 3-D Supersonic Unsteady Potential Flow," (Professor H. T. Yang).

Wang, Kuo-Shong, "Propagation of Finite Amplitude Higher Order Mode Sounds in Rigid and Absorbent Ducts with Particular Application to the Multiple Pure Tone Problem," (Professor P. G. Vaidya).

Youtsos, Anastasius G., "Thermomechanical and Electromagnetic Phenomena in Moving and Deformable Media Via Relativity," (Professor G. Lianis).

1977

O'Hair, Edgar A., "A Study of Turbulent Mixing of Bounded Streams," (Professor S. N. B. Murthy).

Feil, Peter J., "A Methodology for the Reconfiguration of a Severely Damaged Electrical Power System," (Professor A. J. Schiff).

Krieger, William, "Frequency Domain Structural Parameter Estimation," (Professor J. L. Bogdanoff).

Newsom, Donald, "Evaluating Lifeline Response to Earthquakes: A Simulation Methodology," (Professor A. J. Schiff).

Van Allen, Richard, "A Monte Carlo Analysis of Science Instrument Scan Sequence for Unmanned Missions to the Planets," (Professor S. J. Citron).

Yen, Wen Yo, "Effects of Dynamic Aeroelasticity on Handling Qualities and Pilot Rating," (Professor R. L. Swaim).

1978

Baig, Mirza Irfan, "Dynamic Response Analysis of Large Fossil Fuel Plant Structure," (Professor H. T. Yang).

Chen, Chun-Chieh, "Simple Models for Computing Dynamic Responses of Complex Frame Structures," (Professor C. T. Sun).

Shiau, Le-Chung, "Theoretical Study of the Dynamic Response of a Chimney to Earthquake and Wind," (Professor H. T. Yang).

Guthrie, Gregory, "A Mini Computer Based Interactive Computer Graphics System With Illustrative Application to Structural Data Analysis," (Professor A. J. Schiff).

Gran, Carl S. "The Seismic Response of a Column-Supported Cooling Tower," (Professor H. T. Yang).

1979

Hunckler, Charles, "The Dynamic Behavior of an Automobile Tire," (Professor H. T. Yang).

Kendall, Donald, "The Effects of Joule Heat on Turbulent Dispersion of Temperature in High Reynolds Number Flow," (Professor M. M. Stanisic).

Miller, Ronnie K., "Acoustic Emission: An Application to Fracture Mechanics," (Professor C. T. Sun).

Chan, Wen-Sheng, "Finite Element Analysis of Problems Involving Stress Singularities," (Professor C. T. Sun).

Greenburg, Alan M., "A Contribution to the Stability of Electrically Conducting Boussinesq Fluids by the Energy Method," (Professor M. M. Stanisic).

Staab, George, "Singular Finite Element Shape Functions in Estimating Stress Intensity Factors," (Professor C. T. Sun).

Sprandel, James K., "Structural Parameter Identification of Member Characteristics in a Finite Element Model," (Professor A. J. Schiff).

1980

Stiles, Randall, "Time Dependent Analysis of Two-Dimensional, Chemically Reacting, Nonequilibrium Nozzle Flows," (Professor J. D. Hoffman).

Bellagamba, Lawrence, "A Least Squares Method for Constrained Minimization with Application in Structural Design," (Professor H. T. Yang).

Chang, Li Ko, "The Theoretical Performance of High Efficiency Propellers," (Professor J. P. Sullivan).

Chou, Yuan-Fang, "Frequency Effect on Fatigue Crack Propagation in PMMA Panels," (Professor C. T. Sun).

Guruswamy, P, "Aeroelastic Stability and Time Response Analysis of Conventional and Supercritical Airfoils in Transonic Flow by Time Integration Method," (Professor H. T. Yang).

1981

Davis, Donald W. "Application of the Method of Weighted Residuals to the Boundary Layer Momentum and Turbulent Stress Transport Equations," (Professor W. A. Gustafson).

Day, Michael L., "A Mixed Variational Principle for Finite Element Analysis," (Professor H. T. Yang).

Chiu, Di, "Optimal Sensor/Actuator Selection for Linear Stochastic Systems," (Professor R. E. Skelton).

Milling, Robert, "An Experimental Study of Tollmien-Schlicting Wave Cancellation," (Professor J. P. Sullivan).

Yedavalli, Ramakrishna, "Control Design for Parameter Sensitivity Reduction in Linear Regulators: Application to Large Flexible Space Structures," (Professor R. E. Skelton).

Striz, Alfred G., "Application of Harmonic Analysis Method to Aeroelastic Stability Analysis of Conventional and Supercritical Airfoils in Transonic Flow," (Professor H. T. Yang).

Chen, Chia-Hsia, "Flutter and Time Response Analyses of Three Degree of Freedom Airfoils in Transonic Flow," (Professor H. T. Yang).

Yang, Shih-Hsian, "Static and Dynamic Contact Behaviors of Composite Laminates," (Professor C. T. Sun).

1982

Kim, Bong J., "An Approximate Method for Dynamic Analysis of Beam-and Plate-Like Structures," (Professor C. T. Sun).

Tsuchiya, Toshiaki, "Aerothermodynamics of Axial-Flow-Compressors with Water Ingestion," (Professor S. N. B. Murthy).

Wang, Tsorng, "Dynamic Response and Damage Model of a Graphite/Epoxy Laminate to Hard Object Impact," (Professor C. T. Sun).

Chang, Yuan-Bin, "Linear Dynamic Analysis of Revolutional Shells Using Finite Elements," (Professors H. T. Yang and W. Soedel).

Tan, Tein-Min, "Wave Propagation in Graphite/Epoxy Laminate Due To Impact," (Professor C. T. Sun).

Dumanis-Modan, Alon, "Evaluation of the Crack Gage as an Advanced Individual Aircraft Tracking Concept," (Professor A. F. Grandt, Jr.).

Han, An-Dong, "Buckled Plate Vibrations, Large Amplitude Vibrations, and Nonlinear Flutter of Elastic Plates Using High-Order Triangular Finite Elements," (Professor H. T. Yang).

Renie, John, "Combustion Modeling of Composite Solid Propellants," (Professor J. R. Osborn).

1983

DeLorenzo, Michael L. "Sensor and Actuator Selection for the Regulation of Linear Stochastic Systems," (Professor R. E. Skelton).

Hong, Seung Kyu, "Large Eddy Interactions in Curved Wall Boundary Layers — Model and Implications," (Professor S. N. B. Murthy).

Innocenti, Mario, "Cooperative Pilot-Optimal Augmentation System Synthesis for Complex Flight Vehicles," (Professor D. K. Schmidt).

Batina, John, "Transonic Aeroelastic Stability and Response of Conventional and Supercritical Airfoils Including Active Controls," (Professor H. T. Yang).

Moore, Carleton, "A New 48 D.O.F. Quadrilateral Shell Element With Variable-Order Polynomial and Rational B-Spline Geometries with Rigid Body Modes," (Professor H. T. Yang).

1984

Biezad, Daniel J., "Time Series Analysis of Closed-Loop Pilot Vehicle Dynamics," (Professor D. K. Schmidt).

Chen, Jinn-Kuen, "Nonlinear Analysis of Composite Laminates," (Professor C. T. Sun).

Pollock, Peter B., "A Study of Failure in 3-D Carbon Composites," (Professor C. T. Sun).

Sankar, Bhavani, "Contact Behavior, Impact Response and Damage in Graphite Epoxy Laminates Subjected to Initial Stresses," (Professor C. T. Sun).

Yuan, Pin-Jar, "Identification of Pilot Dynamics and Task Objectives from Men-in-the-Loop Simulation," (Professor D. K. Schmidt).

Kamle, Sudhir, "Reflection and Transmission of Flexural Waves at Discontinuities," (Professor J. F. Doyle).

Tsai, Gwo-Chung, "A Study of Frequency Effects and Damage Accumulation During the Fatigue of Graphite/Epoxy and Glass/Epoxy Composite Materials," (Professor C. T. Sun).

Wagie, David, "Model and Controller Reduction of Uncertain Systems Using Covariance Equivalent Realizations," (Professor R. E. Skelton).

Gau, Ching-Fu, "Recursive Partitioning Estimation," (Professor D. Andrisani, II).

Miller, Christopher, "Optimally Designed Propellers Constrained by Noise," (Professor J. P. Sullivan).

1985

Kapania, Rakesh, "Deterministic and Non-Deterministic Response Analysis of Complex Shells," (Professor H. T. Yang).

Sundar, R. M., "An Experimental Investigation of Propeller Wakes Using a Laser Doppler Velocimeter," (Professor J. P. Sullivan).

Joshi, Shiv, "Impact Damage Characterization in Laminated Composites," (Professor C. T. Sun).

Lamberson, Steven, "Equivalent Continuum Finite Element Modeling Of Plate-like Lattice Structures," (Professor H. T. Yang).

Zeiler, Thomas, "An Approach to Integrated Aeroservoelastic Tailoring for Stability," (Professor T. A. Weisshaar).

Grady, Joseph, "Dynamic Delamination Crack Propagation in a Graphite/Epoxy Laminate," (Professor C. T. Sun).

Leewood, Alan, "Numerical Studies of Problems in Anisotropic Plasticity," (Professor J. Doyle and C. T. Sun).

Richard, Ibrahim M., "An Experimental Investigation of the Development and Growth of Counter Rotating Vortices in a Curved Duct," (Professor J. P. Sullivan).

Saigal, Sunil, "Geometric and Material Nonlinear Dynamic Analysis of Complex Shells," (Professor H. T. Yang).

1986

Parpia, Ijaz H. "Multidimensional Time Dependent Method of Characteristics," (Professor C. P. Kentzer).

Perez, Rigoberto, "Stress Intensity Factors for Fatigue Cracks at Holes Under Cyclic Stress Gradients," (Professor A. F. Grandt, Jr.).

Miller, David P., "A Numerical Heat Transfer Model for an Advanced Solar Thermal Propulsion System," (Professor J. R. Osborn).

Chen, Alexander T., "Formulation for Symmetrically Laminated Beam and Plate Finite Elements With Shear Deformation," (Professor H. T. Yang).

Chen, Ta Kang, "An Interaction Synthesis Approach for Robust Active Flutter Suppression Control Law Design," (Professor D. K. Schmidt).

Rechak, Said, "Effect of Adhesive Layers on Impact Damage and Dynamic Response on Composite Laminates," (Professor C. T. Sun).

1987

Collins, Emmanuel G., "State Covariance Assignment of Discrete Systems: Development and Applications," (Professor R. E. Skelton).

Kelly, Scott, "Failure Analysis of Laminated Composite Angels," (Professor C. T. Sun).

Necib, Brahim, "Continuum Modeling and Dynamic Analysis of Large Truss Structures," (Professor C. T. Sun).

Zhou, Shu Gong, "Failure Analysis of Quasi-Isotropic Composite Laminate With Free Edges Under Off-Axis Loading," (Professor C. T. Sun).

Chin, Hsiang, "Effect of Bending-Extension Coupling on the Deformation of Asymmetric Laminates," (Professor C. T. Sun).

Pressley, Homer M. Jr., "In Situ Propellant Burning Rate Determination Using Flash Radiography," (Professor J. R. Osborn).

Ray, Suvendoo K., "A Three-Dimensional Investigation of Steady State Fatigue Crack Closure Behavior for Through-Thickness and Part-Through Flaws," (Professor A. F. Grandt, Jr.).

Chen, Shih-Hsiung, "Panel Method for Counter Rotating Propellers," (Professor M. H. Williams).

Harrison, Garry, "An Experimental Investigation of the Flow Field Around a Counter-Rotating Propeller System Using a Laser Doppler Velocimeter," (Professor J. P. Sullivan).

Hu, Anren, "Modal Cost Analysis of Flexible Structures: Modeling Flexible Structures for Control Design," (Professor R. E. Skelton).

Liu, Richard, "Vibration and Aeroelastic Tailoring of Advanced Composite Plate-Like Lifting Surfaces," (Professor T. A. Weisshaar).

Pope, John Edward, "A Three-Dimensional Model of Fatigue Crack Growth in Incorporating Crack Closure," (Professor A. F. Grandt, Jr.).

1988

Anderson, Mark R., "Actuation Constraints in Multivariable Flight Control Systems," (Professor D. K. Schmidt).

Frederick, Robert A., Jr., "Combustion Mechanisms of Wide Distribution Solid Propellants," (Professor J. R. Osborn).

Garg, Sanjay, "Model-Based Analysis and Cooperative Synthesis of Control and Display Augmentation for Piloted Flight Vehicles," (Professor D. K. Schmidt).

Troha, William Alan, "Influence of Fatigue Crack Closure on the Growth Rate of Surface Flaws," (Professor A. F. Grandt, Jr.).

Shin, Bohyun, "The Development of the Gortler Vortices in a Curved Duct," (Professor M. H. Williams).

Wang, Rong Tyai, "Dynamics and Control of Structures With Extendible Members," (Professor C. T. Sun).

Wu, Yii-Cheng, "A Geometrically Nonlinear Tensorial Formulation of a Skewed Quadrilateral Thin Shell Finite Element," (Professor H. T. Yang).

Blomshield, Frederick S., "Nitramine Composite Solid Propellant Modeling," (Professor J. R. Osborn).

Cho, Jinsoo, "Frequency Domain Aerodynamic Analysis of Interacting Rotating Systems," (Professor M. H. Williams).

Gleason, Daniel, "Analytical Techniques for Tracking Filter Implementation," (Professor D. Andrisani, II).

Jen, Kwang-Chi, "On the Effect of Matrix Cracks on Laminate Strength," (Professor C. T. Sun).

King, Andrew M., "Discretization and Model Reduction for a Class of Nonlinear Systems," (Professor R. E. Skelton).

1989

Kao, Yung-Fu, "A Two Dimensional Unsteady Analysis for Transonic and Supersonic Cascade Flows," (Professor M. H. Williams).

Sawyer, Richard Steven, "Measurement of Lift Development on Rapidly Accelerated Wings," (Professor J. P. Sullivan).

Dawicke, David S., "Three-Dimensional Fatigue Crack Closure Behavior of Metals," (Professor A. F. Grandt, Jr.).

Gates, Thomas, "An Elastic/Viscoplastic Constitutive Model for Fiber Reinforced Thermoplastic Composites," (Professor C. T. Sun).

Le, Dzu Khac, "A Parametric Pole-Matching Approach to Relative Stability and System Integrity," (Professor A. Frazho).

Monoharan, M. G., "Energy Release Rate Analysis of Interfacial Cracks in Composites," (Professor C. T. Sun).

Norman, Timothy Lee, "The Controlled-Damage Concept in Design of Laminated Composites," (Professor C. T. Sun).

Pidaparti, Ramana Murthy V., "Modeling and Fracture Prediction in Rubber Composites," (Professor H. T. Yang).

Rizzi, Stephen P., "A Spectral Approach to Wave Propagation in Layered Solids," (Professor J. Doyle).

Wanthal, Steven P., "Three-Dimensional Finite Element Formulations for Thick and Thin Laminated Plates," (Professor H. T. Yang).

Blair, Maxwell, "Development and Application of the Time Domain Panel Method," (Professor T. A. Weisshaar).

Chien, Lung-Siaen, "Parallel Computational Methods on Structural Mechanics Analysis," (Professor C. T. Sun).

Ku, Chieh-Chang, "Three Dimensional Full Potential Method for the Aeroelastic Modeling of Propfans," (Professor M. H. Williams).

1990

Chen, Chih-Tsai, "Nonlinear Random Vibrations of Structures," (Professor H. T. Yang).

Davis, Roger E., "Secondary Flow and Three-Dimensional Separation in Curved Circular Ducts," (Professor J. P. Sullivan).

Hsieh, Chen, "Control of Second Order Information for Linear Systems," (Professor R. E. Skelton).

Hwang, Ching-Chywan, "Propfan Supersonic Panel Method Analysis and Flutter Predictions," (Professor M. H. Williams).

Nam, Changho, "Aeroservoelastic Tailoring for Lateral Control Enhancement," (Professor T. A. Weisshaar).

Pernicka, Henry John, "The Numerical Determination of Nominal Libration Point Trajectories and Development of a Station-Keeping Strategy," (Professor K. C. Howell).

Mao, Kunming, "A Global-Local Structural Analysis Method," (Professor C. T. Sun).

Augenstein, David L., "A Recursive Solution to the Suboptimal Block Nehari Problem," (Professor A. Frazho).

Liaw, Der-Guang Leslie, "Reliability Study of Uncertain Structures Using Stochastic Finite Elements," (Professor H. T. Yang).

Yoon, Kwang-Joon, "Characterization of Elastic-Plastic and Viscoplastic Behavior of AS4/PEEK Thermoplastic Composite," (Professor C. T. Sun).

1991

Liao, Wei-Chong, "Analysis of Thick Section Composite Laminates," (Professor C. T. Sun).

Caipen, Terry L., "Invariant Boundary Conditions for Cascade Flows," (Professor M. H. Williams).

Ehlers, Steven M., "Aeroelastic Behavior of an Adaptive Lifting Surface," (Professor T. A. Weisshaar).

Heinstein, Martin, "Simulation of Shear Band Phenomenon in Metal Forming and Cutting," (Professor H. T. Yang).

Rui, Yuting, "Elastic-Plastic Behavior of Thermoplastic Composites Under Cyclic Loadings," (Professor C. T. Sun).

Shih, Jua-Min, "Experimental and Finite Element Simulation Methods for Metal Forming and Cutting Processes," (Professor H. T. Yang).

Bacon, Barton J., "Order Reduction for Closed-Loop Systems," (Professors D. K. Schmidt and D. Andrisani, II).

Gordon, Steven C., "Orbit Determination Error Analysis and Station-Keeping for Libration Point Trajectories," (Professor K. C. Howell).

Jih, Chan-Jiun, "Analysis of Delamination in Composite Laminates Under Low Velocity Impact," (Professor C. T. Sun).

Johnston, Robert T., "Unsteady Propeller/Wing Aerodynamic Interactions," (Professor J. P. Sullivan).

Kim, Eung Tai, "The Robust Estimators for Target Tracking," (Professor D. Andrisani, II).

Kim, Jae Hoon, "Model Reduction, Sensor/Actuator Selection and Control System Redesign for Large Flexible Structures," (Professor R. E. Skelton).

Liu, Ketao, "Q-Markov Cover Identification and Integrated MCA-OVC Controller Design for Flexible Structures," (Professor R. E. Skelton).

Luo, Jiayi, "Methods of Analysis for Thick-Section Composite Structures," (Professor C. T. Sun).

Wang, Cong, "Predictive Forming of Advanced Thermoplastic Composite Structures," (Professor C. T. Sun).

Young, Min-Jho, "Analysis of Cracked Plates Subjected to Out-of-Plane Loadings," (Professor C. T. Sun).

1992

Blair, Kim Billy, "Nonlinear Dynamic Response of Shallow Arches to Harmonic Forcing," (Professor T. N. Farris).

Da, Dong, "Robust Control of Uncertain Aerospace and Mechanical Systems With Unmodelled Flexibilities: A Singular Perturbation Approach," (Professor M. J. Corless).

Krozel, James A., "Intelligent Path Prediction for Vehicular Travel," (Professor D. Andrisani, II).

Layton, Jeffrey B., "Multiobjective Control/Structure Design Optimization in the Presence of Practical Constraints," (Professors A. Weisshaar and L. Peterson).

Thornton, Anthony L., "Application of a Solution Adaptive Grid to Flow Over An Embedded Cavity," (Professors C. P. Kentzer and M. H. Williams).

Zhu, Guoming, "L_2 and L_Infinity Multiobjective Control for Linear System," (Professor R. E. Skelton).

Chung, Ilsup, "Inelastic Behavior of Thermoplastic Composites," (Professor C. T. Sun).

Hiday, Lisa Ann, "Optimal Transfers Between Libration-Point Orbits in the Elliptic Restricted Three-Body Problem," (Professor K. C. Howell).

Lee, Shi-Wei Ricky, "Investigation and Modeling of Penetration Process for Composite Laminates Impacted By a Blunt-Ended Projectile," (Professor C. T. Sun).

Liu, Minzhu, "Three-Dimensional Boundary Element Analysis of Contract and Fracture Problems," (Professor T. N. Farris).

Newman, Brett A., "Aerospace Vehicle Model Simplification for Feedback Control," (Professors D. Andrisani, II and D. K. Schmidt).

Tan, Jiak-Kwang, "Simulation of Vortex Bursting," (Professor M. H. Williams).

Weng, Tung-Li, "Numerical Investigations of Fracture Behavior of Ductile Materials," (Professor C. T. Sun).

Wu, Chung-Lin, "Low Velocity Impact of Composite Sandwich Plate," (Professor C. T. Sun).

Chu, Gou-Don, "Free Edge and Size Effects on Failure Initiation and Ultimate Strength of Laminates Containing a Center Hole," (Professor C. T. Sun).

Rausch, Russ David, "Time-Marching Aeroelastic and Spatial Adaptation Procedures on Triangular and Tetrahedral Meshes Using an Unstructured-Grid Euler Method," (Professor H. T. Yang).

Srinivasan, Gopalakrishan, "Spectral Analysis of Wave Propagation in Connected Waveguides," (Professor J. F. Doyle).

1993

He, Chien-Cheng, "Three-Dimensional Finite Element Analysis of Free Edge Delamination in Composite Laminate," (Professor H. T. Yang).

Jiang, Yi-Tsann, "Development of an Unstructured Solution Remising Algorithm for the Quasi-Three-Dimensional Euler and Navier Stokes Equations," (Professor W. J. Usab, Jr.).

Swei, Shan-Min, "Robust Stabilization of Uncertain Systems," (Professors M. J. Corless and M. A. Rotea).

Tsiotras, Panagiotis, "Analytic Theory and Control of the Motion of Spinning Rigid Bodies," (Professor J. M. Longuski).

Jun, Alexander, "Compressive Strength of Unidirectional Fiber Composites with Matrix Nonlinearity," (Professor C. T. Sun).

Kurtz, Russell D., "Analysis of the Fiber Pullout Test and Associated Bimaterial Crack Problems," (Professor C. T. Sun).

Puig-Suari, Jordi, "Aerobraking Tethers for the Exploration of the Solar System," (Professor J. M. Longuski).

Wang, Kuo-Feng, "Simulation of Quenching and Induction Heat Treatment Processes with Experimental Verification," (Professor H. T. Yang).

Blatt, Paul A., "Fatigue Crack Growth Behavior of a Titanium Alloy Matrix Composite Under Thermomechanical Loading," (Professor A. F. Grandt, Jr.).

Chen, Shi-Yew, "Experimental and Boundary Element Analysis of Hertzian Cone Cracking," (Professor T. N. Farris).

Georgiou, Ioannis, "Nonlinear Dynamics and Chaotic Motions of a Singularly Perturbed Nonlinear Viscoelastic Beam," (Professor M. J. Corless).

Iwasaki, Tetsuya, "A Unified Matrix Inequality Approach to Linear Control Design," (Professor R. E. Skelton).

Rodrigues, Eduardo A., "Linear and Nonlinear Discrete-Time State-Space Modeling of Dynamic Systems for Control Applications," (Professor T. A. Weisshaar).

1994

Kherat, Samir Med, "Some Aspects of Gramian Assignment and Interpolation Problems in Control Systems," (Professor A. Frazho).

Berry, Dale Thomas, "A Formulation and Experimental Verification for a Pneumatic Finite Element," (Professor H. T. Yang).

Dixon, Vernon, "Vortex Dynamics in a Viscous Incompressible Fluid," (Professor M. H. Williams).

Fur, Lih-Shing, "Vibration Control of Flexible Structures," (Professor H. T. Yang).

Grigoriadis, Karolos M., "Alternating Projection Techniques for Multiobjective Control," (Professor R. E. Skelton).

Hucker, Scott Alan, "Grinding of Hardened Steel for Tribological Performance," (Professor T. N. Farris).

Martin, Michael, "Inverse Problems in Structural Dynamics," (Professor J. F. Doyle).

Park, Seungbae, "Fracture Behavior of Piezoelectric Ceramics," (Professor C. T. Sun).

Gilbert, Michael Glenn, "Multilevel Control-Structure Design Using LQG Design Sensitivity Analysis Techniques," (Professor T. A. Weisshaar).

Morris, Scott Richard," The Characterization and Modeling of Advanced Thermoplastic Composite Material Forming Processes," (Professor C. T. Sun).

1995

Ankireddi, Seshasayee, "Passive and Active Control of Civil Structures Under Dynamic Wind Loads," (Professor H. T. Yang).

Moulik, Pradipta, "Simulation of Surface Grinding," (Professor H. T. Yang).

Zhang, Shucheng, "Molecular Mixing Measurements and Turbulent Structure Visualizations in Jets With and Without Tabs," (Professor S. P. Schneider).

Chen, Yongliang, "Numerical Approaches for Hydrodynamic Cavitating Flows," (Professor S. D. Heister).

Smith, Michael J. C., "Simulating Flapping Insect Wings Using An Aerodynamic Panel Method: Towards the Development of Flapping Wing Technology," (Professor M. C. Williams).

Zenieh, Salah, "Nonlinear Control of Mechanical and Aerospace Systems With Flexible Elements: Application to Robotics," (Professor M. J. Corless).

Bell, Julia Lea, "Primer Vector Theory in The Design of Optimal Transfers Involving Libration Point Orbits," (Professor K.C. Howell).

Pandey, Ravindra K., "Mechanics of Formation of Wrinkles During the Processing of Thermoplastic Composites," (Professor C.T. Sun).

Potti, Shyam V., "High Velocity Impact and Penetration of Thick Composite Laminates," (Professor C.T. Sun).

Weeks, Craig Andrew, "Nonlinear Rate Dependent Response of Thick-Section Composite Laminates," (Professor C.T. Sun).

1996

Liu, Tianshu, "Applications of Temperature Sensitive Luminescent Paints in Aerodynamics," (Professor J.P. Sullivan).

Prasanth, Ravi Varma R., "Multiple Objective H∞ Control," (Professor M.A. Rotea).

Tao, Jianxin, "Characterization of Matrix Cracking and Delamination in Laminated Composites," (Professor C.T. Sun).

Woodard, Paul, "Analysis of the Carburization and Quenching of Steels," (Professor H.T. Yang).

Wu, Peir-Shin, "Pin-Contact Failure in Composite Laminates," (Professor C.T. Sun).

Zheng, Shunfeng, "Modeling Delamination in Composite Laminates," (Professor C.T. Sun).

Hilbing, James H., "Nonlinear Modeling of Atomization Processes," (Professor S.D. Heister).

Huang, Wen-Liang, "Unsteady Aerodynamics of Advanced Ducted Fans," (Professor M.H. Williams).

Kim, Hong-On, "Elastic-Plastic Fracture Analysis for Small Scale Yielding," (Professor C.T. Sun).

Ely, Todd A., "Dynamics and Control of Artificial Satellite Orbits with Multiple Tesseral Resonances," (Professor K.C. Howell).

Kolonay, Raymond, "Unsteady Aeroelastic Optimization in the Transonic Regime," (Professor H.T. Yang).

McCoy, Robert, "Dynamic Analysis of a Thick-Section Composite Cylinder Subjected to Underwater Blast Loading," (Professor C.T. Sun).

Sims, Jon Andrew, "Delta-V Gravity-Assist Trajectory Design: Theory and Practice," (Professor J.M. Longuski).

Spyropoulos, Evangelos T., "On Dynamic Subgrid-Scale Modeling for Large-Eddy Simulation of Compressible Turbulent Flows," (Professor G.A. Blaisdell).

Su, Xuming, "Three-Dimensional Effects in Elastic-Plastic Fracture," (Professor C.T. Sun).

Zhang, Xiaodong, "Adaptive Sandwich Structures," (Professor C.T. Sun).

1997

Davidson, John B., "Lateral-Directional Eigenvector Flying Qualities Guidelines and Gain Weighted Eigenspace Assignment Methodology," (Professor D. Andrisani II).

Heinimann, Markus, "Analysis of Stiffened Panels with Multiple Site Damage," (Professor A.F. Grandt, Jr.).

Ju, Yongqing, "Thermal Aspects of Grinding for Surface Integrity," (Professor T.N. Farris).

Leeks, Tamara, "Active Aeroelastic Panels with Optimum Self-Straining Actuators," (Professor T.A. Weisshaar).

Qian, Wenqi, "Interfacial Cracks in Isotropic and Anisotropic Media with Friction," (Professor C.T. Sun).

Vaidya, Rajesh S., "Failure Criterion for Notched Fiber Dominated Composite Laminates," (Professor C.T. Sun).

Guo, Cheng, "Dynamic Delamination of Composites," (Professor C.T. Sun).

Hill, Lisa Rene, "Three-Dimensional Piezoelectric Boundary Elements," (Professor T.N. Farris).

Klug, John, "Fracture and Fatigue of Bonded Composite Repairs," (Professor C.T. Sun).

Wernimont, Eric, "Experimental Study of Combustion in Hydrogen Peroxide Hybrid Rockets," (Professor S.D. Heister).

Brockman, Mark, "Quadratic Boundedness of Uncertain Nonlinear Dynamic Systems," (Professor M.J. Corless).

Hoffenberg, Robert, "Simulation of High-Lift Wake Behavior," (Professor J.P. Sullivan).

Lu, Jianbo, "Robust Control and Control System Integration Using Finite-Signal-to-Noise Model," (Professor R.E. Skelton).

Schmisseur, John, "Receptivity of the Boundary Layer on a Mach-4 Elliptic Cone to Laser-Generated Localized Freestream Perturbations," (Professor S.P. Schneider).

Shi, Guojun, "Integrated Identification, Modeling, and Control with Applications," (Professor R.E. Skelton).

Smith, Monty, "Multiple Objective Control System Synthesis Via Lifting Techniques," (Professor A.E. Frazho).

Thiruppukuzhi, Srikanth, "High Strain Rate Characterization and Modeling of Polymer Composites," (Professor C.T. Sun).

Tragesser, Steven, "Analysis and Design of Aerobraking Tethers," (Professor J.M. Longuski).

Zhu, Changming, "Constitutive Modeling of Polymeric Composites Under Various Loading Conditions," (Professor C.T. Sun).

1998

Jiang, Longzhi, "Fracture and Fatigue Behavior of Piezoelectric Materials," (Professor C.T. Sun).

Szolwinski, Matthew, "The Mechanics and Tribology of Fretting Fatigue with Application to Riveted Lap Joints," (Professor T.N. Farris).

Chang, Shih-Hsiang, "Basic Study of Superfinishing of Hardened Steels," (Professor T.N. Farris).

Ladoon, Dale, "Wave Packets Generated by a Surface Glow Discharge on a Cone at Mach 4," (Professor S.P. Schneider).

Lu, Hung-Cheng, "Ballistic Penetration of GRP Composites: Identification of Failure Mechanisms and Modeling," (Professor H.D. Espinosa).

Qin, Jim Hongxin, "Numerical Simulations of a Turbulent Axial Vortex," (Professor G.A. Blaisdell).

Wang, Hsing-Ling, "Evaluation of Multiple Site Damage in Lap Joint Specimens," (Professor A.F. Grandt, Jr.).

Wilson, Roby S., "Trajectory Design in the Sun-Earth-Moon Four Body Problem," (Professor K.C. Howell).

Yen, Chih-Chieh, "Application of Parabolized Stability Equations to Jet Instabilities and the Associated Acoustic Radiation," (Professor M.H. Williams).

1999

Chao, Chien-Chi, "Boundary Element Modeling of 2-D and 3-D Atomization Processes," (Professor S.D. Heister).

Spyropoulos, John T., "A New Scheme for the Navier-Stokes Equations Employing Alternating-Direction Operator Splitting and Domain Decomposition," (Professors A.S. Lyrintzis and G.A. Blaisdell).

Sultan, Cornel, "Modeling, Design, and Control Tensegrity Structures with Applications," (Professor M.J. Corless).

Bunnell, Robert A., "Unsteady, Viscous, Cavitating, Simulation of Injector Internal Flows," (Professor S.D. Heister).

Campbell, Eric T., "Bifurcations from Families of Periodic Solutions in the Circular Restricted Problem with Application to Trajectory Design," (Professor K.C. Howell).

McVeigh, Pamela, "Analysis of Fretting Fatigue in Aircraft Structures: Stresses, Stress Intensity Factors, and Life Predictions," (Professor T.N. Farris).

Beaumont, Matthew, "Characterization of Indalloy 227 Non Lead Solder," (Professor C.T. Sun).

Campbell, Bryan Thomas, "Aerodynamic Losses and Surface Heat Transfer in a Strut/Endwall Intersection Flow," (Professor J.P. Sullivan).

Gick, Rebecca Anne, "Analysis of the Motion of Spinning, Thrusting Spacecraft," (Professor J.M. Longuski).

Li, Haiyun, "An Experimental Investigation of High Pressure Cavitating Atomizers," (Professor S.H. Collicott).

Ryu, Shiyang, "Longitudinal Flying Qualities and Controller Design for Nonlinear Aircraft," (Professor D. Andrisani II).

2000

Viassolo, Daniel, "Design and Implementation of Periodic Digital Controllers," (Professor M.A. Rotea).

Barden, Brian Todd, "Application of Dynamical Systems Theory in Mission Design and Conceptual Development for Libration Point Missions" (Professor K.C. Howell).

Koutsavdis, Evangelos, "On the Development of a Jet Noise Prediction Methodology," (Professor A.S. Lyrintzis).

Kwon, Soonwook, "Characteristics of Three-Dimensional Stress Fields in Cracked Plates Under General Loadings," (Professor C.T. Sun).

Pham, Tuan L., "A Computational Tool for Spray Modeling Using Lagrangian Droplet Tracking in a Homogeneous Flow Model," (Professor S.D. Heister).

Ganapathy, Harish, "Coupled Thermoelastic Analysis of Fretting Contacts," (Professor T.N. Farris).

Han, Chenghua, "Dynamic Response and Failure in Layered Structures and Composites," (Professor C.T. Sun).

Li, Zhiyong, "Determination of Mechanical Properties Using Micro- and Nano-Indentation," (Professors H.T. Yang and S. Chandrasekar).

Pancake, Trent Alan, "Analysis and Control of Uncertain/Nonlinear Systems in the Presence of Bounded Disturbance Inputs," (Professor M.J. Corless).

Taylor, Robert M., "Aerospace Structural Design Process Improvement Using Systematic Evolutionary Structural Modeling," (Professor T.A. Weisshaar).

Yang, Zhengwen, "Fracture Mode Separation for Delamination and Interlaminar Fracture for Composites," (Professor C.T. Sun).

Zavattieri, Pablo D., "Computational Modeling for Bridging Size Scales in the Failure of Solids," (Professor H. D. Espinosa).

2001

Adams, Douglas, "Effects of Lateral Confinement on the Static and Dynamic Strength of Brittle Materials," (Professor C.T. Sun).

Agrawal, Parul, "Micromechanical and Fracture Characteristics of Metal-Ceramic Composites," (Professor C.T. Sun).

Guzmán, José, "Spacecraft Trajectory Design in the Context of a Coherent Restricted Four-Body Problem," (Professor K.C. Howell).

Petropoulos, Anastassios, "A Shape-Based Approach to Automated, Low-Thrust, Gravity-Assist Trajectory Design," (Professor J.M. Longuski).

Smith III, Crawford F., "Response Surface Methods for Optimization," (Professor W.A. Crossley).

Zeng, Qinggang, "A Study on Composite Adhesive Lap Joint," (Professor C.T. Sun).

Chen, Yun, "Analysis of Indentation Cracking in Brittle Solids," (Professor T.N. Farris).

Fanjoy, David, "Aerodynamic Shape and Structural Topology Design of Helicopter Rotor Cross-Sections Using a Genetic Algorithm," (Professor W.A. Crossley).

Hu, Hui-Wen, "Modeling Physical Aging in Polymeric Composites," (Professor C.T. Sun).

Jin, Long, "Simulation of Quenching and Tempering of Steels," (Professor T.N. Farris).

Kim, Haedong, "Computational Techniques for Blade-Vortex Interaction Prediction," (Professor Marc Williams).

Maley, Scott, "Particulate Enhanced Damping of Sandwich Structures," (Professor C.T. Sun).

Tsai, Jia-Lin, "Dynamic Microbuckling Model for Compressive Strength of Polymeric Composites," (Professor C.T. Sun).

Wang, Zeping, "Modeling Micro-Inertia in Composite Materials Subjected to Dynamic Loading," (Professor C.T. Sun).

D'Amato, Fernando J., "Algorithms for the Maximization of Linear Fractional Transformations," (Professor M.A. Rotea).

Ekici, Kivanc, "Parallel Computing Techniques for Rotorcraft Aerodynamics," (Professor A.S. Lyrintzis).

Golden, Patrick J., "High Cycle Fatigue of Fretting Induced Cracks," (Professor A.F. Grandt, Jr.).

Pakalapati, Rajeev T., "Fretting Contact of Dissimilar Isotropic/Anisotropic Materials," (Professor T.N. Farris).

Xu, Changhai, "Simulation of Orifice Internal Flows Including Cavitation and Turbulence," (Professor S.D. Heister).

2002

Salyer, Terry Ray, "Laser Differential Interferometry for Supersonic Blunt Body Receptivity Experiments," (Professor S.H. Collicott).

Choi, Seung-Woo, "Impact Failure Mechanisms of Layered Structures," (Professor J. F. Doyle).

Lee, Dong-Hwan, "Aeroelastic Tailoring and Structural Optimization of Joined-Wing Configuration," (Professor T. A. Weisshaar).

Acikmese, Ahmet, "Stabilization, Observation, Tracking and Disturbance Rejection for Uncertain/Non-linear and Time Varying Systems," (Professor M. J. Corless).

Johnson, Wyatt, "Analysis and Design of Aeroassisted Interplanetary Missions," (Professor J. M. Longuski).

Kang, Un-Taik, "Inverse Method for Static Problems Using Optical Data," (Professor J. F. Doyle).

Kim, Byoung-Do, "Study of Hydrodynamic Instability of Shear Coaxial Flow in a Recessed Region," (Professor S.D. Heister).

Yoon, Suk Goo, "A Fully Nonlinear Model for Atomization of High-Speed Jets," (Professor S.D. Heister).

2003

Kwon, Tae-Jun, "Simulating Collisions of Droplets with Walls and Films Using a Level Set Method," (Professor S.D. Heister).

Park, Jae Seong, "Effect of Contiguity on the Mechanical Behavior of Co-Continuous Ceramic-Metal Composites," (Professor C.T. Sun)

Turaga, Umamaheswar, "A Study of Sandwich T-Joints and Composite Lap Joints," (Professor C.T. Sun).

Chen, Yongkang, "A Study of Capillary Flow in a Vane-Wall Gap in Zero Gravity," (Professor S.H. Collicott).

Eshpuniyani, Brijesh, "Flow Physics of Strained Turbulent Axial Vortices," (Professor G.A. Blaisdell).

Funk, John, "Effects of Impingement and Shear Upon the Ignition of Hypergolic Rocket Bipropellants," (Professor S.D. Heister).

Hoshizaki, Takayuki, "Tactical Strategies for Improving Accuracy in Optical Targeting Systems," (Professor D. Andrisani II).

Park, Chul Young, "Durability of Countersunk Fastener Holes," (Professor A.F. Grandt, Jr.).

Sakaue, Hirotaka, "Anodized Aluminum Pressure Sensitive Paint for Unsteady Aerodynamic Applications," (Professor J.P. Sullivan).

Uzun, Ali, "3-D Large Eddy Simulation for Jet Aeroacoustics," (Professor A.S. Lyrintzis).

2004

Cho, Sung-Man, "Algorithms for Identification of the Nonlinear Behavior of Structures," (Professor J.F. Doyle).

Hassan, Rania, "Genetic Algorithm Approaches for Conceptual Design of Spacecraft Systems Including Multi-Objective Optimization and Design Under Uncertainty," (Professor W.A. Crossley).

Yagci, Baris, "Sinusoid Estimation in Additive Noise and Applications in Experimental Vibration Analysis," (Professor A. Frazho).

Zhang, Haitao, "Mechanical Behavior of Nanomaterials: Modeling and Simulation," (Professor C.T. Sun).

Achuthan, Ajit, "Nonlinear Behavior of Ferroelectric Materials," (Professor C.T. Sun).

Chang, I-Ling, "Simulation of Mechanical Behavior of Nanomaterials by Molecular Statics," (Professor C.T. Sun).

Haradanahalli, Murthy N., "Fretting Fatigue of Anisotropic Materials at Elevated Temperatures," (Professor T.N. Farris).

Kalyanam, Sureshkumar, "Electrical Permeability and Domain Switching Effect on Fracture Behavior of Piezoelectric Material," (Professor C.T. Sun).

Long, Matthew R., "Swirl Injectors for Oxidizer-Rich Stages Combustion Cycle Engines and Hypergolic Propellants," (Professor W.E. Anderson).

Marchand, Belinda G., "Spacecraft Formation Keeping Near The Libration Points of the Sun-Earth/Moon System," (Professor K.C. Howell).

Crafton, Jimmy W., "The Impingement of Sonic and Sub-Sonic Jets Onto a Flat Plate at Inclined Angles," (Professor J.P. Sullivan).

Matlik, John F., "High-Temperature, High-Frequency Fretting Fatigue of a Single Crystal Nickel Alloy," (Professor T.N. Farris).

McConaghy, Thomas Troy, "Design and Optimization of Interplanetary Spacecraft Trajectories," (Professor J.M. Longuski).

2005
Bartha, Bence, "Modeling of Geometry Effects in Fretting Fatigue," (Professor T.N. Farris).

Garcia, Daniel B., "Crack Propagation Analysis of Surface Enhanced Titanium Alloys with Fretting Induced Damage," (Professor A.F. Grandt, Jr.).

You, Huai-Tzu, "Stochastic Model for Ocean Surface Reflected GPS Signals and Satellite Remote Sensing Applications," (Professor J.L. Garrison).

Cho, Jeong-Min, "Effect of Inclusion Size on Failure Mechanism and Mechanical Properties of Polymeric Composites Containing Micro and Nano Particles," (Professor C.T. Sun).

Gregory, James W., "Development of Fluidic Oscillators as Flow Control Actuators," (Professor J.P. Sullivan).

Gao, Guofeng, "Fretting Induced Plasticity in Blade/Disk Contacts," (Professor T.N. Farris).

Mamun, Md-Wahid Al, "Investigation of the Photo Stimulated Luminescence Spectroscopy Technique for Measuring 3-D Stresses," (Professor P.K. Imbrie).

Park, Hongbok, "Flow Characteristics of Viscous High-Speed Jets in Axial/Swirl Injectors," (Professor S.D. Heister).

Suh, Jungjun, "Analysis of Fatigue Crack Growth From Countersunk Fastener Hole," (Professor A.F. Grandt, Jr.).

Namgoong, Howoong, "Airfoil Optimization for Morphing Aircraft," (Professor W.A. Crossley).

Pourpoint, Timothée, "Hypergolic Ignition of a Catalytically Promoted Fuel with Rocket Grade Hydrogen Peroxide," (Professor W.E. Anderson).

Rufer, Shann, "Hot-wire Measurements of Instability Waves on Sharp and Blunt Cones At Mach 6," (Professor S.P. Schneider).

Skoch, Craig, "Disturbances From Shock/Boundary-Layer Interactions Affecting Upstream Hypersonic Flow," (Professor S.P. Schneider).

2006
Garrison, Loren, "Computational Fluid Dynamics Analysis and Noise Modeling of Jets with Internal Forced Mixers," (Professor A.S. Lyrintzis).

Hanna, Ihab M., "Thermal Modeling of Grinding for Process Optimization and Durability Improvements," (Professor T.N. Farris).

Huang, Chihyung, "Molecular Sensors for MEMS," (Professor J.P. Sullivan).

Kwon, Hyukbong, "Buckling and Debond Growth of Partial Debonds in Adhesively Bonded Composite Flanges," (Professor H. Kim).

Tseng, Kuo-Tung, "Fluidic Spray Control," (Professor S.H. Collicott).

Bing, Qida, "Characterizing Compressive and Fracture Strengths of Fiber Reinforced Composites Using Off-Axis Specimens," (Professor C.T. Sun).

Canino, James V., "Numerical Analysis of Coaxial Swirl Injectors," (Professor S.D. Heister).

Lee, Jungmin, "Prediction of Adhesive Stress and Strain Solution and Failure Criterion of Single Lap Bonded Joints," (Professor H. Kim).

Sung, In-Kyung, "Fatigue Life Prediction of Liquid Rocket Engine Combustor with Subscale Test Verification," (Professor W.E. Anderson).

Chandra, Budi Wijaya, "Flows in Turbine Engine Oil Sumps," (Professor S.H. Collicott).

Kim, Jeesoo, "Characterization of Fatigue Crack Growth in Unitized Construction," (Professor A.F. Grandt, Jr.).

Landau, Damon, "Strategies for the Sustained Human Exploration of Mars," (Professor J.M. Longuski).

Okutsu, Masataka, "Design of Human Missions to Mars and Robotic Missions to Jupiter," (Professor J.M. Longuski).

Porras Alonso, Germán, "The Design of System-to-System Transfer Arcs Using Invariant Manifolds in the Multi-Body Problem," (Professor K. C. Howell).

Shimo, Masayoshi, "Multicyclic Detonation Initiation Studies in Valveless Pulsed Detonation Combustors," (Professor S.D. Heister).

2007
Ayoubi, Mohammad Ali, "Analytical Theory for the Motion of Spinning Rigid Bodies," (Professor J.M. Longuski).

Kumari, Shyama, "Contact Geometry, Friction, and Load Variations in Fretting Fatigue of Ti-6Al-4V," (Professor T. N. Farris).

Richardson, Renith, "Linear and Nonlinear Dynamics of Swirl Injectors," (Professor S.D. Heister).

Nusawardhana, A., "Dynamic Programming Methods for Concurrent Design and Dynamic Allocation of Vehicles Embedded in a System-of-Systems," (Professor W. A. Crossley)

Park, Myounggu, "Deformation Dependent Electrical Resistance of MWCNT Layer and MWCNT/PEO Composite Films," (Professor H. Kim)

Sisco, James, "Measurement and Analysis of an Unstable Model Rocket Combustor," (Professor W. E. Anderson)

2008
Corpening, Jeremy, "Computational Analysis of a Pulsed Inductive Plasma Accelerator" (Professor C. Merkle and Professor I. A. Hrbud)

Frommer, Joshua, "System of Systems Design: Evaluating Aircraft in a Fleet Context Using Reliability and Non-Deterministic Approaches" (Professor W. A. Crossley)

Kothari, Rushabh, "Design and Analysis of Multifunctional Structures for Embedded Electronics in Unmanned Aerial Vehicles," (Professor C. T. Sun)

Lana, Carlos, "Constrained State Estimation and Control," (Professor M.A. Rotea)

Millard, Lindsay DeMoore, "Control of Spacecraft Imaging Arrays in Multi-Body Regimes," (Professor K. C. Howell)

Qian, Haiyang, "Study of Failure in Bonded Lap Joints Using Fracture Mechanics," (Professor C. T. Sun)

Raghavan, Seetha, "The Development of Photo-Stimulated Luminescence Spectroscopy for 3-D Stress Measurements," (Professor P. K. Imbrie)

Syn, Chul Jin, "Dynamic Delamination in a Glass Fiber Composite and Interfacial Fracture in a Bi-Material," (Professor Wayne Chen)

Yam, Chit Hong, "Design of Missions to the Outer Planets and Optimization of Low-Thrust, Gravity-Assist Trajectories via Reduced Parameterization," (Professor J. M. Longuski)

Adnan, Ashfaq, "Molecular Simulations of Deformation, Failure and Fracture of Nanostructured Materials," (Professor C. T. Sun)

Gean, Matthew, "Elevated Temperature Fretting Fatigue of Nickel Based Alloys," (Professor T. N. Farris)

Lee, Joon Ho, "Study of an Under-Expanded Sonic Impinging Jet Array," (Professor W .E. Anderson)

Pearson, Nicholas, "Endothermic Catalytic Cracking of JP-10," (Professor W. E. Anderson)

Portillo, Juan, "Nonparallel Analysis and Measurements of Instability Waves in a High-Speed Liquid Jet," (Professor G. A. Blaisdell)

Stein, William Benjamin, "Performance Characterization of a Radio-Frequency Capacitively Coupled Discharge Microthruster," (Professor I. A. Hrbud and Professor A. Alexeenko).

Swanson, Erick O., "Boundary-Layer Transition on Cones at Angle of Attack in a Mach-6 Quiet Tunnel," (Professor S.P. Schneider).

Tang, Ching-Yao, "A Study on the Plume Characteristics and Performance of 2-D Transition Nozzles," (Professor S.D. Heister).

Bhosri, Wisuwat, "Geometric Approach to Prediction and Periodic Systems," (Profesosr A.E. Frazho).

Dong, Zhaoxu, "Mechanical Behavior of Silica Nanoparticle-Impregnated Kevlar Fabrics," (Professor C.T. Sun)

Mane, Muharrem, "Concurrent Aircraft Design and Trip Assignment Under Uncertainty as a System of Systems Problem," (Professor W. A. Crossley)

Subramaniyan, Arun, "Development of Efficient Molecular Simulation Techniques for Engineering Applications," (Professor C. T. Sun)

2009

Guo, Xiaohui, "Investigations of Microscale Fluid-Thermal Phenomena Based on the Deterministic Boltzmann-ESBGK Model," (Professor A. Alexeenko)

Lew, Phoi-tack, "A Study of Noise Generation From Turbulent Heated Round Jets Using 3-D Large Eddy Simulation," (Professor A. S. Lyrintzis)

Merrill, Marriner, "A Nanolaminate Manufacturing Technique for Multifunctional Materials and Devices," (Professor C. T. Sun)

Seah, Chze Eng, "Stochastic Linear Hybrid Systems: Modeling, Estimation and Applications," (Professor I. Hwang)

Sundaram, Narayan K., "Double and Multiple Contacts of Similar Elastic Materials," (Professor T. N. Farris)

Yu, Yen Ching, "Experimental and Analytical Investigations of Longitudinal Combustion Instability in a Continuously Variable Resonance Combustor (CVRC)," (Professor W. E. Anderson)

Borg, Matthew, "Laminar Instability and Transition on the X-51A," (Professor Steven Schneider).

Churchfield, Matthew John, "The Lag RST Turbulence Model Applied to Vortical Flows," (Professor Gregory Blaisdell).

Huang, Hsin-Haou, "Dynamic Characteristics of an Acoustic Metamaterial with Locally Resonant Microstructures," (Professor C .T. Sun).

Jung, Hogirl, "Conjugate Analysis of Asymmetric Heating of Supercritical Fluids in Rectangular Channels," (Professor Charles Merkle).

Medlock, Kristin L. Gates, "Theory and Applications of Ballute Aerocapture and Dual-Use Ballute Systems for Exploration of the Solar System," (Professor James Longuski).

Nugent, Nicholas, "Breakdown Voltage Determination of Gaseous and Near Cryogenic Fluids with Application to Rocket Engine Ignition," (Professor William Anderson).

Tsohas, John, "Hydrodynamics of Shear Coaxial Liquid Rocket Injections," (Professor Stephen Heister).

Foster, John T., "Dynamic Crack Initiation Toughness of High Strength Steel: Experiments and Peridynamic Modeling," (Professor Weinong Wayne Chen).

Iqbal, Liaquat, "Multidisciplinary Design and Optimization (MDO) Methodology for the Aircraft Conceptual Design," (Professor John Sullivan).

Li, Jinhua, "Numerical Methods for Optimal Control of Hybrid Systems and Applications to the National Airspace System," (Professor Inseok Hwang).

Uddin, Mohammed, "The Effect of Nanoparticles Dispersion on The Strength of Nanocomposites: An Experimental and Micromechanical Approach," (Professor C. T. Sun).

APPENDIX D

TEACHING AND RESEARCH AWARD RECIPIENTS

Elmer F. Bruhn Teaching Award

The Elmer F. Bruhn Teaching Award is presented annually to an outstanding teacher in the Purdue University School of Aeronautics and Astronautics, selected by the school's student body for excellence in teaching, and made possible by the interest and generosity of friends and alumni of the school.

Year	Recipient	Year	Recipient
1967	George M. Palmer*	1991	Henry T. Y. Yang
1971	Henry T. Y. Yang	1992	Steven H. Collicott
1972	Ervin O. Stitz	1993	Stephen D. Heister
1973	Francis J. Marshall	1994	Henry T. Y. Yang
1974	William R. Eberle	1995	Kathleen C. Howell
1975	Henry T. Y. Yang	1996	Stephen D. Heister
1976	Michael P. Felix	1997	A. F. Grandt, Jr.
1977	C. T. Sun	1998	W. A. Gustafson
1978	Robert L. Swaim	1999	Stephen D. Heister
1979	Henry T. Y. Yang	2000	William A. Crossley
1980	W. A. Gustafson	2001	Stephen D. Heister
1981	Chin-Teh Sun		and James M. Longuski
1982	Henry T. Y. Yang	2002	Kathleen C. Howell
1983	David K. Schmidt	2003	William A. Crossley
1984	Kathleen C. Howell	2004	James M. Longuski
1985	Henry T. Y. Yang	2005	Kathleen C. Howell
1986	Terrence A. Weisshaar	2006	William A. Crossley
1987	Kathleen C. Howell	2007	Dominick Andrisani
1988	Henry T. Y. Yang	2008	James M. Longuski
1989	Terrence A. Weisshaar	2009	William A. Crossley
1990	Kathleen C. Howell		

Recipient of Sigma Gamma Tau Outstanding Professor Award. (This award was renamed for Professor Elmer F. Bruhn in 1971. Records prior to 1971 are incomplete.)

W. A. Gustafson Teaching Award

The W. A. Gustafson Teaching Award was established by the School of Aeronautics and Astronautics in 1998 to honor Professor Gustafson's distinguished career of teaching and service to the school. Selected by the juniors and seniors of the student body for excellence in teaching, the award is presented annually to an outstanding teacher in the Purdue University School of Aeronautics & Astronautics. The award is made possible by the interest and generosity of friends and alumni of the school.

Year	Recipient
1998	Gregory A. Blaisdell
1999	William A. Crossley
2000	James M. Longuski
2001	William A. Crossley
2002	Anastasios S. Lyrintzis
2003	Martin J. Corless
2004	Marc H. Williams
2005	James M. Longuski
2006	Arthur F. Frazho
2007	Dominick Andrisani
2008	Thomas N. Farris
2009	Martin J. Corless

The C. T. Sun School of Aeronautics & Astronautics Excellence in Research Award

This award was established in 2004 to honor Professor C. T. Sun's distinguished career of teaching and research. It is presented annually to an individual or a team of faculty members in the Purdue University School of Aeronautics & Astronautics to recognize high quality contributions in science and engineering.

Year	Recipient
2004	Chin-Teh Sun
2005	Stephen D. Heister and William Anderson
2006	Mario Rotea
2007	Steven P. Schneider
2008	Weinong Chen
2009	Charles Merkle

APPENDIX E

SUMMARY OF SCHOOL TITLES & SCHOOL HEADS

SCHOOL TITLES

- School of Mechanical and Aeronautical Engineering (1942–45).
  First Purdue program to grant degree in aeronautical engineering (1943).
- School of Aeronautics (1945–56).
  Aeronautical engineering established as independent academic unit on July 1, 1945.
- School of Aeronautical Engineering (1956–60).
  School renamed following dissolution of air transportation program.
- School of Aeronautical and Engineering Sciences (1960–64).
  Aeronautical Engineering merged with School of Engineering Sciences on July 1, 1960.
- School of Aeronautics, Astronautics, and Engineering Sciences (1965–73).
  The engineering sciences portion of the program was terminated January 1, 1973.
- School of Aeronautics and Astronautics (1973–).

SCHOOL HEADS

- Harry L. Solberg (1942–45). Head of combined M.E./A.E. program.
- Elmer F. Bruhn (acting head 1945–47, head 1947–50). *First head of independent aeronautics program.*
- Milton U. Clauser (1950–54).
- Harold M. DeGroff (acting head 1954–55, head 1955–63).
- Paul E. Stanley (interim head 1963–65)
- *Advisory committee:* H. Lo (chair), S. J. Citron, R. J. Goulard, P. E. Stanley, and E. O. Stitz (1965-67).
- Hsu Lo (1967–71).
- John L. Bogdanoff (1971–72).
- Bruce A. Reese (1973–79).
- Henry T. Yang (acting 1979–80, head 1980–84).
- Alten F. Grandt, Jr. (1985–92).
- John P. Sullivan (1993–98).
- Thomas N. Farris (1998-2009)
- Tom I-P. Shih (2009-)

Appendix F

Outstanding Aerospace Engineer Award Recipients

The Purdue University designation Outstanding Aerospace Engineer recognizes the professional contributions of graduates from the School of Aeronautics and Astronautics and thanks them for the recognition that their success brings to Purdue and the school.

Criteria for the award state that recipients must have demonstrated excellence in industry, academia, governmental service, or other endeavors that reflect the value of an aerospace engineering degree. The 101 OAEs represent just over 1% of the more than 7,000 alumni of the school.

1999 Recipients

Richard E. Adams
Neil A. Armstrong
Robert E. Bateman
Donovan R. Berlin
John E. Blaha
Alvin L. Boyd
William W. Brant
Roy D. Bridges, Jr.
Mark N. Brown
Elmer F. Bruhn
John H. Casper
Roger B. Chaffee
Milton U. Clauser
William E. Cooper
Lana Couch
Richard O. Covey
Jack D. Daugherty
Richard L. Duncan
Richard D. Freeman
Guy S. Gardner
Arthur B. Greenberg
Gregory J. Harbaugh
John B. Hayhurst
Walter J. Hesse
Paul T. Homsher
Robert D. Hostetler

Hsichun (Mike) Hua
Richard E. Kasler
Ronald L. Kerber
William C. Kessler
Walker M. Mahurin
Edmund V. Marshall
Garner W. Miller
John I. Nestel
James P. Noblitt
Wendell Norman
William J. O'Neil
Richard M. Patrick
Gary E. Payton
Richard H. Petersen
Mark L. Polansky
James D. Raisbeck
Bruce A. Reese
Herbert F. Rogers
Leonard M. Rose
Richard L. Rouzie
Alfred F. Schmitt
Loren J. Shriver
Shien-Siu Shu
Lester W. Smith
Walter D. Smith
Robert L. Strickler
David O. Swain

Martin W. Taylor
Richard W. Taylor
Janice E. Voss
Charles D. Walker
Henry T. Yang

2000 Recipients
Mark K. Craig
Edward G. Dorsey, Jr.
Leslie A. Hromas
Kenneth O. Johnson
Kenneth G. Miller
Daniel P. Raymer
Charles Robert Saff

2001 Recipients
Alon Dumanis
Michael T. Kennedy
Yasuhiro Kuroda
David McGrath
G. Thomas McKane
Hank Queen
John Rich

2002 Recipients
Paul M. Bevilaqua
Steven E. Lamberson
Jerry L. Lockenour
John H. McMasters
Joseph D. Mason
J. Michael Murphy

2003 Recipients
Raymond R. Cosner
Peggy P. Dedo
Richard L. Fahrner
William H. Gerstenmaier
John L. Hudson
Ronnie K. Miller
Charles E. Taylor
David A. Wagie

2004 Recipients
Bradley D. Belcher
John T. Betts
Lloyd E. Hackman
Anna-Maria R. McGowan
Terrence H. Murphy
David A. Spencer
Anthony L. Thornton

Christopher Whipple
Thomas L. Williams

2005 Recipients
Andrea M. Chavez
Thomas S. Gates
Debra L. Haley
Stephen A. Rizzi
William E. Schmitendorf
Richard E. Van Allen

2006 Recipients
Thomas C. Adamson, Jr.
Steven M. Ehlers
Jerry W. McElwee
Allen Novick
Doris H. Powers
Richard B. Rivir
Norman V. Scurria, Jr.

2007 Recipients
Nancy L. B. Anderson
Thomas J. Beutner
Steven C. Drury
Rune C. Eliasen
Michael W. Hyer
Andrew H. Kasowski
Miroslav A. Simo

2008 Recipients
Frank H. Bauer
Darryl W. Davis
Wayne Eckerle
Walter Eversman
Troy M. Gaffey
Markus B. Heinimann
Timothy A. Kinnan
Kenneth B. Sanger

2009 Recipients
William Ailor III
Charlene Edinboro
Roy Eggink
Andrew King
Thomas Maxwell
Suvendoo Ray
Dennis Warner
John Wheadon, Jr.

Appendix G

Summary of AAE Faculty

This is a summary of individuals who have served on faculty of the School of Aeronautics and Astronautics and/or its predecessors (names in bold print also served a period as school head).

Name	Joined	Left	Name	Joined	Left
Alspaugh, Dale W.	1964	1981	Eringen, A. Cemal	1953	1967
Anderson, Reidar L.	1948	1973	Espinosa, Horacio D.	1992	1999
Anderson, William	2001	current	Fairman, Seibert	1921	1961
Andrisani, Dominick A.	1980	current	**Farris, Thomas N.**	**1986**	**2009**
Augenstein, Bruno W.	1948	1949	Felix, Michael P.	1974	1977
Alexeenko, Alina	2006	current	Feuer, Paula	1954	1974
Basinger, James	1946	1955	Filmer, David	2002	current
Bauers Ca.	1929	?	Fowler, Mart	1947	1955
Bevan, William A.	1926	1929	Frazho, Arthur E.	1980	current
Blaisdell, Gregory A.	1991	current	Garrison, James	2000	current
Bogdanoff, J. L.	**1950**	**1985**	Genin, Joseph	1964	1973
Bollard, R. John	1956	1961	Gersch, Wilbert M.	1963	1970
Briggs, William	1946	1970	Ginsberg, Jerry H.	1969	1973
Bruhn, Elmer F.	**1941**	**1967**	Goulard, Madeline	1958	1975
Burton, Elbert F.	1924	1926	Goulard, Robert J.	1956	1975
Butler, Blaine R.	1972	1983	**Grandt, Alten F., Jr.**	**1979**	**current**
Cargnino, Lawrence T.	1945	1984	Gustafson, Winthrop A.	1960	1998
Chen, Weinong	2004	current	Hancock, J. O.	1936	1983
Citron, Stephen J.	1959	1973	Haskins, George W.	1929	1937
Clauser, Milton	**1950**	**1954**	Heister, Stephen D.	1990	current
Collicott, Steven H.	1991	current	Herrick, Thomas J.	1945	1980
Corless, Martin J.	1984	current	Hoglund, Richard F.	1963	1969
Crosby, Frank	1946	1948	Howell, Kathleen C.	1982	current
Crossley, William A.	1995	current	Hrbud, Ivana A.	2003	2009
Curtis, Richard A.	1969	1975	Hwang, Inseok	2004	current
Cushman, Edward	1945	1947	Jischke, Martin	2000	2007
Cutshall, C. S.	1955	1964	Kayser, Kenneth W.	1974	1977
DeGroff, Harold	**1951**	**1965**	Kentzer, Paul	1959	1990
DeLaurentis, Daniel	2004	current	Kim, Hyonny	2001	2006
Doyle, James F.	1977	current	Klein, Bertram	1947	1950
Drake, John W.	1972	1992	Kline, Leo V.	1956	1961
Eberle, William R.	1970	1977	Koh, Severino L.	1964	1973

Name	Joined	Left	Name	Joined	Left
Kozin, Frank	1958	1967	**Stanley, Paul**	**1945**	**1976**
Lianis, George	1959	1978	Stitz, E. O.	1937	1975
Liston, Joseph	1937	1972	**Sullivan, John P.**	**1975**	**current**
Little, Neil	1926	1966	Sun, C. T.	1968	current
Lo, Hsu	**1949**	**1979**	Sun, Dengfeng	2009	current
Longuski, James M.	1988	current	Sutton, Emmett A.	1962	1966
Lund, Thomas S.	1987	1989	Swaim, Robert L.	1967	1978
Lykoudis, Paul S.	1956	1973	Sweet, Arnold L.	1964	1973
Lyrintzis, Anastasios S.	1994	current	Talavage, Joseph J.	1972	1973
Malone, David W.	1968	1973	Thornburg, Martin L.	1921	1924
Marais, Karen	2009	current	Ting, Edward C.	1968	1973
Marshall, Francis J.	1960	1989	Tomar, Vikas	2009	current
Mason, Joseph D.	1971	1972	Topinka, George	1945	1946
McBride, James W.	1947	1949	Usab, William J., Jr.	1986	1993
Merkle, Charles	2003	current	Valanis, Kyriakos C.	1963	1964
Messersmith, Nathan L.	1992	1997	Vanderbilt, Vern	1947	1948
Miele, Angelo	1955	1959	Vemuri, V. R.	1969	1973
Misener, Walter S.	1947	1949	Webster, Grove	1942	1955
Osborn, J. R.	1980	1989	Weisshaar, Terrence A.	1980	current
Ostoja-Starzewski, Martin	1985	1989	Weske, John R	1947	1948
Palmer, George	1946	1987	Williams, Marc H.	1981	current
Peterson, Clifford C.	1971	1973	Wood, Karl D.	1937	1944
Peterson, Lee D.	1989	1991	**Yang, H. T. Y.**	**1969**	**1994**
Pipes, R. Byron	2004	current	Young, Gilbert A.	1921	1941
Pourpoint, Timothee	2008	current	Zucrow, Maurice	1946	1953
Pritsker, A. Alan	1970	1973			
Qiao, Li	2007	current			
Radavich, John F.	1960	1973			
Ranger, Arthur A.	1968	1973			
Reese, Bruce	**1973**	**1979**			
Rosenberg, Reinhart	1945	1947			
Rotea, Mario A.	1990	2007			
Rottmayer, Earl	1946	1947			
Rusek, John	1998	2002			
Samuels, J. Clifton	1957	1965			
Sanders, W. Burton	1955	1961			
Schapery, Richard A.	1962	1969			
Schiff, Anshel J.	1967	1973			
Schmidt, David K.	1974	1989			
Schmitt, Alfred F.	1953	1955			
Schneider, Steven P.	1989	current			
Serbin, Hyman	1947	1955			
Shanks, Merrill E.	1946	late 60s			
Shih, Tom	**2009**	**current**			
Shu, S. S.	1955	1963			
Skelton, Robert E.	1975	1997			
Solberg, Harry	**1941**	**1945**			
Staley, Alan C.	1924	1926			
Stanisic, M. M.	1956	1985			

INDEX

Index compiled by Marilyn Augst, Prairie Moon Indexing.
Page numbers followed by **f** refer to figures. Page numbers followed by **t** refer to tables.

A

Abbett, Henry, 56
Abbot, Dr., 50
academic programs. *See* curriculum/course listings
accreditation inspection, 155
Adams, Richard E., 331, 454
Adamson, Thomas C. Jr., 455
Ade, George, xxii
administrative changes, 91, 95–96, 149–151, 189–190, 208–210, 260–263
advisory committee on academic affairs, 149–151
Aerodynamics
    1981-83 option, 223t
    1993-94 research, 248–249
    1995 faculty, 238
    2009 faculty research area, 270t
    and Control, 1950 faculty, 96
    graduate courses, 109
    laboratory, 1, 94, 109–111
aerodynamic testing of buildings research, 198–199
Aeroliner student publication, 69, 118
Aeromodelers, 119–121, 120f
Aeronautical Engineering Program, postwar, 59–69, 91–94
Aeronautical Engineering School. *See* School of Aeronautical Engineering
Aeronautics and Astronautics Curriculum Committee, 160, 162, 167, 191–194. *See also* curriculum committees
(a.k.a. Aerospace Curriculum Planning Committee), 150
Aeronautics Committee, 262
aeronautics concentration in curriculum, 274t
Aerospace Corporation, 241t
aerospace engineers from Purdue, 342

aerospace industry
    1950s inspection trips to, 70, 114f, 115
    1965 curriculum for students, 152
    1973 and 1976 employment situation, 190–191, 195
    1980-95 undergraduate design program, 224
    1995 decline of, 240, 242
    in 2010 and future of, 342–343
Aerospace Systems, 2009 faculty research area, 270t, 273, 276
aging aircraft research, 248
Ailor, William III, 455
Air Corp Cadet Program, 31
aircraft
    B-50 Lucky Lady II, 121
    Boeing airplanes, 9, 16
    C-46 Curtiss-Wright twin-engine transport, 85–86, 86f
    Curtiss airplanes, 3f, 5f, 16
    DC-3 airplane, 9, 16, 288
    DC-4 airplanes, 16, 114f, 115
    Ford TriMotor airplane, 9
    Grumman Cougar airplanes, 121
    Martin airplane, 16
    McDonnell Aircraft fighter, 121
    Morrow Victory Trainer, 43, 44f, 45f
    Navy aircraft, 8f
    North American Aviation Mustang fighter, 43
    P-47 Thunderbolt fighter airplane, 85
    P-59 jet-powered airplanes, 86–87, 87f, 282
    Sikorsky airplane, 16
aircraft companies, postwar survey on need for aviation education, 50
aircraft crash safety research, 90
Aircraft Design course, 276
aircraft power plant design, 10, 134
Aircraft Power Plant Option, 60f, 61, 104t, 124
aircraft production, 16, 19, 47, 342–343

Air Force Academy, Colorado Springs, 174–176
Air Force Office of Scientific Research. *See* Office
    of Scientific Research and Development
Airfreight Board, 69–70, 118–119
Airline Engineering Option, 60f, 61
airlines
    inspection trips to, 114f, 115
    postwar survey on need for aviation
        education, 50
airmail delivery in Lafayette, xxi, 341
Airplane and Power Plant Option, 124, 124t–126t,
    131
Airplane Design Curriculum, 59–61, 61f
Airplane Option of 1950 curriculum, 103t
Airport Management & Operations Option, 54
Air Transportation Honorary Society, 70
Air Transportation Program
    about options, 54–59, 223t
    curriculum, 49, 105t, 106t, 107t, 108t
    status of program, 52f, 91–92, 101, 126–128,
        221
Alexeenko, Alina, 267, 269f, 270t, 456
Alford, Joanne, 112f, 113, 118
Allen, John L., 44f, 45, 289, 289f, 345
Allison Gas Turbine, 241t
Alspaugh, Dale W.
    career of, 161, 183, 212
    in faculty list, 173t, 175, 189t, 212t, 456
Alstott, Pamela, 287t
altitude record, 128, 129f
alumni accomplishments, 285–340
    astronaut alumni, 298–308
    class of 1943, 288–298
    Distinguished Engineering Alumni, 308–328
    Honorary Doctorates, 329–340
    Outstanding Aerospace Engineering, 308
    Sigma Gamma Tau Outstanding Senior
        Award, 286–287
    typical graduation class (1986), 286f
American Council on Education for Civil
    Aeronautics Administration, 47
Amrine, Harold, 149
Anderson, Nancy L. B., 455
Anderson, Reidar L., 148t, 167, 180, 456
Anderson, William, 269f
    career of, 265, 273, 280, 452
    in faculty list, 271t, 456
Andrisani, Dominick A., 216f, 232f, 269f
    career of, 217, 227, 228, 238, 252t, 451, 452
    in faculty list, 212t, 225t, 238, 246t, 247t, 270t,
        276, 456
anniversaries of schools
    40th, 209f
    50th golden, xv, xvii, xix, 237, 257, 289
    65th, 342
    Class of 1943's 50th, 288–289
annual conferences, 58–59
Anthes, Victoria, 287t
Anthony, B., 268t
Apollo 1 spacecraft fire, 158, 159f, 228–229, 282f,
    284, 299
Apollo 11 moon landing mission, 130, 190

Aretz, Lawrence I., xxiv
Aretz airport, 117
Armstrong, Neil A., 281f, 307f
    career of, xxiii, 130, 262, 284, 299, 304, 309,
        330, 454
    as Purdue astronaut, 228, 300t, 304, 341–342
    as Purdue student, 117, 120f, 121
Armstrong (Neil) Hall of Engineering. *See* Neil
    Armstrong Hall of Engineering
Army Air Corps officers, 31
Army Ballistic Missile Agency, 135, 136
Army Signal Corps, 136
Army-sponsored research, 280–281
Arnett, Richard, 3f
Arnold, A. M., 82f
Arnold, J. N., 34t
astrodynamics and space applications, 276
astronautics concentration in curriculum, 275t
"astronaut program," 175
astronauts
    beginnings of School of Aeronautical
        Engineering, 128–130
    Challenger explosion and media attention,
        228–229
    lunar sample donations, 284
    Master's Degree in astronautics, 174–176
    Purdue graduates, 298–299, 304–307, 341–342
    space-flight log, 299, 300t–303t
Astrophysics course, 133
Augenstein, Bruno W., 63t, 66–67, 456
auto crash safety research, 90
Aviation Administration Option, 108t
Aviation at Purdue (postwar pamphlet), 50
Aviation Day at Purdue, xxii–xxiii
Aviation Operations Option, 106t
Aviation Week, 259
awards
    A. A. Potter teaching award, 202
    Amelia Earhart awards, 113
    Best Teacher Award, 208, 210
    C. T. Sun School of Aeronautics &
        Astronautics Excellence in Research
        Award, 273, 452
    Dean M. Beverly Stone Award, 263
    Elmer F. Bruhn Teaching Award, 202, 208,
        263, 451
    J. H. Starnes Award, 272
    M. M. Frocht Award, 202
    National Citation of Honor, 113
    Outstanding Aerospace Engineer, 279, 308,
        454–455
    for research, 452
    Sagamore of the Wabash Award, 263
    Sigma Gamma Tau Outstanding Professor
        Award, 451(footnote)
    Sigma Gamma Tau Outstanding Senior
        Award, 286, 287t
    for teaching, 451–452
    Thiokol SPACE Award, 224
    W. A. Gustafson Teaching Award, 263, 452
Ayers, Langdon F., 88
Azzano, Christopher P., 287t

B

Bachelor of Science degrees, 345–385. *See also*
  degrees
    in Aeronautical Engineering, 1, 23, 24t–25t, 49,
      51, 61–62, 113, 139t, 345–355
    in Aeronautics & Astronautics, 355–377
    in Air Transportation, 51, 113, 126, 377–381
    in Engineering Sciences, 382–385
Bader, William, 211f, 237t
Bailey, C. D., 137t
Ball, Edmond F., 231
Ball Aerospace collection, 231
Ball Corporation Aerospace Systems, 241t
Ballistic Missile Defense Organization, 283
Ball State University, 227
Barthel, J. R., 137t
Basic or Airplane Design curriculum, 59–61, 61f
Basinger, James, 86f
    career of, 57, 58, 74f, 85–86, 118, 127
    in faculty list, 92, 96, 456
Bass, Robert, 110f
Bateman, Robert E., 236f, 311–312, 334f, 335,
    454
Battle of Tippecanoe, 341
Bauer, Frank H., 455
Bauers, Mr., 3, 4t, 456
Beachey, Lincoln, xxii–xxiii
Bean, Mr., 49, 74f
Beebe, Robert L, 289, 289f
Beechcraft Company, 115
Beering, Steven C., 260, 278, 299, 308, 333f
Beese, C. W., 33
Belcher, Bradley D., 455
Bell Aircraft Corporation, 86
Benford, R., 226f
Benish, Kerrie, 287t
Bennett, Eric, 227f
Berlin, Donovan R., 329, 454
Bertuccelli, Luca, 287t
Best Teacher awards. *See* awards
Bethel, Howard E., 287t
Betts, John T., 455
Beutner, Thomas J., 455
Bevan, William A., 2, 456
Bevilaqua, Paul M., 325, 455
Binder, Prof., 11t, 46, 62
Blackiston, Harry, 287t
Blaha, John E, 175, 300t, 301t, 302t, 305, 307f, 454
Blaisdell, Gregory A., 211f, 269f
    career of, 220, 238, 252t, 452
    in faculty list, 212t, 238, 246t, 247t, 270t, 456
Blakiston, H. S., 137t
Blue Angels flight performance, 121
Bobillo, Tim, 272
Boeing Company, 111, 115, 235, 236f, 243, 278,
    283
Boeing Lecture Series, 278
Boettcher, Phillip, 287t
Bogdanoff, John L.
    career of, 151, 158, 161, 167, 171, 189t, 193, 197,
      215, 453
    in faculty list, 148t, 173t, 212t, 225t, 456

Bohlman, Jonathon, 287t
Bolds, M. H., 34t, 35t
Bollard, R. John
    career of, 99, 135, 136, 137t
    in faculty list, 138t, 140, 148t, 177, 456
Book of Great Teachers, 273
Borodavchuk, J., 85
Boswinkle, Robert W., 44f, 289–290, 289f, 345
Bower, Donald, 211f, 237t
Bowman, K. J., 280
Boyd, Alvin L., 127, 310–311, 454
Brant, William W., 318, 454
Briden, Amanda, 287t
Bridges, Roy D., 175, 278, 300t, 305, 322, 337, 454
Briggs, William W., 71f, 75f, 86f
    career of, 56–57, 58, 118, 127, 180
    in faculty list, 92, 96, 140, 148t, 456
Brink, Paul L., 44f, 289f, 290, 345
Broglio, L. A., 96
Broughton, Ann, 268t, 269f, 278
Brown, E. R., 21
Brown, Mark N., 300t, 301t, 305, 307f, 454
Brown, Sherwood H., 290, 345
Bruhn, Elmer F., 10f, 82f, 86f, 119f, 213f
    awards received, 273, 330–331, 454
    career of, 9–10, 14–16, 24, 31, 73, 90, 91f, 91–94,
      126, 128, 178–179
    on Curtiss-Wright Cadette Program, 32, 33,
      34t, 35t, 38
    in faculty list, 93, 95, 96, 130, 136, 140, 148t,
      456
    on Lascoe's war surplus securement, 84–85
    Morrow Airplane Project, 42–43, 44f, 45f
    postwar aviation education, 47, 50, 51, 62
    on postwar conferences, 59
    as school head, 46, 63t, 91, 95, 152, 453
    school history & other publications, xv–xvi,
      344
Bruner, Warren, 34
BS degrees. *See* Bachelor of Science degrees
buildings at Purdue. *See also* facilities
    Aeronautical and Engineering Sciences
      Building, 147, 147f, 158
    Aeronautical Engineering Building, 26, 27f,
      28f, 29f
    Aeronautics Building on campus, 70, 77, 79,
      80f
    Aerospace Sciences Building, 243
    Biomedical Engineering Center, 172, 202
    Building Units 1, 2, 3 at Purdue Airport,
      70–72, 76–77, 85–87, 93, 94, 243
    Center for Applied Stochastics, 169, 172
    Chaffee Hall, 158, 160f, 229, 299
    Civil Engineering Building, 157–158
    Engineering Sciences Building, 143, 147
    Grissom Hall, 158, 159f, 229, 231, 235, 242,
      283, 299
    Heavilon Hall, 1, 12, 158, 200f, 281
    Knoy Hall, 162
    Large-Scale Systems Center, 162, 182, 185, 187,
      188, 201
    Michael Golden Hall, 162

Neil Armstrong Hall of Engineering. *See main entry for* Neil Armstrong Hall of Engineering
Nuclear Energy Building, 231, 242–243
Potter Engineering Center, 148, 162
University Computing Center, 229
University Technology Center in High Mach Propulsion, 283
Wood Women's Residence Hall, 32
Burton, Elbert F, 1, 2t, 456
Butler, Blaine R., 187, 189t, 212t, 215, 456
Buttner, Prof., 76

C

Cahill, James, 121
Caldwell, B., 227, 270t
California Institute of Technology, 20t
Calvert, George, 98, 214, 214f
Cameron Copenhaver, Roberta, 216
Campanella, O. H., 280–281
Campbell, Michael, 327–328
Capitol Airlines, 70, 115
Cappellari, J. O., 137t
Cargnino, Lawrence T., xv, xx, 71f, 86f, 213f
    career of, 56–57, 58, 115, 118, 128, 130, 161, 176, 192, 213
    in faculty list, 92, 96, 140, 148t, 189t, 212t, 456
    publications by, xv, xvi, 57
    on turbojet engine hot start, 87
Carmichael, Pres. of Capitol Airlines, 70
Casper, John H., 128, 175, 301t, 302t, 305, 454
Cave, Michael J., 327
Cavoti, Carlos, 137t, 138t
Cernan, Eugene A., 281f, 307f
    career of, xxiii, 284, 299, 304
    as Purdue astronaut, 228, 300t, 342
Cessna Aircraft Company, 115
Chadwell, M., 237t, 268t
Chaffee, Martha, 284
Chaffee, Roger
    Apollo fire, 158, 228, 229, 282f, 284, 299
    career of, 130, 159f, 304, 454
Challenger explosion, 228–229
Chaney, Beth, 211f
Change-Vought Corporation, 115
Chanute Field, Rantoul, Illinois, 31
Chavez, Andrea M., 455
Chen, Tung, 137t
Chen, Weinong, 266, 269f, 271t, 273, 452, 456
Chenea, Paul F., 126, 146
Cheung, L. T, 82f
Christopher, G., 82f
Cicala, Placido, 99
Citron, Stephen J.
    career of, 149, 181, 453
    in faculty list, 146, 148t, 167, 173t, 456
Civil Aeronautics Authority, 17
civilian-pilot training (CPT), 17
Civil Works Administration, 17
Class of 1943, 288–298, 344
Clauser, Milton U., 95f, 110f

career of, 95–98, 329, 454
    in faculty list, 95, 96, 456
    as school head, 91, 95, 121–122, 126, 453
Cold War, 342
College of Engineering, 1996-2009, 259, 261, 264
Collicott, Steven H., 211f, 269f
    career of, 220, 238, 252t, 254t, 273, 451
    in faculty list, 212t, 225t, 238, 246t, 247t, 270t, 456
Colorado School of Mines, 14
commercial airlines, postwar growth, 48
computers, 148–149, 168, 169, 216f, 229
Conner, Robert L., 98
control system design, 249. *See also* Dynamics and Control
control theory, 155
Cooper, William E., 311, 454
Cooperative Education Program, 22, 176, 241t
Copenhaver, Roberta, 216
Córdova, France, 281f, 307f
Cork, M. Joe, 287t
Corless, Martin J., 211f, 269f
    career of, 218, 238, 252t, 452
    in faculty list, 212t, 225t, 238, 270t, 278, 456
Cornell University, and Curtiss-Wright Cadette Program, 32
Corvalan, C. M., 281
Cosby, Ronald, 287t
Cosky, Charissa J., 287t
Cosner, Raymond R., 455
Cottingham, W. B., 195
Couch, Lana M., 319–320, 454
course identification system, 155
course listings. *See* curriculum/course listings
courtesy appointment, 272
Covey, Richard O., 175, 229, 300t, 301t, 302t, 305, 307f, 454
"cradle of astronauts," 128
Craig, Mark K., 323, 455
Crain, L., 268t, 269f
crash safety research, 90
Crepps, R. B., 34t
Crosby, Frank M., 13f, 21, 49, 63t, 65–66, 74f, 456
Crossley, William A.
    career of, 264, 451, 452
    in faculty list, 270t, 271t, 276, 277, 456
Crow, Richard, 35t
curriculum committees, 150, 152, 157, 160, 161
curriculum/course listings
    1924-25, 1–2, 2t
    1930-31, 4t
    1939-40, 11t, 12t
    1941-42 World War II programs, 23–26, 24t–25t, 26t, 31t, 33–35, 40t–42t
    1945-50 postwar programs, 49, 53f, 54–55, 59–61, 73t, 91–93
    1950s, 101–109, 121–126, 130–131, 131t–132t, 133
    1960-73 new courses taught, 173t
    1960s, 140t–143t, 151–162, 153t, 154t, 174–176
    1970s, 162–167, 163t–165t, 190–196, 191t, 192t, 199t
    1980-95, 220–227, 222t, 223t, 225t

1995, 243, 244t
1996-2009, 273–278, 274t, 275t
Curtis, Richard A, 185, 189t, 199t, 200, 456
Curtiss Services Corporation, 3f
Curtiss-Wright Cadette Programs, 32–39, 215, 293
Curtiss-Wright Corporation, Airplane Division, 32–35, 38
Cushman, Edward, 51, 55–56, 57, 85, 86f, 92, 456
Cutshall, C. S., 35t, 148t, 178, 456

D

damage tolerant armor research, 280
dance, dedication for new airport building, 71
Daugherty, Jack D., 312–313, 454
Davis, Darryl W., 328, 455
Davis, Duane, 287t
DeCamp, John R, 118
Dedo, Peggy P., 455
degrees. *See also* Bachelor of Science degrees;
        graduate degrees; Master of
        Science degrees awarded; PhD degrees
        awarded and thesis titles
    awarded by Purdue in aerospace engineering,
        xxi, 342
    online, 277
DeGroff, Harold M., 96f, 110f
    career of, 96, 98–100, 126–127, 130,148, 149
    in faculty list, 130, 138t, 140, 148t, 173t, 456
    as school head, 139–140, 143, 146, 453
DeLaurentis, Daniel, 266–267, 270t, 276, 278, 456
Delta Airlines, 59
demographics of students. *See also* enrollment at
        Purdue
    1994-95, 240t–242t
    1996-2009, 259
Department of Applied Mechanics, 145
Department of Industrial Management &
    Transportation, 127
design courses, 73, 155, 193, 221, 224, 276–277
director of communications and development,
    272
Discovery Park's Center for Advanced Manufac-
    turing, 261
distance education, 259, 277–278
Distinguished Engineering Alumnus, 34, 127,
    308–328
Division of Engineering Mechanics, 145
Division of Engineering Sciences, 139–143,
    145–148, 151
Division of Interdisciplinary Engineering
    Studies, 166
Dobbins, Abigail, 287t
Dorsey, Edward G. Jr., 455
Douglas company, 50
Doyle, James F., 211f
    career of, 204, 239, 252t, 280
    in faculty list, 212t, 225t, 239, 246t, 247t, 271t,
        456
Drake, John W.
    career of, 187–188, 216, 221
    in faculty list, 189t, 199t, 212t, 456
Drury, Steven C., 455

Dumanis, Alon, 455
Duncan, Richard L., 88, 109, 310, 454
Dunn, James R, xvi, 288, 289f, 290–291, 345
Dynamics and Control, 223t, 238, 249–250, 270t–
    271t
dynamometer, 30f

E

Earhart, Amelia, xxiv, xxv–f, 17, 113, 341
earthquake research, 172
Eberle, William R.
    career of, 186–187, 195, 200, 451
    in faculty list, 189t, 199t, 456
Eckerle, Wayne, 455
Edinboro, Charlene, 455
Edmond F. Ball Aerospace Collection, 231
educational philosophy between theory and
    practice, 21–22
Eggink, Roy, 455
Ehlers, Steven M., 455
Ehresman, Charles M., 282
Eliasen, Rune C., 455
Elliott, Edward C., xxiv, 16, 18, 50
Ellis, I., 237t
Elrod, S. B., 34t, 35t
Emmons, P. C., 18, 20–21
endowments by Guggenheim Foundation, 19–20
Engineering Mechanics, Division of, 145
"Engineering News" publication, 162
Engineering Open House, 121
Engineering Option, 153t
Engineering Professional Education, 259,
    277–278
Engineering Sciences, Division of, 139–143,
    145–148, 151
Engineering Sciences Curriculum Committee,
    150, 160, 161. *See also* curriculum committees
engineering sciences program, 145–146, 151,
    154t, 164t–165t, 166–167, 191
Engineers Council for Professional Development
    (ECPD), 155, 157
engines option, Navy V-12 curriculum, 40, 41t
Engle, Marilyn, 211f, 237t
enrollment at Purdue and in schools
    1940-45 during WWII, 28, 32–33, 38, 40, 90–93,
        92f
    1940-2008, 48
    1945-50 postwar, 57–58, 73, 90
    1950s, 98, 112–113, 126, 136, 139t, 146
    1960-73, 143, 156, 157f, 166
    1973-80, 189, 195, 198
    1980-95, 205–207
    1995, 240, 240t–241t, 242t, 245t, 246t–247t
    1996-2009, 259, 260f
environmental concerns and effects on enroll-
    ment, 166
Eringen, A. Cemal, 148t, 173t, 179, 456
Espinosa, Horacio D.
    career of, 220, 239, 252t, 263, 280
    in faculty list, 212t, 239, 246t, 247t, 456
Essig, Robert H., 291, 345

Eversman, Walter, 455
Exponent newspaper, 118

F

facilities. *See also* buildings at Purdue; labora-
    tory facilities; Purdue Airport
    1930s beginnings at Purdue Airport, 16–18
    in 1941-42 during World War II, 24, 26
    in 1945-50 postwar period, 70–87
    in 1950s, 95, 97, 109–112
    in 1980-95, 229–236
    in 1995, 242–243
    in 1996-2009, 281–284
faculty in AAE, summary of, 456–457
faculty/teaching staff
    in 1941-45 during WWII, 34t, 35t, 46
    in 1945-50 postwar, 55–59, 62–63, 63t, 64–69,
        92, 93
    in 1950s, 95–100, 137t
    in 1960-73, 148t, 176–188
    in 1973-80, 189t, 199–204
    in 1980-95, 210–220, 211f, 212t
    in 1995 golden anniversary, 237, 238f–239f
    in 1996-2009, 261–269
Fahrner, Richard L., 455
The Fair Coed (Ade), xxii
Fairman, Seibert, 34t, 35t, 148t, 177, 456
Farris, Glenn, 287t
Farris, Thomas N., 211f, 262, 269f
    career of, 218, 239, 248, 252t, 253t, 262, 452,
        453
    in faculty list, 212t, 225t, 239, 246t, 247t, 271t,
        456
Felix, Michael P., 203, 451, 456
Feuer, Paula B.
    career of, 161, 200–201
    in faculty list, 146, 148t, 173t, 189t, 456
Feustel, Andrew J., 303t, 305, 307f
Fidler, Gary, 287t
field trips to industry, 70
Filmer, David, 266, 269f, 270t, 276, 456
financial matters
    faculty salaries, postwar, 62
    tuition & fees for attending Purdue, 51, 54
Flack, Linda M., 183–184, 183f, 211f, 237t, 268t,
    269f
Fleming, Willliam A., 289f, 291, 345
Flight Operations Option of Air Transportation
    program, 55, 107t
flight research, 20–21, 135, 137t
flight test facilities, 26, 27f. *See also* laboratory
    facilities
flight training at Purdue, xxiii–xxiv
Flying Club, 117–118
Fokker Tri-Motored transport, 9
Fort Ouiatenon, 341
Fountain, P., 344
Fowler, Mart, 119f
    career of, 57–59, 69, 118–119, 127
    in faculty list, 92, 96, 456
Frazho, Arthur E.

career of, 217, 238, 253t, 452
    in faculty list, 212t, 225t, 238, 246t, 247t, 270t,
        456
Freberg, Prof., 46, 62
Frederick, Robert, 234f
Freeman, Richard D, 127, 311, 454
fund raising, 272
Fuqua, Myra, 211f, 237t

G

Gaffey, Troy M., 455
Gaichas, T. A., 137t
Ganong, Prof., 58
Gardner, Guy S., 175, 300t, 301t, 305, 454
Garrison, James L., 265, 269f, 270t, 276, 278, 456
Gates, Thomas S., 455
Gatineau, Robert J., 291
Gaugh, William J., 44f, 291–292, 345
gelled rocket propellants research, 280–281
Gemini-8 flight, 130
General Dynamics, 115
General Electric, 86
Genin, Joseph, 167, 173t, 183, 186, 456
Gentry, Eric, 272
Georgia School of Technology, 20t
Gersch, Wilbert M., 173t, 182, 456
Gerstenmaier, William H., 326, 455
G.I. Bill, 51
Gilbert, Peter, 280f
Gilbreth, Lillian M., 33
Ginsberg, Jerry H., 167, 173t, 183, 186, 456
Glenn L. Martin Co., 115
Glider Club, 115–117, 116f, 117f, 121
gliders, 7f, 115–116, 117f
golden anniversary. *See* anniversaries of schools
Goldman, Jerome M., 289f, 292
Gore, J., 271t
Goulard, Madeline
    career of, 100, 135, 136, 137t, 151, 155, 161,
        192, 201
    in faculty list, 138t, 140, 148t, 173t, 189t, 456
Goulard, Robert J.
    career of, 99, 135, 137t, 149, 161, 171, 192, 201,
        453
    in faculty list, 138t, 140, 148t, 173t, 189t, 456
graduate degrees. *See also* degrees; Master of
    Science degrees awarded; PhD degrees
    awarded and thesis titles
    1937-2009 calendar summary, 206, 207f
    1939-50 in Aeronautical Engineering, 10, 12t,
        23–24, 26t, 88, 92
    1950s in Aeronautics, 109, 136, 139t
    1960s requirements, 169–170, 174–176
    1995 requirements, 245t
    2000 requirements, 276, 277–278
graduate programs and research. *See also*
    research
    1938 vision for research, 22
    1941-45 during World War II, 23–30
    1945-50 postwar, 87–90
    in 1950s, 97, 109, 133t

in 1960-73, 169–173
in 1973-80, 197–199
in 1980-95, 221, 223t
in 1996-2009, 276, 277–278
Grandt, Alten F. (Skip), Jr., xv, xx, 209f, 211f, 234f,
    269f, 333f, 334f
    career of, xv, 204, 209, 224, 226, 227, 239, 248,
        253t, 272, 451, 453
    on Challenger explosion, 228–229
    in faculty list, 212t, 225t, 239, 246t, 247t, 271t,
        277, 456
    letter from Kincheloe's mother, 128
Green, James, 287t
Greenberg, Arthur B., 315–316, 454
Greene, Dennis, 287t
Greenkorn, Robert A., 151, 189
Griffin, Michael, 262, 278, 284
Grissom, Virgil I, 159f. *See also* buildings at
        Purdue: Grissom Hall
    Apollo fire, 130, 228–229, 282f, 284
    career of, 299, 300t, 304, 341
    Grissom Hall named, 158, 159f, 229, 299
Grosh, Richard J., 150–151, 158
Grumman company, 115
Guggenheim Foundation, 19–20
Gustafson, Winthrop A. (Gus), xv, xx, 211f
    career of, xv, 140, 161, 190, 193, 209–210, 238,
        262, 263, 273, 451
    in faculty list, 148t, 173t, 175, 189t, 199t, 212t,
        224, 225t, 238, 246t, 456

H

Haag, C. A., 34t, 35t
Hackman, Lloyd E., 120, 455
Hagenmaier, C., 289f
Hahn, G., 268t
Haines, M. G., 6t
Haley, Debra L., 326–327, 455
Hall, D., 237t
Hall, Prof., 46, 62
Halsema, D., 237t
Hancock, J. O., 213f
    career of, 213
    in faculty list, 148t, 189t, 212t, 456
Hancock John C. (dean), 151, 167, 190
Harbaugh, Gregory J., 158, 301t, 302t, 305, 307f,
    454
Harold Warp Pioneer Village, 87
Harrison, John C., 292–293, 345
Harrison, William Henry, 341
Harston, Gene P., 216, 216f
Hasbe, R. Sergio, 287t
Haskins, George W., xxiii, xxv, 3–4, 5f, 6f, 8f, 9,
    456
Hathaway, Marie, 211f, 237t
Hawkins, G. A., 97, 127, 139, 143, 149, 150, 151,
    156, 158, 167, 175, 189
Hayhurst, John B., 236f, 257, 317, 336, 454
Hedden, C. R., 344
Heinimann, Markus B., 287t, 455
Heister, Stephen D., 211f, 269f

career of, 219, 226, 239, 253t, 272, 273, 280, 282,
    451, 452
in faculty list, 212t, 225t, 239, 246t, 247t, 271t,
    456
helicoptor experiments, 249
Herrick, Robert N., 289f, 293, 345
Herrick, Thomas J., 213
    career of, 64, 88, 151, 190, 192, 212
    in faculty list, 63t, 93, 96, 130, 140, 148t, 189t,
        456
Hertzberg, Steven, 229
Hesse, Walter J., 309, 454
Hill, W. W., 137t
Hillberry, B. M., 248
Hoglund, Richard F., 173t, 182, 456
Hollister, Mr., 5f
Holton, C., 34t
homecoming queen, 112f, 113, 118
Homsher, Paul, 34, 35t, 313–314, 454
Honorary Doctorate Degrees, 329–340
Hopping, Bradley, 287t
Hostetler, Robert D., 314, 454
hot air balloon *Jupiter,* xxi, xxii–f
housing near campus for staff, 62–63
Hovde, Frederick L., 63
Howell, Kathleen C., 211f, 216f
    career of, 218, 238, 253t, 262, 272, 273, 451
    in faculty list, 212t, 225t, 238, 246t, 247t, 270t,
        276, 277, 456
Howland, Merville C., 289f, 293, 345
Howse, Mike, 278
Hrbud, Ivana A., 266, 269f, 271t, 456
Hromas, Leslie A., 91f
    career of, 100, 135, 137t, 455
    in faculty list, 130, 138t, 140, 148t
Hsu Lo Professor of Aeronautical and Astronau-
    tical Engineering, 272
Hua, Hsichun, 317–318, 454
Hudson, John L., 278, 324, 455
Hughes Aircraft Space & Comm., 241t, 243
Hunsacker, Prof., 50
Hwang, Inseok, 267, 269f, 270t, 271t, 276, 456
Hyer, Michael W., 455

I

Imbrie, P., 271t
Indiana 21st Century Research and Technology
    Fund, 282–283
Indiana Institute of Technology, xxii
Indiana Space Grant Consortium, 227–228, 252t
Indiana University, 227
Indians, Wea, 341
Industrial Advisory Council (IAC), 262–263
Industrial Affiliates Program, 243
inspection trips to industry, 70, 114f, 115
Institute for Interdisciplinary Engineering
    Studies at Purdue, 172, 248
Institute of Aeronautical Sciences, Student
    Chapter, 70
Inter-American flight training program, 17–18
interdisciplinary engineering studies, 166, 172
Iowa State University, 32, 280

J

Jackson, J., 268t
James, Robert, 287t
Jamieson, Leah H., 261
Jamrogiewicz, R. A., 287t
Jarvis, Gregory, 229
jet-powered airliners, 48
jet propulsion education & research, 65, 73, 76–77, 86–87
Jischke, Martin, 260–261, 340, 456
John L. Bray Distinguished Professor of Engineering, 272
Johnson, Alan D., 34t, 77
Johnson, K., 268t, 269f
Johnson, Kenneth O., 455
Johnston, Robert, 235f
Jones, Leslie, 137t
Jones, Lewis P., 44f, 293–294, 345

K

Kasler, Richard E., 316, 454
Kasowski, Andrew H., 455
Katehi, Linda, 261
Kayser, Kenneth W., 203, 456
Keller, D. P., 6t
Kennedy, Michael T., 322, 455
Kentzer, Paul, 101f, 214f
    career of, 100–101, 135, 137t, 193, 215
    in faculty list, 140, 148t, 173t, 189t, 212t, 225t, 456
Kerber, Ronald L., 317, 454
Kerkhove, P., 268t, 269f
Kerr, H. Irving, 289f, 294, 345
Kessler, William C., 318–319, 454
Key, Nicole, 271t, 287t
Kherat, Samir, 237, 246t
Kidd, M., 268t, 269f
Kim, Hyonny, 265, 277, 456
Kincheloe, Iven, xxiii, 128–129, 129f, 228, 298
Kincheloe Crain, Frances, 128–129
King, Andrew, 455
Kinnan, Timothy A., 455
Klassen, Diane, 268t, 272
Klein, Bertram, 63t, 67, 93, 456
Kline, J., 268t, 269f
Kline, Leo V., 146, 148t, 177, 456
Knoll, H. B., 344
Koerner, W. G., 82f
Koh, Severino L., 173t, 183, 273, 456
Kordes, E. F, 63t, 88
Korean War, 101, 117
Kozin, Frank, 146, 148t, 171, 173t, 179, 457
Kriebel, R., 344
Kuroda, Yasuhiro, 455
Kvam, E. P., 248

L

laboratories
    aerodynamics, 1, 94, 109–111
    aeronautical engineering, 18–22
    aeronautics, 5f
    aerospace sciences, 147f, 158, 168, 169, 196, 198, 215, 230, 231, 235, 242–243, 249, 250
    aircraft internal combustion engine, 109
    aircraft power plant, 30f, 37f, 71, 73, 74f, 76, 93–94, 121
    Brown chemistry, 158
    composite materials, 169, 230, 231f, 250–251
    dynamic inelasticity, 251
    engine disassembly, 75f
    experimental engineering sciences, 168
    experimental stress, 168, 230
    fatigue and fracture, 251
    flight dynamics & control, 249
    flight simulation, 249–250
    flying, xxiv, 17
    gas turbine, 73, 76
    guidance and control, 230
    in Heavilon Hall, 1, 12, 158, 200f, 281
    high pressure rocket research, 282
    laser, 230
    materials research, 169, 230
    McDonnell Douglas composite materials, 250–251
    metal working, 79, 85f
    photomechanics, 251
    propeller, 74f
    random environments, 169
    rocket, 230–231, 250
    rocket engine, 94, 97, 109
    sophomore aeronautical, 168
    space systems control, 248, 263
    structural dynamics, 251
    structures, 26, 29f, 36f, 42–43, 79, 82f, 84t, 94, 112, 196
    systems, 169
    Thermal Sciences and Propulsion Center, 231, 243, 250
    tribology and materials processing, 251–252
    Zucrow, 279f, 281, 282–283
laboratory facilities. See also facilities; wind tunnels at Purdue
    1938 proposal for, 18–22
    in 1941-45 during World War II, 24, 26, 27f, 29f–30f, 36f–37f
    in 1945-50 postwar, 49, 56, 71f, 72f, 93–94
    in 1960-72, 162–163, 168–169
    in 1973-80, 196
    in 1980-95, 210f, 230–235, 248–252
Lafayette, Indiana, history of air transportation, xxi–xxii, 341
LaGuire, J., 268t
Lamberson, Steven E., 455
land-air transportation system in 1930s, 16
Lappe, Oscar, 44f, 345
large-scale systems option, 161–162, 165, 165t, 167
Lascoe, O. D., 34t, 35t, 56, 58, 84–85
lecture series, 278
L'Ecuyer, M. R., 195, 199t, 246t
Ledington, Matt, 233f
Leland Stanford University, 20t

Lianis, George
    career of, 100, 171, 202–203
    in faculty list, 140, 148t, 173t, 189t, 199t, 457
libraries, 148, 158, 231
Lin, Hua, 150
Lindbergh, Charles, 9
Lindenlaub, Prof., 175
Lindley, R. W., 34t
Lintereur, Louis, 287t
Liston, Joseph, 9f, 15f, 71f, 110f
    career of, 9–12, 134, 180
    in faculty list, 35t, 63t, 93, 95, 96, 140, 148t, 457
    laboratory space development, 18–22, 24, 49,
        73, 76. 109, 281–282
    postwar aviation education, 47, 50, 51, 62, 157
Little, Neil, 34t, 35t, 148t, 178, 457
Lo, Hsu, 63, 132f
    career of, 68, 133, 149, 151, 158, 210, 453
    curriculum development, 160–161, 193
    in faculty list, 63t, 93, 96, 138t, 140, 148t, 189t,
        457
    research work, 89, 135, 136, 137t, 172, 197
Lockenour, Jerry L., 455
Lockheed Electra, xxiv, 17
Lockheed Missiles and Space Company, 224, 243
Lodge, R. A., 176
Longuski, James M., 211f, 269f
    career of, 219, 238, 253t, 273, 451, 452
    in faculty list, 212t, 225t, 238, 246t, 247t, 270t,
        276, 277, 278, 457
Lowy, Stanley H., 294, 345
Lucht, R. P., 280
Ludwieg tube, 235, 236f, 249, 283
Lund, Thomas S., 219, 225t, 457
Lykoudis, Paul S., 99f
    career of, 99, 133, 151, 200
    in faculty list, 138t, 140, 148t, 173t, 457
    research work, 135, 137t, 169, 171
Lyrintzis, Anastasios S., 269f
    career of, 220, 238, 272, 452
    in faculty list, 212t, 225t, 238, 247t, 270t, 277,
        457

M
MacCready, Paul, 278
Mahurin, Walker M., 308–309, 454
Mallett, Frank, 35t
Malmsten, C. R., 216f
Malone, David W, 162, 167, 173t, 185, 457
Marais, Karen, 268, 270t, 457
Marshall, Edmund V., 332–333, 454
Marshall, Francis J., 193f, 209f
    career of, 140, 193, 209, 215, 224, 451
    in faculty list, 148t, 173t, 189t, 199t, 212t, 457
Marshall, Paul, 13f
Mason, Joseph D., 187, 455, 457
Massa, Captain, 119f
Massachusetts Institute of Technology, 20t
Master of Science degrees awarded, 387–426. See
        also graduate degrees
    in Aeronautical Engineering, 88, 388–390
    in Aeronautics & Astronautics, 388–396

in Astronautics, 397
    in Engineering, 398–401
Master's thesis titles, 88, 402–426
mathematics courses, 161
Mathews, J. H., 35t
Maxwell, Thomas, 455
Maybee, J. S., 183
Mayfield, C. E., 35t
Mayhill, Prof., 58
McAuliffe, Christa, 229
McBride, James W., 63t, 67, 76, 78–79, 89, 457
McCabe, G. P., 248
McCulley, Michael J., 301t, 306, 307f
McDermott, Robert F., 175
McDonnell Aircraft Corporation, 34, 78, 135, 136
McDonnell Douglas Corporation, 241t, 243
McElwee, Jerry W., 455
McGowan, Anna-Maria R., 455
McGrath, David, 455
McKane, G. Thomas, 455
McMasters, John H., 455
McNair, Ronald, 229
Mechanical Engineering School. See School of
        Mechanical Engineering
Meikle, G. Stanley, 16, 18, 44f
Mendenhall, Dave, 44f
Mendoros, Dennis, 278
Merkle, Charles
    career of, 266, 272, 273, 280, 452
    in faculty list, 271t, 277, 457
Messersmith, Nathan L., 211f
    career of, 220, 239, 254t, 263
    in faculty list, 212t, 225t, 239, 246t, 247t, 457
Meyer, Scott E., 268t, 269f, 283
Mickelson, Harvey, 137t
Midwest Applied Science Corporation, 149
Mid-West Program for Airborne Television
        Instruction (MPATI), 18
Miele, Angelo, 98f
    career of, 98, 99, 100, 133, 135, 137t
    in faculty list, 138t, 457
military training for men, 101
Miller, E. F., 34t
Miller, Garner W., 316, 454
Miller, Kenneth G., 323, 455
Miller, Ronnie K., 455
Minton, Joseph P., 85
Misener, Walter S., 63t, 67, 76, 457
Mock, C. V., 35t
model airplane building club, 119–121, 120f
Mohr, Russ, 287t
Montgomery, Stephen T., 287t
Moore, Ronald L., 287t
Moore, Terri, 211f, 237t, 268t, 269f
Morris, S., 237t
Morrison, Ruth, 32, 35, 38
Morrow, Howard, 43, 44f, 45f
Morrow Airplane Project, 42–44, 44f–45f
MS degrees. See Master of Science degrees
Multidisciplinary University Research Initiative
        (MURI), 280–281
Mumma, Wade E., 85

Munson, K., 268t
Murphy, J. Michael, 455
Murphy, Terrence H., 455
Myers, Joseph G., 287t

N

NACA Laboratories, 62
NASA National Space Grant College and
    Fellowship Program, 227
National Advisory Committee for Aeronautics
    (NACA), 112, 136
National Aeronautics and Space Administration
    (NASA), 135, 224, 241t, 252t, 256t, 262, 283
National Citation of Honor, 113
National Science Foundation, 136, 169, 171–172,
    176, 197, 252t, 256t
Naval Bureau of Aeronautics, 136
Navy College Training program, 40–42
Navy V-12 Program, 34–42
Neidhold, Carl, 121
Neil A. Armstrong Distinguished Professor of
    Aeronautical and Astronautical Engineering,
    208, 272
Neil Armstrong Hall of Engineering
    dedication ceremony, 281f, 283f, 284, 307f
    displayed items, 128, 282f, 284
    private support of, 272
    school's relocation to, in 2007, xix, 260, 281,
        283–284
Nestel, John I., 294, 312, 454
New York University, 20t
Nobile, A., 268t, 269f
Noblitt, James P., 320–321, 454
Norman, Wendell S., 313, 454
Norris, Stephen R., 287t
North American Aviation Company, 14, 43, 50
Northrop Corporation, 243
Novak, D., 77
Novick, Allen, 325–326, 455

O

O'Bannon, Frank, 263
oceanic air travel, 16
Ochiltree, David W., 289f, 295, 345
Office of Naval Research, 77, 112, 136, 182, 282
Office of Ordnance Research, 112, 136
Office of Scientific Research and Development,
    112, 135, 136, 248, 283
Ohira, H., 137t
oil embargo in 1973, 190
Olds, C. D., 35t
Olsen, E. C., 295
O'Neil, William J., 257, 321–322, 338–339, 454
Onizuka, Ellison, 229
Osborn, J. R.
    career of, 195, 215, 217–218, 228
    in faculty list, 199t, 212t, 457
Ostoja-Starzewski, Martin, 218, 225t, 457
Outstanding Aeronautical Engineer Educator,
    294
Owen, H. F., 34t

P

Palmer, George, 110, 199f, 214f
    career of, 34, 67–68, 135, 137t, 198–199, 215,
        451
    in faculty list, 35t, 63t, 93, 96, 130, 140, 148t,
        173t, 175, 189t, 212t, 224, 457
    laboratory equipment installations, 68, 73, 77,
        79, 81f, 93–94, 111, 168, 235
pamphlets from School of Aeronautics, 51, 54,
    59–61
Panlener, P. J., 34t, 35t
parachute engineering, 130
Paris, F. L., 137t
Patrick, Richard M., 312, 454
Payton, Gary E., 175, 300t, 306, 307f, 454
Pendley, Robert E., 44f, 289f, 295, 345
Pennsylvania State College, and Curtiss-Wright
    Cadette Program, 32
Pernicka, Hank, 216f
Perry, Craig O., 287t
Peters, Christopher, 287t
Petersen, Richard H., 314, 333–334, 454
Peterson, Clifford C, 167, 173t, 187, 457
Peterson, Lee D., 219, 225t, 457
PhD degrees awarded and thesis titles, 427–449.
    See also graduate degrees
Phillips, J., 268t, 269f
physical education for women, 101
Pieri, Gina, 287t
pilot training, 17–18
Pipes, R. Byron, 267, 271t, 272, 457
Piston Joe, 134, 180
Pneuman, G. W., 137t
Polansky, Mark L., 303t, 306, 307f, 454
Pollock, George, 287t
postwar period, 1945-50, 47–94
Potter, Andrey A., 15, 18, 23, 47, 49, 73, 91, 95
Pourpoint, Timothee, 267, 269f, 271t, 280, 457
Powers, Doris H., 455
Pritsker, A. Alan, 162, 167, 173t, 186, 457
Production-Management Option, 60f, 61
professorships, named and chaired, 272
Propulsion graduate courses, 109
Propulsion program, 282, 96, 157, 195, 223t, 239,
    250, 271f
publications. *See also* textbooks
    by faculty and students, 272, 279–280
    references in this book, 344
    from research, 247–248
Purdue Aero Club, xxii
Purdue Aero Flying Club, 117–118
Purdue Aeromodelers, 119–121, 120f
Purdue Aeronautics Corporation, 43, 49, 51, 52f,
    55, 57, 64, 71, 72, 202
Purdue Airlines, 18, 292
Purdue Airport. *See also* facilities
    1930-37 early classes, 7f, 8f
    1930s building of, 16–18, 17f, 18f
    1940s during World War II, 26, 28f
    1945-50 postwar, 49, 50, 70–87
    1950s facilities, 95, 97, 109–112
    1960 facilities after merger, 143

Purdue Engineering Experiment Station, 136
Purdue Glider Club, 115–117, 116f, 117f, 121
Purdue Research Foundation (PRF), 16–17, 89, 136, 282, 341
Purdue Space Day, 278
Purdue Space Shuttle Memorial Fund, 229
Purdue University. *See also* buildings at Purdue
    attraction of high-quality students, 285
    history of aeronautic engineering, xxii–xxv
    reputation of, 55, 175–176, 228
    role in national aerospace engineering, xxi, 48, 49
    tuition & fees for attending, 51, 54
Putnam, Carlton, 59

Q

Qiao, Li, 268, 271t, 457
Queen, Hank, 324, 455

R

Radavich, John F., 148t, 181, 457
radio program, *This Week in Aviation*, 69, 113, 118
Raisbeck, James D., 313, 339–340, 454
Raisbeck Engineering Distinguished Professor of Engineering and Technology Integration, 272
Ranger, Arthur A., 163, 185, 200, 457
Ray, Suvendoo, 455
Raymer, Daniel P., 455
Reagan, David, 211f, 237t, 268t, 269f
reciprocating aircraft engine, 73
Reese, Bruce A.
    career of, 68–69, 190, 193, 195, 196–197, 315, 454
    in faculty list, 63t, 93, 96, 189t, 457
    as school head, 151, 190, 208, 453
Reilly Professor of Engineering, 272
Rennak, R. M., 82f
Rensselaer Polytechnic Institute, and Curtiss-Wright Cadette Program, 32
Republic Aircraft Corporation, 85
Republic Aviation, 115
research. *See also* graduate programs and research
    1938 vision for graduate research, 22
    1949-50 faculty projects and proposals, 88–90
    1950s sponsored support, 112, 134–137, 137t–138t
    1960-73 increased activities, 171–172
    1970s sponsored support, 197–198
    1980s sponsored support, 206, 207f
    1993-94 sponsored support, 247–252, 252t–256t
    1996-2009 sponsored support, 260, 279–281
research awards, 452. *See also* awards
Resnik, Judith, 229
Rialto Airport, 43
Rich, John, 455
Richardson, Robert L., 88
Ripley, Helen, 32
Rivers, W., 6t
Rivir, Richard B., 455

Robinson, R. F., 63t, 77, 93, 96
rocket engine propulsion, 76–77, 250
rocket launch competition, 226
rocket theory courses, 133
Rockwell Defense Electronics, 241t
Rogers, Herbert F., 331–332, 454
Rolls-Royce, 283
Roosevelt, Franklin D., 16
Rose, Leonard M., 44f, 295, 311, 345, 454
Rosenberg, Reinhart, 63t, 64, 88, 457
Ross, David, 16
Ross, Jerry L., 300t, 301t, 302t, 303t, 306, 307f
Ross, Nan Clair, 272
Rotea, Mario A., 211f
    career of, 219, 226, 238, 254t, 255t, 263, 273, 452
    in faculty list, 212t, 225t, 238, 246t, 247t, 457
Rottmayer, Earl, 63t, 66, 457
Rouzie, Richard Lee, 329–330, 454
Rusek, John J., 265, 271t, 457
Rushinsky, John J., 232f
Russian Earth satellite Sputnik I, 133, 137, 299
Ruterbories, Burghard H., 287t

S

Saff, Charles R., 455
Salisbury, Ralph D., 295, 345
Samuels, J. Clifton, 146, 148t, 178, 457
Sanders, Robert, 211f
Sanders, W. Burton, 34t, 148t, 176–177, 457
Sandia National Laboratories, 283
Sanger, Kenneth B., 455
Sayeedi, Naeem, 229
Schaefer, Betty, 36f
Schafer, D., 237t
Schaller, L. F., 34t, 35t
Schapery, Richard A., 161, 173t, 181, 457
Scheuring, Jason, 234f
Schiff, Anshel J., 163, 167, 169, 173t, 184, 457
Schmidt, David K., 216f
    career of, 203, 215, 287t, 451
    in faculty list, 212t, 225t, 457
Schmitendorf, William E., 455
Schmitt, Alfred F.
    career of, 68, 90, 91f, 98–99, 136, 309–310, 454
    in faculty list, 63t, 93, 96, 457
Schneider, Steven P., 211f, 236f, 269f
    career of, 219, 235, 238, 254t, 255t, 273, 283, 452
    in faculty list, 212t, 225t, 238, 246t, 247t, 270t, 457
Schneiter, Leslie E., 44f, 295, 345
Scholer, Walter, 72–73
school heads summary, 453
School of Aeronautical and Engineering Sciences, 139–144, 146–149, 151
School of Aeronautical Engineering, 128–134, 139–144, 146–148
School of Aeronautics, 51–54, 52f, 71, 128, 282
School of Aeronautics, Astronautics, and Engineering Sciences, 152, 158

School of Aeronautics and Astronautics
  astronaut alumni, 299
  facilities in 1980, 231, 235
  historical summary, 342
  name changed, 151
  relocation to Neil Armstrong Hall, 260, 281, 283–284
School of Civil Engineering, 145
School of Engineering Mechanics. *See* Division of Engineering Mechanics
School of Engineering Sciences. *See* Division of Engineering Sciences
School of Industrial Engineering, 149, 158, 231, 235, 242, 251
School of Industrial Engineering and Management, 127
School of Materials Engineering, 280
School of Mechanical and Aeronautical Engineering, 23, 47
  Class of 1943, 288
School of Mechanical Engineering, xxiii, 1–4, 151, 176, 195, 223t, 342
School of Nuclear Engineering, 235, 242–243
Schools of Engineering, name change, 261
school titles summary, 453
Schroeder, Paul M., 295
Schultz, Jeffrey D., 287t
Schwartz, Richard, 261
Science Option, 154t, 155, 156, 160
Scobee, Francis (Dick), 229
Scott, Robert, 211f, 237t
Scudder, Kenneth R., 296
Scurria, Norman V. Jr., 455
Sears, Michael M., 278
seminar participation of staff, 136, 138t
senior aeronautical option, 1, 2t
Serbin, Hyman
  career of, 66, 89, 98
  in faculty list, 63t, 93, 95, 96, 457
Seward, A. L., 137t
Shanks, Merrill E.
  career of, 65, 179–180
  in faculty list, 63t, 93, 96, 140, 148t, 457
Sheadon, John Jr., 455
sheet metal working, 79, 85f
Shields, Randolph, 287t
Shih, Tom I-P., 262, 270t, 453, 457
short-term courses, 130
Shriver, Loren J., 175, 300t, 301t, 306, 307f, 454
Shu, Shien-Siu
  career of, 161, 177–178, 200, 333f, 334, 454
  in faculty list, 146, 148t, 189t, 457
Sigma Alpha Tau Society, 70
Sigma Gamma Tau, 287
Simo, Miroslav A., 455
Sisco, Tim, 287t
Skelton, Robert E., 211f, 216f
  career of, 203–204, 238, 248, 254t, 263
  in faculty list, 199t, 212t, 225t, 238, 247t, 457
Slomski, Joseph F., Jr., 287t
Smith, "Bridge," 4
Smith, Lester W., 310, 454

Smith, Michael, 229
Smith, Walter D., 44f, 296, 311, 454
Snodgrass, R., 268t
Sojka, P. E., 280
Solberg, Harry L., 4, 23, 24, 46, 47, 453, 457
Son, S. F., 271t, 280
songs from Curtiss-Wright Cadette Programs, 39
Southern California University, 48
Spacecraft Design course, 276
Space Day, 278
space exploration, and Purdue's contributions, 228, 342
space-flight log, 299, 300t–303t
space flight research, 171–172
space flight STS-39, 158
space-oriented courses, 133
Space Shuttles, 128, 228–229, 299
Spencer, David A., 216f, 455
sponsored research. *See* research
Spraker, Wilbur A., 296–297
Sputnik I, 133, 137, 299
Squid Research project, 76, 77, 182, 201
Staab, George, 287t
Staley, Alan C, 1, 2t, 457
Stanisic, M. M., 214f
  career of, 214
  in faculty list, 146, 148t, 173t, 189t, 212t, 457
Stanley, Paul E.
  career of, 51, 57–59, 64, 126, 127, 128, 149, 161, 202, 453
  in faculty list, 63t, 92, 93, 96, 138t, 140, 148t, 189t, 457
  research work, 135, 137t, 172
Stein, S., 226f
Steinmetz, H. F., 82f
Stevens, Maxine, 36f
Stewart, Charles R., 287t
Stitz, Ervin O.
  career of, 149, 161, 163, 201–202, 202, 451, 453
  in faculty list, 148t, 189t, 457
Stone, Winthrop E., xxii–xxiii
Strand, H. W., 35t
Streicher, Albert H., 289f, 297
Strickler, Robert L., 321, 454
Stroud, Robert E., 127
Struckel, Margaret, 227f
structural analysis coursework, 42–43
Structural Dynamics Research Center, 241t
Structures and Dynamics, 1950 faculty, 96, 109
Structures and Materials
  1974 major and minor areas, 194, 198
  1981-83 option, 223t
  1993-94 research, 250–252
  1995 faculty, 239
  2009 faculty research area, 271f
structures option, Navy V-12 curriculum, 40, 42t
Stuart Field, Purdue U., xxiii, xxiv–f
student activities, 69–70, 112–121
Sullivan, John P., 210f, 211f
  career of, 204, 227, 238, 254t, 255t, 261–262
  in faculty list, 212t, 225t, 238, 246t, 270t, 276, 457

as school head, 209, 257, 261, 453
Sumcad, Stephanie, 287t
Sun, Chin-Teh, 211f, 231f, 269f
    career of, 169, 184–185, 239, 272, 273, 451, 452
    in faculty list, 173t, 189t, 199t, 212t, 225t, 239,
        246t, 247t, 277, 457
    research work, 198, 248, 253t, 255t, 280
Sun, Dengfeng, 268, 270t, 271t, 457
surveys
    1968 student concerns, 160–161
    postwar plans & need for aviation education,
        47–48, 49–50
Sutton, Emmett A., 173t, 181–182, 457
Swaim, Robert L., 184f, 189t, 199t
    career of, 137t, 184, 190, 193, 200, 451, 454
    in faculty list, 173t, 457
Swain, David O., 319, 336–337
Sweet, Arnold L., 161, 167, 182, 457
Syracuse University, 20t

T

Talavage, Joseph J., 167, 188, 457
Tally, L. H., 344
A Taste of Aerospace, 228
Taylor, Charles E., 337–338, 455
Taylor, Martin W., 127, 313, 455
Taylor, Richard W., 82f, 236f, 330, 455
Taylor, Ronald, 287t
teaching assistant positions, 197
teaching awards, 202, 208, 272–273, 451–452. *See*
    also awards
textbooks. *See also* publications
    on aerodynamic loads by Yang, 186
    on aircraft propulsion engines by Cargnino,
        57
    on airplane design by Wood, 10
    on applied mathematics by Genin and
        Maybee, 183
    on creative engineering by Liston, 134
    on creative product evolvement by Liston and
        Stanley, 180
    on mathematical theory of turbulence by
        Stanisic, 214
    on mechanics of materials by Fairman and
        Cutshall, 177
    on statics and dynamics by Genin and
        Ginsberg, 183, 186
    on structural design by Bruhn, 14–16
Theoretical Aeronautics Option, 122, 122f–123f,
    131
Thiokol Corporation, 224, 226f, 241t, 243
This Week in Aviation, radio program, 69, 113, 118
Thompson, F. H., 34t, 35t
Thompson, H. C., 176
Thornburg, Martin L., 1, 457
Thornton, Anthony L., 455
Ting, Edward C., 167, 185, 457
Tingle, Scott, 306
Tinius Olson test machine, 83f
Tipmore, Claud, 214
Tomar, Vikas, 269, 271t, 277, 457
Topinka, George, 63t, 64–65, 457

Topping, R. W., 344
torque stands, 30f
Trabant, E. A., 139, 146
Traffic Administration Option, 55
Troy, Christine, 287t
Trueblood, Ralph B., 44f, 297, 345
Trumble, K. P., 280
TRW, 243
T-shirts with Top Ten list, 242
turbojet engines, 48, 76
Turner, R. H., 82f
Turpin, J. Clifford, xxiii, 298

U

undergraduate curriculum. *See* curriculum/
    course listings
United Airlines Company, 58
United Airlines DC- planes, 114f, 115
University Faculty Scholars program, 272–273
University of Akron, 20t
University of Colorado, 45
University of Dayton Research Institute, 280
University of Massachusetts, 280
University of Michigan, 20t
University of Minnesota, 32
University of Notre Dame, 227
University of Texas, 32
University of Washington, 20t
University Space Research Association (USRA),
    224
U.S. Coast and Geodetic Survey, 17
U.S. Department of Commerce, 17
U.S. Department of Defense, 256t
Usab, William J., Jr., 218, 225t, 287t, 457
USAF Academy, 174–176
US News & World Report, 259

V

Valanis, Kyriakos C., 182, 457
Van Allen, Richard E., 455
Vanderbilt, Vern
    career of, 66, 93–94, 97
    in faculty list, 63t, 93, 96, 457
Veal, Cicero B., xxii
vehicular dynamics, 195–196
Vellinger, A. J., 34t, 35t
Vemuri, V. R., 162, 167, 186, 457
Ventre, Brian, 287t
veterans in postwar aviation education, 48, 54,
    117, 126, 206
Voss, Janice E., 301t, 302t, 303t, 306, 307f, 455
Vought-Sikorsky Aircraft Company, 14

W

Wagie, David A., 455
Wagner, S., 269f
Wakefield Trophy, 121
Walker, Charles D., 300t, 306, 307f, 455
Walker, Neil, 287t

Walker, Rhea, 72f, 98, 200, 200f
Warner, Dennis, 455
Warren, R., 6t
war surplus equipment, 77–79, 84–87
War Training Division, Purdue, 33
War Training Service (WTS) program, 17
Washburn, Karen, 287t
wave propagation research, 251
WBAA radio station, 118
Wea Indians, 341
Weber, James A., 287t
Weber, Jürgen, 278
Weber, Mary E., 302t, 303t, 307
Webster, Grove
    career of, 43, 47, 49, 50, 55, 57, 126, 127
    in faculty list, 51, 58, 92, 96, 457
Weeks, Lawrence C., 287t
Weiler, E. T., 126
Weisshaar, Terrence A., 211f, 226f, 269f
    career of, 217, 239, 255t, 256t, 451
    in faculty list, 212t, 224, 239, 270t, 271t, 276, 457
Werthman, Lance, 287t
Weske, John R., 63t, 65, 76, 78, 457
West, Frank E., 297–298, 345
Westinghouse, 86
Wheaton, Brad, 280f
Whipple, Christopher, 324–325, 455
White, Edward M., 158, 159f, 228, 282f, 284
Whittenberg, John, 229
Wight, Nathan, 272
William E. Boeing Lecture Series, 278
Williams, Donald E., 300t, 301t, 307, 307f
Williams, Marc H., 211f, 269f
    career of, 218, 238, 256t, 262, 452
    in faculty list, 212t, 225t, 238, 246t, 270t, 457
Williams, Thomas L., 455
wind tunnels at Purdue, 280f, 283. See also
        laboratories: aerospace sciences
    in 1930-37, 4, 5f
    in 1937-41, 12, 14f
    in 1941-45 war years, 26, 27f, 37f
    in 1945-50, 77–79, 80f, 94
    in 1950s, 109–111
    in 1960s, 168
    in 1973-80, 198–99
    in 1980-95, 230, 232f, 235, 235f, 236f
    in 1983-84, 249
    Boeing/AFOSR Mach 6 Quiet Tunnel, 280f, 283
    Boeing Wind Tunnel, 81f, 111, 235, 236f, 249
    display at Engineering Open House, 121
    hypersonic, 235, 283
    Japanese variable-density, 78, 81f, 111
    low turbulence, 78
    subsonic, 78, 81f, 109, 110f, 111f, 168, 198, 232f, 233f, 235, 249
    supersonic, 68, 94
Wingert, R. J., 82f
Wise, John, xxi, xxii–f
Wise, Sharon, 211f, 237t
Witkowski, David P., 233f

Witmer, C. C., xxii
Wittig, Sigmar, 278
Wolf, David A., 301t, 303t, 307, 307f
women
    in Curtiss-Wright Cadette Programs, 32–39
    "first woman" distinctions, 100, 201
    outreach efforts toward, 227–228
    pilots at Purdue, 112f, 113
Wood, Karl D., 9f, 24
    career of, 9–10, 15–16, 18–22, 33, 34t, 45–46, 77, 80f, 91
    in faculty list, 11f, 12f, 457
    quoted on flying a brick, 288
Wood, Thomas, xxi
Woodward, W. E., 34t
Wooten, Breanne, 287t
World War II, 23–46
Wright, Orville, xxiii
Wright brothers, 2
Wright-Patterson Air Force Base, 241t
Wurster, Dorothy, 36f

X

Xia, G., 268t

Y

Yang, Henry T. Y., 211f, 333f, 334f
    awards received, 208, 273, 335, 451, 455
    career of, 186, 193, 196, 197, 217, 256t, 264
    as dean of engineering, 209, 264, 320f
    in faculty list, 173t, 189f, 212t, 457
    as school head, 190, 208, 261, 453
Yarbrough, Edwin, 287t
Yeager, Chuck, 119f
Yoder, T., 269f
Yorn, Kendall, 287t
Yost, Mark, 211f, 217
Young, Gilbert A., 1, 18, 23, 457
Younts, J., 268t

Z

Zachary, J., 237t
Zak, A. R., 137t
Zakem, Steve, 287t
Zarrora, F., 35t
Zimney, Charles M, 298, 345
Zucrow, Maurice, 97f, 109
    career of, 65, 69, 73, 76–77, 93–94, 97, 133
    in faculty list, 63t, 93, 95, 96, 457